Imagined Civilizations

Roger Hart

Imagined Civilizations

China, the West, and Their First Encounter

The Johns Hopkins University Press
Baltimore

© 2013 The Johns Hopkins University Press
All rights reserved. Published 2013
Printed in the United States of America on acid-free paper
9 8 7 6 5 4 3 2 1

The Johns Hopkins University Press
2715 North Charles Street
Baltimore, Maryland 21218-4363
www.press.jhu.edu

Library of Congress Control Number: 2010924546

ISBN 13: 978-1-4214-0606-0
ISBN 10: 1-4214-0606-3

A catalog record for this book is available from the British Library.

*Special discounts are available for bulk purchases of this book. For more information,
please contact Special Sales at 410-516-6936 or specialsales@press.jhu.edu.*

The Johns Hopkins University Press uses environmentally friendly book materials,
including recycled text paper that is composed of at least 30 percent post-consumer
waste, whenever possible.

With enormous gratitude to all my teachers,
with special thanks to Benjamin Elman
and Mario Biagioli

Contents

Imagined Civilizations

Chapter 1
Introduction

THIS BOOK takes as its starting point an observation that one might hope would be uncontroversial: "civilizations" are no less imagined than "nations." During the later decades of the twentieth century, the term "imagined communities" gained considerable prominence through critical studies of various forms of nationalism. Yet many of these same critical studies reinforced the assumed reality of "the West." This book proposes to extend the approach taken in critical studies of nations by applying the term "imagined" to civilizations, in a similarly critical fashion. To say that civilizations, such as "China" and "the West," are imagined is not to dismiss them as somehow merely fictive and therefore without significance, but rather to suggest an alternative starting point for historical inquiry.

World history has usually been studied and understood within a framework of civilizations, and it is in this context that the transmission, by a handful of Jesuits, of specific European sciences, technologies, religion, and culture into seventeenth-century China has been accorded considerable significance as the asserted "first encounter" of "China" and "the West." Many recent studies have argued that it was the recognition of the superiority of Western science that led a select group of concerned Chinese officials to convert to Catholicism. Perhaps the most celebrated example of this asserted first encounter between Western science and Chinese tradition is the translation of Euclid's *Elements* in 1607 by the Italian Jesuit Matteo Ricci (1552–1610) and the Chinese official Xu Guangqi 徐光啟 (1562–1633). Xu Guangqi has been portrayed as a polymath and the greatest scientist of the Ming dynasty (1368–1644). These previous studies, based on the prolific writings of the Jesuits themselves, have focused primarily on the Jesuits and have taken them as the historical protagonists.

In this book I re-examine these events from the point of view of the Chinese and reach conclusions strikingly different from those of the previous studies. More specifically, this book presents a critical history of the dissemination of what was called Western Learning (*Xi xue* 西學) into seventeenth-century China, by drawing heavily upon Chinese primary sources, and taking the Chinese—who in fact wielded considerable power over their Jesuit collaborators—as the

1

central protagonists. I show that whereas the Jesuits claimed that the Chinese officials who collaborated with them were their converts, these Chinese officials represented the Jesuits as tributary officials who had come to China to serve the Chinese emperor. The writings of the Jesuits, these Chinese officials argued, recovered lost doctrines from ancient China, which had been preserved in the West. Adopting these doctrines would help the dynasty return to the perfected moral order of ancient China, which they imagined existed in "the West," where, for over a thousand years, there had been no wars, rebellions, or changes in dynasty. The extravagant assessments of the superiority, newness, and practical efficacy of Western Learning made by these Chinese officials, who had in fact little knowledge of current Chinese sciences, were in historical context their bids for patronage through memorials presented to the emperor, in which they fashioned themselves as statesmen with novel solutions to the ongoing crises of the Ming empire.

The approach taken here is deflationary, critical, and microhistorical. I propose to re-examine these events not as an encounter of two great civilizations, but simply as a collaboration of a small group of Chinese scholar-officials and Jesuit missionaries. A central argument of this book, then, is that rather than viewing this as a "first encounter" of two great civilizations—"China" and "the West"—we should instead critically historicize these terms as actors' categories. That is, we should analyze how the historical protagonists themselves imagined "the West," by way of furthering their own interests in the context of seventeenth-century China. Similarly, I will argue that we should critically analyze how narratives about this "first encounter" contributed to imagining "China" and "the West" during the twentieth and twenty-first centuries.

Preliminary Considerations

I will be offering more detailed and more technical arguments in this book, but here I would like to offer at least a preliminary critique of some terminology that I will want to question, such as the conventional usage of the terms "China," "the West," "modernity," "science," and "modern Western science": although these terms certainly have legitimate uses, and although some of them are employed by the historical actors themselves, I want to question the utility of adopting them as analytic categories in historical explanations.

"China" and "the West"

One of the central theses of this book is the need to question the common adoption in historical studies of the terms "China" and "the West" as explanatory or analytic categories. I will present arguments in this book to suggest that these two terms, adopted from the popular imaginary, have no place in rigorous historical analysis. Attempts in academic studies throughout the twentieth century to define "China" and "the West" rigorously have failed. To begin with, whatever these two terms are supposed to mean, they cannot simply be the names of places. Instead, they are asserted to refer to cultures or "traditions" that have remained somehow continuous over several millennia. Unfortunately, attempts to offer more precise characterizations of them have produced little more than a series of oppositions purported to distinguish them: "modern" versus "traditional," "democratic" versus "despotic," "scientific" versus "mystical," "theoretical" versus "practical," "rational" versus "intuitive," "capitalistic" versus "pre-capitalistic," and "Christian" versus "heathen," to name only a few. Not only are these characterizations mutually incongruous, they accord so poorly with the historical record that their proponents quickly retreat into lengthy evasions claiming them to be only "propensities" or "generalizations." In making these characterizations, proponents typically adopt one part of the West as representing the whole. On a moment's reflection, it is easily seen that any characterization meant to remain valid over an area as vast as the West, and over several centuries or even millennia, is likely to be far too clumsy to tell us anything specific about history.

Ironically, despite the critical turn in the scholarship of the late twentieth century, claims about something termed "the West" have remained commonplace. For it has been a continued credulity toward "the West" that has allowed poststructuralists and postcolonialists to inflate the importance of their academic (and often arcane) criticisms to prophecy a fall of the West. For example, claims of the deconstruction of Western metaphysics, which began with assertions of inconsistencies in a close reading of Plato's texts, were plausible only through an insistent credulity toward something called "the West" and an all-encompassing totality called "Western metaphysics." Similarly, for too many postcolonialist writers, a continued credulity toward "the West" allowed them to borrow the simplistic story of the universal triumph of Western civilization while inverting the moral: Western civilization is now all but universal not because it was true or better, but because it was hegemonic, dominating, colonizing, disciplining, or normalizing.

This is not to say that we can eliminate the terms "China" and "the West." Rather, we need to observe that these are actors' categories, not components of a historical explanation. Indeed, it is the ideological manipulation of these categories that the historian must explain. That is, whereas the historical figures studied in this book (along with later writers) do make various claims about "China" and "the West," instead of adopting their claims as historical conclusions, we must instead seek to understand them in their historical context—who made these claims, how and why they were made, and so forth. In short, instead of asking what happened when China first encountered the West, we must ask a series of very different questions: How were China and the West imagined by the Jesuits, the Chinese collaborators, and their detractors? How were these representations manipulated to further their agendas? How did later historians imagine China and the West, and what is the relation of these representations to earlier imaginings of China and the West by the Jesuits?

This distinction I am suggesting—between the ideologies of the historical actors themselves and the analytic explanations offered by historians—is crucial in historical research. Many, perhaps most, of the ideas and beliefs that historians analyze are ones that we may not believe in ourselves. This distinction causes no particular problem in cases where the historian is studying systems of thought that are no longer accepted. There is little danger that, in a modern study of the divine right of kings, a historian will adopt as his or her own conclusion the claims of the historical actors, for example that a particular historical figure became king because it was in fact his divine right. Instead, historians present a critical study in which claims of divine right are contextualized as an ideology that plays an important role in the legitimation of the king.

A problem arises, then, when historians (and readers) adopt as an explanatory framework the claims made by the historical actors. Here it is important that the historian explain these ideologies in context, without presenting the ideologies themselves as historical conclusions. For example, the historian should explain the crucial historical role that Marxist ideology played in the formation of revolutionary movements, without adopting as analytic historical explanations the Marxist claims that history is a progression of stages of economic development determined by contradictions between classes, with revolutions resulting from those contradictions. Similarly, the historian should contextualize the Jesuits' belief that historical events are the working of God's will, without adopting this belief as a historical explanation. And the historian should critically analyze the role that claims of an Aryan race and its glorious history played in the rise of National Socialism, without accepting this pernicious fiction as an explanation for historical developments. Similarly, I will argue, the historian must critically examine the claims that historical actors made about "China" and "the West," without adopting these categories as a framework for historical analysis. Unfortunately, too many good historians, even at present, adopt "China" and "the West" as analytic categories in historical explanations—some even argue that the central problem addressed by historians today should be why the West is unique.

"Ancient" and "Modern"

In questioning grand narratives about "China" and "the West," one false alternative I wish to avoid is presenting equally grand narratives about radical breaks. Grand narratives about such breaks share much with those about "China" and "the West." Indeed, claims of a temporal "great divide" between the ancient and the modern have been one of the principal devices through which a spatial "great divide" between the West and the Rest has been imagined. As I will argue in more detail, attempts to offer a concrete historical definition of the terms "modern" or "modernity" have failed as badly as attempts to define "the West." Although there are arguably styles within specific forms of art that are rightly termed "modern," these styles vary from one form to another, and have little in common with the asserted radical break in thought, science, society, institutions, economics, and politics that is termed the "modern." Indeed, attempts to distinguish the modern from its Other, like attempts to distinguish the West from the Rest, have produced little more than equally unending, incongruous, and fruitless lists of purported oppositions: Protestantism versus Catholicism (Hegel), positive science versus religion and metaphysics (Comte), capitalism versus the ancient and feudal (Marx), association versus community (Tönnies), bureaucratic rationality versus tradition (Weber), status versus contract (Maine), organic versus mechanical solidarity (Durkheim), commodity versus gift exchange (Mauss), territory versus kinship (Morgan), hot versus cold societies (Lévi-Strauss), pre-logical primitive mentality versus science (Lévy-Bruhl), complex versus simple societies (Spencer), individual versus holistic societies (Dumont), and printing versus pre-literacy, to name just a few.

Much of the work by historians of science in much of the twentieth century was focused on the purported break termed the "Scientific Revolution," which was asserted to have ushered the West into modernity. Again, the historical subjects of the period did often promote themselves and their projects through claims to being modern; historical explanations must critically contextualize these ideologies as acts of collective self-fashioning rather than accepting them as historical fact.

"Traditional China" and "Modern West"

In making these arguments, I seek also to avoid the vague disputes about continuities (or traditions) versus discontinuities. On the one hand, one can always posit continuities, even within the most radical breaks. On the other hand, one can just as easily posit discontinuities and breaks in precisely what is claimed to be a continuous "tradition." To explain this more concretely, assume that we sought to "compare" four entities: early China, the early West (whether conceived of as early Greece or seventeenth-century Europe), modern China, and the modern West. One can produce a lengthy list of similarities between early China and the early West, just as easily as between the early West and the modern West. One can just as easily produce lists of dissimilarities between early China and the early West as between early China and modern China. The resulting assertions of "essential" similarity and dissimilarity, of continuity and discontinuity, are so vague that they encourage endless manipulation into clumsy (and usually mutually contradictory) theses.

For example, one type of thesis resulting from claims of similarities has been that the development of the economic base determines the superstructure, and thus different civilizations are supposed to be essentially similar to each other at the same stage of economic development; in this view, early China would be fundamentally similar to the early West. Another thesis has argued for radical breaks between different periods in the history of the West (whether a radical break from ancient to modern, or a series of shifts of epistemes); in this view, the early West and the modern West differ radically. And yet another thesis (often asserted by employing the biological metaphors of development and teleologies) argues for essential continuities between the early West and the modern West; in this view, China and the West differ radically. Each of these theses relies on fictive constructs—whether mentalities, epistemes, or teleologies—through which the asserted unity of civilizations is imagined. Rather than remaining trapped in the resulting disputes about which of these vague, conflicting conceptualizations has merit, I seek to historicize these claims—to show how the manipulations of asserted continuities and discontinuities presented in historical narratives constitute one part of the very process through which these unities are imagined.

Imagined Civilizations

By way of criticizing the ideological operations behind the use of the term "civilizations," I will analyze them as "imagined." Obviously, I hope to borrow, suggestively, from Benedict Anderson's widely used term "imagined communities." That is, I wish to draw on the considerable scholarship that has analyzed the ways in which nations are not natural, pre-given categories but rather the products of a complex process of state-building and imagination: "imagined communities" refers specifically to the building of nations and in particular to the apparatus of the modern state—including bureaucracies, militaries, judiciaries, administrations, and other institutions—along with the creation and establishment of ideologies through which this amalgam is imagined as one. I propose to extend this criticism to civilizations, such as "the West."

   Civilizations are arguably far more illusory than nation-states. For example, "the West"—with no formal or official bureaucracy, military, judiciary, administration, or other institutions—has little if any of the apparatus or concrete embodiments of the nation-state. Furthermore, as I argue later in this book, it has proven impossible to offer any rigorous explanation for what the basis of the purported unity of a "civilization" might be, especially given that this unity must (allegedly) persist over a history of several millennia. I also hope to avoid the impulse—so evident in Ernest Gellner's work but also found in Benedict Anderson's—to critique non-Western nationalism as derivative, imagined, or rooted in forms of tribalism, in contrast to "the West," which remains uncriticized and even exalted. I will also use this same critical approach to analyze the asserted break between the "ancient" and the "modern": we should be skeptical of such asserted divisions, whether imagined spatially or temporally.

"Science" and "Western Science"

Another term that requires some preliminary discussion is "science," and in particular the combination "modern Western science." In its current usage, the term "science" denotes a very broad category. There are certainly many legitimate uses of the term "science" in this broad sense, such as in library classificatory systems, for organizing university departments into divisions (such as liberal arts, sciences, engineering), or in the titles of textbooks. The problem arises, however, when we rely on a term as vague as "science" as the basis for historical analysis. Unfortunately, as I argue in more detail in chapter 2, there is no precise definition for the term "science," and more important for this study, there is no rigorous way to distinguish historically that which is science from that which is not. That is, there are no methodologies, sets of beliefs, institutions, or attitudes that can be demonstrated to be common to *all* science in *all* historical periods. When we study earlier historical periods, we find that work in science often incorporated elements that seem familiar (for example, the application of mathematics, empirical observations, or experiment) alongside elements that may seem quite strange (including, for example, various ideologies, philosophies, cosmologies, or causal systems). Until the most recent several decades, studies in the history of science too often focused on those aspects of science in earlier periods that seemed familiar, ignoring aspects of science that seemed unfamiliar. Sometimes, these studies became little more than praise-and-blame narratives in which views that could be made to appear similar to modern views were extolled as scientific, and those that appeared to differ from modern views were derided as superstition. Such studies extolled Newton for his physics while ignoring his extensive work in alchemy as an embarrassment to him and to science itself.

This simplistic approach has been almost universally rejected by historians of science. More recent work in the history of science seeks to understand science in its cultural context, rather than to pass judgment using anachronistic standards not available in the period under study. Historians now seek to understand this work in the context of the science of the time. Thus the term "science" remains useful as a broad category, but we should not expect too much from it in terms of doing the precise work of historical explanation.

One helpful alternative to the term "science" is the use of the term "sciences." In this book, when I use the term "science," I will use it in this very broad manner, as a plurality of "sciences," as an inclusive term to cover a broad range of investigations, such as Pythagorean mathematics together with number mysticism, the astronomy and numerology of Kepler, and Newton's alchemy and physics. What counts as a "true" science or a "false" science is a controversy between the historical protagonists themselves, using standards very different from our own. In other words, I use the term "science" in a manner similar to the way one uses another inclusive term, "humanities," to categorize a broad range of academic disciplines. With these caveats in mind, I see no particular reason to avoid the term "science"; I will use "sciences" when I wish to emphasize their plurality.

Although I trust that these assertions will not be too controversial, at least for historians of science, I raise this issue because of the widespread use of the terms "Western science" and "modern Western science." Were we not so habituated to using terms such as "Western science," "Western philosophy," "Western literature," and "Western culture," it would require little more than a moment's reflection to realize that any attempt to characterize all the science, philosophy, literature, or culture of the West over several hundreds or thousands of years is impossible. Again, like the term "science," these terms do have legitimate uses, for example in the titles of textbooks or courses, but again, not as an analytic category in historical explanations.

"Patrons," "Collaborators," and "Converts"

I would also like to present at least a preliminary explanation for my choice of "patrons" and "collaborators" for the Chinese who worked together with the Jesuits, instead of the more conventional "converts." In the historical context of seventeenth-century China, Xu Guangqi was the Jesuits' patron: Xu held enormous power over his Jesuit collaborators; Xu devoted his later career to promoting Western Learning, and achieved success within the imperial court by doing so; Xu defended the Jesuits against attacks that could have resulted in their exile or execution. Similarly, Li Zhizao 李之藻 (1565–1630) and Yang Tingyun 楊廷筠 (1557–1628), who, together with Xu, have been called the "Three Pillars" (*san da zhushi* 三大柱石) of the Jesuit mission in seventeenth-century China, were also patrons of the Jesuits.

"Collaborators" is a more inclusive term, more effectively encompassing the wide range of Chinese who worked with the Jesuits along with those who in different ways may have supported various aspects of their mission. The term "collaborators" also reminds us of their vulnerability to possible charges of treason for aiding foreigners. As officials of the Ming dynasty, their ultimate loyalty was to be to the emperor and, of course, not to the Jesuits. I argue in this book that one of the fundamental "problems" of translation was not differences in languages or cultures, but rather the necessity for the collaborators to create and preserve the ambiguities that allowed them to profess loyalty to the Chinese emperor while producing documents that the Jesuits could claim were evidence of their success in proselytization. There is, however, one drawback in using the term "collaborators"—its connotation of passivity. I certainly do not wish to imply that the collaborators were passive: one of the central criticisms I offer of the received historiography is in fact its conceptualization of the collaborators as passive translators; indeed, this book focuses on the agency of the Chinese literati in their promotion of Western Learning.

"Convert," the term conventionally used for the Chinese who collaborated with the Jesuits, refers specifically to someone who adopted the Catholic faith. Assertions that Xu was a convert are based primarily on the Jesuits' claims, which later historians have too often repeated as fact. Xu venerated *Tianzhu* 天主 [Lord of Heaven], who Xu imagined belonged to the pantheon of sage kings of ancient China. Erik Zürcher has more accurately described the views of Xu as "Tianzhu-ism" or "Confucian monotheism."[1] To assert that Xu was in fact a convert, just because the Jesuits claimed that he was, is no more justified than asserting that the Jesuits were in fact tributary officials who traveled to China to serve the Ming dynasty, simply because Xu claimed so in a memorial to the emperor.

Another problem with the term "convert" is that it reverses power relations in seventeenth-century China: it then seems as if the Jesuits' Chinese patrons were subservient to the Jesuits; it also seems as if the Jesuits' interpretations of their doctrines should be the standard by which we must judge the views promulgated by their Chinese patrons. Although the Jesuits may have asserted as much, this does not help us understand the dissemination of Western Learning during the Ming dynasty. Xu, for example, was under no compulsion to accept any more of the Jesuits' doctrines than he pleased, especially because of his considerable power over the Jesuits. And Xu, by insisting that the doctrines of the Jesuits were lost doctrines from ancient China, retained the ultimate power to interpret these doctrines as he wished.

More broadly speaking, another fundamental problem with using the term "convert" is that it is best understood as an actors' category. That is, throughout the history of the Catholic mission in imperial China, who was to be seen as a convert was in fact a matter of considerable dispute among the historical pro-tagonists. Factions that supported the recruiting of Chinese collaborators were quick to claim that they had successfully converted the Chinese, glossing over differences in viewpoints and offering rationalizations and apologies for aspects of Catholic doctrines that individual Chinese collaborators did not accept. Con-versely, opposing factions sought to demonstrate that the Chinese had not in fact converted. Deciding who (if any) among the Chinese were true converts requires a belief in the correctness of one particular version of modern doctrinal standards (that is, we should presumably be applying standards that are not simply those of one group of historical protagonists or another). In general, whether or not specific collaborators fit a particular set of modern criteria for conversion is not central to my focus in this book on the history of science.

---

[1] Zürcher correctly observes that "[a]s in the case of some other late Ming converted literati, Xu Guangqi's creed can best be characterized as 'Confucian monotheism': it is almost exclusively focused on the belief in a single, all-powerful Creator-God, the controller of human destiny in life, and the stern judge in the hereafter. The whole complex of Rebirth and Redemption is hardly ever touched upon." Erik Zürcher, "Xu Guangqi and Buddhism," in *Statecraft and Intellectual Renewal in Late Ming China: The Cross-Cultural Synthesis of Xu Guangqi (1562–1633)*, ed. Catherine Jami, Gregory Blue, and Peter M. Engelfriet (Leiden: Brill, 2001), 162. See also Nicolas Standaert, "Erik Zürcher's Study of Christianity in Seventeenth-Century China: An Intellectual Portrait," *China Review International* 15 (2008): 476–502. I thank Nicolas Standaert for bringing Zürcher's term "Tianzhu-ism" and the latter article to my attention. Indeed, as we will see, we should not assume that Xu was a Christian: there is little if any evidence that Xu believed in the central doctrines of Christianity—salvation through Jesus Christ, his sacrifice on the cross, and his resurrection.

Origins of This Book

This book began as a macrohistory, and in fact many of the ideas that this book proposes to critique once served as the original framework for my research. And among these, the most important was the implicit assumption of two suprahistorical entities, imagined as "China" and "the West," that were to be a fundamental starting point for historical analysis. Because the conclusions of the research that I present here differ considerably from my original expectations—the changes that historical materials forced on my earlier hypotheses remain surprising to me even now—the genesis, then, of the research presented in this book perhaps deserves an explanation.

I was originally trained in mathematics—as a result of my interest in philosophy and a philosophical tradition in which mathematics was seen as the foundation for the study of philosophy (a short list of examples might include the early Greeks, Descartes, Leibniz, and Spinoza, up to Husserl, Russell, Whitehead, and Wittgenstein). I was particularly interested in questions related to theories of knowledge—questions of truth, certainty, scientific method, and proof. Work on the logical and axiomatic foundations of mathematics seemed the most rigorous formulation of these problems. I was fascinated by the foundations of modern mathematics (for example, as a sophomore in college I experimented with translating the set-theoretic foundations and theorems of real analysis into logical notation). I took courses on projective geometry (a system of geometry constructed on the axiom—contrary to Euclid's *Elements*—that two parallel lines meet at infinity) and the non-Euclidean geometry of general relativity. And of course, I was particularly intrigued by the stunning results that had brought into question the entire project of establishing the logical foundations and certainty of mathematics: Kurt Gödel's famous theorem on undecidability in which he showed, roughly speaking, that any finite axiom system is incomplete as the foundation of mathematics; the construction of axiomatic systems for modern mathematics in which the axioms are neither intuitively obvious nor necessarily true; the controversies surrounding the axiom of choice and its equivalence to the well-ordering theorem; and Ludwig Wittgenstein's criticisms of rule-following in mathematics as a "form of life," together with his criticisms of axioms as not unquestionable but only unquestioned.

At the same time, I pursued a somewhat naive interest in Chinese philosophy, and in particular Daoism and the radical skepticism of *Zhuangzi*:[2] to me, this seemed to offer a radical alternative to the Western philosophical tradition. In graduate school I became involved with Volunteers in Asia,[3] and went to China with them to teach for a year. Fascinated by Chinese culture, I eventually stayed for five years, immersing myself in Chinese philosophy, literature, history, and

---

[2] The translation that I read was *Chuang Tzu: Basic Writings*, trans. Burton Watson (New York: Columbia University Press, 1964).

[3] VIA (formerly Volunteers in Asia) is a private, nongovernmental, nonreligious, nonprofit organization based at Stanford University. See www.viaprograms.org.

culture. At the universities where I taught I took more than twenty courses and completed the readings in subjects ranging from classical poetry to literary criticism, political economy, and Maoist thought; on my own I read everything from classical philosophy to modern literature to dictionaries of modern and classical Chinese. I became fascinated with Chinese culture, living in Chinese teachers' dorms, and insisting on speaking in Chinese even with Westerners who had a modicum of Chinese. Although Chinese thought could no longer be for me an exoticized, radical Other, and although I could not have articulated what could possibly link together as "Chinese" the diversity of writings that included Daoism, Confucianism, Legalism, Chinese historical writings, Chinese poetry, literature, and Maoism, I continued to accept uncritically the viewpoint presented in the courses I took in China—that there was something called "the West," and something called "China," and that each differed fundamentally from the other.

When I returned to the United States, I continued my studies of Chinese thought at UCLA. Again, as in China, there was a pervasive assumption among scholars of a fundamental opposition between "China" and "the West." Comparative studies of literature, philosophy, society, politics, and economics often sought to explain the differences between the two, along with the consequences for their respective development. One recurrent theme in this scholarship was the absence of modern science, capitalism, or democracy in traditional China, as well as the impact of their respective introduction from the West. Other themes included the validity of applying Western theory to China: whether economic and sociological theories applied to China; whether China necessarily had to progress along the same path as the West to attain modernity; and the validity of the use of Western terminology as a translation of Chinese thought.

I chose to write my master's thesis on the Daoist text *Zhuangzi* 莊子 (350? to 50? BCE, attributed to Zhuang Zhou 莊周, fl. 320? BCE). Initially, I had hoped to compare Chinese and Western thought: in particular, I had hoped to compare the radical Daoist skepticism in *Zhuangzi* with Western deconstructive criticism. Yet as my research progressed, the very terms of comparison—Chinese Daoism and Western deconstruction—began to disintegrate, as categories too clumsy to be useful in analyzing these texts. My readings in what has been termed deconstruction, post-structuralism, or post-modernism soon convinced me of the futility of seeking essential features or even common impulses shared by these diverse writings. Instead of adopting as terms of analysis the vague, inflationary characterizations of deconstruction (such as relativism, the radical rejection of truth, or the undermining of Western metaphysics—characterizations too often preferred in polemics both for and against deconstruction), I learned to appreciate and analyze the specificity of these arguments in their historical context. Similarly, both philosophical analysis and philological research convinced me of the futility of trying to discover a kind of coherence within *Zhuangzi* that could serve as a basis for comparison with deconstruction.[4] In retrospect, I regret only that

---

[4] To be somewhat more precise, it has long been known that *Zhuangzi* is a compilation of writings of different writers from different periods. But research on *Zhuangzi* has often sought to demonstrate that it is a compilation by declaring certain parts (roughly the "Inner

the implications arising from the failure of this attempt to take the categories "China" and "the West" as the basis for comparing thought had not become more apparent to me in the initial formulations of my subsequent work.

During this period I read Liang Qichao's 梁 啟 超 (1873–1929) enthusiastic description of the introduction of Euclid's *Elements* into China by Matteo Ricci and Xu Guangqi.[5] Roughly speaking, I was interested in this topic because at that time it seemed to me that Western forms of reason, and in particular axiomatic deduction and logic, had played an essential role in the development of Western science, and especially mathematics. The origin of the axiomatic method could presumably be traced to Euclid's *Elements*. This then seemed to rigorously define a form of reasoning that was uniquely Western and had been crucial in the development of Western thought from its beginnings in early Greece up to the present. The secondary historical studies I read argued that the Chinese mathematical tradition had not developed an axiomatic method; many of these studies claimed (though without offering much supporting historical evidence) that the absence of Euclidean deduction had severely inhibited the development of Chinese science. The date of the introduction of the axiomatic method into China could be placed with reasonable certainty at the translation of the *Elements* in 1607; there were sufficient extant primary historical sources and secondary historical studies to support the writing of a history of the introduction of Euclid's *Elements* into China.

While I was in China I had the privilege of teaching a course on real and functional analysis; I knew from having read Chinese textbooks in preparation for that course that Chinese mathematics eventually became "axiomatic"—that is, current work in pure mathematics is written in a style of modern mathematical proof that seems (at least in appearance) to be axiomatic. It seemed to me, then, that this essential form of Western reason, fundamental to science and exemplified by the axiomatic method of Euclid's *Elements*, had originated in Greece but eventually was adopted in China. Wishing to understand this historical process, I chose this topic for my dissertation.

Although my interest was in thought—scientific and philosophical—my studies had also convinced me that the concrete results of comparative studies provide more insight into philosophical problems than do purely philosophical reflections.[6] I had thus hoped to follow in the tradition of applying historical

---

Chapters") to be the authentic work of Zhuang Zhou while asserting that the rest (roughly the "Miscellaneous" and "Outer" chapters) are the inauthentic product of later writers. That is, proof of the inauthenticity of the later chapters has been used, ironically, to reinforce the purported authenticity of the "Inner Chapters." For an analysis of the compilation of *Zhuangzi*, see Michael Loewe, ed., *Early Chinese Texts: A Bibliographical Guide* (Berkeley: Society for the Study of Early China, Institute of East Asian Studies, University of California, Berkeley, 1993).

[5] Liang Qichao, *Intellectual Trends in the Ch'ing Period*, trans. Immanuel C. Y. Hsü (Cambridge, MA: Harvard University Press, 1959).

[6] Perhaps this is most vividly explained in the following anecdote noted by Jacques Derrida: according to Maurice Merleau-Ponty, Edmund Husserl (whose phenomenological reduction of contingent fact was to allow the reflective mind to reveal all possiblities by a process of imaginary variation) late in life recognized the limits of a purely philosophical approach: "In a

research to philosophical problems,[7] an effort perhaps similar to what Pierre Bourdieu calls "fieldwork in philosophy."[8]

The early formulations of my work were marked by contradictions between conventional assumptions shared by the historical literature and more recent critical research controverting those assumptions—it is perhaps because of the disciplinary divisions of contemporary scholarship that even the most recent work in one field often incorporates outdated assumptions that have already been shown to be untenable by neighboring fields. For example, I continued to accept the assumption that axiomatic deduction and logic played a crucial role in the historical development of mathematics. I also assumed that axiomatic deduction differed radically from Chinese thought, and that its introduction into China was, as secondary studies claimed, of considerable importance. Yet at the same time I knew the problems attendant upon the asserted tie between mathematics, certainty, truth, and foundationalism. That is, in my research into the history of Chinese science I continued to assume—following the secondary historical literature on the subject—that axiomatic reason and logic were the foundation of mathematics, and that logic and mathematics were the foundation of scientific thinking. I assumed all this despite—and not out of ignorance of— the work in mathematics by Gödel demonstrating the limitations of axiomatic systems, the work in philosophy by Wittgenstein critiquing claims of certainty in mathematics, and more recent critical work in the history of science.

As I struggled to reconcile the conventional assumptions that framed my initial hypotheses with more recent research that controverted them, the new directions I explored continued to retain much of their initial formulations—and in particular, the assumption of a radical divide between China and the West. At this time I had become interested in questions about incommensurability through reading the works of Thomas Kuhn, Paul Feyerabend, and Benjamin Lee Whorf, together with related philosophical work on conceptual schemes by Willard Van Orman Quine, and anthropological work on the differences between primitive

---

letter to Lévy-Bruhl which has been preserved, Husserl seems to admit that the facts go beyond what we imagine and that this point bears a real significance. It is as if the imagination, left to itself, is unable to represent the possibilities of existence which are realized in different cultures. . . . [Husserl] saw that it is perhaps not possible for us, who live in certain historical traditions, to conceive of the historical possibility of these primitive men by a mere variation of our imagination." Maurice Merleau-Ponty, "Phenomenology and the Sciences of Man," in *The Primacy of Perception*, ed. James Edie, trans. John Wild (Evanston: Northwestern University Press, 1964), 90–91; quoted in Jacques Derrida, *Edmund Husserl's "Origin of Geometry": An Introduction*, trans. John P. Leavey, Jr. (Lincoln: University of Nebraska Press, 1978), 111. See also Merleau-Ponty's "The Philosopher and Sociology," in *Signs*, trans. Richard C. McCleary (Evanston, IL: Northwestern University Press, 1964), 98–113.

[7] For example, see Thomas S. Kuhn, "Introduction: A Role for History," in *The Structure of Scientific Revolutions*, 2nd ed. (Chicago: University of Chicago Press, 1970), 1–9.

[8] Pierre Bourdieu, "'Fieldwork in Philosophy,'" in *In Other Words: Essays Towards a Reflexive Sociology*, trans. Matthew Adamson (Stanford, CA: Stanford University Press, 1990), 3–33. Bourdieu attributes the phrase "fieldwork in philosophy" to J. L. Austin, *How to Do Things with Words*, 2nd ed. (Cambridge, MA: Harvard University Press, 1975).

and scientific thought. Incommensurability seemed to be a promising way to conceptualize the problems I was working on.

Historical work at that time on the Jesuits in China seemed to confirm further the importance of incommensurability. In the most philosophically sophisticated study of the Jesuits in China, Jacques Gernet's thesis asserted a fundamental linguistic incommensurability between Christian theological and Chinese philosophical concepts, which he linked to differing worldviews.[9] More specifically focused on Euclid's *Elements*, Jean-Claude Martzloff, one of the most eminent historians of Chinese mathematics, had argued that it was because of linguistic incommensurability that the Chinese had failed to comprehend the deductive structure of the *Elements*, precisely because of the impossibility of translating the copula (in English, the verb "to be") into classical Chinese language.[10] Yet after reading these works I had, as had others, remained unconvinced by their arguments.

At this point, my initial reaction was not to reject the notion of incommensurability (and the assumed dichotomy of China and the West), but to try to formulate ways to render it more rigorous. One possibility was suggested by Mario Biagioli's analysis of incommensurability as both social and diachronic: whereas Kuhn had taken what he interpreted as the incommensurability between scientific paradigms as a historical explanation for the non-dialogue between conflicting scientific communities, Biagioli argued that what appeared to be incommensurability emerged from willful misunderstandings in the process of the formation of competing scientific groups.[11] Following this and other attempts to differentiate forms of incommensurability,[12] I began to entertain what I considered to be a more careful application of incommensurability in the analysis of the introduction of Euclid's *Elements* into China. For at the core of the traditional formulations of incommensurability, it seemed to me, was a conflation of differing notions of incommensurability: Thomas Kuhn had explained the incommensurability of scientific paradigms using examples of the incommensurability of languages;[13] yet Benjamin Whorf had cited examples of differing scientific worldviews to explain the incommensurability of languages and the dependence of thought on language.[14] The translation of the *Elements* into Chinese seemed an ideal historical case of incommensurability between two distinct scientific traditions as well as between two linguistic systems.

---

[9] Jacques Gernet, *China and the Christian Impact: A Conflict of Cultures*, trans. Janet Lloyd (New York: Cambridge University Press, 1985).

[10] Jean-Claude Martzloff, *A History of Chinese Mathematics* (New York: Springer, 2006), first published as *Histoire des mathématiques chinoises* (Paris: Masson, 1988).

[11] Mario Biagioli, "The Anthropology of Incommensurability," *Studies in History and Philosophy of Science* 21 (1990): 183–209.

[12] For another proposal, see David B. Wong, "Three Kinds of Incommensurability," in *Relativism: Interpretation and Confrontation*, ed. Michael Krausz (Notre Dame, IN: University of Notre Dame Press, 1989), 140–58.

[13] See Kuhn, *Structure of Scientific Revolutions*.

[14] For example, Benjamin Lee Whorf, *Language, Thought, and Reality: Selected Writings* (Cambridge, MA: MIT Press, 1959), 214.

Roughly speaking, I had hoped to distinguish between what I termed linguistic, scientific, and social incommensurability. Linguistic incommensurability was to be seen between systems of languages that were already given, in this case the Chinese and European languages. Scientific incommensurability was to be seen between two systems of scientific thought, in this case traditional Chinese mathematical thought and Western mathematical thought exemplified by the axiomatic deduction of Euclid's *Elements*. (The underlying assumption, which I had failed to recognize, was the presumed existence of two entire systems posed against each other—one labeled Chinese and one labeled Western—as the terms of analysis.) And finally, social incommensurability was to explain the disputes between the Chinese converts and their opponents who, although they had shared in common the Chinese language, became speciated into two groups with two competing, incommensurable discourses.

If linguistic incommensurability had in some sense been overcome and Xu Guangqi had become an advocate of the axiomatic method, could I offer a sociological analysis of scientific incommensurability in the resistance to Western science by competing social groups? What I had in mind was undertaking a study that would follow the example of Marc Bloch's "The Advent and Triumph of the Watermill," which makes historically intelligible the seemingly inexplicably slow adoption of watermills.[15] That is, I hoped to write a social history of the introduction and acceptance of the axiomatic method that neither naturalized this acceptance nor simply caricatured resistance to it as unwarranted. Could the problematic of incommensurability itself be inverted by establishing that linguistic incommensurability, which originated in physical separation, found a solution in bilingualism? And yet, had scientific incommensurability re-emerged as a result of conflicts between social groups competing for legitimacy? It was within this theoretical framework that most of my early research was formulated, and the result was a set of questions remarkable only for their incongruence with my current project: How were mathematical truths determined in the absence of an axiomatic method? Did Chinese mathematics employ an alternative form of proof? Was the translation accurate? Which edition had been translated? Why were only the first six chapters translated? What were the Chinese reactions to the translation? Did the Chinese understand the importance of the axiomatic method? How did the early translation compare with later versions? Could this

---

[15] Bloch argues that although the advent of watermills dated from before 0 CE in Greece and 700–900 in Europe, the "triumph" of watermills did not occur until much later. He argues that it was neither a lack of technical inventiveness nor the necessity of employing labor that later slowed the implementation of the watermill, but rather the following factors: (1) the lack of rivers in some geographical locations; (2) ministerial cautions that besieged fortresses having watermills could succumb to famine if water was cut off; and (3) a preference for the mobility of portable mills. The success of the watermill was then not at "one stroke," Bloch argues, but rather the result of bitter, protracted social conflicts. Marc Bloch, "The Advent and Triumph of the Watermill," in *Land and Work in Mediaeval Europe: Selected Papers*, trans. J. E. Anderson (Berkeley: University of California Press, 1967), 136–68.

comparison of forms of Western and Chinese thought provide insight into the development of Chinese mathematics, and thus all of Chinese science?[16]

My approach to the translation of the *Elements*, however, encountered fundamental problems in the rigorous formulation and analysis of each of the three aspects of incommensurability—linguistic, social, and scientific. First was the problem of linguistic incommensurability, and the lack of any linguistic evidence to support the claim. Martzloff had presented a translation of classical Chinese into modern European languages to demonstrate that translation was impossible; but it seemed to me that the alleged difficulties in translation that he found resulted from his own mistranslation rather than any real impossibility. Indeed, secondary research that had not explicitly adopted the framework of linguistic incommensurability seemed to suggest that there was none: several sources maintained that the translation was accurate; the possibility of overcoming incommensurability through bilingualism seemed to find support in the number of terms created for the translation and still in use; and many articles repeatedly asserted (incorrectly, I later discovered) that Xu Guangqi had mastered the deductive system embodied in the *Elements* and had applied it to a wide variety of problems. And then the very claims of incommensurability made by Gernet and Martzloff seemed to be self-contradictory: their demonstrations of incommensurability required a framework within which the purportedly incommensurable schemes were both intelligible, and yet this was a framework that did not exist for the historical protagonists four hundred years ago. Ultimately, the most important evidence controverting the thesis of incommensurability lay in the primary historical documents themselves: in my reading of the *Jihe yuanben* 幾何原本 (the Chinese translation of Euclid's *Elements*), I failed to uncover any of the difficulties in translation that I had been led to expect. I found no particular difficulties in expressing the copula "to be," nor in expressing the concept of axioms, nor in deducing conclusions from them. I could find no particular problems in the proofs. I had somehow thought that proofs by contradiction would be a problem, yet even those seemed clear enough. In short, I saw no evidence to indicate that the translation of Euclid's *Elements* presented any more of a problem than does the translation of modern Western mathematical texts into modern Chinese.

The second problem was the difficulty of documenting social incommensurability: despite the considerable secondary literature on Ricci, Xu Guangqi, their translation, and the anti-Christian attacks against them, among these there were no materials documenting the reaction of Chinese mathematicians.[17] In this

---

[16] Although these questions all seemed reasonable enough at the outset of my research, I now see that they derived their significance from a set of implicit anachronistic assumptions about the importance of axiomatic deduction in the history of the development of mathematics, about the importance of the introduction of Euclid's *Elements* into China, and about the fundamental differences between China and the West. That is, these questions diverted attention from an understanding of these events in their historical context.

[17] The very few extant treatises by Ming dynasty authors responding to Euclid's *Elements* evince little knowledge of Chinese mathematics.

secondary literature there were no comprehensive studies of Ming mathematics, and in fact very few studies of any facet of Ming mathematics whatsoever. Instead, the historiography had focused almost entirely on the work of the Jesuits and their collaborators. The extant work of Chinese mathematicians had been ignored in these historical accounts—both Western and Chinese—which often were more concerned with extolling the virtues of Western science than with any analysis of the Chinese sciences of the period.

The most fundamental problem confronting my original project was, however, its assumption of scientific (or conceptual) incommensurability between Chinese and Western mathematics. I realize now that I had posited this supposed difference not as the result of a comparative analysis of Western and Chinese scientific treatises: rather, I had implicitly assumed incommensurability because of the pervasive presumption of radical differences between Chinese and Western thought, and in particular the assumption that something called "science," or "reason," was uniquely Western. And yet rarely had any of the studies making such claims attempted to document the absence of "science" or "reason" in China, or even to offer a plausible explanation of what such an assertion might mean; these grandiose conclusions were not, of course, based on any examination of Chinese scientific treatises. In sum, my choice of axiomatic deduction thus represented—not a questioning of the assumption of a fundamental opposition between China and the West—but an unintentional attempt to *preserve* this purported opposition by rendering it more rigorous and plausible, by narrowing it to one of the few remaining areas where historical research had not yet exposed the absurdity of this assumption.[18]

My hypotheses had been formulated with a certain blindness. For I knew the problems inherent in each link in the chain of reasoning in the received teleological account that asserted that axiomatic deduction, unique to the West and originating with the Greeks, was the foundation of mathematics and thereby Western science. I knew that the apparent continuity implied by the anachronistic use of the terms "science," "logic," and "axiom" masked important historical changes; I knew how profoundly the "informal" axiom system of Euclid differed from the "formal" axioms of the present.[19] I realized the implausibility of ahistorical claims that axiomatic deduction played a historically causal role in the development of most of mathematics—it had not been possible to axiomatize even the number system until considerably later. I was also familiar with the problems attending claims that mathematics was the foundation of all sciences.

Most important, my approach required an interpretive blindness in reading historical materials: I had dismissed entire passages as irrelevant, or simply accidents of historical circumstance. Thus my response was to see each of these

---

[18] G. E. R. Lloyd presents axiomatic-deductive demonstration as one of the central differences between Chinese science and Greek science. Lloyd, *Adversaries and Authorities: Investigations into Ancient Greek and Chinese Science* (Cambridge: Cambridge University Press, 1996), 62–63, 211ff.

[19] I have borrowed this terminology from S. C. Kleene, *Introduction to Metamathematics* (Princeton: Van Nostrand, 1952).

problems separately, all the while trying to correct each piece of the received story, rendering it more precise; I did not yet understand that what I should be questioning was the story as a whole.

In the face of these contradictions, it was the insistence of the historical sources that ultimately resulted in a change in my interpretive strategies.[20] I had at first—as had many others—allowed myself to read Xu Guangqi's preface to the translation of Euclid's *Elements* as advocating the axiomatic approach to mathematics. I finally noticed—and began to take seriously—his statements at the conclusion of his preface. There Xu explicitly claims that the import of the *Elements* is *not* mathematical, stating that the *Elements* has "lesser uses and greater uses" 小用大用. Xu explains that "the lesser [use] is to investigate things to fathom principle; one extremity of the principles of things is also called the numerical arts" 小者格物窮理，物理之一端別為象數. For Xu, the greater use is ultimately "self-cultivation to serve heaven" 修身事天. The reason, then, that Xu explicitly presents for translating the *Elements* is to show that Western Learning is beyond doubt: "I urgently transmitted the lesser, desiring to quickly first put forth what is easily believed, so that people could understand [Mr. Ricci's] writings and wish to see his intent and principles, and so understand that his learning can be believed without doubt" 余乃亟傳其小者，趨欲先其易信，使人繹其文，想見其意理，而知先生之學，可信不疑.[21] I began to see that Euclid's *Elements* belonged to a historical context very different from those projected through an anachronistic view of science,[22] yet one that was entirely intelligible to us today. This started the slow process of critically reformulating my earlier assumptions.

As I read more Chinese primary sources, a very different picture began to emerge. In particular, I was astonished by how little work had been done on the numerous extant Chinese mathematical treatises from the period immediately prior to the introduction of Euclid's *Elements*. Historians had more or less accepted as fact Xu Guangqi's pronouncements that Chinese mathematics was in a state of decline, and that what remained of it was vulgar and corrupt. According

---

[20] Hans-Georg Gadamer describes the process of reading as follows: "A person who is trying to understand a text is always projecting. He projects a meaning for the text as a whole as soon as some initial meaning emerges in the text. Again, the initial meaning emerges only because he is reading the text with particular expectations in regard to a certain meaning." Gadamer, *Truth and Method*, 2nd ed., trans. Joel Weinsheimer and Donald G. Marshall (New York: Crossroad, 1989), 267.

[21] This has been largely ignored by historians of science, but both Willard Peterson and Wang Ping have noted this passage. Peterson, "Why Did They Become Christians? Yang T'ing-Yun, Li Chih-Tsao, and Hsu Kuang-Ch'i," in *East Meets West: The Jesuits in China, 1582–1773*, ed. Charles E. Ronan and Bonnie B. C. Oh (Chicago: Loyola University Press, 1988), 129–52; Wang, *Xifang lisuanxue zhi chuanru* 西方曆算學之輸入 [The introduction of Western astronomical and mathematical sciences into China] (Taibei: Zhongyang yanjiuyuan jindaishi yanjiusuo 中央研究院近代史研究所, 1966), 15. I thank Nathan Sivin for the latter reference.

[22] See Thomas Kuhn's discussion of the "mental transposition" necessary to understand Aristotle. Kuhn, however, formulates this transposition in terms of incommensurability. Kuhn, *The Copernican Revolution: Planetary Astronomy in the Development of Western Thought* (Cambridge, MA: Harvard University Press, 1957), 95–99.

to Xu, Western mathematics was in every way superior, and this alleged loss of Chinese mathematics was no more to be regretted than discarding "tattered sandals." Yang Tingyun and Li Zhizao, who, along with Xu, were the most important patrons of the Jesuits, made similar claims. For historians, then, having accepted Xu's claims, there was little reason to investigate Ming mathematics.

As I studied extant Chinese mathematical treatises of the period, I became increasingly skeptical of Xu's claims. The most compelling evidence against his claims was to be found in Chinese work on linear algebra—arguably the most sophisticated and recognizably "modern" mathematics of this period in China. General solutions to systems of n linear equations in n unknowns were not commonly recorded in European mathematical treatises at the time. My research on the subject rapidly became not just a section, not just a chapter, but a book-length research monograph on linear algebra in imperial China.[23] It became apparent that there was considerable work on mathematics during the Ming dynasty. Thus, I will argue, it was not that the Jesuits, by introducing Western Learning to the Chinese, sparked a renaissance of interest in science in China. Rather, the Jesuits' Chinese patrons recognized the Ming court's interest in science and used it as a way to advance themselves by promoting Western Learning.

By tracing these mathematical practices preserved in extant treatises, I discovered something that provided a key insight into the strategies of the Jesuits' Chinese patrons. To support their claim that Western Learning was in every way superior, they purloined linear algebra problems from the very Chinese mathematical treatises they derided as vulgar, and published them in what they purported to be a translation of European mathematics. Chinese readers of the time had no way of knowing that these problems were not from European treatises. It thus became clear to me that it is not the case, as has previously been argued, that the Jesuits' Chinese patrons were first attracted to Western Learning because of its scientific superiority and then converted to Christianity. Rather, purported translations of European texts provided the Jesuits' Chinese patrons with an ingenious means to falsely assert the superiority of Western Learning, which they promoted in order to advance their careers.

In the course of writing this book, probably one of the most startling realizations for me was that it is mistaken to take, as previous studies had, the Jesuits as the principal historical actors. Like many of the mistaken assumptions made in previous studies, this assumption is in part the result of adopting the viewpoint of the Jesuit sources, and in part the result of anachronistically projecting history of the nineteenth and twentieth centuries onto seventeenth-century China. I came to see, as I noted above, that instead it was the Chinese patrons who were in positions of enormous power over their Jesuit collaborators. I can hardly overemphasize the importance of understanding this power relation: it is the opposite of what we expect in "encounters" of the West and non-West; it is the opposite of that described in most historical studies of colonialism and post-colonialism. The implications are profound: the Jesuits were not in a position

---

[23] Roger Hart, *The Chinese Roots of Linear Algebra* (Baltimore: Johns Hopkins University Press, 2011).

to dictate to their Chinese patrons what they should believe or write; we cannot project European historical context onto seventeenth-century China; and in particular, we cannot project European interpretations of texts—whether Euclid, Aristotle, or Catholicism—onto their Chinese translations. It therefore makes little sense to speak of any "misunderstanding": the Chinese patrons appropriated those aspects that were useful to them for their own purposes. Jesuit claims of the failure of their Chinese collaborators to understand their doctrines must be contextualized as alibis for failures of their mission.

Finally, I should mention important insights related to the history of mathematics that emerged through my research on linear algebra in imperial China. First, I came to understand that we must distinguish between mathematical practices and extant mathematical treatises, which preserve only fragmentary records of these practices. In the case of linear algebra in imperial China, through my research, I found that the literati who compiled mathematical treatises often understood only the rudiments of the practices they recorded. Indeed, throughout imperial China, since at least the Han dynasty (206 BCE–220 CE), aspiring literati in pursuit of imperial patronage presented to the imperial court treatises on a wide variety of arts—including mathematics, astronomy, medicine, metaphysics, and philology, to name only a few—along with prefaces claiming these arts to be essential to ordering the empire and intimating that the compiler should be appointed as an imperial advisor. It is to this long tradition that Xu's writings on mathematics belong (even though Xu's writings on mathematics are inferior to earlier treatises, this did not significantly hinder Xu's rise within the Ming bureaucracy). Second, evidence I discovered only after this book was essentially completed shows that distinctive linear algebra problems recorded in early Chinese mathematical treatises from the first century are also recorded and solved in European mathematical treatises beginning in the thirteenth century. Mathematical practices of this period—even the most sophisticated—were what might be termed non-scholarly: they did not require literacy, and they were not based on or transmitted primarily by texts. Together with commerce, art, and religion, these mathematical practices likely circulated throughout Eurasia by way of traders, travelers, merchants, and missionaries. It is to this long tradition that the Jesuits' purveyance of European mathematics in seventeenth-century China belongs.

Notes on Sources and Conventions

Because this book is intended for an interdisciplinary audience, I will briefly explain here some of the editorial decisions I have made in writing it, along with the conventions I have followed.

Background Reading

Following scholarship that takes a microhistorical and critical approach, I have not attempted to offer larger, descriptive syntheses of secondary historical literature that summarize the broader historical background (for example, of the social and economic history or intellectual trends in China during the Ming dynasty). Readers interested in larger overviews should consult the historical literature already available on the subject. For a general overview of Chinese history, Jacques Gernet's *A History of Chinese Civilization* is perhaps best.[24] One of the best collections of translations of primary sources on Chinese thought is Wm. Theodore de Bary's *Sources of Chinese Tradition*: volume 1 covers philosophical, religious, political, and historical writings from antiquity up to 1600; chapter 27 of volume 2, "Chinese Responses to Early Christian Contacts," includes several short translations from the Chinese collaborators and their detractors.[25] A good general bibliography of more specialized secondary historical studies can be found in John Fairbank and Merle Goldman's *China: A New History.*[26]

Beyond these more introductory works, the single most important study for my purposes herein is Benjamin Elman's *Cultural History of Civil Examinations in Late Imperial China,*[27] a critical analysis of Chinese thought in its social and institutional context; much of my work is based on his pioneering research. The best overview of the development of science, technology, medicine, and mathematics in China during this period is Elman's *On Their Own Terms: Science in China, 1550–1900;*[28] particularly important for this study are Parts I, "Introduction," and II, "Natural Studies and the Jesuits" (pp. 1–223), along with the extensive list of secondary sources in "Notes" (pp. 437–525), the "Bibliography of Chinese and Japanese Sources" (pp. 527–40), and Appendix 1, "Tang Mathematical Classics" (pp. 423–24). Joseph Needham's encyclopedic *Science and Civilisation in China* remains the most comprehensive resource on Chinese sciences.[29] And for an excellent introduction to work in the history of science, I would highly recommend Mario Biagioli's *The Science Studies Reader,* which also includes an

---

[24] Jacques Gernet, *A History of Chinese Civilization,* 2nd ed., trans. J. R. Foster and Charles Hartman (Cambridge: Cambridge University Press, 1996).

[25] Wm. Theodore de Bary et al., eds., *Sources of Chinese Tradition,* 2nd ed., 2 vols. (New York: Columbia University Press, 1999–2000).

[26] John King Fairbank and Merle Goldman, *China: A New History,* enl. ed. (Cambridge, MA: Belknap Press of the Harvard University Press, 1998).

[27] Benjamin A. Elman, *A Cultural History of Civil Examinations in Late Imperial China* (Berkeley: University of California Press, 2000).

[28] Idem, *On Their Own Terms: Science in China, 1550–1900* (Cambridge, MA: Harvard University Press, 2005).

[29] *Science and Civilisation in China,* ed. Joseph Needham, 7 vols. (Cambridge: Cambridge University Press, 1954–2008).

extensive list of further readings.[30] For important examples of a critical approach to civilizations, I recommend recent studies by Haun Saussy and Lydia Liu.[31]

For overviews of various aspects of the Ming Dynasty, I would highly recommend *The Cambridge History of China*: Part 1 presents the political history; Part 2 presents overviews of government, law, foreign relations, social and economic developments, communication, intellectual thought, interaction with the missionaries, and religion (including Buddhism and Daoism).[32] For biographical sketches of important historical figures of the period, see L. Carrington Goodrich and Chaoying Fang's *Dictionary of Ming Biography*.[33]

There have been numerous historical studies on the Jesuit mission in late Ming China, without which this book would not have been possible. Particularly important among Western-language studies are the pioneering works by Jonathan Spence, Willard Peterson, and Jacques Gernet, to name only a few, together with Jean-Claude Martzloff's research into Chinese mathematics, and Peter Engelfriet's research on the translation by Xu Guangqi and Matteo Ricci of Euclid's *Elements*.[34] Catherine Jami, Peter Engelfriet, and Gregory Blue have published an important collection of essays on Xu Guangqi.[35] Matteo Ricci's introduction to Catholic doctrines written for his Chinese audience, *The True Meaning of the Lord of Heaven*, has been translated into English.[36] Further translations of short passages from the debates between the collaborators and their enemies can be found within the text in Gernet's *China and the Christian Impact* and Nicolas Standaert's study of Yang Tingyun.[37] A reference book of particular importance is Standaert's encyclopedic *Handbook of Christianity in China*.[38] I also recommend four recent studies of various aspects of early Catholic missions in China: Liam Brockey's *Journey to the East*, Florence Hsia's *Sojourners in a*

---

[30] Mario Biagioli, ed., *The Science Studies Reader* (New York: Routledge, 1999).

[31] For example, see Haun Saussy, *Great Walls of Discourse and Other Adventures in Cultural China* (Cambridge, MA: Harvard University Asia Center, 2001); Lydia H. Liu, *The Clash of Empires: The Invention of China in Modern World Making* (Cambridge, MA: Harvard University Press, 2004).

[32] Frederick W. Mote and Denis Twitchett, eds., *The Ming Dynasty, 1368–1644*, 2 vols. (New York: Cambridge University Press, 1988, 1998).

[33] L. Carrington Goodrich and Chaoying Fang, eds., *Dictionary of Ming Biography, 1368–1644*, 2 vols. (New York: Columbia University Press, 1976).

[34] Peter M. Engelfriet, *Euclid in China: The Genesis of the First Chinese Translation of Euclid's Elements, Books I–VI (Jihe Yuanben, Beijing, 1607) and Its Reception up to 1723* (Boston: Brill, 1998).

[35] Jami et al., *Statecraft and Intellectual Renewal*.

[36] Matteo Ricci, SJ, *The True Meaning of the Lord of Heaven (T'ien-Chu Shih-I)*, trans. Douglas Lancashire and Peter Kuo-chen Hu, SJ (St. Louis, MO: Institute of Jesuit Sources, 1985).

[37] Gernet, *China and the Christian Impact* (see n. 9 on p. 14 of this book); Nicolas Standaert, *Yang Tingyun, Confucian and Christian in Late Ming China: His Life and Thought* (Leiden: E. J. Brill, 1988).

[38] Nicolas Standaert, ed., *Handbook of Christianity in China*, 2 vols. (Leiden: Brill, 2001).

Strange Land, Eugenio Menegon's *Ancestors, Virgins, and Friars*, and R. Po-chia Hsia's *A Jesuit in the Forbidden City*.[39]

Crucial evidence presented in this book is mathematical. My first book, *The Chinese Roots of Linear Algebra*, presents important evidence for the arguments presented here. For a broader overview of Chinese mathematics, see either Li Yan and Du Shiran's *Chinese Mathematics: A Concise History* or Jean-Claude Martzloff's *History of Chinese Mathematics*,[40] which includes an extensive bibliography on secondary historical research on Chinese mathematics.

Chinese Conventions

Several different systems are used to transcribe Chinese characters phonetically into English, and this may be a source of confusion for readers not trained in Chinese studies. I have chosen to follow the pinyin system that is currently used in the People's Republic of China and is now becoming the standard in Western scholarship. Older works in Western scholarship often used the Wade-Giles system, which is similar to pinyin, but uses somewhat altered spellings. For example, Xu Guangqi (the pinyin transliteration for the Chinese characters 徐光啟) is transliterated as Hsu Kuang-ch'i in the Wade-Giles system. In Chinese, one written Chinese character represents one spoken syllable. Because there are over 50,000 of these characters represented by just over 400 distinct spoken syllables, there can be hundreds of distinct characters sharing the same transliteration.[41] In the pinyin system, when two or more Chinese characters combine to form one word, the transliterations of the characters are also combined. For example, *shuxue* 數學 (mathematics) is a single word consisting of the two characters *shu* (number) and *xue* (learning).[42] Following standard conventions, I have silently

---

[39] Liam Matthew Brockey, *Journey to the East: The Jesuit Mission to China, 1579–1724* (Cambridge, MA: Belknap Press of Harvard University Press, 2007); Florence C. Hsia, *Sojourners in a Strange Land: Jesuits and Their Scientific Missions in Late Imperial China* (Chicago: University of Chicago Press, 2009); Eugenio Menegon, *Ancestors, Virgins, and Friars: Christianity as a Local Religion in Late Imperial China* (Cambridge, MA: Harvard University Press, 2009); R. Po-chia Hsia, *A Jesuit in the Forbidden City: Matteo Ricci 1552–1610* (New York: Oxford University Press, 2010).

[40] Li Yan and Du Shiran, *Chinese Mathematics: A Concise History*, trans. John N. Crossley and Anthony W.-C. Lun (Oxford: Clarendon Press, 1987); Martzloff, *History of Chinese Mathematics* (see n. 10 on p. 14 of this book).

[41] I am describing here modern Chinese in the standard Mandarin dialect; these numbers vary over different dialects and historical periods. In Mandarin, the spoken syllables are further distinguished by four distinct tones for each syllable. I have followed the standard practice of not including tonal markings in pinyin because of the typographic difficulties in doing so.

[42] It should be noted that the primary historical documents from seventeenth-century China used in this study were written in what is often referred to as classical or literary Chinese. Although there is a considerable overlap of characters, classical Chinese differs substantially from modern Chinese in syntax, and often in the meaning and usage of the individual characters. In comparison to modern Chinese, classical Chinese uses fewer multi-character words.

changed all other transliterations into pinyin, except in facts of publication such as the title of a book or an author's name, where one must follow the original transliteration no matter how unconventional. It should be noted that Needham's *Science and Civilisation in China* uses its own system of transliteration, as does the Brooks' translation of the Confucian *Analects*.[43]

Another source of confusion is Chinese names, which are written with the family name (usually one but sometimes two characters) first, followed by the given name (usually two but sometimes one character). In the pinyin transliteration, the transliterations of the characters that constitute each name are combined and capitalized. For example, Sima Qian 司馬遷 is a name with the two-character family name Sima and the single-character given name Qian. Chinese literati often used several different given names: their name given at birth (*ming* 名), their courtesy names (*zi* 字), their sobriquets (*hao* 號), along with posthumous titles (*shi* 諡) and various other honorifics. For example, Xu Guangqi's family name is Xu, his given name is Guangqi, his courtesy name is Zixian 子先, his sobriquet is Xuanhu 玄扈, his Christian name is Baolu 保禄 (Paul), his posthumous title is Taibao 太保, and his posthumous name is Wending 文定. For another example, Li Zhizao 李之藻 (1565–1630), who also collaborated with the Jesuits, used the courtesy names Zhenzhi 振之 and Wocun 我存, and the sobriquets Liang'anjushi 涼庵居士 and Cunyuanjisou 存圓寄叟, among others. I have followed the practice of using the most commonly used name. Usually, this is the given name, but, for example, the philosopher and statesman Wang Shouren 王守仁 (1472–1528) is more commonly known as Wang Yangming 王陽明. It should be noted that courtesy names, sobriquets and official or posthumous titles are sometimes used in the titles of collections of their writings. In unusual cases, names were changed to avoid taboo characters, usually the names of emperors. For example, one stroke was removed from the given name (second character) of Li Zhi 李治 to become Li Ye 李冶. In the Bibliography, names are listed with the family name first, but in the case of Chinese names I have omitted the comma because the names themselves have not been inverted.

Translation

In consideration of readers who are not in the field of Chinese studies, I have attempted to translate as much as possible into English. The difficulties of translation are vexing but not insurmountable. As I argue in later chapters, exaggerated claims of radical differences between cultures are often made through a

---

In my transliterations of classical Chinese into pinyin, I have followed the standard sources, such as the dictionary *Ciyuan* 辭源 (Beijing: Shangwu yinshuguan 商務印書館, 1991) and the convention of trying to avoid combining pairs of characters unless they are words in classical Chinese.

[43] E. Bruce Brooks and A. Taeko Brooks, *The Original Analects: Sayings of Confucius and His Successors; A New Translation and Commentary* (New York: Columbia University Press, 1998).

mystification of the problems of translation. In particular, it is sometimes argued that Chinese concepts differ radically from Western counterparts, and therefore the Chinese terms (in transliteration) must be used. In this book I argue that such claims are based on a naive view of language, simplistic assumptions of a correspondence between words and concepts, and ultimately, a credulity toward a radical divide between "China" and "the West." The problem of the use of translated terms here does not, I believe, fundamentally differ in nature from the problem of the use of contemporary English to discuss concepts from seventeenth-century Latin, Italian, or French. The principle I have followed is pragmatic—simply trying to ensure maximum intelligibility and precision.

For translating concepts, in cases where a Chinese term is well-known and I know of no particularly suitable English equivalent, I have used the conventional pinyin transliteration. One example is the philosophical term *qi* 氣 (*chi* in Wade-Giles), which is sometimes translated as "material force," sometimes as "ether," and sometimes as "energy." (I wish to emphasize, however, that the problem in translation here is not a radical incommensurability between Chinese and Western concepts. There is no "Chinese" concept of *qi*: the meanings assigned to this abstract term have been a matter of heated debate throughout the history of imperial China.) In cases where there is no acceptable equivalent, I have chosen the closest reasonable equivalent and noted possible differences. In cases where a term more accurate than the conventional one exists, I have used the better term. For example, I have chosen to translate the Chinese mathematical term *tianyuan* 天元 as "celestial origin algebra": because no suitable alternative exists, I have used the term "algebra" even though, as one scholar noted, celestial origin algebra does not meet the criteria of a certain very narrow definition of the term "algebra," since there is no explicit written symbol that represents the unknown variable; I have used the term "origin" instead of the more commonly used term "element," following Jean-Claude Martzloff's argument that the former is a better choice for translation of *yuan*.[44] In my translations of Chinese texts I have attempted when possible to consult existing translations, and, when appropriate, to cite them in the notes for the convenience of readers not trained in Chinese who might be interested in further reading; I have emended and silently corrected these translations where appropriate. I explain in the footnotes such important difficulties as may arise in the translations. For short phrases I have followed the convention of including both pinyin and Chinese characters, but for longer passages I have not transliterated into pinyin. In transcribing Chinese texts, for the sake of precision I have noted in the footnotes places where I have emended mistakes and idiosyncratic or variant characters in the text.

Again, for the convenience of the broader intended audience, for titles of Chinese primary sources and secondary historical research articles, I have used translations in the main text, instead of following the sinological tradition of using pinyin transliterations. This is one of the styles recommended by the

---

[44] These issues are further discussed in chapter 4, "Mathematical Texts in Historical Context."

Chicago Manual of Style,[45] and is especially helpful in languages where, unlike Latin or French, the reader of English cannot be expected to guess from cognates the meaning of the title. Indeed, the examples provided in the *Chicago Manual of Style* are for Russian titles; as an occasional reader of works in this field with no training in Russian, I have found the use of translated titles in these works particularly helpful. Furthermore, even for the specialist trained in Chinese, pinyin transliterations are no substitute for Chinese characters—it is often impossible to guess the actual title when titles are given only in transliteration without characters. I have used Chinese pinyin for titles only when no reliable translation is possible. Full titles with pinyin and characters are, of course, given in the first instance of each citation, and in the bibliography. For the sake of clarity and uniformity, I have silently changed transliterations of titles in quotations of secondary sources to translations of those titles. On the other hand, in footnotes and technical appendices intended only for specialists in the field, I have followed the usual sinological conventions and have not provided translations into English.

For Chinese primary historical sources, I have followed the standard bibliographic conventions (including, for example, when appropriate, recording the number of *juan* 卷).[46] When these materials are translated, I use the term "chapter" for *juan* 卷, "volume" for *ben* 本, and "collectanea" for *congshu* 叢書. Treatises published during this period often do not include important publication facts. Publication dates are unreliable; I have used the dates recorded in prefaces where they are available. We rarely know the number of copies printed. Treatises are often preserved in collectanea compiled in later periods; the abbreviations I have used for these collections can be found at the beginning of the Bibliography.

Because this book is concerned with historiography, and is intended for intellectual historians among others, the most important date is sometimes when a work was originally published. Therefore, in the Bibliography I also include facts about the first publication of a work, when they can be determined; the publication facts given in the footnotes are for the edition cited.

For translations of governmental offices and official titles, I have followed Charles Hucker's *A Dictionary of Official Titles in Imperial China.*[47] I have consulted standard sources to convert to the Western calendar.

---

[45] *The Chicago Manual of Style*, 16th ed. (Chicago: University of Chicago Press, 2010).

[46] For example, see *Tushuguan guji bianmu* 图书馆古籍编目 [Cataloging ancient Chinese books], ed. Beijing daxue tushuguan xuexi 北京大学图书馆学系 and Wuhan daxue tushuguan xuexi 武汉大学图书馆学系 (Beijing: Zhonghua shuju 中华书局, 1985).

[47] Charles O. Hucker, *A Dictionary of Official Titles in Imperial China* (Stanford, CA: Stanford University Press, 1985).

A Microhistorical Approach to World History of Science

My approach is microhistorical,[48] as I noted at the outset. That is, as an alternative to macrohistories of civilizations, this book adopts an approach to history that contributed to the transformation of studies in the history of science, from broad hagiographic and teleological narratives to more historical analyses of scientific practices in social context. Two seminal examples of this approach are Steven Shapin and Simon Schaffer's analysis of Boyle's experimental program, and Mario Biagioli's analysis of Galileo's patronage.[49]

From Hagiography to Science in Context

This shift in the history of science from hagiography to historical analyses of scientific practices in social context is perhaps best summarized in Steven Shapin's recent *Never Pure*.[50] To explain how this approach differs from its predecessors, Shapin outlines the formative views of the founder of the history of science, Harvard's George Sarton:

> It was *right* for the historian to be a hagiographer: "above all," Sarton wrote, "we must celebrate heroism whenever we come across it. The heroic scientist adds to the grandeur and beauty of every man's existence." Although Sarton at times observed that scientists had the full range of human vices as well as virtues, at other times he insisted that great scientists represented fallible human nature at its highest stage of development: "truth itself is a goal comparable with sanctity . . . the disinterested and fearless search of truth is the noblest human vocation." Science is "the very anchor of our philosophy, of our morality, of our faith," and it was our proper calling as historians to make that foundational role visible to the wider culture.[51]

---

[48] I will call the approach I describe below microhistorical, following Carlo Ginzburg's pioneering work. See especially Ginzburg, *The Cheese and the Worms: The Cosmos of a Sixteenth-Century Miller*, trans. John Tedeschi and Anne Tedeschi (Baltimore: Johns Hopkins University Press, 1980); idem, "Checking the Evidence: The Judge and the Historian," *Critical Inquiry* 18 (1991): 79–92; idem, "Microhistory: Two or Three Things that I Know about It," *Critical Inquiry* 20 (1993): 10–35.

[49] Steven Shapin and Simon Schaffer, *Leviathan and the Air-Pump: Hobbes, Boyle, and the Experimental Life* (Princeton, NJ: Princeton University Press, 1985); Mario Biagioli, *Galileo, Courtier: The Practice of Science in the Culture of Absolutism* (Chicago: University of Chicago Press, 1993). See also idem, *Galileo's Instruments of Credit: Telescopes, Images, Secrecy* (Chicago: University of Chicago Press, 2006); idem, "Postdisciplinary Liaisons: Science Studies and the Humanities," *Critical Inquiry* 35 (2009): 816–33.

[50] Steven Shapin, *Never Pure: Historical Studies of Science as If It Was Produced by People with Bodies, Situated in Time, Space, Culture, and Society, and Struggling for Credibility and Authority* (Baltimore: Johns Hopkins University Press, 2010).

[51] Shapin, *Never Pure*, 3–4.

Shapin then offers as a contrast the following "selective list" of the central points of departure for more recent studies of the history of science:

- You could say that science happens within, not outside of, historical time, that it has a deep historicity, and that whatever transcendence it possesses is itself a historical accomplishment.
- You could say that science similarly belongs to place, that it bears the marks of the places where it is produced and through which it is transmitted, and that whatever appearance of placelessness it possesses is itself a spatially grounded phenomenon.
- You could say that science is not one, indivisible, and unified, but that *the sciences* are many, diverse, and disunified.
- You could . . . say that there is *no* single, coherent, and effective Scientific Method that does the work that genius was once supposed to do, and even that there are no supposedly special cognitive capacities found in science that are not found in other technical practices or in the routines of everyday life.
- You could say that scientists are morally and constitutionally diverse specimens of humankind, that extraordinarily reliable knowledge has been produced by morally and cognitively ordinary people, and, further, that the ordinariness of *individual* scientists was not effectively repaired by any special virtues said to attach to their *communal* way of life.
- You could say that Truth (in any precise philosophical sense) is not a product of science, or that it is not a unique product. Or you could say that the historian is not properly concerned with Truth but with credibility, with whatever it is that *counts as Truth* in a range of historical settings.
- You could say that science is not pure thought but that it is *practice*, that the hand is as important as the head, or even that the head follows the hand.
- You could say that the making and warranting of scientific knowledge are *performances*, that those producing scientific knowledge can and do use a full range of cultural resources to produce these performances, and that these include displaying the marks of integrity and entitlement: expertise, to be sure, but also signs of dedication and selflessness. . . .[52]

This represents, Shapin asserts, "a short-list of the leading edges of change in the historical understanding of science over the past several decades."

I propose to extend this approach to world history: as an alternative to narratives about a purported encounter of China and the West, I analyze these events simply as a local collaboration between the members of a small group of Chinese scholar-officials and the Jesuit missionaries. In historical context, the conflict was not between China and the West, but between these collaborators

---

[52] Shapin, *Never Pure*, 5, emphasis in original.

and their enemies—for example, in China, the Confucian officials who sought their suppression, and in Europe, competing orders of the Catholic church.

Evidence from Mathematics

A distinctive feature of the microhistorical approach I employ in this book is my focus on mathematical evidence preserved in extant treatises. This approach develops from Carlo Ginzburg's work on clues, evidence, and historical methods. More specifically, Ginzburg writes about what has been called "Morelli's method" in art history, attributed to the Italian art critic Giovanni Morelli (1816–1891). Ginzburg notes that Morelli sought to distinguish original paintings from copies by examining "the most trivial details that would have been influenced least by the mannerisms of the artist's school: earlobes, fingernails, shapes of fingers and toes."[53] Extant mathematical treatises from seventeenth-century China provide a similar opportunity: the mathematics in these documents is unlikely to have been tampered with; it is possible to reconstruct and analyze the mathematical practices recorded in these documents and compare them with the practices recorded in other documents of the period.

In this respect, mathematics offers considerably more reliable evidence than other records of the period, for example, military, agricultural, technological, or even astronomical records. Previous accounts have often taken the statements of the Jesuits and the Chinese collaborators more or less as historical fact. Take the military, for example. Was Xu Guangqi expert on military matters? Were the Jesuit missionaries expert in casting cannons? Was it true, as claimed by Sun Yuanhua 孫元化 (1581–1632, baptized Ignatius in 1621), a disciple of Xu who attempted to implement Xu's military proposals, that "one of those [Western] cannons is worth a thousand [others]," "because if 'they are supported by equipment, the use of telescopes, and if the measurements are done according to the art of *gougu* [right triangles],' then it would be possible to make 'every shot a hit.'"[54] Was Sun's

---

[53] Carlo Ginzburg, "Clues: Roots of an Evidential Paradigm," chap. 5 in *Clues, Myths, and the Historical Method*, trans. John Tedeschi and Anne C. Tedeschi (Baltimore: Johns Hopkins University Press, 1989), 97. Ginzburg notes that this method resulted in Morelli "proposing many new attributions for works handing in the principal European museums. . . . Some of the new identifications were sensational" (p. 97). An earlier version of this chapter is published as Carlo Ginzburg and Anna Davin, "Morelli, Freud, and Sherlock Holmes: Clues and Scientific Method," *History Workshop* 9 (Spring 1980): 5–36.

[54] Huang Yi-Long, "Sun Yuanhua: A Christian Convert Who Put Xu Guangqi's Military Reform Policy into Practice," in Jami et al., *Statecraft and Intellectual Renewal*, 235; quotations are from Sun Yuanhua. I have chosen to use Huang's study as an example here because his very detailed analysis of Sun Yuanhua's military endeavors offers perhaps the most well-researched prima facia case in support of the asserted superiority of Western Learning. The central flaw in Huang's historical analysis—a flaw shared by most previous studies of the Jesuits in China— is that his argument often depends on accepting the claims of the Jesuits and their Chinese collaborators more or less as fact, while dismissing considerable evidence against Sun and the Chinese collaborators as politically motivated.

public beheading, after a military debacle, due to wrongful accusations? Or were perhaps some of the many charges against Sun, made by numerous ministers, credible? We cannot determine with certainty whether Sun's forays into training troops or producing cannons were successes or failures, nor can we reconstruct battles of the past. But if we dismiss as false the considerable evidence in numerous memorials accusing Sun of incompetence, we might at least countenance critical skepticism toward the scant evidence that seems to support Sun, such as claims of "each cannon shot killing 100 men."[55] Similarly, to take a second example, previous accounts have often taken as fact the claims of successful astronomical predictions made by the Jesuits and the Chinese collaborators. Was Xu Guangqi also expert on astronomy? Were the Jesuit missionaries also expert at predicting eclipses? Astronomical documents of the period, including records of eclipse predictions, are not only inconclusive but were subject to political manipulation.[56] In sum, what is certain is that the Jesuits and their Chinese patrons were experts in propaganda, eager to proclaim victory in their endeavors; we should no more accept their claims as fact than the Jesuits' claims about the superiority of their exorcisms and rain-making rituals over those of the Chinese.

In contrast, extant mathematical treatises preserve evidence from which we can reach definitive conclusions: Xu and his collaborators, as I will demonstrate in this book, were not expert on mathematics; the European mathematics brought by the Jesuits was not uniformly superior, and the Jesuits' Chinese patrons and collaborators knew it.

---

[55] Huang, "Sun Yuanhua," 237, quoting Peng Sunyi 彭孫貽 (1615–1673), *Shanzhong wenjian lu* 山中聞見錄 [Record of things seen and heard amidst the mountains], in *Xian Qing shi liao* 先清史料 [Historical materials of the early Qing] (Changchun: Jilin wenshi chubanshe 吉林文史出版社, 1990), hereinafter *XQSL*. Huang similarly reports without any expression of skepticism the Jesuits' claim that before Sun Yuanhua was executed, Johann Adam Schall von Bell (1592–1666, a German Jesuit missionary to China, who used the Chinese name Tang Ruowang 湯若望), "disguised as a laborer delivering mortar, managed to enter the prison and administer last rites to Sun." Huang, "Sun Yuanhua," 253.

[56] F. R. Stephenson and L. J. Fatoohi, "Accuracy of Solar Eclipse Observations Made by Jesuit Astronomers in China," *Journal for the History of Astronomy* 26 (1995): 227–36; L. J. Fatoohi and F. R. Stephenson, "Accuracy of Lunar Eclipse Observations Made by Jesuit Astronomers in China," *Journal for the History of Astronomy* 27 (1996): 61–67; John M. Steele, "Predictions of Eclipse Times Recorded in Chinese History," *Journal for the History of Astronomy* 29 (1998): 275–85; idem, *Observations and Predictions of Eclipse Times by Early Astronomers* (Dordrecht: Kluwer Academic Publishers, 2000); Yunli Shi, "Eclipse Observations Made by Jesuit Astronomers in China: A Reconsideration," *Journal for the History of Astronomy* 31 (2000): 135–47; Lingfeng Lü, "Eclipses and the Victory of European Astronomy in China," *East Asian Science, Technology, and Medicine* 27 (2007): 127–45. For a recent study of methods of eclipse prediction, see Clemency Montelle, *Chasing Shadows: Mathematics, Astronomy, and the Early History of Eclipse Reckoning* (Baltimore: Johns Hopkins University Press, 2011). I thank Lim Jongtae for bringing Lü's article to my attention.

An Outline of the Chapters

This book, then, presents a critical history of what has been called the first encounter of China and the West. It comprises seven chapters that critically explore different aspects of previous accounts of the Jesuits in seventeenth-century China.[57] The remaining six chapters are outlined below:

Chapter 2, "Science as the Measure of Civilizations," critically analyzes accounts that imagine "the West" by identifying it with "Science" (or, alternatively, "Modern Science"). My approach is historical. More specifically, I examine accounts of the history of science to analyze the various ways in which claims about science were framed within debates about civilizations. In the early twentieth century, science was identified exclusively with the West, and the West was imagined in opposition to China through claims about science. By the middle of the twentieth century, Joseph Needham, whose encyclopedic *Science and Civilisation in China* documented the considerable development of sciences in China, proposed an alternative configuration: premodern science, Needham asserted, was equally shared by the premodern world; it was modern science, beginning with the Scientific Revolution, that developed only in the West; and it was modern Western science that was universal. Nathan Sivin countered that scientific revolutions were not unique to the West, yet he still accepted the claim that only the West had produced *the* Scientific Revolution. Other writers continued to identify science as exclusively Western, ignoring the technical details specific to the sciences themselves, offering philosophical, social, linguistic, logical, and political explanations for the absence of science in China. Despite their differences, what these accounts share is the view that civilizations are the fundamental subject of world history, and science serves as their measure.

Chapter 3, "From Copula to Incommensurable Worlds," examines how differences in language are inflated into differences between civilizations, and more specifically, how studies of such seemingly abstruse aspects of language as the function of the copula (in English, the verb "to be") are inflated to assertions of incommensurability between China and the West. Differences in languages have often been adduced as powerful ways to imagine both unity and difference. Although differences in languages were important in imagining differences between nations (for example, English versus French), in the twentieth century these differences were often elided to assert that the languages of the West differed radically from those of the Rest. These asserted differences in languages then served as a stepping-stone for dramatic conclusions about the fundamental differences between civilizations. This chapter examines how translation, or more precisely, the alleged impossibility of translation, is used to imagine radical differences.

---

[57] My purpose here is not to present an overarching story of the Jesuits in China or of the life and times of Xu Guangqi.

Chapter 4, "Mathematical Texts in Historical Context," analyzes three mathematical treatises from the Yuan and Ming dynasties. In previous accounts of the "first encounter" of China and the West, the asserted decline of China served to highlight the West's ascendence. Chinese science has often been held to have been in decline during the Ming dynasty, and in particular, it has been asserted that during the Ming dynasty Chinese mathematics "had fallen into oblivion." Against this backdrop, the introduction of Euclid's *Elements* into China is asserted to have sparked a Chinese renaissance. This chapter analyzes three mathematical treatises in historical context to better understand the mathematics recorded in these treatises—Yuan dynasty Celestial Origin (*tian yuan* 天元) algebra, popular mathematics during the Ming dynasty, and mathematical work on precision calculations of the equal temperament of the musical scale.

Chapter 5, "Tracing Practices Purloined by the Three Pillars," focuses on reconstructing mathematical practices from evidence preserved in extant treatises from imperial China. Mathematics is special among the sciences in that we have textual records from which we can reconstruct the mathematical practices of the period. The historical evidence will show that claims for the superiority of Western mathematics made by the Jesuits' Chinese patrons are in historical context little more than propaganda.

Chapter 6, "Xu Guangqi, Grand Guardian," re-examines Xu Guangqi and his writings in historical context. Xu has often been viewed as a prescient advocate of nascent Western modernity: because of the putative decline in Chinese science and mathematics, he translated treatises on Western science and technology, including mathematics, agriculture, and military technologies; he tirelessly promoted reforms of the calendar, astronomy, agriculture, and the military; against orthodox Confucianism and Buddhist superstition, he advocated practical studies; and perhaps most famously, he translated Euclid's *Elements* into Chinese. This chapter provides a close reading of some of Xu's most important writings: his civil examination essays, his earliest extant patronage letter, his earliest extant work on Christianity, his earliest work related to mathematics, his prefaces to mathematical treatises, and most important, his memorial to the emperor in defense of the Jesuits. I will argue that Xu was a gifted scholar of the Confucian arts, who wrote prefaces, treatises, and memorials on a wide range of issues— ranging from mathematics and astronomy to agriculture and the military—in which he had little or no expertise, indiscriminately promoting Western Learning in his pursuit of imperial patronage.

Chapter 7, "Conclusions," presents a summary of the findings, and of the proposed microhistorical approach to world history.

Chapter 2
Science as the Measure of Civilizations

IMAGINING "CIVILIZATIONS" to be the central actors in historical drama, writers since the eighteenth century have devised a variety of defining features that were supposed to distinguish between them. Terms such as "modernity," "science," and "capitalism" headed the list of mutually incongruous candidates invoked to portray stark differences: the "non-West" was identified often by mere absence (for example, a lack of science, of capitalism, or of modernity) or else designated by pejoratives (for example, practical, intuitionistic, despotic). In this literature, with the Great Explanandum—the known uniqueness of the West— as the given starting point, practically any study of language, thought, society, institutions, or politics could be called into service as the explanation of what makes the West unique. Anthropomorphized through the assignment of personality traits (pride, xenophobia, conservatism, fear), civilizations—"the West," "China," and "Islam"—became the central protagonists in praise-and-blame histories. Simplistic teleologies of science provided universal benchmarks by which to measure the progress of civilizations toward modernity; the purported radical break between the ancient and the modern in the West was transposed onto a fictive Great Divide between the primitive non-West and the modern West. Ignorance of the sciences of non-Western civilizations was mistaken for ignorance of science in those civilizations.

Yet despite overwhelming historical evidence to the contrary, claims that science is uniquely Western have continued to appear, even in the most respected scholarly publications in the history of science. Important examples include the works of Toby Huff and A. C. Crombie.[1] The view that science is uniquely Western

---

[1] Toby E. Huff, *The Rise of Early Modern Science: Islam, China, and the West* (Cambridge: Cambridge University Press, 1993); idem, *Intellectual Curiosity and the Scientific Revolution: A Global Perspective* (Cambridge: Cambridge University Press, 2011); Alistair C. Crombie, *Styles of Scientific Thinking in the European Tradition: The History of Argument and Explanation Especially in the Mathematical and Biomedical Sciences and Arts*, 3 vols. (London: Duckworth, 1994); see also idem, "Designed in the Mind: Western Visions of Science, Nature and Humankind," and idem, "The Origins of Western Science," in *Science, Art and Nature in Medieval and Modern Thought* (London: Hambledon Press, 1996).

is most clearly articulated by Crombie, who, in his review of one of Huff's earlier books, quotes from the introduction to his own multivolume work, presenting the following theses: "the history of science as we have it is the history of 'a vision and an argument initiated in the West by ancient Greek philosophers, mathematicians and physicians'"; "it is 'a specific vision, created within Western culture'"; it was based on the Greeks' "two fundamental conceptions of universal natural causality matched by formal proof"; "'from these two conceptions all the essential character and style of Western philosophy, mathematics, and natural science have followed.'" In contrast, Crombie asserts, "China had no Euclid, and did not adopt his scientific style when that became available."[2]

Rather than returning to take sides within these debates, this chapter will take the framework that has preconditioned these controversies as itself the object of historical analysis. That is, I shall analyze here what these accounts share: the assumption that civilizations—"China" and "the West"—are to be fundamental starting points in analyses of the history of science; that to the West and China we can then rigorously assign antithetical pairs of attributes (e.g., scientific versus intuitive, theoretical versus practical, causal versus correlative thinking, adversarial versus irenic, or geometric versus algebraic) that remain valid across historical periods, geographic locales, social strata, gender identifications, economic and technological differentials, and domains of scientific research, along with their subdomains and competing schools; and that ultimately, studies of the history of science can contribute to the further assignment of normative attributes in praise-and-blame historiographies of civilizations (e.g., the uniqueness of the West in producing universal science, the xenophobia of China, or the equality of all civilizations).

Perhaps it was the vagueness inherent in concepts as broad as "civilization" and "science" that encouraged such easy answers to questions about their relationship. With the existence of suprahistorical entities called civilizations established by assumption, the central question became how to determine what exactly characterized them. What were the essential features that distinguished one civilization from another? Given an assumed Great Divide that existed between the West and its Other, this question was often posed in a quite specific form: What was it that made the West unique? Yet the very requirements that these essential defining features were supposed to fulfill presented something of a paradox: these features were to be transhistorical, holding sway across spans of hundreds or thousands of years; they were also to be unique—confined within the boundaries of a single civilization (for example the West)—and thus their antitheses were to characterize other civilizations, again over hundreds or thousands of years. To be sure, transhistorical continuity was often constructed through metaphors of growth and development; uniqueness was often defended through claims of propensities and emphasis. Moreover, although these distinguishing features were to define a single civilization, there was to

---

[2] Alistair C. Crombie, review of *Rise of Early Modern Science*, by Toby E. Huff, *Journal of Asian Studies* 53 (1994): 1213–14; quoted from Crombie, *Styles of Scientific Thinking* 3–5. It should be noted that the number of specialists who adhere to such viewpoints is decreasing.

be a sense in which it remained possible to compare them: all civilizations were by definition in some sense unique, yet the ultimate conclusions reached were often comparisons of civilizations along normative teleologies of moral, political, scientific, or economic progress. If all civilizations are by definition unique, one civilization—the West—is asserted to be more unique than others.

To respond to this conundrum, almost any field of research purporting to discover essential features of the West could be called into service. For some writers, the fundamental differences were linguistic: alphabetic versus allegedly ideographic scripts, the existence versus nonexistence of the copula, scientific versus poetic, theoretical versus practical, or abstract versus concrete; these traits were then linked to the development of rigorous scientific language or efficient bureaucracies.[3] In other accounts, the fundamental difference was capitalism, which itself ushered in modernity. In yet other accounts, the key was religion: Max Weber improbably connected the differences he alleged to have discovered between Protestantism and Chinese religions to the development of capitalism.[4] For still others, the fundamental differences were philosophic: conceptions of natural law, causal versus correlative thinking, the ordering of time and space, demonstrative logic versus consensus.[5] For others, the fundamental difference was political—democracy versus Oriental despotism.[6] This list represents but a fraction of claims for the key features distinguishing the West from the Rest; the search continues to this day.[7]

Remarkably, these were not just conclusions emerging from studies of the West—virtually every subdiscipline of sinology has made its own contribution to the imagining of fundamental oppositions between China and the West. In philosophy, in the 1920s Homer Dubs assessed various explanations for what he assumed, as his title indicates, was "the failure of the Chinese to produce philosophical systems."[8] In economics, it was capitalism. In literature, in the 1950s, John Bishop argued that in comparison with the standards set by Western fiction, Chinese fiction was unoriginal, lacking in personality, monotonously

---

[3] For example, see Jack Goody, *The Logic of Writing and the Organization of Society* (Cambridge: Cambridge University Press, 1986); Jacques Gernet, *China and the Christian Impact: A Conflict of Cultures*, trans. Janet Lloyd (New York: Cambridge University Press, 1985).

[4] Max Weber, *The Religion of China: Confucianism and Taoism*, trans. Hans H. Gerth (Glencoe, IL: Free Press, 1951). Weber also invoked double-entry bookkeeping for good measure.

[5] For example, see Joseph Needham, *Human Law and the Laws of Nature in China and the West* (London: Oxford University Press, 1951); Derk Bodde, "Evidence for 'Laws of Nature' in Chinese Thought," *Harvard Journal of Asiatic Studies* 20 (1959): 709–27; and idem, "Chinese 'Laws of Nature': A Reconsideration," *Harvard Journal of Asiatic Studies* 39 (1979): 139–55.

[6] The classic is Karl August Wittfogel, *Oriental Despotism: A Comparative Study of Total Power* (New Haven, CT: Yale University Press, 1957).

[7] Samuel P. Huntington, *The Clash of Civilizations and the Remaking of World Order* (New York: Simon & Schuster, 1996); Niall Ferguson, *Civilization: The West and the Rest*, 1st American ed. (New York: Penguin Press, 2011).

[8] Homer Dubs, "The Failure of the Chinese to Produce Philosophical Systems," *T'oung Pao* 26 (1929): 96–109.

focused on plot, and immoral;[9] other studies argued that Chinese literature had no allegory.[10]

Among these themes, the most persistent, most debated, and purportedly the most important of these differences between China and the West has been in the realm of science. Work in the early twentieth century often argued that China had no science whatsoever.

Teleologies of Science and Civilizations

The contention that science is uniquely Western has never been presented as a thesis to be demonstrated historically—that is, stated explicitly, formulated rigorously, evaluated critically, and documented comprehensively. Rather, throughout much of the twentieth century, variants on this theme frequently appeared in panegyrics for Western civilization ("Science . . . is the glory of Western culture"),[11] in the forgings of exalted origins for the West in Greek antiquity ("science originated only once in history, in Greece"),[12] and in accounts that confidently offered explanations for the absence of science in other civilizations—accounts thus unencumbered by any requirement to examine sciences already known to be absent. As presented, these were hardly simple assertions of differential developments of specific sciences in particular geographic areas during particular historical periods. Instead these accounts asserted a Great Divide between civilizations, the West and its Others.[13] One particularly dramatic formulation was Ernest Gellner's "Big Ditch" symbolizing the enormous differences separating "traditional" societies from the scientific "Single World or Unique Truth" produced by "one *kind* of man."[14] Such assertions—although apparently about the West—should have depended for their validity on investigations of other, non-

---

[9] John Bishop, "Some Limitations of Chinese Fiction," *Far Eastern Quarterly* 15 (1956): 239–47. For a critical discussion of these assertions see Eugene Eoyang, "A Taste for Apricots: Approaches to Chinese Fiction," in *Chinese Narrative: Critical and Theoretical Essays*, ed. Andrew Plaks (Princeton: Princeton University Press, 1977), 53–69.

[10] For a summary and critique of these claims see Haun Saussy, *The Problem of a Chinese Aesthetic* (Stanford: Stanford University Press, 1993).

[11] Henry E. Kyburg, Jr., *Science & Reason* (New York: Oxford University Press, 1990), 3.

[12] Lewis Wolpert, *The Unnatural Nature of Science* (Cambridge, MA: Harvard University Press, 1994), 35.

[13] For examples of debates over this kind of a "Great Divide," see Robin Horton and Ruth Finnegan, eds., *Modes of Thought: Essays on Thinking in Western and Non-Western Societies* (London: Faber, 1973). For criticisms of assertions of a "Great Divide," see Clifford Geertz, "Anti Anti-Relativism," in *Relativism: Interpretation and Confrontation*, ed. Michael Krausz (Notre Dame, IN: University of Notre Dame Press, 1989), 19; and Bruno Latour, *We Have Never Been Modern*, trans. Catherine Porter (Cambridge, MA: Harvard University Press, 1993), 97–109.

[14] Ernest Gellner, *Spectacles & Predicaments: Essays in Social Theory* (Cambridge: Cambridge University Press, 1979), 145–47; and idem, "Relativism and Universals," in *Rationality and Relativism*, ed. Martin Hollis and Steven Lukes (Cambridge, MA: MIT Press, 1982), 188–200.

Western cultures. However, the historical evidence accompanying such claims related only to the uncontroversial half of the assertion—the existence of sciences in the West. The substantive half—the assertion of the absence of sciences everywhere else—rested on little more than the writers' own ignorance of the sciences of other cultures, mistaken for other cultures' ignorance of science. The most important historical counterexample was China—research beginning in the 1940s increasingly provided considerable evidence that there were in China many practices similar to those that have been labeled "science" in the West.[15] This, then, is the reason that "Chinese science" became a problem.

In the context of this broader literature, "science" was then but one possible solution among many that have been offered to explain the purported differences between "civilizations." Comparative studies, however, rarely even attempted to offer any precise criteria for defining what they might mean by the term "science." Often, "science" was simply left undefined in these studies (this was sometimes rationalized by scientistic claims that "science" was to be a primitive undefined concept); elsewhere "science" was defined by invoking equally amorphous terms such as "reason" or "rationality."[16]

In other accounts, the mere existence of lexical terms (most often the word "science" itself, under proper translation, in Latin, Italian, or English) was supposed to suffice to guarantee the existence of sciences; and then—again under suitable translation—the lack of the lexical term "science" in Chinese seemed to demonstrate that China lacked sciences. Overlooked is that the Latin term *scientia* referred to the study of universal, necessary truths of Aristotelian philosophy; because many of the diagrams used in the proofs in Euclid's *Elements* were considered to be not necessary but accidental, many of the results were held not to be necessary truths. And as Steven Shapin notes, studies of things in nature were termed "natural history" and investigations into the natural world were termed "natural philosophy"; "the term 'scientist' was invented only in the nineteenth century and was not in routine use until the early twentieth."[17]

In other accounts still, it was scientific methodology that defined science. At their least sophisticated, these asserted methodologies were little more than the familiar ideologies of scientists culled from selective readings of the early Greeks, Galileo, Bacon, or Newton. Elsewhere, it was axiomatization that was

---

[15] The most important example is the pioneering work of Joseph Needham. See Needham's multivolume series *Science and Civilisation in China*, ed. Joseph Needham, 7 vols. (Cambridge: Cambridge University Press, 1954–2008). For criticisms of Needham's work, see Nathan Sivin, *Science in Ancient China: Researches and Reflections* (Brookfield, VT: Ashgate, 1995). For Needham's publications (up to 1973), see "Bibliography of Joseph Needham," in Mikulas Teich and Robert Young, eds., *Changing Perspectives in the History of Science: Essays in Honour of Joseph Needham* (Boston: D. Reidel, 1973).

[16] For example, Lukes' attempt to offer a transcultural definition of the term "rationality" ends up with little more than the principle of non-contradiction. See Steven Lukes, "Some Problems about Rationality," in *Rationality*, ed. Bryan R. Wilson (Evanston: Harper & Row, 1970), 194–213.

[17] Steven Shapin, *The Scientific Revolution* (Chicago: University of Chicago Press, 1996), 6–7, n. 3.

held to differentiate scientific traditions;[18] sometimes it was deduction or logic, originating with the Greeks, or Descartes, or Galileo; sometimes it was preferred analogies—natural law, the book of nature, or the mechanical universe.

There are, however, good reasons to believe that the required definition of "science" cannot in fact be made, for it would require rigorous philosophical criteria capable of demarcating science from non-science—criteria that would remain both transhistorically and transculturally valid.[19] First, it has proven impossible to offer criteria sufficient to demarcate "science" from "non-science":[20] the term "science" is not a "natural kind,"[21] nor can it be defined by a simple description or disjunction of descriptions; it has been impossible to offer a definition of "science" that could claim both to encompass all of the sciences and to exclude what is not science.[22] The difficulty with defining "science" by some unified methodology is suggested by Feyerabend's criticism that "*the events, procedures and results that constitute the sciences have no common structure*; there are no elements that occur in every scientific investigation but are missing elsewhere."[23] Second, such a definition for "science" would have to be transhistorical—the definition would have to apply equally well to the present and to ancient Greece (in accounts that place the origin of science there) or seventeenth-century Europe (in the case of modern science).[24] Third, such a definition for "science" would have to be transcultural if it were to avoid the charge of simply circularly invoking particular

---

[18] This claim at least had the merit of specifying a seemingly more concrete, identifiable methodology, and one that apparently can be found in ancient Greece, early modern Europe, and modern mathematics.

[19] For recent critiques, see Peter L. Galison and David J. Stump, eds., *The Disunity of Science: Boundaries, Contexts, and Power* (Stanford, CA: Stanford University Press, 1996).

[20] There are enough important philosophical arguments against adopting naive views to warrant demanding a precise definition. Important examples include Ludwig Wittgenstein, *Philosophical Investigations*, trans. G. E. M. Anscombe (Oxford: Blackwell, 1958); Paul Feyerabend, *Against Method* (London: Verso, 1988); Thomas S. Kuhn, *The Structure of Scientific Revolutions*, 2nd ed. (Chicago: University of Chicago Press, 1970); Willard Van Orman Quine, "Two Dogmas of Empiricism," in *From a Logical Point of View: 9 Logico-Philosophical Essays*, 2nd ed. (Cambridge, MA: Harvard University Press, 1961); idem, "2 Dogmas in Retrospect," *Canadian Journal of Philosophy* 21 (1991): 265–74.

[21] For an overview of these arguments see Richard Rorty, "Is Science a Natural Kind?" In *Objectivity, Relativism, and Truth: Philosophical Papers* (Cambridge: Cambridge University Press, 1991), 24–30.

[22] In a more general form, the problem here may be that linguistic terms often lack central, essential meanings, but instead seem to share what Wittgenstein called a family resemblance of uses.

[23] Feyerabend, *Against Method*, 1, emphasis in the original.

[24] And again, we encounter a more general problem of transhistorical definitions: radical changes often lie behind the apparent nominal continuity of a word. In contrast to the term "science," it is much easier to offer precise definitions of specific scientific or mathematical terms, such as calculus. The existence of terms such as logic or axiomatic deduction can hardly assure us of the existence of science.

sciences in particular localities (e.g., ancient Greece, early modern Europe) as its essential defining forms.[25]

These writers of the twentieth century, then, sought to provide answers to questions about the relationship between science and civilizations without any particularly clear formulation of either of the concepts that would serve as the framework for their inquiry. But rather than attempting to provide for these authors the definitions that they themselves never used—definitions that arguably do not exist—we must instead chart the history of debates in which it was the very ambiguities of these terms that became the site of ideological contest. How did differing visions of civilizations and their relationships inflect conceptions of "science" and the writing of the history of science? What role did changing accounts of the fundamental defining features of "science" play in narrating the histories of civilizations?

The Scientific West and the Intuitive East

Accounts written in the early twentieth century contrasted the scientific West and intuitive East; and China, part of an undifferentiated "East," lacked science entirely.[26] Bertrand Russell, after lecturing in China, wrote in 1922 in a chapter entitled "Chinese and Western Civilization Contrasted," that "comparing the civilization of China with that of Europe, one finds in China most of what was to be found in Greece, but nothing of the other two elements of our civilization, namely Judaism and science. . . . Except quite recently, through European influence, there has been no science and no industrialism."[27] In the 1940s, Filmer Northrop, a professor of philosophy at Yale University, posited suprahistorical differences between "Eastern intuitive" and "Western scientific" philosophical systems representative of entire civilizations, arguing that "a culture which admits only concepts by intuition is automatically prevented from developing science of the Western type."[28] Similarly, Wilmon Sheldon, also a professor of philosophy at

---

[25] There is no reason to assume that political or even linguistic boundaries are congruent to those of the networks of scientific knowledge. And more generally, such a claim would seem to demand some rationale for the choice of geopolitical constructs—terms such as the West and China—a rationale seemingly tangential to the study of questions of cognition and epistemology if it were to escape the charge that it was a transparent attempt at glorification.

[26] These claims, as noted above, were just one part of a literature of the period asserting Chinese absence of modernity, capitalism, science, and philosophy. For criticisms of claims that China lacked mathematics, see John [Jock] Hoe, *Les systèmes d'équations polynômes dans le Siyuan yujian (1303)* (Paris: Collège de France, l'Institut des hautes études chinoises, 1977), 7–8.

[27] Bertrand Russell, "Chinese and Western Civilization Contrasted," in *The Basic Writings of Bertrand Russell, 1903–1959*, ed. Robert E. Egner and Lester E. Denonn (New York: Simon & Schuster, 1961), 551; reprinted from Russell, *The Problem of China* (London: Allen & Unwin, 1922).

[28] Filmer S. C. Northrop, "The Complementary Emphases of Eastern Intuitive and Western Scientific Philosophy," in *Philosophy—East and West*, ed. Charles A. Moore (Princeton, NJ: Prince-

Yale, contrasted Eastern and Western philosophy, asserting bluntly that "the West generated the natural sciences, as the East did not."[29] Albert Einstein, in a casual letter frequently quoted by later historians, stated in 1953 that the "development of Western science is based on two great achievements: the invention of the formal logical system (in Euclidean geometry) by the Greek philosophers, and the discovery of the possibility to find out causal relationship by systematic experiment (Renaissance). In my opinion one has not to be astonished that the Chinese sages have not made these steps. The astonishing thing is that these discoveries were made at all."[30] These accounts offered no analysis of Chinese sciences, presumably because there were not supposed to be any to analyze.

It was to disprove such claims that Joseph Needham began his *Science and Civilisation in China*, which soon developed into a multivolume series documenting the developments in China in chemistry, mathematics, astronomy, physics, and other sciences.[31]

---

ton University Press, 1946), 168–234, quoted in Hu Shih, "The Scientific Spirit and Method in Chinese Philosophy," in *The Chinese Mind: Essentials of Chinese Philosophy and Culture*, ed. Charles A. Moore (Honolulu: University of Hawaii Press, 1967), 104. A shorter version of Hu's article appeared as "The Scientific Spirit and Method in Chinese Philosophy," *Philosophy East and West* 9 (1959): 29–31. See also Filmer S. C. Northrop, *The Meeting of East and West: An Inquiry Concerning World Understanding* (New York: Macmillan, 1946), chaps. 8–10. Northrop asserted a "unity of Oriental culture" (a "single civilization of the East" that included China, Japan, and India) and distinguished an entire Western scientific philosophy from an Eastern intuitive philosophy (p. 312).

[29] Wilmon Henry Sheldon, "Main Contrasts between Eastern and Western Philosophy," in *Essays in East-West Philosophy: An Attempt at World Philosophical Synthesis*, ed. Charles A. Moore (Honolulu: University of Hawaii Press, 1951), 291; quoted in Hu, "Scientific Spirit and Method," 104. The assertions by Northrop and Sheldon are criticized by Hu, who presents the views of two Chinese philosophers, Wang Chong (27–c. 100) and Zhu Xi (1130–1200). But Hu never cites evidence from Chinese scientific treaties.

[30] Quoted in Arthur F. Wright, review of *Science and Civilisation in China*, Vol. 2, *History of Scientific Thought*, by Joseph Needham, *American Historical Review* 62 (1957): 918 and Wolpert, *Unnatural Nature of Science*, 48. Cited in Charles C. Gillispie, *The Edge of Objectivity: An Essay in the History of Scientific Ideas* (Princeton, NJ: Princeton University Press, 1960), 9; and Robert M. Hartwell, "Historical Analogism, Public Policy, and Social Science in Eleventh- and Twelfth-Century China," *American Historical Review* 76 (1971): 722–23.

[31] *Science and Civilisation in China* comprises the following volumes: vol. 1, *Introductory Orientations*; vol. 2, *History of Scientific Thought*; vol. 3, *Mathematics and the Sciences of the Heavens and the Earth*; vol. 4, *Physics and Physical Technology*—pt. 1, *Physics*, pt. 2, *Mechanical Engineering*, pt. 3, *Civil Engineering and Nautics*; vol. 5, *Chemistry and Chemical Technology*—pt. 1, *Paper and Printing*, pt. 2, *Spagyrical Discovery and Invention: Magisteries of Gold and Immortality*, pt. 3, *Spagyrical Discovery and Invention: Historical Survey, from Cinnabar Elixirs to Synthetic Insulin*, pt. 4, *Spagyrical Discovery and Invention: Apparatus, Theories, and Gifts*, pt. 5, *Spagyrical Discovery and Invention: Physiological Alchemy*, pt. 6, *Military Technology: Missiles and Sieges*, pt. 7, *Military Technology: The Gunpowder Epic*, pt. 9, *Textile Technology, Spinning and Reeling*, pt. 11, *Ferrous Metallurgy*, pt. 12, *Ceramic Technology*, pt. 13, *Mining*; vol. 6, *Biology and Biological Technology*, pt. 1, *Botany*, pt. 2, *Agriculture*, pt. 3, *Agro-Industries and Forestry*, pt. 5, *Fermentations and Food Science*, pt. 6, *Medicine*; vol. 7, pt. 1, *Language and Logic*, pt. 2, *General Conclusions and Reflections*.

Within this civilizational dispute, historians of [Western] science often perceived Needham's research not as one part of a larger project of the study of the history of science, but as a project in opposition to their own work. In the late fifties and early sixties they continued to insist that science was exclusively Western: in response to studies of the sciences of other civilizations (and Needham's in particular), the criteria *defining* "science" changed; but the defining *boundaries* of science as exclusively Western did not. For example, in A. C. Crombie's account, the Orient became differentiated into distinct civilizations, but the "achievements" of these distinct civilizations were undifferentiatedly dismissed as technologies. Western science was no longer defined solely in stark opposition to Oriental intuition. In its place, "Western science" was defined by an incongruous amalgam of "essential elements" culled from the tradition claimed for the West, including noncontradiction, empirical testing, Euclid, and logic:

> Impressive as are the technological achievements of ancient Babylonia, Assyria, and Egypt, of ancient China and India, as scholars have presented them to us they lack the essential elements of science, the generalized conceptions of scientific explanation and of mathematical proof. It seems to me that it was the Greeks who invented natural science as we know it, by their assumption of a permanent, uniform, abstract order and laws by means of which the regular changes observed in the world could be explained by deduction, and by their brilliant idea of the generalized use of scientific theory tailored according to the principles of noncontradiction and the empirical test. It is this essential Greek idea of scientific explanation, "Euclidean" in logical form, that has introduced the main problems of scientific method and philosophy of science with which the Western scientific tradition has been concerned.[32]

In another account—even though the defining features of the "mainstream" of science were new—the defining boundaries of science as Western remained stubbornly constant. For de Solla Price, this "mainstream" was the emblem of scientific modernity—mathematical astronomy; instead of attributing the development of science to a scientific method, he appealed to "inspiration" as historically causal. Thus mathematical astronomy differentiated "our own high civilization" from its Other:

> What is the origin of the peculiarly scientific basis of our own high civilization? ... Of all limited areas, by far the most highly developed, most recognizably modern, yet most continuous province of scientific learning, was mathematical astronomy. This is the mainstream that leads through

---

[32] Alistair C. Crombie, "The Significance of Medieval Discussions of Scientific Method for the Scientific Revolution," in *Critical Problems in the History of Science*, ed. Marshall Clagett (Madison: University of Wisconsin Press, 1959), 81; quoted in Joseph Needham, "Poverties and Triumphs of the Chinese Scientific Tradition," in *The Grand Titration: Science and Society in East and West* (London: George Allen & Unwin, 1969), 41–42. Needham's response to this view is discussed below. For a detailed presentation of Crombie's views, see his *Styles of Scientific Thinking*.

the work of Galileo and Kepler, through the gravitation theory of Newton, directly to the labours of Einstein and all mathematical physicists past and present. In comparison, all other parts of modern science appear derivative or subsequent; either they drew their inspiration directly from the success-ful sufficiency of mathematical and logical explanation for astronomy, or they developed later, probably as a result of such inspiration in adjacent subjects.[33]

Primitive versus Modern Science

Needham too shared the assumption that civilizations were to be a fundamental starting point in studies of the history of science: in place of science versus non-science, he offered his own set of four major contrasts between China and the West—organic versus mechanical philosophies, algebra versus Euclidean geom-etry, wave versus particle theories, and practical versus theoretical orientations;[34] his "grand titration" was to redistribute credit for scientific discoveries among civilizations;[35] he proposed to restore for China its pride, correcting its slighting by making it an equal contributor among the tributaries that flowed into the river of modern science; ultimately, he sought to discover the social and economic reasons why Chinese civilization was more advanced than the West before the sixteenth century and later fell behind.[36]

Needham's project was from its inception formulated not as one component of, but rather in opposition to, mainstream history of science that asserted that science was unique to Western civilization. Yet Needham's project adopted many of the features of these histories of [Western] science of the period: against the catalogues of scientific achievements claimed for the glory of the West, he offered achievements now claimed for the Chinese; against the exaggerated claims of

---

[33] Derek J. de Solla Price, *Science since Babylon* (New Haven, CT: Yale University Press, 1961), 2–5; quoted in Needham, "Poverties and Triumphs," 42.

[34] Needham, "Poverties and Triumphs," 20–23. This essay was first published in Alistair C. Crombie, ed., *Scientific Change: Historical Studies in the Intellectual, Social and Technical Con-ditions for Scientific Discovery and Technical Invention, from Antiquity to the Present* (London: Heinemann, 1963). Needham (like G. E. R. Lloyd) affirms these oppositions with some reserva-tions, for example stating that he does "not want to disagree altogether with the idea that the Chinese were a fundamentally practical people" (p. 23). See also the comments and discussion on Needham's article in Crombie, *Scientific Change*.

[35] Needham proposed a retrospective competition between the West and China, fixing dates of discovery through a "grand titration" that compared "the great civilizations against one another, to find out and give credit where credit is due." Needham, *Grand Titration*, 12.

[36] Needham sought to "analyse the various constituents, social or intellectual, of the great civilizations, to see why one combination could far excel in medieval times while another could catch up later on and bring modern science itself into existence." Ibid.

Western contributions to other civilizations, Needham asserted Chinese influence where the evidence was incomplete.

The two views presented above—those of Crombie and de Solla Price—were written in response to the discoveries of the sciences of other civilizations, and to Needham's work in particular; these views themselves elicited a response from Needham.[37] Needham noted that discoveries of the sciences of other cultures resulted not in the rejection of claims of European uniqueness but rather in the deprecation of the sciences of other cultures: "As the contributions of the Asian civilizations are progressively uncovered by research, an opposing tendency seeks to preserve European uniqueness by exalting unduly the role of the Greeks and claiming that not only modern science, but science as such, was characteristic of Europe, and of Europe only, from the very beginning. . . . The counterpart of this is a determined effort to show that all scientific developments in non-European civilizations were really nothing but technology."[38]

But as attention to Needham's phrase "not only modern science, but science as such" suggests, this criticism was not Needham's central thesis. Needham presented two problems that became the central "Needham questions" defining the field of the history of Chinese science:

Why did modern science, the mathematization of hypotheses about Nature, with all its implications for advanced technology, take its meteoric rise *only* in the West at the time of Galileo? This is the most obvious question which many have asked but few have answered. Yet there is another which is of quite equal importance. Why was it that between the second century BCE and the sixteenth century CE East Asian culture was much *more* efficient than the European West in applying human knowledge of nature to useful purposes?[39]

Critics who saw in Needham an exaggerated attempt to rehabilitate Chinese science ignored his ultimate reaffirmation of modern science as uniquely Western—Needham did not dispute the radical break between the scientific and nonscientific, but only the manner in which the boundary was drawn. For Needham, this break derived directly from accounts that asserted a radical divide in the West between the ancient and modern by adding to "science" the even more amorphous term "modern": "When we say that modern science developed only in Western Europe at the time of Galileo in the late Renaissance, we mean surely that there and then alone there developed the fundamental bases of the structure of the natural sciences as we have them today, namely the application of mathematical hypotheses to Nature, the full understanding and use of the

---

[37] Needham, "Poverties and Triumphs." Needham presents views similar to those of Crombie and de Solla Price from Gillispie, *Edge of Objectivity*, and from J. D. Bernal (Needham provides no reference for his quotation from Bernal).

[38] Needham, "Poverties and Triumphs," 41.

[39] Ibid., 16, emphasis in original. Needham proposed the following solution: "Only an analysis of the social and economic structures of Eastern and Western cultures, not forgetting the great role of systems of ideas, will in the end suggest an explanation of both these things." Ibid.

experimental method, the distinction between primary and secondary qualities, the geometrisation of space, and the acceptance of the mechanical model of reality."[40] And indeed Needham's central concern is this supplemental term "modern": "Hypotheses of primitive or medieval type distinguish themselves quite clearly from those of modern type. Their intrinsic and essential vagueness always made them incapable of proof or disproof, and they were prone to combine in fanciful systems of gnostic correlation. In so far as numerical figures entered into them, numbers were manipulated in forms of 'numerology' or number-mysticism constructed *a priori*, not employed as the stuff of quantitative measurements compared *a posteriori*."[41] Thus against schemes that posited a radical difference between civilizations East and West, Needham insisted on preserving the uniqueness of modern Western science by claiming that the premodern world—including China and Greece—"must be thought of as a whole";[42] the radical break for Needham was the boundary between the modern and the primitive.[43]

Notwithstanding Needham's brief list of the characteristics of modern science—experimentalism, mathematization, geometrization, and mechanism—these were hardly the central features animating his discussion of modern science. Instead, for Needham the central distinction between primitive and modern science was its universality: "Until it had been universalized by its fusion with mathematics, natural science could not be the common property of all mankind. The sciences of the medieval world were tied closely to the ethnic environments in which they had arisen, and it was very difficult, if not impossible, for the people of those different cultures to find any common basis of discourse."[44] Needham then incorporated science into this universal teleology: "the river of Chinese science flowed, like all other such rivers, into this sea of modern science."[45] And by the concluding paragraph, Needham's "science" has become nothing more than an impoverished signifier in a teleology purely utopian: "Let us take pride enough in the undeniable historical fact that *modern* science was born in Europe and only in Europe, but let us not claim thereby a perpetual patent thereon. For what was born in the time of Galileo was a universal palladium, the salutary enlightenment of all men without distinction of race, colour, faith or homeland, wherein all can qualify and all participate. Modern universal science, yes; Western science, no!"[46]

The view Needham adopted, of a radical break in Western thought between the ancient and modern, was a central theme of work in the history of science

---

[40] Needham, "Poverties and Triumphs," 14–15.

[41] Ibid., 15. Needham's characterization of "primitive" (i.e., premodern) science is not true for all of either the early Greek sciences or the early Chinese sciences.

[42] Ibid., 16.

[43] Needham states, "Galileo broke through its walls"; "the Galilean break-through occurred only in the West." Ibid., 15.

[44] Ibid.

[45] Ibid., 16.

[46] Ibid., 54, emphasis in original.

in the first half of the twentieth century; precisely what that shift was and when it occurred was a central topic of debate. For Pierre Duhem, it was debates over Aristotle and Averroes in Paris in the fourteenth century; for Alexandre Koyré, it was the shift from the closed world of the ancients to the infinite universe of the moderns; for Thomas Kuhn, it was the change from a geocentric to a heliocentric model in astronomy; for A. C. Crombie, it was the experimental work of Robert Grosseteste; and this is to name only a few of the most important theses from the history of science.[47]

Scientific Revolutions versus *The* Scientific Revolution

The most important response to the two Needham questions—why China was more proficient at technology before the sixteenth century and why modern science arose only in the West—was a series of criticisms presented by Nathan Sivin.[48] Against the former, Sivin argued that in the period from the first century BCE to the fifteenth century CE, science and technology were separate, and thus Chinese superiority in technology was not indicative of more advanced science;[49] he also criticized attempts to compare the science and technology of civilizations in their entirety.[50] In response to the latter—Needham's "Scientific Revolution problem"—Sivin critiqued several assumptions underlying the question why China lacked a scientific revolution and pointed out fallacies of historical reasoning that discovered conditions asserted to have inhibited the

---

[47] Pierre Duhem, *The Aim and Structure of Physical Theory* (Princeton, NJ: Princeton University Press, 1991); Alexandre Koyré, *From the Closed World to the Infinite Universe* (Baltimore: Johns Hopkins University Press, 1957); Thomas S. Kuhn, *The Copernican Revolution: Planetary Astronomy in the Development of Western Thought* (Cambridge, MA: Harvard University Press, 1957); Alistair C. Crombie, *Robert Grosseteste and the Origins of Experimental Science, 1100–1700* (Oxford: Clarendon Press, 1958).

[48] Sivin's criticisms of the Needham questions begin on the first page of his first book—see Nathan Sivin, *Chinese Alchemy: Preliminary Studies* (Cambridge, MA: Harvard University Press, 1968), 1. Some of Sivin's more notable critiques first appeared in the following articles: Sivin, "Copernicus in China," in *Colloquia Copernica II: Études sur l'audience de la théorie héliocentrique*, vol. II (Warsaw, 1973), 63–122; idem, "Wang Hsi-Shan" and "Shen Kua," in *Dictionary of Scientific Biography*, ed. Charles C. Gillispie (New York: Scribner, 1970–1980), s.v.; idem, "Why the Scientific Revolution Did Not Take Place in China—or Didn't It?" *Chinese Science* 5 (1982): 45–66. Revised versions of these essays have been collected in Sivin, *Science in Ancient China*; unless otherwise noted, I have used the revised versions. See also idem, "Max Weber, Joseph Needham, Benjamin Nelson: The Question of Chinese Science," in *Civilizations East and West: A Memorial Volume for Benjamin Nelson*, ed. Eugene Victor Walter (Atlantic Highlands, NJ: Humanities Press, 1985), 37–49.

[49] Sivin, "Why the Scientific Revolution Did Not Take Place," 46.

[50] Sivin argues that although before the fifteenth century Europe was technologically less advanced, Chinese astronomy even in the fourteenth century was less accurate than Ptolemaic astronomy which predated it by a thousand years. Ibid., 46–47.

growth of Chinese science.[51] Sivin's ultimate response, however, was to assert that "by conventional intellectual criteria, China had its own scientific revolution in the seventeenth century."[52] This revolution was not, Sivin argued, as sweeping as the Scientific Revolution in Europe.[53]

Sivin's claim was part of his criticism of the received accounts of the rejection of modern Western science by a xenophobic, conservative, traditional China.[54] Against portrayals of the Jesuits as having introduced modern science, Sivin argued that the Jesuits withheld the Copernican system, instead presenting the Tychonic system as the most recent, and misrepresenting the history of Western astronomy to disguise this.[55] Against claims that the lack of Chinese acceptance of early modern science was due to intellectual, linguistic, or philosophical impediments, Sivin argued that it was contradictions in the Jesuits' presentation of Western astronomy—including these misleading characterizations of Copernican astronomy that the Jesuits were by decree forbidden to teach—that made it incomprehensible.[56] And against caricatures of the Chinese as xenophobic and conservative, Sivin concluded that the Chinese did accept Western astronomical techniques, resulting in a "conceptual revolution in astronomy."[57]

---

[51] Sivin, "Why the Scientific Revolution Did Not Take Place," 51–59.

[52] Ibid., 62.

[53] In his analysis of the introduction of European astronomy into China by the Jesuits, Sivin asserts that Wang Xishan 王錫闡 (1628–1682), Mei Wending 梅文鼎 (1633–1721), and Xue Fengzuo 薛鳳祚 (c. 1620–1680) were responsible for what Sivin terms a scientific revolution— a fundamental conceptual change from numerical to geometric methods in astronomy. Sivin states, "They radically and permanently reoriented the sense of how one goes about comprehending the celestial motions. They changed the sense of which concepts, tools, and methods are centrally important, so that geometry and trigonometry largely replaced traditional numerical or algebraic procedures. Such issues as the absolute sense of rotation of a planet and its relative distance from the earth became important for the first time. Chinese astronomers came to believe for the first time that mathematical models can explain the phenomena as well as predict them. These changes amount to a conceptual revolution in astronomy." Ibid.

[54] For example, George C. Wong argued that Chinese scholar-officials' opposition to Western science was due to their traditional beliefs. Wong, "China's Opposition to Western Science during the Late Ming and Early Ch'ing," *Isis* 54 (1963): 29–49; see also Sivin's critical response, "On China's Opposition to Western Science during the Late Ming and Early Ch'ing," *Isis* 56 (1965): 201–5.

[55] Sivin asserts that the cosmology presented by the Jesuits from 1608 to 1642 was written from an Aristotelian point of view: Johannes Schreck (1576–1630), Giacomo Rho (1593–1638), and Johann Adam Schall von Bell (1592–1666) did not mention Galileo. Sivin states that Schall's "quasi-historical treatise" *On the Transmission of Astronomy in the West* included Copernicus and Tycho with Ptolemy in the ancient school; the modern school comprised Schall, Rho, and their collaborators; the summary of Copernicus's *De revolutionibus* did not mention the motion of the earth; and Schall's conclusion asserted that astronomy had not progressed beyond Ptolemy. For the Jesuits in China, Sivin notes, "heliocentricism was unmentionable." Sivin, "Copernicus in China," 19–22.

[56] European cosmology was discredited, Sivin argues, by its internal contradictions, and only in the mid-nineteenth century were contemporary treatises introduced that resulted in acceptance of the heliocentric system. Ibid., 1, 50.

[57] Sivin, "Why the Scientific Revolution Did Not Take Place," 62 (see n. 53 above).

Sivin's response, however, incorporated many of the assumptions within which the claims he critiqued had been framed. And like Needham, Sivin also incorporated into his criticisms many of the assumptions of teleological histories of Western science. For example, although Sivin in his work emphasizes the importance of studying science in cultural context, his analysis often implicitly assigns to internalist scientific criteria an explanatory causal role. The inevitability of the acceptance of Western astronomy is naturalized by appeal to anachronistic modern scientific criteria, such as accuracy and explanatory power: it was the power of Western models not only to predict but also to exhibit inherent patterns, he asserts, that attracted the Chinese. Social context is then incorporated into Sivin's account primarily as a distorting factor: European cosmology was rendered incomprehensible by the misleading characterizations of Copernican astronomy, which the Jesuits were forbidden to teach, and was ultimately discredited by its own internal contradictions. This naturalizing of scientific truth, and the sociology of error, are elements of a conventional teleology of scientific development implicit in his account.

Seventeenth-century European astronomy thus remained "modern science" posed against "traditional" Chinese science,[58] for example in Sivin's assertion that Wang and his contemporaries did not succeed "in a mature synthesis of traditional and modern science."[59] The West remained the source of modern science for the Chinese: "The character of early modern science was concealed from Chinese scientists, who depended on the Jesuit writings. Many were brilliant by any standard. As is easily seen from their responses to the European science they knew, they would have been quite capable of comprehending modern science if their introduction to it had not been both contradictory and trivial."[60] The limited extent of the transformation of the scientific revolution in China remained the result of distorting nonscientific influences, blamed now not on the Chinese but on the Jesuits: "In short, the scientific revolution in seventeenth-century China was in the main a response to outmoded knowledge [transmitted by the Jesuits] that gave little attention to, and consistently misrepresented, the significance of developments in the direction of modern science."[61]

The key to Sivin's argument was thus his redistribution of "scientific revolutions" among civilizations: by asserting that there was not one but two scientific revolutions—one Chinese and one European—Sivin implied that differences between China and Europe were of degree rather than kind.[62] But by the criteria he used for scientific revolutions—shifts in a disciplinary matrix—there

[58] See also Sivin's study of the thought of Shen Gua, from which he concludes that there was no unified conception of science. Ibid., 47–51.

[59] Idem, "Wang Hsi-Shan," in Gillispie, *Dictionary of Scientific Biography*, 14:163.

[60] Sivin, "Copernicus in China," 1.

[61] Sivin, "Wang Hsi-Shan," 14:166, note "n." Sivin refers the reader to his article "Copernicus in China" for further documentation of this thesis.

[62] Sivin, "Why the Scientific Revolution Did Not Take Place," 65. Sivin also mentions a third scientific revolution "that didn't take place in Archimedes' lifetime."

certainly have been many others.[63] Sivin's account adopted from the histories of Western science the conflation of scientific revolutions (in this technical sense) with the mythologies of *the* Scientific Revolution—a difference that Sivin implicitly incorporates in his use of a capitalized "Scientific Revolution" for Europe.[64] This conflation was itself rooted in attempts by these histories to offer scientific revolutions as the historical cause of the radical break between the ancient and modern that *the* Scientific Revolution emblematized.[65] This radical break had then been translated to a radical difference between the modern scientific West (unique among civilizations in having had the Scientific Revolution) and traditional China. Sivin documented a scientific revolution in China—a change in the disciplinary matrix in Chinese astronomy that was itself a limited copy of the Scientific Revolution of Europe. But he denied to this scientific revolution the miraculous transformative powers claimed in the mythologies of *the* Scientific Revolution of the West.

Praise-and-Blame Histories of Civilizations

On the question of Chinese science, much of the scholarly literature written during the period when Sivin and Needham were publishing their work offered no study of any aspect of it. Instead, these works (sometimes presented against Needham by borrowing from earlier claims that science was uniquely Western, and sometimes following Needham's call to find the social causes that modern science was uniquely Western) purported to offer explanations for the absence of science in China—explanations that were philosophical, social, linguistic, logical, or political.[66] For example, Mark Elvin offered the metaphysical thought

---

[63] The term "disciplinary matrix" is introduced in Kuhn's postscript to *Structure of Scientific Revolutions* to mean that which is shared by practitioners and accounts for the unanimity of their judgment; the term replaces his earlier concept of paradigm. Kuhn's disciplinary matrix consists of four major components: (1) symbolic generalizations (statements that can be formalized into logical or mathematical symbols, functioning to define terms and express relations); (2) the metaphysical conceptions involved in scientific interpretations and models that provide preferred analogies; (3) values that are shared by a wider community than simply one group of scientific practitioners; and (4) "exemplars"—concrete solutions to scientific problems. Kuhn, *Structure of Scientific Revolutions*, 182–87.

[64] Sivin, "Why the Scientific Revolution Did Not Take Place," passim.

[65] Perhaps the most dramatic emblem of the alleged radical shift between the worldviews of the ancients and those of the moderns is the shift from the geocentric to the heliocentric picture of the solar system (Kuhn, *Copernican Revolution*). Alexandre Koyré offers a similarly dramatic symbolization of the radical break between the "closed world" of the ancients and the modern "infinite universe" (Koyré, *From the Closed World*). Kuhn's example of the duck/rabbit diagram is yet another example of an emblem of this alleged radical shift (Kuhn, *Structure of Scientific Revolutions*, 114).

[66] For an overview of the literature predating this work, see Sivin, "Why the Scientific Revolution Did Not Take Place," 54 n. 7. Elsewhere, Sivin also repeatedly criticizes sinologists for ignoring scientific texts. Sivin, "Copernicus in China," 13.

that developed from Wang Yangming 王陽明 (1472–1529) as "the reason why China failed to create a modern science of her own accord."[67] Joseph Levenson explained the purported absence of a Chinese scientific tradition as the result of an "amateur ideal."[68] Alfred Bloom asserted that the Chinese language had inhibited the ability of the Chinese to think theoretically.[69] Robert Hartwell argued that the major impediment was the absence of the formal logical system embodied in Euclidean geometry.[70] And Wenyuan Qian provided a "politico-ideological" explanation.[71]

Yet this literature was not about science. Levenson failed to cite a single primary source on Chinese science in his bibliography; instead, he drew his conclusions on the nondevelopment of science and modernity by universalizing the ethics of Ming painting as exemplary of all of Ming culture, and comparing this with stereotypes of Western science and modern values.[72] Bloom made no pretense of citing historical materials, much less scientific materials from China or the West, in justifying his leap from measuring the testing skills of students in present-day China—formulated in the language of the Sapir-Whorf hypothesis—to the development of Chinese science in the past.[73] Elvin cited only one intellectual, Fang Yizhi 方以智 (1611–1671).[74] Hartwell's explanation of the nondevelopment of Chinese science was appended to a study of trends in Chinese historiography.[75] And Qian's dialogic narrative contained its own admissions of the historical falsity of the central theses of the book.[76] These accounts—because the absence of science in China was assumed—could ignore the technical details specific to the sciences themselves, and instead derive lessons on topics deemed more vital—whether political despotism, philosophical orthodoxy, linguistic inadequacies, or cultural stagnation.

---

[67] Mark Elvin, *The Pattern of the Chinese Past: A Social and Economic Interpretation* (Stanford, CA: Stanford University Press, 1973), 234.

[68] Joseph R. Levenson, *The Problem of Intellectual Continuity*, vol. 1 of *Confucian China and Its Modern Fate* (Berkeley: University of California Press, 1958), chaps. 1 and 2.

[69] Alfred H. Bloom, *The Linguistic Shaping of Thought: A Study in the Impact of Language on Thinking in China and the West* (Hillsdale, NJ: Lawrence Erlbaum, 1981).

[70] Hartwell, "Historical Analogism," 723.

[71] Wen-yuan Qian, *The Great Inertia: Scientific Stagnation in Traditional China* (Dover, NH: Croom Helm, 1985), 26.

[72] Levenson, *Problem of Intellectual Continuity*, 204–9.

[73] Bloom, *Linguistic Shaping of Thought*.

[74] Elvin, *Pattern of the Chinese Past*, 227–34.

[75] Hartwell, "Historical Analogism." Hartwell's central thesis is that historical analogism "was dominant among policy makers during most years of the eleventh, twelfth, and thirteenth centuries" (p. 694); it was displaced by "an orthodoxy of the Zhu Xi classicist-moral-didactic compromise" from the fourteenth to nineteenth centuries (p. 717).

[76] Qian, *Great Inertia*, 12–14, 96–97. As Sivin notes, Qian's book is "a shallow 'answer' to the Scientific Revolution Problem uninformed by acquaintance with the primary literature." Sivin, "Selected, Annotated Bibliography of the History of Chinese Science: Sources in Western Languages," in Sivin, *Science in Ancient China*, 7.

Chapter 3
From Copula to Incommensurable Worlds

B Y ADOPTING certain naive presuppositions, studies of the asserted problems encountered in translations between languages have often reached dramatic conclusions about the fundamental differences between civilizations. These presuppositions are naive in the sense that they circumvent many of the questions that should properly confront historical inquiry, adopting instead simple formulas. For example, on what level of social organization should historical explanation concentrate—what are the significant units of society in analyses of historical change? Instead of determining the complex networks of alliances that dynamically constitute groupings within societies, these studies consider the boundaries already given—drawn along lines of languages, or, more often, systems of languages that mark the purported divides between civilizations. What are the fracture lines in societies underlying antagonisms and conflict? Instead of analyzing complicated divisions along the dimensions of class, gender, status, allegiances, or competing schools of thought, all such differences are collapsed into a unity predetermined by the sharing of a single language (the same, that is, once all historical, regional, educational, and status differences are effaced). What kinds of relationships should historical analysis elucidate? With civilizations as the given units of analysis, such studies are typically content with assertions of similarities and differences.

What, then, is the relationship between thought and society? Instead of historicizing the role of ideologies, self-fashioned identities, and performative utterances in the formation of social groupings, individuals are instead reduced to representatives or bearers of entire civilizations. How does one understand thought through the transcriptions preserved in historical documents? Instead of explaining the dissemination of copies, commentaries, and interpretations of texts in their cultural context, such studies fix an original against which the correspondence of the translation can be compared. And what is the relationship between thought and language? Too often, such studies implicitly presuppose a correspondence between words and concepts. After such a series of simplifying reductions have been adopted, the conventional conclusions about civilizations are an almost inevitable result.

Rather than critiquing in a general fashion the paradoxes that inhere in claims made about civilizations in studies of translations, this chapter will illustrate these paradoxes through the analysis of specific claims. To accomplish this, I have chosen the two most important studies that have used incommensurability to analyze the translations by the Jesuit missionaries and their collaborators in seventeenth-century China. I first outline the claims, presented in these two studies, of linguistic and conceptual incommensurability between China and the West, claims that are based on the asserted difficulties of translating the copula (the verb "to be" used to connect subject and predicate) and the concept of existence. I then turn to the theories of incommensurability that underwrite these studies, along with several related philosophical theories: Emile Benveniste's analysis of the copula "to be," Jacques Derrida's critique of Benveniste, Willard Van Orman Quine's arguments on the indeterminacy of translations, and Donald Davidson's criticisms of assertions of conceptual schemes. Finally, as an alternative to incommensurability, I present an analysis of the translations in cultural context.

China, the West, and the Incommensurability That Divides

As we saw in the preceding chapter, throughout much of the twentieth century an anthropomorphized "China" and "the West" were imagined to be two central actors in historical drama. In this context, studies of the "first encounter" of these two great civilizations acquired a particular urgency. Interpretive approaches have often been limited to two alternative models—conflict, opposition, and misunderstanding, or synthesis, accommodation, and dialogue.[1] But in the 1980s, relativism—again formulated within the context of an assumed plausibility of a divide separating China and the West—became yet another important approach.[2] Theories of linguistic and conceptual incommensurability often

---

[1] Important studies include Jonathan D. Spence, *The Memory Palace of Matteo Ricci* (New York: Viking Penguin, 1984); Charles E. Ronan and Bonnie B. C. Oh, eds., *East Meets West: The Jesuits in China, 1582–1773* (Chicago: Loyola University Press, 1988); John D. Young, *East-West Synthesis: Matteo Ricci and Confucianism* (Hong Kong: Centre of Asian Studies, University of Hong Kong, 1980); idem, *Confucianism and Christianity: The First Encounter* (Hong Kong: Hong Kong University Press, 1983); David E. Mungello, *Leibniz and Confucianism: The Search for Accord* (Honolulu: University Press of Hawaii, 1977); idem, *Curious Land: Jesuit Accommodation and the Origins of Sinology* (Stuttgart: Steiner, 1985). For important criticisms of the received historiography, see Lionel M. Jensen, *Manufacturing Confucianism: Chinese Traditions and Universal Civilization* (Durham, NC: Duke University Press, 1997).

[2] For example, see *Journal of Asian Studies* 50, no. 1 (February 1991), which is dedicated to the issue of relativism, and especially David D. Buck's summary, "Forum on Universalism and Relativism in Asian Studies: Editor's Introduction," 29–34. See also Philip C. C. Huang, "Theory and the Study of Modern Chinese History: Four Traps and a Question," *Modern China* 24 (April 1998): 183–208. For Huang, "culturalism" and the resulting relativism constitute one of his "four traps" (pp. 192–201). For collected essays on the debates on relativism, see Martin Hollis and Steven Lukes, eds., *Rationality and Relativism* (Cambridge, MA: MIT Press, 1982); and Michael

underwrite this appeal to relativism, providing for relativism perhaps its most rigorous formulation. It should be noted that these claims of relativism and incommensurability have played an important role in encouraging the analysis of Chinese sources and viewpoints by positing a special Chinese worldview protected from pretentious dismissal by a historiography mired in universalism. Yet they have done so at the cost of further reifying "China" and "the West," and further radicalizing the purported divide that separates them. To elucidate the role played by claims about translation in theories of incommensurability, this chapter will examine two important studies: Jacques Gernet's *China and the Christian Impact*[3] and a related analysis of the translation of Euclid's *Elements* by Jean-Claude Martzloff in his *History of Chinese Mathematics*.[4]

Gernet's China and the Christian Impact

In probably the most sophisticated study of the Jesuits in China during the seventeenth century, Gernet's *China and the Christian Impact* adopts incommensurability between Western and Chinese concepts as the philosophical framework that is to explain the history of the translation and introduction of Christianity.[5] Against previous studies of the introduction of Christianity into China that had been based primarily on Western sources, Gernet proposes as a new approach the study of the "Chinese reactions to this religion." Previous approaches were often universalistic, assuming that "one implicit psychology—our own—valid for all periods and all societies is enough to explain everything." Gernet asserts that for the missionaries, the rejection of Christianity "could only be for reasons that reflected poorly on the Chinese." Later interpreters similarly have "a tendency to see the enemies of Christianity as xenophobic conservatives" while praising the collaborators as open minded. This thesis, Gernet asserts, "is contradicted by the facts."[6]

Krausz, ed., *Relativism: Interpretation and Confrontation* (Notre Dame, IN: University of Notre Dame Press, 1989).

[3] Jacques Gernet, *China and the Christian Impact: A Conflict of Cultures*, trans. Janet Lloyd (New York: Cambridge University Press, 1985); originally published as *Chine et christianisme: Action et réaction* (Paris: Gallimard, 1982).

[4] Jean-Claude Martzloff, *A History of Chinese Mathematics* (New York: Springer, 2006); originally published as *Histoire des mathématiques chinoises* (Paris: Masson, 1988). It should be noted that more than twenty years after it was first published, Martzloff's book is still one of the most authoritative general studies of the history of Chinese mathematics.

[5] Gernet, *China and the Christian Impact*, and Spence, *Memory Palace of Matteo Ricci*, are generally recognized as the two seminal works describing this period. For two important critiques of Gernet's book, see Paul A. Cohen, review of *China and the Christian Impact: A Conflict of Cultures*, by Jacques Gernet, *Harvard Journal of Asiatic Studies* 47 (1987): 674–83; and Howard L. Goodman and Anthony Grafton, "Ricci, the Chinese, and the Toolkits of Textualists," *Asia Major*, 3rd ser., 3 (1990): 95–148.

[6] Gernet, *China and the Christian Impact*, 1–2.

Gernet's defense of the Chinese rejection of Christianity is based on a claim of the fundamental incommensurability of languages and the associated Chinese and Western worldviews: "The missionaries, just like the Chinese literate elite, were the unconscious bearers of a whole civilisation. The reason why they so often came up against difficulties of translation is that different languages express, through different logics, different visions of the world and man."[7] Gernet outlines this theoretical framework in the final sections of his concluding chapter. He offers several examples of the difficulties in bridging "mental frameworks": for example, "In trying to assimilate the Chinese Heaven and the Sovereign on High to the God of the Bible, the Jesuits were attempting to bring together concepts which were irreconcilable."[8] He discovers radical differences between Chinese and Western thought: "The Chinese tendency was to deny any opposition between the self and the world, the mind and the body, the divine and the cosmic. . . . For Chinese thought never had separated the sensible from the rational, never had imagined any 'spiritual substance distinct from the material,' never had conceived of the existence of a world of eternal truths separated from this world of appearances and transitory realities."[9]

These differences (although still often conceptualized by Gernet as absences) are adduced as evidence that demonstrates the "radical originality" of China: "Ultimately, what the Chinese criticisms of Christian ideas bring into question are the mental categories and types of opposition which have played a fundamental role in Western thought ever since the Greeks: being and becoming, the intelligible and the sensible, the spiritual and the corporeal. Does all this not mean that Chinese thought is quite simply of a different type, with its own particular articulations and its own radical originality?"[10]

The philosophical framework of conceptual incommensurability that Gernet employs in this work is based on the linguistic theory of Emile Benveniste: "Benveniste writes: 'We can only grasp thought that has already been fitted into the framework of a language. . . . What it is possible to say delimits and organises what it is possible to think. Language provides the fundamental configuration of the properties that the mind recognises things to possess.'"[11] More specifically, Gernet asserts that the two fundamental differences between Chinese and Western languages are categories of thought that derive from language and the concept of *existence*:[12] "Benveniste's analysis illuminates two characteristics of

---

[7] Gernet, *China and the Christian Impact*, 2. For criticisms of assertions that individuals are the bearers of whole cultures, see Johannes Fabian, *Time and the Other: How Anthropology Makes Its Object* (New York: Columbia University Press, 1983).

[8] Gernet, *China and the Christian Impact*, 193.

[9] Ibid., 201.

[10] Ibid., 208.

[11] Ibid., 240. Gernet cites Emile Benveniste, "Catégories de pensée et catégories de langue"; translated in Emile Benveniste, *Problems in General Linguistics*, trans. Mary Elizabeth Meek (Coral Gables, FL: University of Miami Press, 1971).

[12] These two claims are in fact related, since Benveniste's central example of categories of thought is based on the copula.

Greek—and, more generally, Western—thought, both of which are closely related to the structure of Greek and Latin: one is the existence of categories the obvious and necessary nature of which stems from the use to which the language is unconsciously put. The other is the fundamental importance of the concept of being in Western philosophical and religious thought."[13] As I will argue below, Gernet's examples—the translation of Christian terms—present special philosophical problems. So before exploring these, we will examine Martzloff's analysis of the translation of Euclid's *Elements* as a more concrete but related example for the comparison of Chinese and Greek thought and language.

Martzloff's History of Chinese Mathematics

The translation of Euclid's *Elements* into Chinese by Xu Guangqi and Matteo Ricci in 1607 would seem ideal for an examination of linguistic incommensurability, given the extant historical documents. Jean-Claude Martzloff, one of the most respected historians of Chinese mathematics, has written extensively on the translation.[14] He adopts Gernet's incommensurability in his explanation of the history of the translation, arguing that the Chinese had failed to comprehend the deductive structure of the *Elements* precisely because of linguistic incommensurability. Martzloff argues that the central problem was the difficulty of translating the copula, because of its absence in classical Chinese:

> In addition to the terminology, the even more formidable problem of the difference between the Chinese syntax and that of European languages had to be faced. The main difficulty was the absence of the verb "to be" in classical Chinese. The translators were unable to find better substitutes for it than demonstratives or transitive verbs such as *you, wu,* and *wei.* . . . But often, the verb "to be" disappeared altogether, as in the following case:

---

[13] Gernet, *China and the Christian Impact,* 240. Gernet quotes Benveniste's assertion that the existence of the verb "to be" in Greek made possible the philosophical manipulation of the concept, with the result that the concept of being became central in Greek thought.

[14] Martzloff, *History of Chinese Mathematics,* 111–22, 273–77, 371–89. In addition, see idem, "La compréhension chinoise des méthodes démonstratives euclidiennes au cours du XVIIe siècle et au début du XVIIIe," in *Actes du IIe colloque international de sinologie: Les rapports entre la Chine et l'Europe au temps des lumières* (Paris: Les Belles Letters, 1980), 125–43; idem, "La géométrie euclidienne selon Mei Wending," *Historia Scientiarum* 21 (1981): 27–42; idem, "Matteo Ricci's Mathematical Works and Their Influence," in *International Symposium on Chinese-Western Cultural Interchange in Commemoration of the 400th Anniversary of the Arrival of Matteo Ricci, S. J. In China. Taibei, Sept. 11–16, 1983* (Taibei: Furen daxue chubanshe, 1983), 889–95; idem, "Eléments de réflexion sur les réactions chinoises à la géométrie euclidienne à la fin du XVIIe siècle: Le *Jihe lunyue* de Du Zhigeng vu principalement à partir de la préface de l'auteur et de deux notices bibliographiques rédigées par des lettres illustres," *Historia Mathematica* 20 (1993): 160–79. Again, for a more detailed study of the translation, see Peter M. Engelfriet, *Euclid in China: The Genesis of the First Chinese Translation of Euclid's Elements, Books I–VI (Jihe Yuanben, Beijing, 1607) and Its Reception up to 1723* (Boston: Brill, 1998).

圜者。一形於平地居一界之間。自界至中心作直線。俱等。

[The] circle: [a] shape situated on flat ground (*ping di*) [sic] within [a] limit.
[The] straight strings (*xian*) constructed from [the] limit to [the] centre: all
equal.[15]

Martzloff then offers for comparison Clavius's original:

"Circulus, est figura plana sub una linea comprehensa, quae peripheria
appelatur, ad quam ab uno puncto eorum, quae intra figuram sunt posita,
cadentes omnes rectae linae, inter se sunt aequales."[16]

Martzloff then links the copula to questions of existence, asserting that "one
might think that this type of phenomenon contributed to a masking of the conception, according to which geometric objects possess inherent properties, the
existence or non-existence of which is objectifiable."[17] Although Martzloff apparently borrows this framework from Gernet's *China and the Christian Impact* and
Benveniste,[18] in his argument he cites primarily A. C. Graham's "'Being' in Western Philosophy" as asserting that neither *you* 有, *wu* 無, nor *wei* 為 are equivalent
to the copula.[19]

In addition to Gernet's and Martzloff's assertions, based on Benveniste, a wide
variety of arguments on the relation of language to thought have been presented
in historical studies of China. In some cases, suggestions about the role of language have been quite productive. For example, Peter Boodberg suggests that
"the great semantic complexity of *tao* may have predetermined the rich system

[15] Martzloff, *History of Chinese Mathematics*, 116–18. Chinese punctuation and interpolations
in the English translation (in brackets) are Martzloff's. Martzloff's French original: "Le cercle?
Une forme située sur la terre plate (*ping di* 平地) (sic) entre de la limite! Les 'fils' (*xian* 線) droits
construits de la limite au centre? Tous égaux!" Martzloff, *Histoire des mathématiques chinoises*,
103. "Sic" is in the originals, both English and French.

[16] Martzloff, *History of Chinese Mathematics*, 118; Martzloff quotes from Clavius, *Euclidis Elementorum*, bk. 1, def. 7.

[17] Martzloff, *History of Chinese Mathematics*, 118. In the original French edition, Martzloff states
directly that the central problem is of "the concept of existence": "Ce type de phénomène
rendait pour le moins hasardeuse la transmission du concept d'existence si important en
mathématiques (parallèles, constructions géométriques, raisonnements par l'absurde dans
lesquels on prouve qu'un certain objet mathématique n'existe pas)." Martzloff, *Histoire des
mathématiques chinoises*, 103.

[18] It should be noted that a subsequent article by Martzloff instead asserts that it was a commensurability between the European and Chinese concepts of space and time that led to the rapid
acceptance of Jesuit astronomy (p. 67). Yet the incommensurability thesis remains in Martzloff's
claim that the Chinese and Europeans had "fundamentally different orientations" (p. 82–83).
Gernet, in the following article, argues that the commensurability that Martzloff argues for is
only apparent. Martzloff, "Space and Time in Chinese Texts of Astronomy and Mathematical
Astronomy in the Seventeenth and Eighteenth Centuries," *Chinese Science* 11 (1993): 66–92;
Jacques Gernet, "The Encounter between China and Europe," *Chinese Science* 11 (1993): 93–102.

[19] A. C. Graham, "'Being' in Western Philosophy Compared with Shih/Fei Yu/Wu in Chinese
Philosophy," *Asia Major* 7 (1959): 79–112. Graham's conclusions are in fact in many ways the
opposite of Martzloff's. Martzloff also cites Gilles Gaston Granger, *La théorie aristotélicienne de
la science* (Paris: Aubier Montaigne, 1976).

of associations surrounding *Tao* in its metaphysical and literary career."[20] In other cases, theses about the role of language are misguided. For example, Alfred Bloom notoriously asserts that the lack of counterfactuals and universals in the Chinese language inhibited the ability of the Chinese to think theoretically.[21] Many authors have presented claims that the Chinese language inhibited the development of science.[22] Until recently, such studies have rarely critically analyzed any of the details of the theories they cite;[23] the following section, then, will examine Benveniste's claims about the copula.

The Philosophy, Language, and Translation of Existence

Benveniste's central thesis is that language and thought are coextensive, interdependent, and indispensable to each other. "Linguistic form is not only the condition for transmissibility," Benveniste asserts, "but first of all the condition for the realization of thought";[24] the structure of language "gives its *form* to the content of thought."[25] Benveniste examines Aristotle's categories of thought to assess whether we have "any means to recognize in thought such characteristics as would belong to it alone and owe nothing to linguistic expression."[26] He concludes that Aristotle's categories were simply the fundamental categories of the language in which Aristotle thought—"the ten categories can . . . be transcribed in linguistic terms."[27] "Unconsciously," Benveniste argues, Aristotle "took as a cri-

---

[20] Peter A. Boodberg, "Philological Notes on Chapter One of the Lao Tzu," *Harvard Journal of Asiatic Studies* 20 (1957): 601.

[21] Alfred H. Bloom, *The Linguistic Shaping of Thought: A Study in the Impact of Language on Thinking in China and the West* (Hillsdale, NJ: Lawrence Erlbaum, 1981). For a critical review, see Kuang-ming Wu, "Counterfactuals, Universals, and Chinese Thinking," review of *The Linguistic Shaping of Thought: A Study in the Impact of Language on Thinking in China and the West* by Alfred Bloom, *Philosophy East and West* 37 (1987): 84–94.

[22] For one further example of these arguments, see Derk Bodde, *Chinese Thought, Society, and Science: The Intellectual and Social Background of Science and Technology in Pre-Modern China* (Honolulu: University of Hawaii Press, 1991).

[23] Important pioneering studies include Haun Saussy, *The Problem of a Chinese Aesthetic* (Stanford: Stanford University Press, 1993); and Lydia H. Liu, *Translingual Practice: Literature, National Culture, and Translated Modernity—China, 1900–1937* (Stanford, CA: Stanford University Press, 1995). In their first chapters, both works provide critical discussions of much of the literature on philosophy and translation. See also Haun Saussy, *Great Walls of Discourse and Other Adventures in Cultural China* (Cambridge, MA: Harvard University Asia Center, 2001); and Lydia H. Liu, *The Clash of Empires: The Invention of China in Modern World Making* (Cambridge, MA: Harvard University Press, 2004).

[24] Benveniste, *Problems in General Linguistics*, 56; originally published as *Problèmes de linguistique générale* (Paris: Gallimard, 1966).

[25] Benveniste, *Problems in General Linguistics*, 53.

[26] Ibid., 56.

[27] Ibid., 60.

terion the empirical necessity of a distinct *expression* for each of his predications.
. . . It is what one can *say* which delimits and organizes what one can think."[28]

In "The Supplement of Copula: Philosophy *Before* Linguistics," as an example of the paradoxes in claims that language governs thought, Jacques Derrida critiques Benveniste's assertion that the Greek language determined Aristotle's categories.[29] It is Benveniste's own writings, Derrida asserts, that offer a "counter-proof" against the assertion that *being* is nothing more than a category linguistically determined by the copula *to be*: Benveniste himself asserts that there is a meaning of the philosophical category *to be* beyond that expressed in grammar. For Benveniste argues, Derrida asserts, that (1) "the function of 'the copula' or 'the grammatical mark of equivalence' is absolutely distinct from the full-fledged use of the verb *to be*" in the sense of *existence*; and (2) "in all languages, a certain supplementary function is available to offset the lexical 'absence' of the verb 'to be,'" used grammatically as a mark of equivalence.[30]

Derrida then links this conflation of these two uses of "to be"—grammatical and lexical—to themes in the history of Western philosophy.[31] It is the "full-fledged" use that Heidegger claims to seek to recover when he suggests that *being* has become both compromised and effaced: "'Being' remains barely a sound to us, a *threadbare* appellation. If nothing is left to us, *we must seek at least* to grasp *this last vestige of a possession*."[32] This nostalgia for a return to the use of *to be* as *existence* is echoed, Derrida asserts, by Benveniste: "It must have had a definite lexical meaning before *falling*—at the end of a long historical development—to the rank of 'copula.' . . . We must *restore its full force and its authentic function to the verb 'to be'* in order to measure the distance between a nominal assertion and an assertion with 'to be.'"[33] The copula thus transcends the grammatical categories of particular languages: in some languages it is denoted by only a lexical absence; on the other hand, the "full fledged" notion of *to be* cannot be a category determined by language if it is to be possible to return from the effaced use of *being* to its "full force" and "authentic function."

Derrida's critique of Benveniste exemplifies the paradoxes that Derrida suggests inhere in assertions that philosophic discourse is governed by the constraints of language. For the oppositions of linguistics—"natural language/formal language, language system/speech act, insofar as they are productions of philo-

---

[28] Benveniste, *Problems in General Linguistics*, 61, emphasis in original.

[29] Jacques Derrida, "The Supplement of Copula: Philosophy *Before* Linguistics," in *Textual Strategies: Perspectives in Post-Structuralist Criticism*, ed. Josué V. Harari (Ithaca, NY: Cornell University Press, 1979).

[30] Ibid., 114. Compare this with Boodberg's discussion: "(d)jwer (modern wei, graphs: 唯, 惟, or 維), a copula-like particle common in the language of the Shih and Shu. If this be the lost Chinese verb 'to be' . . . ." Boodberg, "Chapter One of the Lao Tzu," 603.

[31] In this sense, grammatical peculiarities of the Indo-European languages have influenced the development of philosophy within those languages, by the conflation of "to be" and "existence."

[32] Martin Heidegger, *An Introduction to Metaphysics*, trans. Ralph Manheim (New York: Anchor Books, 1961), 58–61; quoted in Derrida, "Supplement of Copula," 118, with Derrida's italics.

[33] Benveniste, *Problems in General Linguistics*, 138; quoted in Derrida, "Supplement of Copula," 119, with Derrida's italics.

sophical discourse, belong to the field they are supposed to organize."[34] Derrida thus inverts Benveniste's claim, asserting that "philosophy is not only *before* linguistics in the way that one can be *faced* with a new science, outlook, or object; it is also before linguistics in the sense of preceding, providing it with all its concepts."[35]

Among the claims for radical differences between languages, the absence of the copula has seemed to be both the most concrete and the most significant, for the copula has seemed to be the most plausibly connected with philosophically important consequences. Derrida's criticisms point to two fundamental problems in Gernet's and Martzloff's applications of the theories of Benveniste: (1) although the existence of the copula seemed to mark an important difference between the Indo-European and Chinese languages, Benveniste's claim is, rather, that the copula exists in all languages, and (2) the correlation between the copula and the metaphysics of existence results from the conflation of two separate uses of the same lexical term.[36]

If the linguistics of the copula does not demonstrate that radical differences exist between languages—much less between thoughts expressed in those languages—the general question remains: To what extent can differences in thought be shown to result from differences in language? Derrida offered one answer: the concepts from linguistics that are to provide the basis for comparison are themselves constituted by the philosophy they purport to analyze. Quine offers a different critique: to compare systems of thought, we must first have solved the problem of translation. Against the views of Ernst Cassirer, Edward Sapir, and Benjamin Whorf that differences in language lead to fundamental differences in thought, Quine objects that we cannot, in principle, provide translation rigorous enough to assess such grand philosophical theses. He concludes that it is not that "certain philosophical propositions are affirmed in the one culture and denied in the other. What is really involved is difficulty or indeterminacy of correlation. It is just that there is less basis of comparison— less sense in saying what is good translation and what is bad—the farther we get

---

[34] Ibid., 82.

[35] Ibid., 98.

[36] It should also be noted that elsewhere in his argument asserting that language determines categories of thought, Benveniste uses the example of science and the Chinese language to argue *against* linguistic incommensurability: "Chinese thought may well have invented categories as specific as the *dao*, the *yin*, and the *yang*; it is nonetheless able to assimilate the concepts of dialectical materialism or quantum mechanics without the structure of the Chinese language proving a hindrance. No type of language can by itself alone foster or hamper the activity of the mind. The advance of thought is linked much more closely to the capacities of men, to general conditions of culture, and to the organization of society than to the particular nature of a language. But the possibility of thought is linked to the faculty of speech, for language is a structure informed with signification, and to think is to manipulate the signs of language." Benveniste, *Problems in General Linguistics*, 63–64.

away from sentences with visibly direct conditioning to nonverbal stimuli and the farther we get off home ground."[37]

If Quine's argument, based on the indeterminacy of correlation, concludes that it is impossible to rigorously compare differing conceptual schemes, Davidson, in "On the Very Idea of a Conceptual Scheme," warns that the notion of a conceptual scheme is itself ultimately unintelligible.[38] Davidson's argument is presented against the conceptual relativism of Quine, Whorf, Thomas Kuhn, and Paul Feyerabend.[39] Davidson notes the following paradox: the demonstration that two conceptual schemes are incommensurable requires the solution of the purported incommensurability in a frame of reference that incorporates both; assertions of conceptual schemes are always framed in a language that purports to explain that which cannot be explained.[40] The differences, he notes, "are not so extreme but that the changes and the contrasts can be explained and described using the equipment of a single language."[41] Davidson thus argues that "we cannot make sense of total failure" of translation, whether based on a plurality of imagined worlds or on incommensurable systems of concepts employed to describe the same world.[42] Davidson concludes that "no sense can be made of the idea that the conceptual resources of different languages differ dramatically."[43]

---

[37] Willard Van Orman Quine, "Meaning and Translation," in *On Translation*, ed. Reuben A. Brower (Cambridge, MA: Harvard University Press, 1959), 171–72.

[38] Donald Davidson, "On the Very Idea of a Conceptual Scheme," *Proceedings and Addresses of the American Philosophical Association* 47 (1974): 5–20.

[39] Benjamin Lee Whorf, "The Punctual and Segmentative Aspects of Verbs in Hopi," in *Language, Thought, and Reality: Selected Writings* (Cambridge, MA: MIT Press, 1959), 51–56; Willard Van Orman Quine, "Two Dogmas of Empiricism," in *From a Logical Point of View: 9 Logico-Philosophical Essays*, 2nd ed. (Cambridge, MA: Harvard University Press, 1961); Paul Feyerabend, "Explanation, Reduction, and Empiricism," in *Scientific Explanation, Space, and Time*, ed. Herbert Feigl and Grover Maxwell (Minneapolis: University of Minnesota Press, 1962), 28–97; Thomas S. Kuhn, *The Structure of Scientific Revolutions*, 2nd ed. (Chicago: University of Chicago Press, 1970). Davidson argues that Quine's conceptual schemes constitute yet a "third dogma." For Quine's response, see "On the Very Idea of a Third Dogma," in *Theories and Things* (Cambridge, MA: Harvard University Press, 1981).

[40] This is precisely the contradiction found in Martzloff—his assertion of incommensurability between Chinese and Latin itself requires a solution to the problem of incommensurability on a firm ground for philosophical translations between Greek, Latin, English, Chinese, and French, a solution that could not have existed four hundred years ago, at the time of the translation.

[41] Davidson, "Very Idea of a Conceptual Scheme," 184.

[42] Ibid., 185.

[43] Idem, *Inquiries into Truth and Interpretation* (New York: Oxford University Press, 1984), xviii. More precisely, Davidson states, "Our general method of interpretation forestalls the possibility of discovering that others have radically different intellectual equipment. But more important, it is argued that if we reject the idea of an uninterpreted source of evidence no room is left for a dualism of scheme and content. Without such a dualism we cannot make sense of conceptual relativism" (p. xviii).

From Words to Worlds via Translation

Davidson's central argument stands against the assertion of conceptual schemes by showing that they are unintelligible—that is, the very formulation of such schemes is paradoxically circular. But in the course of his argument, Davidson also offers the following deflationary aside: "Instead of living in different worlds, Kuhn's scientists may, like those who need Webster's dictionary, be only words apart."[44] Davidson is correct that assertions of incommensurable conceptual schemes cannot be formulated coherently; the question remaining for us to answer becomes, How are claims of different worlds constructed from differences in words? It is precisely because of the impossibility, suggested by Davidson, of describing radically different conceptual worlds that such claims must defer by analogy to *other* radical differences. Thus, for example, Kuhn explains the purported incommensurability of scientific paradigms through analogies to differing linguistic taxonomies.[45] Feyerabend explains incommensurability between "cosmologies" through analogies in forms of art.[46] Whorf explains differing linguistic systems through examples from science.[47] Quine explains differing conceptual schemes through analogies in physics.[48] And Martzloff and Gernet explain the incommensurability of thought through analogies to linguistic differences, via Benveniste's copula.

How are incommensurate worlds, then, created from words? The process of translation would at first seem an unlikely tool, since the putative goal of translation is to establish equivalences between two languages. But it is this assumption

---

[44] Davidson, "Very Idea of a Conceptual Scheme," 189.

[45] See Kuhn, *Structure of Scientific Revolutions*; for a later formulation, see idem, "Second Thoughts on Paradigms," in *The Structure of Scientific Theories*, ed. Frederick Suppe (Urbana: University of Illinois Press, 1974), 459–82.

[46] Citing Emanuel Löwy, *Die Naturwiedergabe in der älteren griechischen Kunst* (Rom: Loescher, 1900), Paul Feyerabend, in *Against Method* (London: Verso, 1988), 170–226, offers as an example of incommensurability two contrasting forms of Greek art that he identifies as cosmologies A and B. A is the "archaic style" in which a "paratactic aggregate" forms a "visual catalogue" through the placement of standard figures in varying symbolic positions—for example, death is the standard figure drawn horizontally. In B, art is arranged so that the underlying essence is grasped through representations that trigger illusions (e.g., two-dimensional drawings). Feyerabend then defines incommensurability: "A discovery, or a statement, or an attitude [is called] incommensurable with the cosmos (the theory, the framework) if it suspends some of its universal principles" (p. 215).

[47] For example, Whorf asserts that the adoption of Western science entails the adoption of the Western system in its entirety: "That modern Chinese or Turkish scientists describe the world in the same terms as Western scientists means, of course, only that they have taken over bodily the entire Western system of rationalizations, not that they have corroborated that system from their native posts of observation." Whorf, *Language, Thought, and Reality*, 214.

[48] Quine's famous analogies are from boundary-value problems in physics: "The totality of our so-called knowledge or beliefs ... is a man-made fabric which impinges on experience only along the edges"; "total science is like a field of force whose boundary conditions are experience." Quine, "Two Dogmas of Empiricism," 42, quoted in Davidson, "Very Idea of a Conceptual Scheme," 191.

that provides for these claims a crucial resource: the purported impossibility of finding equivalent words then itself serves as a sign of radically different worlds.[49]

The differences that appear in Martzloff's examples are not between Clavius's Latin version and Ricci's Chinese translation—no unmediated comparison is possible—but instead between the untranslated Latin, the translation that Martzloff provides of Ricci's Chinese into English, and ordinary expressions of English. Clavius's Latin represents an uncorrupted original by remaining untranslated and dehistoricized: effaced are the problems of translation from Greek to Latin to French and English, the complex history of the translation and editions of this text,[50] and in particular Clavius's redaction, which altered and deleted much of the structure of the proofs.[51] Martzloff's translations convey the radical otherness of classical Chinese by employing techniques of defamiliarization similar to those used elsewhere to demonstrate (again, in English) the purported awkwardness of Chinese monosyllabism: "King speak: Sage! not far thousand mile and come; also will have use gain me realm, hey?"[52] Indeed, Martzloff argues against "more elegant, more grammatical" renderings, stating that "English grammaticality tends to obliterate the structure of the Chinese and the connotation of the specialised terms."[53]

The differences Martzloff presents are the artifacts of the choices he makes in his translation. First, he insists on an extreme literalism in his selections of possible equivalents, for example "straight strings" for *xian* 線 and "flat ground" for *pingdi* 平地, with the latter marked by *sic* to emphasize the inappropriateness of what was, after all, his own choice. He marks articles with brackets. However, Martzloff's most jarring technique is his omission of the copula in English, a language in which the copula is denoted lexically. Although Martzloff cites Graham as asserting that classical Chinese lacks the equivalent of the copula, Martzloff notes that "we shall not retain [Graham's] English translations, since these translations introduce numerous elements which do not exist at all in classical

---

[49] Arguably, it is the purported correlations of words with things that underwrites claims of deep ontological differences. Claims of radical differences are based on the assertion that one language fails to have a word for a particular thing, not that the thing cannot possibly be named or described.

[50] John E. Murdoch, "Euclid: Transmission of the Elements," in *Dictionary of Scientific Biography*, ed. Charles C. Gillispie (New York: Scribner, 1970–1980), 4:437–59.

[51] On Clavius's mathematics and science, see James M. Lattis, *Between Copernicus and Galileo: Christoph Clavius and the Collapse of Ptolemaic Cosmology* (Chicago: University of Chicago Press, 1994).

[52] The example is from August Schleicher, *Die Sprachen Europas in systematischer Uebersicht* (Bonn: H.B. König, 1850), 51; quoted in William Dwight Whitney, *Language and the Study of Language: Twelve Lectures on the Principles of Linguistic Science* (London: N. Trübner, 1867), 331; quoted and criticized in J. R. Firth, "Linguistic Analysis and Translation," in *Selected Papers of J. R. Firth, 1952–59*, ed. F. R. Palmer (Bloomington: Indiana University Press, 1968), 76. For analysis of this translation, see Haun Saussy, "Always Multiple Translation, or, How the Chinese Language Lost Its Grammar," in *Tokens of Exchange: The Problem of Translation in Global Circulations*, ed. Lydia H. Liu (Durham, NC: Duke University Press, 1999), 107–23. I thank Haun Saussy for bringing this passage to my attention.

[53] Martzloff, *History of Chinese Mathematics*, 118.

Chinese (for example, the verb 'to be')."[54] Martzloff, however, then supplements English with a nonlexical symbol, the colon, to denote the absent copula: "since ordinary words are not sufficient, we also use a punctuation mark" (in his original translations into French, Martzloff employed a question mark paired with an exclamation mark).[55] It is then senseless to correct Martzloff's translation. While ordinary translation seeks to establish equivalences, Martzloff seeks to convey radical differences; but following these principles, an English translation of Clavius's Latin would be rendered equally bizarre, and thus Martzloff must leave the Latin untranslated. The extremes to which Martzloff takes the translation are necessary to evoke the linguistic differences that are to serve as an analogy for radical differences in thought.

If Martzloff's claimed linguistic incommensurability relating to *existence* is an artifact of his jarring omission of the copula in insistently preserving this "lexical absence" of classical Chinese in modern English and French, which denote the copula lexically, is there any evidence to support claims of a radical incommensurability on the conceptual level? As noted above, Gernet's claims of conceptual incommensurability were based on his assertion, following Benveniste, of differences between China and the West in the fundamental "concept of being" and "the existence of categories" that unconsciously stem from the use of language;[56] Gernet adopts the claims of abstract differences in the concepts of *existence* and *categories* as a philosophical theory within which to frame his description of historical events. The critiques by Derrida, Quine, and Davidson suggested general philosophical problems with these claims; here instead I will seek historical explanations. That is, we must return to the debates on *existence* and *categories* not abstracted from but, rather, resituated within their historical context, and not as philosophy explaining history, but as philosophy inseparable from the history it was to explain. To do so, I will re-examine one of the central texts analyzed by Gernet, *The True Meaning of the Lord of Heaven* (*Tianzhu shi yi* 天主實義, c. 1596, attributed to Matteo Ricci), for evidence of the problem of the translation into Chinese of notions of *existence* and *categories*.[57]

The problem that confronted the Jesuits and their Chinese patrons in the *Tianzhu shi yi*, it turns out, is not one of an impossibility of expressing the philosophical concept of *existence* in the abstract but, rather, debates about existence

---

[54] Ibid., 273.

[55] Ibid., 274. For the French, see n. 15 on page 56 of this book.

[56] Gernet, *China and the Christian Impact*, 240.

[57] *Tianzhu shi yi* 天主實義 [True meaning of the lord of heaven], in *Tianxue chu han* 天學初函 [Learning of heaven, first series] (Taibei: Taiwan xuesheng shuju 臺灣學生書局, 1965), hereinafter *TXCH*. Although I will follow the conventional attribution of this work to Ricci, it is likely that the content was determined by his Chinese patrons. *Tianzhu shi yi* is translated into English as Matteo Ricci, SJ, *The True Meaning of the Lord of Heaven (T'ien-Chu Shih-I)*, trans. Douglas Lancashire and Peter Kuo-chen Hu, SJ (St. Louis, MO: Institute of Jesuit Sources, 1985). Several of the examples that I analyze are in fact cited or translated in Gernet, *China and the Christian Impact*, and provide important evidence against the conclusions Gernet draws in the final chapter.

with specific referents—in particular, spirits and God. Chapter 4 begins with the following summary of the previous chapter:

> Chinese scholar: Yesterday after I took my leave and reviewed your distinguished instruction, sure enough [I] understood that it all follows true principles. I do not know why the deluded scholars of my country should accept as the orthodox truth denials of the existence of spirits.
>
> 中士曰：昨吾退習大誨，果審其皆有真理。不知吾國迂儒，何以攻折鬼神之實為正道也。
>
> Western scholar: I have comprehensively examined the ancient classics of your esteemed country. Without exception [these texts] take sacrifices to the spirits as momentous occasions for the Son of Heaven and the feudal lords; thus [they] revered these spirits as being above them and all around them. How then could it possibly be that there is no such thing and thus in this [they] acted deceitfully![58]
>
> 西士曰：吾遍察大邦之古經書，無不以祭祀鬼神為天子諸侯重事，故敬之如在其上、如在其左右，豈無其事而故為此矯誣哉？

In the *Tianzhu shi yi*, there is no shortage of ways to attribute to "spirits" the predicate of "existence": "Tang['s soul] continued to exist without dissipating" 湯為仍在而未散矣; "souls of the deceased exist eternally without extinction" 死者之靈魂為永在不滅; "the human soul does not dissipate after death" 人魂死後為不散泯.[59] Nor is there any shortage of debates on questions of existence of spirits. In *Tianzhu shi yi*, the Chinese scholar summarizes contemporary Chinese positions on the existence of spiritual beings as follows:[60]

> Chinese scholar: Among contemporaries who discuss the spirits, each has his own viewpoint. Some state that nowhere in the world are there things such as spirits. Others state if one believes in them then they exist; if one does not believe in them then they do not exist. Others state that to assert that they exist is incorrect; to assert that they do not exist is also incorrect;

---

[58] *Tianzhu shi yi*, 450; Ricci, *True Meaning of the Lord of Heaven*, 34–36. In the following translations I have consulted Lancashire and Hu's translations, borrowing and making alterations where appropriate.

[59] *Tianzhu shi yi*, 451–52; Ricci, *True Meaning of the Lord of Heaven*, 176–77. Just as the asserted absence of a word precisely equivalent to the English verb "to be" is hardly evidence for broader philosophical conclusions such as conceptual incommensurability (as I have argued above), so too the translations I offer in this chapter of various Chinese phrases into English terms such as existence, soul, and forms of the verb "to be" are not meant to suggest a transparency of translation or to imply the exact equivalence of Chinese and English words or concepts (I thank Marta Hanson for her suggestions on this point). The possibility or impossibility of translation cannot demonstrate either radical incommensurability or exact correspondence; these are questions not of linguistics but of philosophical interpretation and intellectual history.

[60] For an important earlier example, see the chapter "Gui shen" 鬼神 [Ghosts and spirits] in Zhu Xi 朱熹 (1130–1200), *Zhuzi yu lei* 朱子語類 [Conversations with Master Zhu, arranged topically], ed. Li Jingde 黎靖德 and Wang Xingxian 王星賢, 8 vols. (Beijing: Zhonghua shuju 中華書局, 1986).

only to assert that they both exist and do not exist is to attain it [the correct viewpoint]![61]

中士曰：今之論鬼神者，各自有見。或謂天地間無鬼神之殊[62]。或謂信之則有，不信之則無。或謂如說有則非，如說無則亦非，如說有無，則得之矣。

The reply of the Western scholar proceeds from the claim that "all affairs and things that exist indeed do exist, [those that] do not exist indeed do not exist" 凡事物，有即有，無即無, and proceeds to the conclusion that spirits exist. The dispute with commonplace Chinese views is thus not that the spirits of ancestor worship do not exist; on the contrary, they exist eternally. Rather, there is a more important spirit that the Chinese must worship if they seek fortune and salvation. His defense of the existence of spirits is based on his own translation of an enigmatic phrase attributed to Confucius, interpreted through his knowledge of God:

> Thus Confucius states: "Respect the spirits, and distance them."[63] Happiness, fortune, and forgiveness of sin are not within the powers of the spirits, but up to the Lord of Heaven alone. Yet the trendy curry favor [with the spirits], desiring to receive this [happiness, fortune, and forgiveness] from them; but it is not the true way to obtain this. The meaning of "distance them" is the same as "if one sins against Heaven, there is no one to pray to."[64] How is it possible that "distance them" can be explained as "they do not exist," snaring Confucius in the deceit that spirits do not exist?[65]

---

[61] *Tianzhu shi yi*, 452; Ricci, *True Meaning of the Lord of Heaven*, 178–79. This passage is also translated in Gernet, *China and the Christian Impact*, 68: "Some say that they definitely do not exist. Others say that they do exist if one believes in them and do not if one does not believe in them. Others say that it is just as false to say that they do exist as to say that they do not and that the truth is that they both do and at the same time do not exist." Gernet concludes that these debates were "hardly . . . to the missionaries' liking," but does not explain why this translation does not controvert his claims in his concluding section, "Language and Thought," and in particular his central assertion that "there was no word to denote existence in Chinese, nothing to convey the concept of being or essence, which in Greek is so conveniently expressed by the noun *ousia*" (p. 241).

[62] Reading *shu* 屬 for *shu* 殊 in the translation.

[63] The entire passage states, "Fan Chi inquired about wisdom. Confucius stated: 'Endeavor to make the people righteous, respect the spirits and distance them; this can be termed wisdom.' [Fan Chi also] inquired about humaneness. [Confucius] stated: 'The humane face difficulties first and obtain afterwards; this can be called humaneness.'" 樊遲問知。子曰：「務民之義，敬鬼神而遠之，可謂知矣。」問仁。曰：「仁者先難而後獲，可謂仁矣。」 *Analects* 6.22, *Lunyu zhushu* 論語註疏 [*Analects*, with commentary and subcommentary], in *Shisanjing zhushu: fu jiaokan ji* 十三经注疏：附校勘记 [Thirteen Classics, with commentary and subcommentary, and with editorial notes appended] (Beijing: Zhonghua shuju 中华书局, 1980), hereinafter *SSJZS*, 2:2479. Thus, in the comparison Confucius appears to be exhorting Fan Chi to concentrate on the affairs of the living rather than those of spirits. See also *Analects* 11.12, *Lun yu zhushu*, 2:2499.

[64] *Analects* 3.13, *Lun yu zhushu*, 2:2467.

[65] *Tianzhu shi yi*, 468; Ricci, *True Meaning of the Lord of Heaven*, 202–3.

故仲尼曰：「敬鬼神而遠之。」彼福祿免罪，非鬼神所能，由天主
耳。而時人謟瀆66，欲自此得之，則非其得之之道也。夫「遠之」意與
「獲罪於天，無所禱」同。豈可以「遠之」解「無之」，而陷仲尼于
無神之惑哉。

What is superstitious about Chinese ancestor worship, then, is not belief in these spirits themselves but seeking favor by worshiping the spirits of ancestors rather than salvation by worshiping the Lord of Heaven. Ricci's answer indicates that the problem is not a lack of belief but, rather, that the Chinese concept of God has been effaced. Contrary to Gernet's claims, the translation of European treatises into Chinese provides no convincing evidence to demonstrate difficulties in expressing the concept of *existence*, whether of tangible or intangible objects. In historical context, these problems were not about the impossibility of translating the abstractions of modern philosophy into Chinese; they were debates about the existence of spirits and, in particular, the primal noun (God) and the primal verb (Being).

Similarly, differences in categories of thought—whether based on Benveniste's view that language dictates categories through an unconscious process or Kuhn's view that incommensurability results from mismatched taxonomies—do not provide a framework outside history available only to modern commentators. Instead, the claim of differences in taxonomies was itself a strategy in Jesuit propaganda. Ricci argues,

> In dividing things into categories, the learned men of your noble country state: some [things] attain form, such as metal and stone; some in addition attain the energy of life and grow, such as grass and trees; some also attain the senses, such as birds and beasts; and some are more refined and attain consciousness and intelligence, such as man.
>
> 分物之類，貴邦士者曰：或得其形，如金石是也；或另得生氣而長大，如草木是也；或更得知覺，如禽獸是也；或益精而得靈才，如人類是也。
>
> Our learned men of the West seem to have made even more detailed [categories], as can be seen from the following chart. Only its classes of accidents are most numerous and difficult to list completely, and therefore they are only summarized, emphasizing their nine major classes.[67]
>
> 吾西庠之士猶加詳焉，觀後圖可見。但其依賴之類最多，難以圖盡，故略之，而特書其類之九元宗云。

Linguistic differences and conceptual differences do exist—as they do within civilizations, cultures, and subcultures—but this is very different from a theory of radical conceptual incommensurability split along a purported China/West divide. As an explanatory framework, conceptual incommensurability is at best

---

[66] Substituted for idiosyncratic variant.

[67] *Tianzhu shi yi*, 461; Ricci, *True Meaning of the Lord of Heaven*, 190; Gernet, *China and the Christian Impact*, 243.

inflationary: Does the rejection of Ricci's claims asserting the existence of spirits really require for its explanation a theory of incommensurability alleging the impossibility of the translation of the concept of existence? Does contesting Ricci's God and the evidence that he adduces for Him from the *Analects* really require for its explanation a theory of two radically different philosophical worldviews? Does the disbelief in the soul really require for its explanation assertions of the incomprehension of Western concepts and scholastic philosophy? Worse, Gernet's thesis of incommensurability "is contradicted by the facts." The above examples suggest no insurmountable difficulties in expressing concepts of existence.

Gernet's claims themselves, like those of Martzloff, create the otherness of the world he purports to describe. These claims of differences derive their plausibility in the first place from the assumption of a Great Divide between China and the West. The complex similarities and dissimilarities between Jesuit doctrines and Legalism, Buddhism, Confucianism, Daoism, and popular religions are collapsed along a single radical China/West divide, which is then reinforced through the insistent, repeated assertion of difference: "The conclusion that emerges from the various texts which I have just cited is that Chinese conceptions are in *every regard* the opposite of those taught by the missionaries."[68] Counterexamples—the adoption of Western Learning [*Xi xue* 西學] by Chinese collaborators—are themselves appropriated to further reproduce this divide: "What appeared to fit these traditions—or rather what could be easily integrated—was accepted; the rest was unacceptable";[69] "Chinese who sympathized with Christianity praised it in terms of Chinese philosophy and attributed their own conceptions to it."[70] Through links forged by tracing linguistic continuities, this break between cultures then becomes the suprahistorical break between civilizations;[71] the rejection of Jesuit doctrines is inflated to represent the rejection of "the European vision of the world."[72]

The import, for research on China, of this barrier constructed by incommensurability was that—against dismissals common in the received historiography that narrated the inexorable triumph of Western universals—relativism protected the assertions of the Chinese protagonists. Relativism insisted not just on equivalence but on the impossibility in principle of any comparison. Elevated to the status of an equal, the claims and allegations of the Jesuits' Chinese opponents came to merit historical analysis. This, then, is the central contribution resulting from Gernet's adoption of relativism and incommensurability—his insistence on understanding the arguments of the Chinese opponents of the Jesuits and their

---

[68] Gernet, "Christian and Chinese Visions of the World in the Seventeenth Century," *Chinese Science* 4 (1980): 14, emphasis added.

[69] Ibid., 15.

[70] Ibid., 16.

[71] For example, Gernet notes, "The difference between the philosophical ideas inherited from Ancient Greece and those of the Chinese emerges clearly here." Gernet, *China and the Christian Impact*, 210.

[72] From the concluding sentence to Gernet, "Christian and Chinese Visions of the World," 17.

collaborators through the analysis of Chinese primary materials. For this, *China and the Christian Impact* has been justly recognized as seminal.

Yet one consequence of this very relativism was that it effaced the social and political context it should have analyzed—it is important here to see what this framework of incommensurability must, by construction, leave out. When social and historical context is shifted into the realm of discourse, rejections of Jesuit doctrines are explicable only by assertions of the impossibility of translation between two radically different worldviews; individuals are deprived of agency in deciding what is determined by these conceptual structures. Yet how can we be assured that the difficulty is one of translation if the existence of the referent of the term—whether *spirits* or *God*—is itself in question? How does one determine whether the notion of *God* was correctly translated? Without a grounding in a truth external to the text, the only criterion for assessing the translations is comparisons with explanations and glosses in other texts; in short, we have returned to doctrinal disputes within Christianity, of which the Jesuits were themselves but one faction. And perhaps the most serious problem in Gernet's framework is the collapsing of the complex interactions between individuals and subcultures to two mutually exclusive poles, China and the West; the Chinese collaborators, belonging to neither, can only be minimized as transparent, passive translators.

The Contextual Turn

Historical studies framed within the linguistic turn have often displaced social context into the realm of discourse;[73] theories of incommensurability have been but one variant on this trend. One important alternative that replaces discourse back into its social context is suggested by Mario Biagioli's proposal of a "*diachronic* approach" to incommensurability by analyzing "its *emergence* in relation to the internal structure, external boundaries, and relative power or status of the socio-professional groups involved in the non-dialogue."[74] Biagioli's concern is to explain how phenomena that have been interpreted as linguistic incommensurability are in fact strategies adopted in social conflicts between groups that share the same language.[75] I will take a related approach, showing

---

[73] The most sustained attempt to formulate such a theory is Michel Foucault, *The Archaeology of Knowledge & the Discourse on Language*, trans. A. M. Sheridan Smith (New York: Pantheon Books, 1972). For an analysis, see Hubert L. Dreyfus and Paul Rabinow, *Michel Foucault: Beyond Structuralism and Hermeneutics*, 2nd ed. (Chicago: University of Chicago Press, 1983). The theoretical reformulation presented in *Archaeology of Knowledge* should be contrasted with Foucault's earlier historical works.

[74] Mario Biagioli, "The Anthropology of Incommensurability," *Studies in History and Philosophy of Science* 21 (1990): 184.

[75] Incommensurability, Biagioli asserts, is often associated with instances of trespassing professional or disciplinary boundaries. He concludes against Kuhn that while the possibility of developing bilingualism is not logically wrong, the assumption is unwarranted in ignoring "the fundamental relation between social groups and cognitive activity." Ibid., 207.

how claims of purported difficulties in translations between different languages themselves served as resources in social conflicts. Claims about translation will not be conceptualized as disembodied philosophical theories that provide a framework external to history within which events are analyzed; instead, claims about translation will themselves be examined as in need of historical explanation. That is, instead of accepting claims made by historical protagonists about translation, whether about its accuracy, difficulty, or impossibility, we must analyze, contextualize, and render historically intelligible those claims themselves. For example, in seventeenth-century Europe, theories about the translation and accommodation of the spoken word of God to the languages of the illiterate masses provided both important resources and alibis in debates over heliocentric theories;[76] theories about translations of the Bible and the Book of Nature were central sites of contest in debates over the legitimacy of Galileo's claims.[77]

Translation Historicized

How, then, did claims about translation provide for the Jesuits and their Chinese patrons important opportunities? To answer this question, we must first examine the translations, conceptualized not as an impossibility demonstrated by posing, against the translations, objections from biblical hermeneutics but, rather, analyzed as historical events.[78] Perhaps the simplest approach to translation, and the one adopted by the Jesuits and their Chinese patrons most frequently (except for the translation of important theological terms) was the creation of neologisms. In some cases, they resorted to loan-words—neologisms created by transliteration (here, and throughout this book, phrases typeset in a smaller font are commentaries, which in Chinese are distinguished from the original text by the use of half-size characters):[79]

> The study of *anima* translated, means "soul" or "nature of the soul" within *philo-sophia* translated, means the "study of investigating things and exhausting principles" is the most beneficial and the most respected.

---

[76] Robert S. Westman, "Proof, Poetics, and Patronage: Copernicus's Preface to De Revolutionibus," in *Reappraisals of the Scientific Revolution*, ed. David C. Lindberg and Robert S. Westman (Cambridge: Cambridge University Press, 1990), esp. 90–91.

[77] Mario Biagioli, "Stress in the Book of Nature: The Supplemental Logic of Galileo's Realism," *MLN* 118 (April 2003): 557–85. Biagioli argues that "Galileo's so-called mathematical realism was, in fact, a form of scriptural fundamentalism" (p. 559).

[78] For an important theoretical analysis of the problem of translation between Chinese and Western languages and a critical overview of theories on translation, see Liu, *Translingual Practice*, esp. chap. 1.

[79] For an analysis of translation and a discussion of terminology, see Theodora Bynon, *Historical Linguistics* (Cambridge: Cambridge University Press, 1977), chap. 6; see also John Lyons, *Semantics*, 2 vols. (Cambridge: Cambridge University Press, 1977).

亞尼瑪譯言靈魂亦言靈性之學。於費祿蘇非亞譯言格物窮理之學中，為最
益，為最尊。[80]

Transliteration provided one possible translation for the term *God*. For example, the first chapter of *Tianzhu shi yi*, titled "Lun Tianzhu shi zhi tian di wanwu, er zhuzai anyang zhi" 論天主始制天地萬物而主宰安養之 [Showing that the Lord of Heaven created heaven, earth, and the myriad things, and controls and sustains them], argues that there must be a creator of the heavens, and "Thus this is the Lord of Heaven, the one our Western nations term *Deus*" 夫即天主，吾西國所稱『陡斯』是也.[81] A transliteration for *Deus* was, in fact, the choice for the translation of the term "God" into Japanese. This historical possibility of creating neologisms by semantically neutral transliteration undermines theories of incommensurability that assert radical impossibilities based on nothing more than the absence of lexical terms.[82]

Another approach was loan translations—the creation of semantic neologisms by combining characters. This was the approach most often employed by the Jesuits and their Chinese patrons. The following examples will serve as a very small sample: in theology, terms for omniscience (*zhi zhi* 至智), omnipotence (*zhi neng* 至能), and infinite goodness (*zhi shan* 至善);[83] in Aristotelian philosophy, terms for cause (*suoyiran* 所以然), active cause (*zuozhe* 作者), formal cause (*mozhe* 模者), material cause (*zhizhe* 質者), final cause (*weizhe* 為者), the four types of causes—proximate, distant, universal, and special (respectively, *jin* 近, *yuan* 遠, *gong* 公, *si* 私),[84] substance (*zilizhe* 自立者), and accident (*yilaizhe* 依賴者);[85] and in Euclidean geometry, point (*dian* 點), line (*xian* 線), and surface (*mian* 面).[86] A very different strategy of translation adopted by the Jesuits and their patrons was the selective omission of doctrines that would have subjected them to even harsher attacks: their writings often omitted mention of the Trinity, revelation, and, with the exception of baptism, the sacraments: presumably, the most difficult of the sacraments to explain would have been the Eucharist—that

---

[80] *Ling yan li shao yin* 靈言蠡勺引 [Preface to *A Preliminary Discussion of Anima*], in *TXCH*, 2:1127.

[81] *Tianzhu shi yi*, 1.381; Ricci, *True Meaning of the Lord of Heaven*, 70.

[82] For examples of the translation of Christian doctrines into European languages, see Bynon, *Historical Linguistics*, chap. 6.

[83] *Tianzhu shi yi yin* 天主實義引 [Preface to *True Meaning of the Lord of Heaven*], in *TXCH*, 1:370; Ricci, *True Meaning of the Lord of Heaven*, 62.

[84] *Tianzhu shi yi*, 1:390–91; Ricci, *True Meaning of the Lord of Heaven*, 84–86.

[85] *Tianzhu shi yi*, 1:406; Ricci, *True Meaning of the Lord of Heaven*, 108. Hu and Lancashire note that the current terms are *ziliti* 自立體 and *yifuti* 依附體; see Ricci, *True Meaning of the Lord of Heaven*, 108, n. 18.

[86] For a comprehensive table of Euclidean terms translated by the Jesuits, along with their modern equivalents, see Engelfriet, *Euclid in China: The Genesis of the First Chinese Translation*.

bread and wine were in actuality the body and blood of Christ. Their writings also frequently failed to mention the crucifixion of Jesus.[87]

For the translation of the most important terms in Christian theology, the Jesuits and their Chinese patrons used semantic extension, borrowing and redefining terms from Buddhism and Confucianism. From Buddhism they appropriated terms that had been employed as equivalents for Sanskrit: for Heaven, *tiantang* 天堂 (Sanskrit *devaloka*, mansion of the gods); for Hell, *diyu* 地獄 (Sanskrit *naraka*); for Devil, *mogui* 魔鬼 (*mo* for the Sanskrit *mara*); for angels, *tianshen* 天神 (Sanskrit *deva*; Protestant translators later employed *tianshi*); for soul, *linghun* 靈魂.[88] But it was from Confucianism that the Jesuits and their Chinese patrons borrowed their most crucial terms. As Gernet notes, "The first missionaries were especially delighted to find in the Classics—the works venerated above all others among the literate elite—the term 'Sovereign on High' (*shangdi*), invocations to Heaven and expressions such as 'to serve Heaven' (*shi tian*), 'to respect' or 'fear Heaven' (*jing tian, wei tian*)."[89] For example, the overall project of the Jesuits was described by Xu Guangqi as "self-cultivation to serve Heaven" (*xiu shen shi tian* 修身事天).[90]

Translating "God"

Among the terms translated, the one of the utmost significance for the Jesuits and their Chinese patrons was *God*. Instead of the transliteration of *Deus*, early translations made use of several choices: "Lord of Heaven" (*Tianzhu* 天主, more literally, "Master of Heaven," a term that appears in Buddhist texts), "Emperor on High" (*Shangdi* 上帝, a term that appears in several early Chinese texts, including the *Li ji* 禮記, *Shi jing* 詩經, *Shu jing* 書經, *Mozi* 墨子, and *Shi ji* 史記), "Heavenly Emperor" (*Tiandi* 天帝, a term that appears primarily in Buddhist texts but also in *Zhan guo ce* 戰國策), and "Most Venerated" *Shangzun* 上尊.[91]

---

[87] Yang Guangxian 楊光先 (1597–1669) later accused the Jesuits of frequently failing to mention the crucifixion, arguing that the Jesuits wished to deliberately conceal that the Lord they worshiped was nothing more than a criminal during the period of the Han dynasty.

[88] These examples are from Douglas Lancashire and Peter Kuo-chen Hu, SJ, "Introduction," in Ricci, *True Meaning of the Lord of Heaven*, 35–36.

[89] Gernet, *China and the Christian Impact*, 25.

[90] Xu Guangqi uses this term to characterize the Jesuits' mission in China in a memorial to the Ming court. The phrase *shi tian* appears in *Liji zhushu* 禮記註疏 [*Record of Rites*, with commentary and subcommentary], in *SSJZS*, 2:1612. *Xiu shen* is a central concept in Cheng-Zhu Neo-Confucianism. See, for example, Zhu Xi 朱熹 (1130–1200), *Da xue zhang ju* 大學章句 [*Great Learning*, separated into chapters and sentences], in *Sishu zhangju jizhu* 四書章句集注 [Four Books, separated into chapters and sentences, with collected commentaries] (Beijing: Zhonghua shuju 中華書局, 1983), hereinafter *SZJ*, 3.

[91] *Tianzhu shi yi*, 1:359, 359, 366, and 369, respectively. Other terms were also used for God in more specific contexts, such as *Dayuan* 大元 (*Tianzhu shi yi*, 1:399; Ricci, *True Meaning of the Lord of Heaven*, 96). Lancashire and Hu note that Yuan 元 was sometimes used, along with Yuan 原, "to express 'source' or God as Creator" (Ricci, *True Meaning of the Lord of Heaven*, 96, n. 32). As noted by Lancashire and Hu, these terms were sometimes preceded by a blank space to

Different choices in translation provided the Jesuits and their Chinese patrons with different opportunities. Phonemic and semantic neologisms necessitated lengthy explanations and commentaries; examples include treatises explaining the concept of *anima* and the soul.[92]

Borrowings from Buddhism and Confucianism provided important additional opportunities. The borrowing of terms from Buddhism was the result of an attempt by the Jesuits in the early years of the mission in China to represent themselves as similar to the Buddhists. But in later years, their use of Buddhist terms also provided the Jesuits with the claim that their doctrines corrected Buddhist distortions. For the Jesuits, Buddhist theories were doubly false: Buddhist doctrines were perversions of Indian beliefs that were no longer accepted in India, doctrines that had in fact originated in the false beliefs of Pythagoras.[93]

The Jesuits' strategy in China was focused—as in Europe—on gaining patronage with the ruler. The official Confucian orthodoxy provided crucial opportunities for this project and, at the same time, for the elite literati-officials who collaborated with the Jesuits. For the Jesuits and their Chinese patrons, the problem in the choice of the proper term for *God* was not a lack of possible equivalents but, rather, the opportunities offered by each that entailed complex strategic implications and consequences—social, political, philosophical, and philological. (The conflict over these choices led to the Rites Controversy; the use of the terms *Tian* and *Shangdi* was forbidden by the pope, in 1704.)[94] Claims made by the Jesuits and their Chinese patrons about translation served as an important means of legitimation. As Gernet notes,

> The missionaries also often resort to another idea: namely that part of the ancient Chinese tradition had disappeared in the Burning of the Books ordered by the first of the Qin emperors in 213 BCE and that it was precisely that part that set out the thesis of an all-powerful, creator God, the existence of heaven and hell and the immortality of the soul; the teaching of the missionaries fortunately made it possible to complete what had been lost in the classical traditions of China. This is, indeed, pretty well the thesis put forward by Ricci in *The True Meaning of the Master of Heaven*, where he explains to a Chinese man of letters why it is that the Classics make no mention of paradise and hell.[95]

---

denote respect; the terms *Zhongguo* 中國 and *Zhonghua* 中華 were also preceded with a blank space (*Tianzhu shi yi*, 1:367). To denote this, I have capitalized these terms in pinyin; it should be noted, however, that the addition of a blank space in the Chinese text is not used uniformly throughout.

[92] *Ling yan li shao* 靈言蠡勺 [A preliminary discussion of anima], in *TXCH*.

[93] For examples of Jesuit claims of Buddhist distortions, see Erik Zürcher, "The Jesuit Mission in Fujian in Late Ming Times: Levels of Response," in *Development and Decline of Fukien Province in the 17th and 18th Centuries*, ed. E. B. Vermeer (Leiden: Brill, 1990), 418 n. 3.

[94] In 1704 Pope Clement XI banned the use of *tian* and *shangdi* as translations for God. Lancashire and Hu, "True Meaning of the Lord of Heaven," 20.

[95] Gernet, *China and the Christian Impact*, 28.

For the Jesuits and their Chinese patrons, translation was not theorized as introducing new knowledge but, rather, as a recovery of knowledge that had been lost from the Chinese tradition.

The most important opportunity offered by translation was in the selection, among possible equivalents, for the term *God*: ambiguities in translation were a crucial resource in camouflaging ambiguities in the loyalty of elite literati-official collaborators toward the Jesuits. These elite collaborators were central in the dissemination of Western Learning in China, not merely as the object of Jesuit proselytism strategies, but in translating Western Learning into the language and literary style of the elite, helping to legitimate it, and building their careers on its advocacy.

The success of Xu Guangqi—the most important among the Jesuits' Chinese patrons—exemplifies how problems of translation permitted the Chinese collaborators to produce documents that could be read by both the Jesuits and the Ming imperial court as expressions of loyalty and faith.[96] Xu had been trained in the Hanlin Academy to write memorials to the Ming court on issues ranging from taxes to water conservancy, from military proposals to astronomy. In his memorials Xu repeatedly risked his career for the Jesuits, while at the same time fashioning himself as a statesman with novel practical solutions to Ming dynasty crises. Probably the most important example is Xu's explicit defense of the Jesuits in "Bian xue zhang shu" 辨學章疏 [Memorial on distinguishing learning] (I will analyze this in detail in the section titled "Xu Guangqi's 'West,'" on pages 245–253 of this book). For Gernet, this passage is evidence of a Chinese "ancient mental framework" incommensurable with Western thought: "For Zhang Xingyao and Xu Guangqi, the ancient mental frameworks remain unchanged despite their conversions: orthodoxy must contribute towards the universal order and is recognisable by its beneficial moral and political effects."[97]

Yet the content of this passage reflects not the workings of a mental framework but, rather, its intended purpose; it is, after all, a memorial on political policy presented to the Ming imperial court. What Gernet ignores is evidence from other sources that can equally support the claim that Xu adopted aspects of Christianity: Xu's extant letters show, as Fang Hao notes, that "there are many places that express the sincerity of his religious beliefs."[98] In particular, Xu's eleventh extant letter to his family shows the importance he attached to Catholic rituals—he was concerned that his father-in-law had not been given absolution before

---

[96] It should be noted that the specific mechanisms of patronage in China differ in crucial ways from those found in Europe. Perhaps the most important differences are the examination system and the official bureaucracy, and the way that these entities structured patronage in the Ming court.

[97] Gernet, *China and the Christian Impact*, 110–11.

[98] Fang Hao 方豪, *Zhongguo Tianzhujiao shi renwu zhuan* 中國天主教史人物傳 [Biographies of persons in the history of Chinese Catholicism] (Beijing: Zhonghua shuju 中華書局, 1988), 103.

his death.[99] Questions about Xu's beliefs, then, cannot be answered a priori on the basis of differences in language or worldview; instead, it is precisely these questions that were central matters of debate among the historical protagonists. If instead of theorizing translation as an impossibility, we seek to understand the process by which translation did occur, this passage is representative of that process. Xu's novel solutions—military, moral, mathematical, and astronomical—appealed to a desperate Ming court that promoted him to one of the highest posts;[100] in turn, Xu's success helped legitimate the Western Learning that he advocated. And in this passage, Xu's most dramatic evidence that this imagined "West" offered solutions to Ming dynasty crises was a lexical absence—the word for "rebellion."[101] Here again—as was the case with Martzloff and Gernet—fantastic claims about the radically different Other are imagined via assertions about the absence of words.

Conclusions

Relativism and the incommensurability (linguistic or conceptual) on which it was based required for its initial formulation the assumption of a radical divide between two imagined unities, China and the West. Differences in languages then served as a natural symbol for this presumed divide—both by paralleling political boundaries and by providing compelling metaphors for the suprahistorical continuity of civilizations. This use of languages as emblems for civilizations could, however, with equal ease be used to support either universalistic or relativistic conclusions: on the one hand, differences in languages, through purported hierarchies constructed for languages (e.g., their precision and scientificity, or the development of alphabetization) have been linked to other civilization-defining teleologies (e.g., science, or capitalism);[102] on the other hand, differences in language have also provided important metaphors for constructing relativism's radically different Other.

---

[99] Xu Guangqi, *Jia shu* 家書 [Family letters], in *Xu Guangqi ji* 徐光啟集 [Collected works of Xu Guangqi] (Shanghai: Shanghai guji chubanshe 上海古籍出版社, 1984), 2:492, hereinafter *XGQJ*. For an English translation of these letters, see Gail King, "The Family Letters of Xu Guangqi," *Ming Studies* 31 (1974): 1–41.

[100] Xu's career followed the pattern of exile and returns to power typical of the period; with the ascendance of Sizong in 1628, Xu returned to power. In the last year of his life, 1633 (Chongzhen 6), he reached one of the highest posts in government, the Grand Guardian of the Heir Apparent, and Grand Secretary of the Hall of Literary Profundity (Taizi taibao wenyuange daxueshi 太子太保文淵閣大學士).

[101] See the translation of this memorial in chapter 6, "Xu Guangqi, Grand Guardian," on pages 245–253 of this book; for this particular claim, see footnote 196 on page 248.

[102] Important examples include Jack Goody, *The Logic of Writing and the Organization of Society* (Cambridge: Cambridge University Press, 1986), and Bodde, *Chinese Thought, Society, and Science*.

For the purposes of this latter relativism, features of languages were further essentialized and radicalized: the difficulties of translation came to represent impervious barriers; mutual intelligibility came to represent an essentialized, systemic unity; diachronic continuities came to represent suprahistorical self-identity. The enormous diversity of schools of thought collapsed into an essentialized China provided crucial alibis in the forging of continuities and discontinuities on which the claims of incommensurability depended. On the one hand, this diversity provided a wealth of examples to demonstrate the opposition between China and the West; on the other hand, this same diversity could always provide examples to explain the acceptance of Western doctrines in China as nothing more than the acceptance of the Chinese tradition.

Historically contextualized, in their translations the Jesuits and their Chinese patrons adopted for their own religious concepts terminology appropriated from Buddhism. They created neologisms from terms in the Confucian tradition. Their neologisms provided the opportunity for explanations and commentary. Jesuit doctrines were then claimed to be the recovery of the lost meanings of the Confucian classics destroyed in the Qin burning of the books. But most important, problems in translation served as an important strategy of the Chinese patrons and collaborators. By introducing ambiguities in the translation of terms such as "Lord of Heaven," "Emperor on High," and "serving heaven," the Jesuits' Chinese patrons produced documents that could be read by both the Chinese court and the Jesuit missionaries as expressions of allegiance. Translation was thus not an obstacle to dialogue but in fact a crucial resource; the Jesuits' Chinese patrons were not transparent scribes but active agents manipulating these translations.

Chapter 4
Mathematical Texts in Historical Context

THE SONG (960–1276) and early Yuan (1260–1368) mark the apex that traditional Chinese mathematics never again attained, the received historiography has maintained: during the period from the mid-Yuan until the arrival
of the Jesuits at the end of the Ming dynasty (1368–1644), Chinese mathematical treatises were lost, earlier discoveries forgotten, and mathematics disdained.
According to this view, it was only the introduction of Western mathematics by
the Jesuits in the late Ming that created a revival of interest in Chinese mathematics.[1]

The Ming dynasty has often been portrayed as a period of decline—whether
ritual, moral, intellectual, political, or scientific—and it is within this narrative
that studies of Ming mathematics have consistently been framed. These claims
of mathematical decline have been self-reinforcing—the lack of studies of Ming
mathematics has confirmed the claim that there was little mathematics to study.
In sum, for historical accounts of Chinese mathematics, the term "decline" has
often seemed to be all that was needed to describe mathematics from the early
Yuan to the late Ming.

This chapter proposes an alternative approach to the study of mathematics
during the Ming dynasty, one that avoids simply choosing between "decline" and
its inverse, "development." These two choices—"decline" and "development"—
tell us little about the mathematics of any period, differing only in representing
opposite directions in an ahistorical "internal" teleology offered as an explanatory framework to fill the void that remains after the effacement of the social and
intellectual context.[2] Instead, mathematical practices will be conceptualized as

---

[1] For an important corrective of this view, see Willard J. Peterson, "Calendar Reform Prior to the
Arrival of Missionaries at the Ming Court," *Ming Studies* 21 (1986): 45–61.

[2] This is not to deny that historical research has often benefited from assertions of "development": claims of "development" impart to historical study an urgency and significance by
the asserted connections to the present; claims of "development" also provide a framework
for historical study—ordering new discoveries along a teleology, constructing hagiographies
of the discoverers, and investigating priority and influence. In contrast, however, assertions of
"decline" suggest that a period may be ignored and the work of the period safely dismissed.

local forms of knowledge inseparable from the historical context in which they were produced.[3]

This chapter will attempt to reconceptualize mathematics during the Yuan and Ming by examining three of the most well-known extant mathematical treatises from the period. The first mathematical treatise is Li Ye's *Sea Mirror of Circle Measurement*, which has been considered perhaps the most important representative of Yuan dynasty Celestial Origin (*tian yuan* 天元) algebra,[4] and has also served as an important example of the mathematics "lost" during the Ming. The second is Cheng Dawei's *Comprehensive Source of Mathematical Methods*, a Ming dynasty mathematical treatise that was frequently reprinted during the Ming and Qing, and was transmitted to Japan and Korea as well. The third mathematical treatise is Zhu Zaiyu's *New Explanation of the Theory of Calculation*, a work on precision calculations of the equal temperament of the musical scale, which is rarely even mentioned in histories of Chinese mathematics. It is this text for which we can offer the most substantive analysis of the historical context—the debates over the role of music in ritual. But before we examine these three mathematical treatises, we will review characterizations of the Ming Dynasty as a period of "decline."

---

[3] My use of the term "local" to describe forms of knowledge derives from work in the history of science. By its use I do not wish to endorse Clifford Geertz's formulations, and certainly not Joseph Needham's characterization of premodern science as essentially local, which he then contrasted with universal modern science. My emphasis on locality is microhistorical and meant to question tautological claims of "universality": when beliefs or practices asserted to have persisted over limited historical periods are proclaimed to be universal, this universality is then offered in place of a historical explanation for their persistence. Instead of adopting claims of universality made by the historical protagonists as fact, we must critically contextualize these claims as performative and ideological. That is, in contrast to studies in which mathematical techniques have been removed from social context (and often even mathematical context) to be placed into a teleology with modern mathematics as its endpoint, knowledge is viewed as comprehensible only in its social and historical context. In particular, we must avoid adopting as our starting point in historical studies the ideology that mathematics is pure and distinct from other areas of intellectual endeavor and social experience. And yet, social context itself cannot be a sufficient explanation—we cannot know (certainly, at least, for the period under study) a priori which contexts will result in the creation of specific forms of mathematics. That is, instead of taking beliefs or metaphors current during a particular period as the transparent cause of mathematical development, it is these beliefs that must be contextualized as but one constituent of the knowledge constructed. For a historical explanation, we must analyze both the context and the mathematics itself. See Clifford Geertz, *Local Knowledge: Further Essays in Interpretive Anthropology* (New York: Basic Books, 1983); idem, *The Interpretation of Cultures: Selected Essays* (New York: Basic Books, 1973); idem, "Local Knowledge and Its Limits," *Yale Journal of Criticism* 5 (1991): 129–35.

[4] I will follow Martzloff's translation of *tian yuan* here, rather than the more traditional translation "Celestial Element," since *tian yuan* refers to a "reference place (or origin) on the counting surface." Similarly, *si yuan* will be translated as "Four Origins." See Jean-Claude Martzloff, *A History of Chinese Mathematics* (New York: Springer, 2006), 258–59.

Ming Dynasty "Decline" in Search of a Teleology

As we have seen, historical accounts have often portrayed the Ming Dynasty—and the late Ming in particular—as a period of decline. What has changed over the centuries in these accounts is not so much the assessment of the Ming but rather the framework invoked to narrate that decline.[5] For scholars of the early Qing dynasty (1644–1911), the Ming was a period of moral and intellectual decline: Gu Yanwu 顧炎武 (1613–1682) and Wang Fuzhi 王夫之 (1619–1692) placed the blame for the ensuing Manchu conquest on Ming intellectuals.[6]

It was only later that comparisons with the West were incorporated into the framework of dynastic collapse, articulated perhaps most eloquently by Liang Qichao 梁啟超 (1873–1929). Liang still linked the demise of the Ming Dynasty to intellectual decadence,[7] but extolled Xu Guangqi 徐光啟 (1562–1633) and the Jesuit collaborators for their important contributions to science during this period.[8] In Liang's view, the Qing Dynasty represented a turn toward science and a decisive break with the subjective metaphysics of inner cultivation of Neo-Confucianism—the dominant orthodoxy over the preceding 600 years—and its concomitant syncretism with Zen Buddhism and Daoism.[9] Decline for Liang is thus no longer simply the political failure of a single dynasty or a failure to adhere to the institutions and rituals of the ancient past.

Perhaps nowhere is the decline of the Ming more compellingly portrayed than in Ray Huang's *1587, a Year of No Significance*.[10] Though Huang cleaves to the

---

[5] Claims of decline (and changing frameworks used to explain it) are hardly new to the Ming Dynasty. For example, Ouyang Xiu's 歐陽修 (1007–1072) *Xin Tang shu* 新唐書 [New History of the Tang] begins, Peter Bol notes, "with a new definition of the decline from antiquity in terms of unity and duality, in contrast to the early Tang view of decline in terms of *wen* and to the later view in terms of the decline of the *dao*." Peter K. Bol, *"This Culture of Ours": Intellectual Transitions in T'ang and Sung China* (Stanford: Stanford University Press, 1992), 195; see also 308–9.

[6] For a discussion of these views, see Liang, *Zhongguo jin sanbai nian xueshushi* 中國近三百年學術史 [History of Chinese intellectual thought during the last 300 years] (Yangzhou: Jiangsu Guangling guji keyinshe 江蘇廣陵古籍刻印社, 1990), 8; and idem, *Intellectual Trends in the Ch'ing Period*, trans. Immanuel C. Y. Hsü (Cambridge, MA: Harvard University Press, 1959).

[7] Liang Qichao argues that scholars of the period fought among themselves while the Manchus prepared for invasion; the concept of "investigating affairs" had become, he asserts, nothing more than a meaningless Buddhist slogan. It should be noted that, in contrast to other critiques of Ming intellectuals, Liang's critique of the Ming is not directed at Wang Yangming 王陽明 (1472–1529), who Liang asserts did make important contributions, but at his followers. Liang, *Zhongguo jin sanbai nian xueshushi*, 3.

[8] Liang notes Euclid's *Elements* in particular as a central work of the period (ibid., 8–9). This passage is also quoted in the most important bibliography of Jesuit works translated into Chinese. See Xu Zongze 徐宗澤, *Ming-Qing jian Yesu huishi yi zhu tiyao* 明清間耶穌會士譯著提要 [Annotated bibliography of the translations and writings of the Jesuits during the Ming and Qing dynasties] (Taibei: Zhonghua shuju 中華書局, 1989), 3–4.

[9] Liang, *Zhongguo jin sanbai nian xueshushi*, 1–10.

[10] Ray Huang, *1587, a Year of No Significance: The Ming Dynasty in Decline* (New Haven, CT: Yale University Press, 1981).

most traditional of frameworks—the political decadence of the emperor and his advisors, which led to dynastic collapse—his study begins and ends with brief, implicit hints that the ultimate framework of comparison and analysis is the rising, early modern West.

This narrative of Ming dynasty decline has also been transfigured into Western historiographic conventions, as exemplified by Mark Elvin's *The Pattern of the Chinese Past*.[11] For Elvin, decline inheres not in moral, ritual, or political decadence, but rather in the loss of economic, scientific, and technological innovation.[12] And the three explanations that Elvin offers for this Chinese decline in the 14th century—the closing of the frontier, increasing isolation from the rest of the world, and the "disastrous" consequences of Wang Yangming's moral intuitionism for Chinese science—derive from the three central emblems of European early modernity: expansion, openness, and science.[13]

Other studies focused on specific aspects of the period (as opposed to Ming China in its totality) have demonstrated the inadequacy of "decline" as a historical explanation within their respective fields; but this scholarship has often reversed assertions of decline by invoking explicitly developmental metaphors that appeal to equally anachronistic teleological frameworks. For example, in his studies of Ming philosophy, Theodore de Bary's central metaphor is of Neo-Confucianism "unfolding" toward an apparently modern "liberal view of the self," as well as pragmatism and humanitarianism.[14] Similarly, the assertions by economic historians of burgeoning "sprouts of capitalism" in the late Ming economy invoke a biological metaphor to assure—against the purported stagnancy of traditional China—motion in the direction of development along a Marxist teleology.[15]

---

[11] Mark Elvin, *The Pattern of the Chinese Past: A Social and Economic Interpretation* (Stanford, CA: Stanford University Press, 1973).

[12] The "dynamic quality of the medieval Chinese economy disappeared," Elvin argues, in about the middle of the fourteenth century and recovered only slowly; but more importantly, "the internal logic of Chinese historical development began to change"—there were "no advances in science to stimulate advances in productive technology." Ibid., 203–4.

[13] Ibid., 203ff.

[14] In the introduction to Wm. Theodore de Bary and the Conference on Ming Thought, eds., *Self and Society in Ming Thought* (New York: Columbia University Press, 1970), de Bary notes that Ming thought had fallen into "bad repute": it was viewed as a period of "general decline and aimless drifting" in which Wang Yangming was perceived as the only bright spot (p. 1). De Bary asserts, however, that Chinese scholars had unfairly deprecated Ming thought, citing Gu Yanwu, Cui Shu 崔述 (1740–1816), and Liang Qichao as examples. Against these caricatures, de Bary argues that the Wang Yangming school developed into a "new humanitarianism" with an "optimistic and liberal view of the self" generating "a new 'pragmatism' which gave increasing attention to 'practical' realities" (p. 23). This was part of a "'near-revolution' in thought during the latter part of the Ming" (p. 24). See also idem, "Individualism and Humanitarianism in Late Ming Thought," in de Bary et al., *Self and Society in Ming Thought*, 145–248; and de Bary, *The Liberal Tradition in China* (New York: Columbia University Press, 1983).

[15] For an overview of these claims, see Foon Ming Liew, "Debates on the Birth of Capitalism in China during the Past Three Decades," *Ming Studies* 26 (1988): 61–75. Dividing Chinese historiography on the sprouts of capitalism in China into three periods (pre-1955, 1955–76, and

The "Decline" of Chinese Mathematics

There have been, however, few attempts to rehabilitate Ming mathematics or science in general; indeed, nowhere have we been more secure in the belief in Chinese decline than in its failure to develop that which is claimed to be uniquely Western—Science, and *the* Scientific Revolution.

The consensus conclusion reached in historical studies of Ming mathematics has, at least until very recently, been fairly unanimous. Prior to the arrival of the Jesuits, Ming mathematics, along with science and thought in general, was in a state of decline.[16] Chinese mathematics, it is often held, reached its apex in the Song and Yuan dynasties: the thirteenth century has been deemed by Li Yan the "zenith of the development of mathematics";[17] Ulrich Libbrecht states bluntly that the "later phase of the Song marks both the apogee of the development of mathematics in China and its terminal point."[18] During the Ming, this received view asserts, crucial mathematical treatises and techniques from the Chinese tradition were lost.[19] The creativity necessary for mathematical development was stultified by the pervasive Cheng-Zhu orthodoxy and the civil service examinations; mathematics and science were disdained by the inwardly focused followers of Wang Yangming.[20]

More recent studies have often reconfirmed these assertions of "decline." Guo Shuchun, one of the leading scholars of Chinese mathematics, asserts that dur-

---

1976–80s), Liew argues that the stagnancy of China's economy resulted not from internal causes but rather foreign imperialism in the 1800s.

[16] See Joseph Needham, *Mathematics and the Sciences of the Heavens and the Earth*, vol. 3 of *Science and Civilisation in China* (Cambridge: Cambridge University Press, 1959), 50–52; Ulrich Libbrecht, *Chinese Mathematics in the Thirteenth Century: The Shu-Shu Chiu-Chang of Ch'in Chiu-Shao* (Cambridge, MA: MIT Press, 1973), 13; Yoshio Mikami, *The Development of Mathematics in China and Japan*, 2nd ed. (repr., New York: Chelsea, 1974), 108, 112; Li Yan and Du Shiran, *Chinese Mathematics: A Concise History*, trans. John N. Crossley and Anthony W.-C. Lun (Oxford: Clarendon Press, 1987), 175; Qian Baocong 钱宝琮, *Zhongguo shuxue shi* 中国数学史 [History of Chinese mathematics] (Beijing: Kexue chubanshe 科学出版社, 1964), 234ff.; and Mei Rongzhao 梅荣照, "Ming-Qing shuxue gailun" 明清數學概論 [Outline of Ming-Qing mathematics], in *Ming-Qing shuxueshi lunwenji* 明清數學史論文集 [Collected essays on the history of Ming-Qing mathematics], ed. Mei Rongzhao 梅荣照 (Nanjing: Jiangsu jiaoyu chubanshe 江苏教育出版社, 1990), 1–20.

[17] Li and Du, *Chinese Mathematics*, title of chapter 5, p. 109.

[18] Libbrecht, *Chinese Mathematics in the Thirteenth Century*, 2.

[19] Li and Du, *Chinese Mathematics*, 175; Mei, "Ming-Qing shuxue gailun," 1–7; Mikami, *Mathematics in China and Japan*; Qian, *Zhongguo shuxue shi*, 234ff.

[20] Mei, "Ming-Qing shuxue gailun," 5; Qian Baocong 钱宝琮, "Song-Yuan shiqi shuxue yu daoxue de guanxi" 宋元时期数学与道学的关系 [Relationship between mathematics and Song-Yuan Learning of the Way], in *Qian Baocong kexueshi lunwen xuanji* 钱宝琮科学史论文选集 [Selected essays by Qian Baocong on the history of science], ed. Zhongguo kexueyuan ziran kexueshi yanjiusuo 中国科学院自然科学史研究所 (Beijing: Kexue chubanshe 科学出版社, 1983), 238–39; Mei Rongzhao 梅荣照 and Wang Yusheng 王渝生, "Xu Guangqi de shuxue sixiang" 徐光启的数学思想 [Mathematical thought of Xu Guangqi], in *Xu Guangqi yanjiu lunwenji* 徐光启研究论文集 [Collected essays on research into Xu Guangqi], ed. Xi Zezong 席泽宗 and Wu Deduo 吴德铎 (Shanghai: Xuelin chubanshe 学林出版社, 1986), 41.

ing the late Yuan and Ming Dynasties, "classical Chinese mathematics sharply declined" 中國古典數學急劇衰落: not only did this period fail to produce works comparable to the *Mathematical Treatise in Nine Sections* (*Shu shu jiu zhang* 數書九章)[21] or the *Jade Mirror of the Four Origins* (*Si yuan yu jian* 四元玉鑑),[22] but earlier Song achievements such as the Four-Origin technique (*si yuan shu* 四元術) and root extraction were not understood.[23] Liu Dun argues that "except for a few subjects (e.g., commercial arithmetic using the abacus and calculations concerning harmonics), mathematics was in a stagnant or even retrograde position in the Ming Dynasty."[24] Nathan Sivin asserts that "there were few important innovations at the highest level of mathematics from the mid-fourteenth century until the seventeenth century, when the Jesuit missionaries prompted an efflorescence of interest in European geometry, true trigonometry, logarithms, and so on. This hiatus may have been part of the price paid for by the abacus."[25] This picture of decline is starkly rendered by Jean-Claude

---

[21] Qin Jiushao 秦九韶 (13th c.), *Shu shu jiu zhang* 數書九章 [Mathematical treatise in nine sections], in *Zhongguo kexue jishu dianji tonghui: Shuxue juan* 中國科學技術典籍通彙：數學卷 [Comprehensive collection of the classics of Chinese science and technology: Mathematics volumes] (Zhengzhou: Henan jiaoyu chubanshe 河南教育出版社, 1993), hereinafter *ZKJDT*. For a translation and analysis, see Libbrecht, *Chinese Mathematics in the Thirteenth Century*.

[22] Zhu Shijie 朱世傑 (1249–1314), *Si yuan yu jian* 四元玉鑑 [Jade mirror of the four origins], in *ZKJDT*. For a translation and analysis, see Jock Hoe, *The Jade Mirror of the Four Unknowns by Zhu Shijie: An Early Fourteenth Century Mathematics Manual for Teaching the Derivation of Systems of Polynomial Equations in up to Four Unknowns; A Study* (Christchurch, N.Z.: Mingming Bookroom, 2007).

[23] Guo Shuchun 郭书春, "Introduction," *Zhongguo kexue jishu dianji tonghui: Shuxue juan* 中國科學技術典籍通彙：數學卷 [Comprehensive compilation of the Chinese technological and scientific classics: Mathematics], ed. Guo Shuchun 郭书春 (Zhengzhou: Henan jiaoyu chubanshe 河南教育出版社, 1993), 1:18–19.

[24] Liu lists the causes as the following: (1) "the lack of creative mathematicians"; (2) "the scarcity of paradigmatic works such as the *Nine Chapters of Mathematical Arts* (*Jiu zhang suan shu* 九章算術) and the *Mathematical Treatise in Nine Sections* (*Shu shu jiu zhang* 數書九章)"; and (3) "the lack of problems with a heuristic significance." Citing Gu Yingxiang 顧應祥 (1483–1565) as an example, Liu further argues that "Ming Dynasty scholars had only a superficial knowledge of earlier mathematics, and most important mathematical works, along with their outstanding achievements, were virtually unknown, and some had been lost." Liu continues, "By the early Qing Dynasty, Western mathematical knowledge had already reached China and began to display its superiority in both logical reasoning and practice." Liu's assessment of the Ming is nonetheless one of the most positive of any historian; it should be noted that, in contrast to his overall assessment of the Ming dynasty, Liu later calls Cheng Dawei a "pioneer." Liu Dun, "400 Years of the History of Mathematics in China: An Introduction to the Major Historians of Mathematics since 1592," *Historia Scientiarum* 4 (1994): 103–4. (The mathematical work of Gu Yingxiang and Cheng Dawei will be discussed below.)

[25] Sivin further argues that the abacus was "extremely well suited to the routine needs of the growing urban merchant class. Because the abacus could only represent a dozen or so digits in a linear array, it was useless for the most advanced algebra until it was supplemented by pen-and-paper notation." Nathan Sivin, "Science and Medicine in Chinese History," in *Heritage of China: Contemporary Perspectives on Chinese Civilization*, ed. Paul S. Ropp (Berkeley: University of California Press, 1990), 172; republished in idem, *Science in Ancient China: Researches and Reflections* (Brookfield, VT: Ashgate, 1995). Sivin's view is cited by John K. Fairbank as an

Martzloff, one of the foremost historians of Chinese mathematics, in *Companion Encyclopedia of the History and Philosophy of Mathematical Sciences*. In a section entitled "The Ming Decline and Contacts with Western Mathematics under the Qing," Martzloff describes in some detail the Jesuit contributions to Chinese mathematics in the late Ming and early Qing:

> The Jesuits offered their services and undertook a successful programme of translations of mathematical works containing knowledge they believed indispensable for astronomy: elementary arithmetic, calculating instruments such as Napier's rods and Galileo's proportional compass, geometry, plane and spherical trigonometry, and logarithms. The sources of the translations were manuals used in European Jesuit colleges. In particular, in 1607 the Italian Jesuit Matteo Ricci (1552–1610) and the Chinese high official Xu Guangqi (1562–1633) published the *Jihe yuanben* ("Elements of Geometry") based on Clavius's commentary on the first six books of Euclid's *Elements*.

In contrast, on the subject of Chinese mathematics during the Ming, Martzloff offers only the following two sentences:

> Under the Ming, the major achievements of the Song and Yuan sank into oblivion. Mathematics had centred then on the abacus, an instrument whose origin is obscure but which is not known to have existed before the fourteenth century.[26]

Catherine Jami echoes this view in the recent *Oxford Handbook of the History of Mathematics*, stating:

> It is widely admitted that by 1600, the most significant achievements of the Chinese mathematical tradition had fallen into oblivion. The *Jiu zhang suan shu* 九章算術 "Nine Chapters on mathematical procedures" (first century AD), regarded as the founding work of the Chinese mathematical tradition (Chemla and Guo 2004) and included in the *Suan jing shi shu* 算經十書 "Ten mathematical classics" (656) had effectively been lost. Furthermore the sophisticated *tian yuan* 天元 "celestial element" algebra developed in the 13th century had been forgotten. The calculating device on which both were based, the counting rods, had fallen into disuse; the abacus had become the universally used calculating device.[27]

---

example of "how China's early precocity in invention could later hold her back." John King Fairbank and Merle Goldman, *China: A New History*, enl. ed. (Cambridge, MA: Belknap Press of the Harvard University Press, 1998), 3.

[26] Jean-Claude Martzloff, "Chinese Mathematics," in *Companion Encyclopedia of the History and Philosophy of the Mathematical Sciences*, ed. Ivor Grattan-Guinness (New York: Routledge, 1994), 99–100. The remainder of his section discusses the introduction of Western mathematics.

[27] Catherine Jami, "Heavenly Learning, Statecraft, and Scholarship: The Jesuits and Their Mathematics in China," in *The Oxford Handbook of the History of Mathematics*, ed. Eleanor Robson and Jacqueline A. Stedall (Oxford: Oxford University Press, 2009), 60. The only mathematical treatise from the Ming Dynasty discussed by Jami is Cheng Dawei's *Suan fa tong zong* 算法

These assertions of the decline in Ming mathematics are not, however, conclusions derived from exhaustive studies of Ming mathematics itself: there have been, in fact, few general studies of Ming mathematics;[28] there are few research articles concerning any aspect of Ming mathematics;[29] there have been few systematic analyses of Ming mathematicians and few studies of individual practitioners;[30] there has been little research into extant Ming mathematical treatises;[31]

統宗. Jami's view of Chinese mathematics is not new, as can be seen from this quote from the early twentieth century: "After the noteworthy achievements of the 13th century, Chinese mathematics for several centuries was in a period of decline. The famous 'celestial element method' in the solution of higher equations was abandoned and forgotten. Mention must be made, however, of Ch'eng Tai-wei [Cheng Dawei], who in 1593 issued his *Suan-fa T'ung-tsung* ('A Systematised Treatise on Arithmetic'), which is the oldest work now extant that contains a diagram of the form of abacus, called *suan-pan*, and the explanation of its use. ... The 'Systematised Treatise on Arithmetic' is famous also for containing some magic squares and magic circles." Florian Cajori, *A History of Mathematics*, 2nd ed. (New York: Macmillan, 1919), 76.

[28] Books that cover Ming mathematics usually focus on translations of European mathematics, the Qing dynasty, or both. For example, Mei, *Ming-Qing shuxueshi lunwenji*, offers only a very brief overview of Ming mathematics; *Ming mo dao Qing zhong qi* 明末到清中期 [Late Ming to mid-Qing], vol. 7 of *Zhongguo shuxueshi daxi*, 中国数学史大系 [Compendium of the history of Chinese mathematics], ed. Li Di 李迪 (Beijing: Beijing shifan daxue chubanshe 北京师范大学 出版社, 2000), focuses on the introduction of Western mathematics and the Qing.

[29] General histories of Chinese mathematics such as Mikami, *Mathematics in China and Japan*, Li and Du, *Chinese Mathematics*, and Needham, *Mathematics and the Sciences of the Heavens* devote only a few pages to the Ming; their primary focus is the introduction of Western mathematics.

[30] An important exception is Dai Nianzu 戴念祖, *Zhu Zaiyu: Ming dai de kexue he yishu juxing* 朱載堉：明代的科學和藝術巨星 [Zhu Zaiyu: A giant of the sciences and the arts in the Ming dynasty] (Beijing: Renmin chubanshe 人民出版社, 1986). Some articles have also examined the work of Cheng Dawei. One of the best works is Yan Dunjie 严敦杰 and Mei Rongzhao 梅荣照, "Cheng Dawei ji qi shuxue zhuzuo" 程大位及其數學著作 [Cheng Dawei and his mathematical works], in Mei, *Ming-Qing shuxueshi lunwenji*, 26–52. Yan and Mei argue (incorrectly) that Cheng was the greatest of Ming mathematicians; his work is important only against the background of the "backwardness of traditional mathematics in the Ming dynasty." For an interesting analysis of Cheng's work from the point of view of art history, see Craig Clunas, "Text, Representation and Technique in Early Modern China," in *History of Science, History of Text*, ed. Karine Chemla (Boston: Kluwer Academic, 2004), 107–21.

[31] General bibliographies in Western languages of Chinese mathematics cite only a fraction of the extant Chinese mathematical treatises. The bibliography in Libbrecht, *Chinese Mathematics in the Thirteenth Century* is not comprehensive because of its stated focus on the Song, after which, Libbrecht asserts, Chinese mathematics declined. He refers to only one Ming treatise. Catherine Jami, "Western Influence and Chinese Tradition in an Eighteenth Century Chinese Mathematical Work," *Historia Mathematica* 15 (1988): 311–31, contains a reference to only the treatise studied, without edition or location; the bibliography in Frank J. Swetz and Ang Tian Se, "A Brief Chronological and Bibliographic Guide to the History of Chinese Mathematics," *Historia Mathematica* 11 (1984): 39–56 provides little information not found in general histories. Other articles have focused primarily on Song and pre-Song. General histories, such as Martzloff, *History of Chinese Mathematics*, contain few references to Ming mathematics; Li and Du, *Chinese Mathematics* and Mikami, *Mathematics in China and Japan* contain no bibliography of sources but only mention titles within the text. For a bibliography, the reader of Li and Du, *Chinese Mathematics* is referred to Needham, *Mathematics and the Sciences of the*

and the only attempt at a bibliography of Ming mathematics—compiled in the 1930s—has been largely overlooked.[32]

In short, the mathematics of the Ming has not been deemed worthy of study: because there were no important texts, there has been little reason to compile a bibliography of mathematical texts written during the Ming; because mathematical texts from early dynasties were lost, there has been little need to document precisely which texts were extant during the Ming. And because there has seemed to be little mathematics deserving of study, these accounts often revert to the larger lessons that the failure of the Ming has to offer: indeed the derision for Ming mathematics borrows terms from the narrative of moral decline that seem hardly appropriate for mathematics—Needham speaks of "decay," and Mikami calls Ming scholars "degenerate."[33]

Chinese Mathematics in the Late Song and Early Yuan Dynasties

Conventional histories, as noted above, have held that Chinese mathematics reached its apex with the Celestial Origin and Four Origin algebras of the late Song and early Yuan.[34] What we know of these algebras is primarily from four texts:[35] Li Ye's 李冶 (1192–1279) *Sea Mirror of Circle Measurement* (*Ce yuan hai*

Heavens, which is not specialized enough to list more than several titles—the primary focus is on astronomy and the Jesuit impact.

[32] Li Yan 李儼, "Ming dai suanxue shu zhi" 明代算學書志 [Bibliography of mathematical works of the Ming Dynasty], in *Zhong suan shi lun cong* 中算史論叢 [Collected essays on the history of Chinese mathematics], 5 vols. (Beijing: Zhongguo kexueyuan 中國科學院, 1954–55), 2: 86–102, originally published in *Tushuguanxue jikan* 圖書館學季刊 1, no. 4, 667–82; idem, "Zengxiu Mingdai suanxue shuzhi" 增修明代算學書志 [Additions and revisions to the bibliography of mathematical works of the Ming Dynasty], in *Zhongguo suanxue shi luncong* 中國算學史論叢 [Collected essays on the history of Chinese mathematics] (Taibei: Zhengzhong shuju 正中書局, 1954), originally published in *Tushuguanxue jikan* 1, no. 4, 667–82. I have seen little discussion of the contents of these bibliographies in Western or Chinese studies. Presumably, the primary reason that they have been ignored is the general dismissal of Ming mathematics. It is for this reason that I compiled a table of mathematical treatises up through the Ming dynasty that are listed in Chinese bibliographies. See Roger Hart, *The Chinese Roots of Linear Algebra* (Baltimore: Johns Hopkins University Press, 2011), "Appendix B: Chinese Mathematical Treatises," pp. 213–54. For a compilation of extant bibliographies of treatises from the Ming and Qing dynasties, see *Zhongguo suanxue shumu huibian* 中国算学书目汇编 [Compilation of bibliographies of Chinese mathematical treatises], ed. Li Di 李迪 (Beijing: Beijing shifan daxue chubanshe 北京师范大学出版社, 2000).

[33] "Ming scholars were so degenerate that they were little able to understand the celestial element [celestial origin] algebra and that the calendrical reform was left unaffected although the prevalent calendar had become very loose and inaccurate with the lapse of years." Mikami, *Mathematics in China and Japan*, 112.

[34] For a discussion of the Celestial Origin and Four Origin algebras, see Li and Du, *Chinese Mathematics*, 135–48, and Martzloff, *History of Chinese Mathematics*, 258–71.

[35] Qian, *Zhongguo shuxue shi*, 168.

jing 測圓海鏡, 1248);[36] Li Ye's more elementary *Elaboration on the Yigu Collection* [*Yi gu yan duan* 益古演段, 1259];[37] Zhu Shijie's 朱世傑 (1249–1314) exposition of Four Origin algebra, *Jade Mirror of the Four Origins* [*Si yuan yu jian* 四元玉鑑, 1303];[38] and Zhu Shijie's more elementary *Mathematical Primer* (*Suanxue qi meng* 算學啟蒙, 1299),[39] which contains examples of the Celestial Origin algebra. Yet there are relatively few extant texts from this period: others (in addition to the four texts listed above) include Ding Ju's 丁巨 *The Methods of Calculation of Ding Ju* (Ding Ju suanfa 丁巨算法一卷, 1355);[40] Zhao Youqin's 趙友欽 *A New Book on the Original Gexiang* (*Yuan ben ge xiang xin shu wu juan* 原本革象新書五卷);[41] An Zhizhai 安止齋 and He Pingzi's 何平子 *A Detailed Explanation of Methods of Calculation* (*Xiang ming suanfa er juan* 詳明算法二卷); and *Collection of the Complete Powers of Calculational Methods* (*Suanfa quan neng ji er juan* 算法全能集二卷).[42]

The *Sea Mirror of Circle Measurement* is a remarkable mathematical work, and there are several additional reasons for examining this text in some detail:

1. It is the earliest extant text containing examples of the thirteenth-century Chinese algebras.
2. It was widely available during the Ming—it is listed in most bibliographies of Ming works that record mathematical treatises, suggesting that it was known to both mathematicians and literati.[43] The *Elaboration on the Yigu Collection*

---

[36] Editions include *Ce yuan hai jing shier juan* 測圓海鏡十二卷 [The sea mirror of circle measurement, 12 *juan*] in *JFQS* and *SKQS*; *Ce yuan hai jing xi cao shier juan* 測圓海鏡細草十二卷 [The sea mirror of circle measurement with detailed explanations, 12 *juan*], Ming *chaoben* 明抄本, also in *CJCB*, *ZBZZ*, and *BTSC*. See also Zhang Chuzhong 張楚鍾, *Ce yuan hai jing shi bie xiang jie yi juan* 測圓海鏡識別詳解一卷 [The sea mirror of circle measurement with distinctions and detailed explanations, one *juan*], in *QSZSX*. Important studies of this text include Karine Chemla, "Étude du livre Reflets des mesures du cercle sur la mer de Li Ye (1248)" (Thèse de mathématiques, Université Paris XIII, 1982).

[37] Editions include Li Ye 李冶, *Yi gu yan duan* 益古演段 (1259); and Zhu Shijie 朱世傑 *Suanxue qi meng* 算學啟蒙 (1299). Other texts by Li Ye include *Yi gu yan duan er juan* 益古演段二卷 in *SKQS*; *Yi gu yan duan san juan* 益古演段三卷, Qing *chaoben* 清王萱鈴抄本, also in *CJCB*, *ZBZZ*, *BTSC*, and *SKQS*.

[38] Editions include: *Si yuan yu jian san juan* 四元玉鑑三卷 [Jade mirror of the four origins] in *BTSC* and *WWBC*; *Xin bian si yuan yu jian san juan* 新編四元玉鑑三卷 [A new edition of the Jade mirror of the four origins, three *juan*], Qing *chaoben* 清嘉慶二十四年王萱鈴家抄本; Luo Shilin 羅士琳 and Zhu Shijie, *Si yuan yu jian xi cao san juan* 四元玉鑑細草三卷 [The jade mirror of the four origins, with detailed explanations, 3 *juan*] in *GSHG* and *CFZXS*. See also Yi Zhihan 易之瀚, *Si yuan shi li yi juan* 四元釋例一卷 [Explanations and examples of the four origins], in *GSHG* and *CFZXS*; Cui Chaoqing 崔朝慶 *Du Si yuan yu jian ji yi juan* 讀四元玉鑑記一卷 [Notes on reading the Jade Mirror of the Four Origins], in *ZYLZ* and *NJZJ*.

[39] Editions include: Zhu Shijie and Luo Shilin 羅士琳, *Suanxue qimeng san juan* 算學啟蒙三卷 [A primer on calculation studies, 3 *juan*], in *GSHG* and *CFZXS*; Zhu Shijie and Luo Shilin, *Xin bian suanxue qimeng san juan fu shi wu yi juan* 新編筭學啟蒙三卷附識誤一卷 in *GSHG*.

[40] In *CJCB* and *ZBZZ*.

[41] Originally in *YLDD*; the meaning of the title of this treatise is not clear.

[42] In *XLTCS*.

[43] The *Sea Mirror of Circle Measurement* is listed in the following bibliographies: *Wenyuan ge shumu* 文淵閣書目 [Bibliography of the Hall of Literary Depth], in *Congshu jicheng chu bian*

was also widely available but less influential;[44] the *Jade Mirror of the Four Origins* is listed only in the *Bibliography of the Thousand Qing Hall* (*Qianqing tang shumu* 千頃堂書目); and the *Mathematical Primer* is not recorded at all.

3. The *Sea Mirror of Circle Measurement* was the focus of study by several Ming literati: it was copied by Tang Shunzhi 唐順之 (1507–1560);[45] it was later recompiled by Gu Yingxiang 顧應祥 (1483–1565), who solved the problems with simpler calculational techniques and omitted the Celestial Origin algebra that he confessed he was unable to understand;[46] and it is mentioned briefly by Xu Guangqi, who proposed to recover its "meanings."[47]

For these reasons, the *Sea Mirror of Circle Measurement* has been a central marker in historical reconstructions of Chinese mathematical development. In these reconstructions, the *Sea Mirror of Circle Measurement* has served as the exemplar

---

叢書集成初編 [Compendium of collectanea, first edition] (Beijing: Zhonghua shuju 中华书局, 1985–1991), hereinafter *CJCB*; Qian Pu 錢溥 (*jinshi* 1439), *Bige shumu* 秘閣書目 [Bibliography of the Imperial Pavilion], in *Siku quanshu cunmu congshu bu bian* 四庫全書存目叢書補編 [Collection of works preserved in the catalogue of the Four Treasuries, supplemental series] (Jinan: Qilu shushe 齊魯書社, 2001), hereinafter *SQCCB*; Cheng Dawei 程大位 (1533–1606), *Suanfa tongzong* 算法统宗 [Comprehensive source of mathematical methods], in *ZKJDT*; Jiao Hong 焦竑 (1541–1620), *Guo shi jing ji zhi* 國史經籍志 [Record of books for the dynastic history] (Taibei: Taiwan shangwu yinshuguan 臺灣商務印書館, 1965); Sun Nengchuan 孫能傳 (*juren* 1582) and Zhang Xuan 張萱, *Neige cangshu mulu* 內閣藏書目錄 [Catalogue of books of the Grand Secretariat], in *Xu xiu Siku quanshu* 續修四庫全書 [Continuation of the complete collection of the Four Treasuries] (Shanghai: Shanghai guji chubanshe 上海古籍出版社, 1995–2002), hereinafter *XXSQ*; Zhao Qimei 趙琦美 (1563–1624), *Maiwang guan shumu* 脈望館書目 [Bibliography of the Maiwang Hall], in *Congshu jicheng xu bian* 叢書集成續編 [Compendium of collectanea, continuation] (Shanghai: Shanghai shudian 上海書店, 1994), hereinafter *CJXB*; *Jingu tang shumu* 近古堂書目 [Bibliography of the Hall of Approaching the Ancients], in *CJXB*; Dong Qichang 董其昌 (1555–1636), *Xuanshang zhai shumu* 玄賞齋書目 [Bibliography of the Studio of Occult Rewards] (Beijing: Guoli Beiping tushuguan 國立北平圖書館, 1932–1937); Huang Yuji 黃虞稷 (1629–1691), *Qianqing tang shumu* 千頃堂書目 [Bibliography of the Thousand Qing Hall], in *CJXB*; and Fu Weilin 傅維鱗 (d. 1667), *Ming shu* 明書 [Book of the Ming], in *Siku quanshu cunmu congshu* 四庫全書存目叢書 [Collection of works preserved in the catalogue of the Four Treasuries] (Jinan: Qilu shushe 齊魯書社, 1995–97), hereinafter *SQCC*. See Hart, *Chinese Roots of Linear Algebra*, "Appendix B: Chinese Mathematical Treatises," 213–54.

[44] The *Elaboration on the Yigu Collection* is listed in the following bibliographies: Yang et al., *Wenyuan ge shumu*; Qian, *Bige shumu*; Jiao, *Guo shi jing ji zhi*; Sun and Zhang, *Neige cangshu mulu*; Dong, *Xuanshang zhai shumu*; Huang, *Qianqing tang shumu*; and Fu, *Ming shu*. It is also possible that the *Yi gu suan fa* 益古算法 listed in Cheng, *Suanfa tongzong*, is the same treatise.

[45] This treatise is no longer extant, and is not listed in any bibliographies; Gu Yingxiang's preface notes the existence of this text. Gu, *Ce yuan hai jing fen lei shi shu* 測圓海鏡分類釋術 [Sea mirror of circle measurement, arranged by categories, with explanations of the methods], in *Yingyin Wenyuan ge Siku quanshu* 影印文淵閣四庫全書 [Complete collection of the Four Treasuries, photolithographic reproduction of the edition preserved at the Pavilion of Literary Erudition] (Hong Kong: Chinese University Press, 1983–1986), hereinafter *SKQS*.

[46] Ibid.

[47] Xu states, "I wished to explain its meanings, but lacked the time" 余欲為說其義，未遑也. *Gougu yi xu* 勾股義序 [Preface to *Right Triangles, Meanings*], in *Xu Guangqi ji* 徐光啟集 [Collected works of Xu Guangqi] (Shanghai: Shanghai guji chubanshe 上海古籍出版社, 1984), hereinafter *XGQJ*, 1:84.

of the Celestial Origin algebra of the thirteenth-century; placed into a transcultural framework for the development of mathematics, it represents a benchmark that the Chinese reached first. It has also been important in the negative evaluations of Chinese mathematics of the Ming—both as evidence of decline (the Celestial Origin algebra has served as the standard that Ming mathematics failed to reach) and the incompetence of Ming mathematicians (as judged from Gu's statement of incomprehension).

This teleology, based on the development of solutions for polynomial equations—for which the thirteenth-century Chinese algebras represent the endpoint—has then become one of the central themes in the received historiography of Chinese mathematics. In this teleology, the origins of development are to be found in the *Nine Chapters of Mathematical Arts*,[48] which is often portrayed in these accounts as presenting solutions to polynomial equations of the second degree, citing the following problem 20 from the "Gou gu" chapter (*juan*) as an example of the solution of $x^2 + 34x = 71000$:

> Given is a city square of unknown size, with gates opening in the middle of each [side]. Exiting from the North Gate, there is a tree at twenty paces. Exiting from the South Gate fourteen paces, turning and proceeding West one thousand seven hundred seventy-five paces, the tree comes into view. What is the length [of one side] of the city square?

> Technique: Multiply the number of paces taken from the North Gate by the number of paces taken toward the West [20×1775], double that [2×35500], and take that as the constant [71000]; [take the number of paces taken from the North Gate][49] added to the number of paces taken from the South Gate as the coefficient[50] [$14 + 20$], extract the root [250], namely, the [length of the] city wall.

> 今有邑方不知大小，各中開門。出北門二十步有木。出南門十四步，折而西行一千七百七十五步見木。問邑方幾何。術曰：以出北門步

---

[48] For information on the origins, compilation, and editions of the *Nine Chapters of Mathematical Arts*, see Michael Loewe, ed., *Early Chinese Texts: A Bibliographical Guide* (Berkeley: Society for the Study of Early China, Institute of East Asian Studies, University of California, Berkeley, 1993). The *Nine Chapters on Mathematical Arts* is in many collectanea, including *SKQS* and *ZKJDT*; I have used *ZKJDT*. For analysis of the text, see Bai Shangshu 白尚恕, *Jiu zhang suanshu zhushi* 《九章算术》注释 [*Nine Chapters on the Mathematical Arts* with commentary and explanations] (Beijing: Kexue chubanshe 科学出版社, 1983); Guo Shuchun 郭书春, *Jiu zhang suanshu huijiaoben* 九章算術匯校本 [Critical edition of the *Nine Chapters on Mathematical Arts*] (Shenyang: Liaoning jiaoyu chubanshe 辽宁教育出版社, 1990); Karine Chemla and Guo Shuchun, *Les neuf chapitres: Le classique mathématique de la Chine ancienne et ses commentaires* (Paris: Dunod, 2004); and Shen Kangshen, Anthony W.-C. Lun, and John N. Crossley, *The Nine Chapters on the Mathematical Art: Companion and Commentary* (New York: Oxford University Press, 1999).

[49] Interpolated on the basis of the mathematical content.

[50] In modern terminology, the coefficient of x. Libbrecht translates *cong fa* 從法 as "derived element"; the phrase *dai cong kai fang fa* 帶從開方法 is opaquely rendered as "corollary to the square root method" in Li and Du, *Chinese Mathematics*, 53.

數乘西行步數，倍之，為實。并出南門步數為從法，開方除之，即邑
方。[51]

This problem can be transformed, under suitable translation, into an equivalent polynomial equation: the problem seems to have been solved through what in modern notation would be expressed as the equality of the ratios of the sides of similar triangles,

$$\frac{\left(\frac{x}{2}\right)}{20} = \frac{1775}{x+14+20},$$

which, when multiplied out, yields

$$x^2 + 34x = 71000.$$

In modern historical accounts, this translation into modern notation is often accompanied by the further claim that in fact this provides a general solution of polynomial equations of the second degree with integral coefficients.[52]

The teleology of polynomial equations within which this problem is interpreted then provides these accounts with an epistemological framework in which the following axes gauge increasing progress toward modern mathematics: (1) the increasing size of the exponents; (2) the additional use of negative exponents; (3) the use of negative and fractional coefficients; (4) the increasing number of variables; and (5) the increasing generalization. The history of this development then consists of discovering the first origins of, and plotting progress along, each of these axes. Polynomials of the third degree, for example, appear first in the Tang; however, before the Northern Song, this view holds, progress was "limited" by the geometric problems to which the algebras were applied. In the Northern Song, Jia Xian 賈憲 and Liu Yi 劉益 (fl. 1113) solved higher-order equations with positive roots.[53] Although there are records of works predating the *Sea Mirror of Circle Measurement* that apparently contain Celestial Origin algebra, these works are no longer extant.[54] The importance of the *Sea*

---

[51] Li Ye 李冶 (1192–1279), *Ce yuan hai jing* 測圓海鏡 [Sea mirror of circle measurement], in *ZKJDT*, 1:199; Bai, *Jiu zhang suanshu zhushi*, 337.

[52] Li and Du, *Chinese Mathematics*, 53.

[53] Little is known about either of these authors. Fragments of Liu Yi's work have been preserved in the *Yang Hui suanfa* 楊輝算法; see Lay Yong Lam, *A Critical Study of the Yang Hui Suan Fa: A Thirteenth-Century Chinese Mathematical Treatise* (Singapore: Singapore University Press, 1977). For a reconstruction of Jia Xian's *Huang di jiu zhang suanfa xi cao* 黃帝九章算法細草, based on the *Jiu zhang suanfa zuan lei* 九章算法纂類 appended to Yang Hui's *Xiang jie jiu zhang suan fa* 詳解九章算法, see Qian Baocong 钱宝琮, "Zeng cheng kaifang fa de lishi fa zhan" 增乘開方法的歷史發展 [Historical development of the *zeng cheng* method for root extraction], in *Song Yuan shuxueshi lunwenji* 宋元數學史論文集 [Collected essays on the history of mathematics during the Song and Yuan Dynasties], ed. Qian Baocong 钱宝琮 et al. (Beijing: Kexue chubanshe 科学出版社, 1966), 36–59.

[54] Several texts apparently on the Celestial Origin algebra are listed in the preface by Zu Yi 祖頤 to Zhu Shijie's *Jade Mirror of the Four Origins*: Shi Xindao's 石信道 *Qian jing* 鈐經 (among the Ming bibliographies, this work is recorded only in Cheng, *Suanfa tongzong*); Jiang Zhou's 蔣周 *Yi gu* 益古; Li Wenyi's 李文一 *Zhao dan* 照膽; and Liu Ruxie's 劉汝諧 *Ru ji shi suo* 如積釋鎖. Li

Mirror of Circle Measurement derives from the benchmarks it provides along these axes of development, including: (1) examples of higher-order equations—among 170 problems, 19 solve polynomials of the third degree, 13 solve problems of the fourth degree, and one solves a polynomial of the sixth degree; (2) the use of both positive and negative constant coefficients; and (3) equations with fractional coefficients.[55]

This view not only takes the current state of mathematics as the teleological endpoint, naturalizing development toward that telos, but the gridwork framing this teleology reflects ideologies that have removed mathematics from the world: social context only distorts, and cannot be a constitutive element in the formation of knowledge; applications are irrelevant; and even the mathematical context is ignored—the properties of triangles serve only as a practical distraction, or even limitation. This teleology of polynomial equations was not, however, known to the practitioners; as the next sections will show, the developments presented in the *Sea Mirror of Circle Measurement* are in an entirely different direction.

Li Ye's Sea Mirror of Circle Measurement

The *Sea Mirror of Circle Measurement*—the entire treatise—is based on a single diagram (Figure 4.1 on the next page).[56] The text begins with a series of definitions (see Figure 4.2 on page 92). Contrary to expectations for a text acclaimed for its exposition of polynomial equations, these definitions derive from this single geometric diagram—a triangle with a circle inscribed, a square circumscribed about the circle, and fourteen further triangles constructed from line segments including the diameters of the circle and other points of intersection. And contrary to possible expectations for a text of Greek geometry, no procedure is given for the construction of the diagram, no demonstration is offered that the diagram can in fact be drawn, no proof is presented that the objects of study exist, and no definitions are offered for the elements of the diagram—points, lines, circles, or triangles.

Ye apparently obtained a treatise in Dongping 東平 (in present day Shandong 山東) that used the following notation for expressing polynomial degrees: *xian* 仙, *ming* 明, *xiao* 霄, *han* 漢, *lei* 壘, *ceng* 層, *gao* 高, *gao* 上, *tian* 天, *ren* 人, *di* 地, *xia* 下, *di* 低, *jian* 減, *luo* 落, *shi* 逝, *quan* 泉, *an* 暗, and *gui* 鬼, where the term *ren* is the constant term, those before represent the equivalent of positive exponents, and those after the negative. Qian, *Zhongguo shuxue shi*, 172–73.

[55] See, for example, Kong Guoping 孔国平, "Li Ye" 李冶 [Li Ye], in *Zhongguo gudai kexuejia zhuanji* 中国古代科学家传记 [Biographies of scientists in ancient China], ed. Du Shiran 杜石然 (Beijing: Kexue chubanshe 科学出版社, 1992–1993), 631.

[56] Throughout this analysis I have drawn heavily on Bai Shangshu 白尚恕, *Ce yuan hai jing jin yi* 測圓海鏡今譯 [*Sea mirror of circle measurement*, translated into modern Chinese] (Shandong: Shandong jiaoyu chubanshe 山东教育出版社, 1985); I have followed his notation when possible (and in particular, in the diagram and in the equations in the following text).

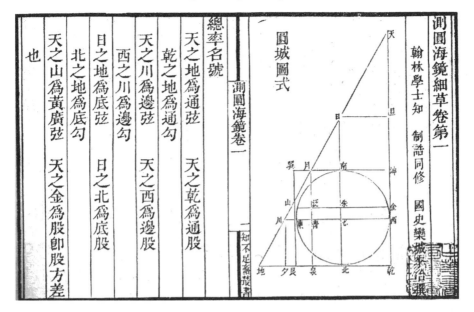

Fig. 4.1: The diagram of the triangle in Li's *Sea Mirror of Circle Measurement.*

Instead, the points of the diagram are labeled with terms that resonate with metaphysical significance. The first sentence of the text describes the largest triangle △*ABC* (see Figure 4.1 and Figure 4.3 on page 93) in the following terms:

> Heaven-to-Earth [*AB*] is the Through Connecting Hypotenuse; Heaven-to-Caelum (*qian* 乾)[57] [*AC*] is the Connecting Long-Side; and Caelum-to-Earth [*BC*] is the Connecting Short-Side.
>
> 天之地為通弦，天之乾為通股，乾之地為通勾。[58]

These terms are not just names used as labels. There appear to be two diagrams superimposed one on the other: the first is the circle which, as the next chapter of this text explains, represents the wall around a city, with the center labeled Heart (*xin* 心) and four gates, which are labeled North, South, East, and West. In the second diagram, the labels seem to be assigned systematically—for example, Heaven (*tian* 天), Sun (*ri* 日), Moon (*yue* 月), Mountains (*shan* 山), Rivers (*chuan* 川) and Earth (*di* 地) are placed along the same line in that order—but the source for the system is not known.[59]

This first section of the *Sea Mirror of Circle Measurement* continues by generating—from the twenty-two labeled points of intersection—fourteen more triangles. The inscribed circle and the circumscribed square, along with the various

---

[57] *Qian* 乾 is one of the sixty-four hexagrams; it represents heaven. I have translated *qian* as "Caelum" to distinguish it from "Heaven," which I have used for the translation of *tian* 天.

[58] Li, *Ce yuan hai jing,* 1:732.

[59] One possible source for comparison is Li Ye's *Jing zhai gu jin tou* 敬齋古今黈 (in *SKQS*); but I have not found any direct references to this system.

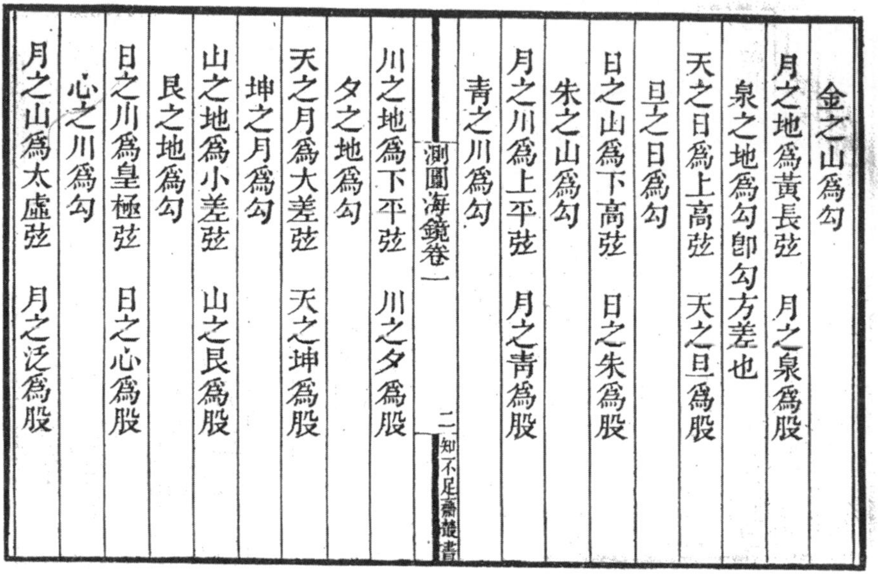

月之山為太虛弦　月之泛為股
心之川為勾
日之川為皇極弦　日之心為股
艮之地為勾
山之地為小差弦　山之艮為股
坤之月為勾
天之月為大差弦　天之坤為股
夕之地為勾
川之地為下平弦　川之夕為股

測圓海鏡卷一
二 知不足齋叢書

青之川為勾
月之川為上平弦　月之青為股
朱之山為勾
日之山為下高弦　日之朱為股
旦之日為勾
天之日為上高弦　天之旦為股
泉之地為勾即勾方差也
月之地為黃長弦　月之泉為股
金之山為勾

Fig. 4.2: The terms for the triangles in Li's *Sea Mirror of Circle Measurement.*

arcs, rectangles, squares, and areas are all completely ignored. Each of these triangles is assigned a prefix. Table 4.1 on page 94 lists the triangles to which Li assigns names: the first column lists the character (or sometimes two) that is assigned as the name of the triangle; the third column lists the vertices of that triangle, starting with the uppermost vertex and proceeding counterclockwise (so that the first two vertices form the hypotenuse, the second two vertices form the shorter side, and the first and last vertices form the longer side); and the final column lists the length of the hypotenuse, short-side, and long-side in that order. Again, the source and metaphysical significance of much of this terminology is unknown.[60]

The assignment of these prefixes for the fourteen triangles is made in parallel sentences:

> Heaven-to-River is the Border Hypotenuse, Heaven-to-West is the Border Long-Side, West-to-River is the Border Short-Side. Sun-to-Earth is the Lower Hypotenuse, Sun-to-North is the Lower Long-Side, North-to-Earth is the Lower Short-Side. Heaven-to-Mountain is the Yellow Hypotenuse, Heaven-to-Metal is the Long-Side, namely the Square Difference[61] of the Long-Side, and Metal-to-Mountain is the Lower Short-Side. . . .

---

[60] Bai Shangshu offers brief explanations for some of these terms, which I have followed where appropriate. For example, for $\triangle MNG$, I have used the character *zhuan* 專 as a substitute for a character not found in standard references, which Bai interprets to mean small, since $\triangle MNG$ is the smallest triangle.

[61] It is not clear what the term *fangcha* means here.

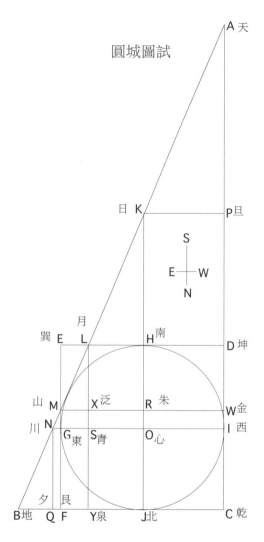

Fig. 4.3: The diagram of the triangle in Li's *Sea Mirror of Circle Measurement.*

天之川為邊弦，天之西為邊股，西之川為邊勾。日之地為底弦，日之北為底股，北之地為底勾。天之山為黃弦，天之金為股，即股方差也，金之山為勾。[62]

This procedure is then repeated fourteen times, once for each triangle. It is these terms that are used to identify each triangle, and subsequently each element of the triangle—the sides, differences, second differences, etc.—will all

---

[62] Li, *Ce yuan hai jing,* 1:732–33.

Table 4.1: Classification of triangles in *Sea Mirror of Circle Measurement*.

Prefix	(English)	Vertices	(English)	Lengths of sides
通	Connecting	天地乾	ABC	680, 320, 600
邊	Border	天川西	ANI	544, 256, 480
底	Lower	日地北	KBJ	425, 200, 375
黃廣	Yellow-Wide	天山金	AMW	510, 240, 450
黃長	Yellow-Long	月地泉	LBY	272, 128, 240
上高	Upper-High	天日旦	AKP	255, 120, 225
下高	Lower-High	日山朱	KMR	255, 120, 225
上平	Upper-Level	月川青	LNS	136, 64, 120
上平	Lower-Level	川地夕	NBQ	136, 64, 120
大差	Greater-Difference	天月坤	ALD	408, 192, 360
小差	Lesser-Difference	山地艮	MBF	170, 80, 150
皇極	Imperial-Ultimate	日川心	KNO	289, 136, 255
太虛	Great-Void	月山泛	LMX	102, 48, 90
明	Bright	日月南	KLH	153, 72, 135
專	Small	山川東	MNG	34, 16, 30

be referenced via these terms. Thus these names and the triangles organize the objects created.

Although in Euclidean geometry these triangles are all similar, here they are not conceptualized as such; in fact, there is no mention made of this similarity. It is not the triangles' geometric similarity but their metrical differences that are generative—it is the differences in lengths that merit calculation. Two functions are introduced: Combination (*he* 和)—the addition of the lengths of two different sides; and Difference (*jiao* 較)—the length of the one side subtracted from another. Both functions take two variables, and the results for six combinations of two sides from each triangle are given: if we denote the length of the short side by a, the long side by b, and the hypotenuse by c, we have $(b \pm a)$, $(c \pm a)$, and $(c \pm b)$.

Two more functions are then introduced. These functions are not merely a further application of Combination and Difference, but functions with three variables—all three sides—yielding four possibilities: Difference-Combination $(c + b - a)$, the Difference-Difference $(c - b + a)$, Combination-Combination $(c + b + a)$, and the Combination-Difference $(a + b - c)$. The results for the Connecting Triangle are as follows:

The Connecting Hypotenuse is 680, the Short-Side 320, the Long-Side 600. The Short-Side—Long-Side Combination is 920, the [Short-Side—Long-Side] Difference is 280. The Short-Side—Hypotenuse Combination is 1000, the Difference is 360. The Long-Side—Hypotenuse Combination is 1280, the Difference is 280. The Hypotenuse Difference-Combination is 960, the

Difference is 400. The Hypotenuse Combination-Combination is 1600, the Difference is 240.

通弦六百八十，勾三百二十，股六百。
勾股和九百二十，較二百八十。
勾弦和一千，較三百六十。
股弦和一千二百八十，較二百八十。
弦較和九百六十，較四百。
弦和和一千六百，較二百四十。 63

These results are listed, in parallel form, for each of the fifteen triangles (with two exceptions),[64] as for example, the Great-Void triangle:

The Great-Void Hypotenuse is 102, the Short-Side 48, the Long-Side 90. The Short-Side—Long-Side Combination is 138, and the Difference is 42. The Short-Side—Hypotenuse Combination is 150, the Difference is 54. The Long-Side—Hypotenuse Combination is 192, the Difference is 12. The Hypotenuse Difference-Combination is 144, the Difference is 60. The Hypotenuse Combination-Combination is 240, the Difference is 36.

太虛弦一百〇二，勾四十八，股九十。
勾股和一百三十八，較四十二。
勾弦和一百五十，較五十四。
股弦和一百九十二，較一十二。
弦較和一百四十四，較六十。
弦和和二百四十，較三十六。 65

If the addition and subtraction of the lengths of the various sides of triangles was used to generate a set of numbers, the next section seeks to find equalities among these, and to define new terms.

Heaven-to-Sun and Sun-to-Heart are equal [*AK = KO*]. Heart-to-River and River-to-Earth are equal [*ON = NB*]. Sun-to-Heart and Sun-to-Mountain are equal [*KO = KM*]. Therefore Mountain-to-River is called the "Lesser Difference" (*xiao cha* 小差). River-to-Heart and River-to-Moon are equal. Therefore Moon-to-Sun is called the "Greater Difference" (*da cha* 大差). The Bright Short-Side and the Small Long-Side are mutually obtained, termed the "Inner Ratio" (*nei lü* 内率), to seek the Void-Product. The Bright Long-Side and the Small Short-Side are mutually obtained, termed the "Outer Ratio" (*wai lü* 外率), to seek the Void-Product. The Void Short-Side and the Void Long-Side are mutually obtained, termed the "Void Ratio" (*xu lü* 虛率), to seek the Void-Product. Each Short-Side—Long-Side Combination is just the Hypotenuse-Yellow Combination. Each Greater

---

[63] Li, *Ce yuan hai jing*, 1:733.

[64] The two exceptions are cases where the triangles are the same size (*shang xia gao* 上下高 and *shang xia ping* 上下平)—the calculations fail to produce different results and are combined under *gao* 高 and *ping* 平.

[65] Li, *Ce yuan hai jing*, 1:735.

Difference is just the Long-Side-Yellow Difference. Each Lesser Difference is just the Short-Side-Yellow Difference. The High-Long-Side—Level-Short-Side Difference is called the "Corner Difference" (*jiao cha* 角差), or also the "Distant Difference" (*yuan cha* 遠差).

天之于日與日之于心同。心之于川與川之與地同。日之于心與日之于山同。故以山之川為小差。川之于心與川之于月同。故以月之日為大差。明勾專股相得，名為內率，求虛積。明股專勾相得，名為外率，求虛積。虛勾虛股相得，名為虛率，求虛積。凡勾股和即弦黃和。凡大差即股黃較。凡小差即勾黃較。高股平勾差，名角差，又名遠差。 [66]

And so on. Again, the origins and usage of this terminology are lost, beyond the definitions offered in the text itself;[67] but the complexity and multiplicity of terms indicates the importance attached to determining the ratios of the various sums and differences of the sides of the triangles.

The Celestial Origin Algebra

The first example of the application of Celestial Origin algebra in the *Sea Mirror of Circle Measurement* does not appear until the thirteenth problem of the second chapter. The Celestial Origin algebra is introduced without any explanation of terminology or how to calculate using it and, in fact, it appears without any explanation whatsoever:

It is asked: Exiting the West gate [*I*] and proceeding South 480 paces, there is a tree [*A*]; exiting from the North gate [*J*] and walking 200 paces toward the East [*B*], it [the tree at *A*] can be seen. The question and answer are the same as before.

或問：出西門南行四百八十步有樹，出北門東行二百步見之。問答同前。

Method: Take the product of the distances walked by the two [persons] as the constant, the sum of the distances walked by the two as the first term, and one as the second term, obtaining the radius [of the inscribed circle].

法曰：以二行步相乘為實，二行步并為從，一步常法，得半徑。 [68]

Translated into modern notation, this problem can be made equivalent to the following polynomial equation:

$$x^2 + 680x = 96000.$$

The solution is 120, the radius of the inscribed circle.

---

[66] Li, *Ce yuan hai jing*, 1:736.

[67] Bai, *Ce yuan hai jing*, and the annotations on which his work is based, offer little explanation of these definitions.

[68] Li, *Ce yuan hai jing*, 1:767.

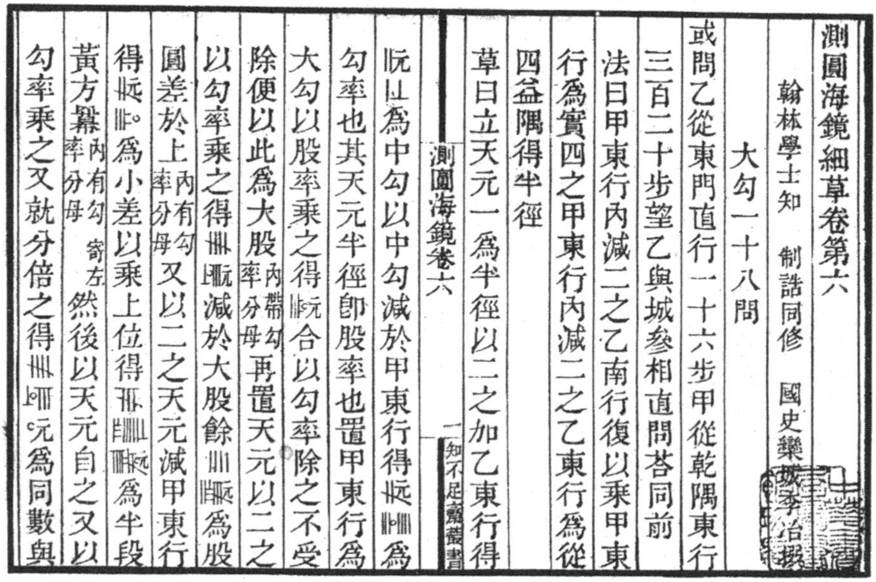

測圓海鏡細草卷第六

翰林學士知 制誥同修 國史樂城李冶撰

大勾一十八問

或問乙從東門直行一十六步甲從乾隅東行三百二十步壑乙與城參相直問荅同前

法曰甲東行內減二之乙南行復以乘甲東行為實四之甲東行內減二之乙東行為從

四益隅得半徑

草曰立天元一為半徑以二之加乙東行得

測圓海鏡卷六

忨凵為中勾減於甲東行得忨卌為勾率也其天元半徑即股率也置甲東行為大勾以股率乘之得卌帶勾再置天元以二之除便以此為大股率內減於大股餘卌為股以勾率乘之得卌減於大股餘卌為股圓差於上率內有勾又以二之天元減甲東行得忨為小差以乘上位得卅卌為半段黃方羃率內有勾寄左然後以天元自之又以勾率乘之又就分倍之得卌卌元為同數與

Fig. 4.4: A celestial-origin problem from Li's *Sea Mirror of Circle Measurement*.

Problem thirteen states that "the question and answer are the same as before." The previous problems, it turns out, have also been concerned with finding the radius (or diameter) of the inscribed circle. The introduction to the second chapter states,

> Suppose there is a round city wall with unknown diameter, gates opening on the four sides, and outside each gate there are horizontal and vertical intersecting roads. The Northwest corner of the intersecting roads is Caelum [C], the Northeast corner of the intersecting roads is *Gen* [F], the Southeast corner of the intersecting roads is *Xun* [E], and the Southwest corner of the intersecting roads is Earth [D].[69] All methods of measurement will be set up, one by one, in what follows.
>
> 假令圓城一所，不知周徑，四面開門，門外縱橫各有十字大道。其西北十字道頭定為乾地，其東北十字道頭定為艮地，其東南十字道頭定為巽地，其西南十字道頭定為坤地。所有測望法，一一設如後。[70]

Given the diagram, the modern reader familiar with the properties of triangles and inscribed circles can solve for r by means of the properties of right triangles. Writing x for the length of BC, y for the length of AC, z for AB, and r for the

---

[69] The four corners, *qian, gen, xun,* and *kun* all represent hexagrams in the *Yi jing.*

[70] Li, *Ce yuan hai jing,* 1:762.

radius, we obtain

$$r^2 + \left(\frac{x}{y}r\right)^2 = \left(\frac{x}{y}(y-r)-r\right)^2.$$

Then, solving this equation for r in terms of x, y and z, we get the surprisingly simple formula,

$$r = \frac{x+y-z}{2}.$$

Multiplying the numerator and denominator by $(x+y+z)$ yields an equivalent formula,

$$r = \frac{xy}{x+y+z}.$$

This latter formula is in fact found in *Sea Mirror of Circle Measurement* as the solution to the first problem of the second chapter, which, like the first Celestial Origin problem, is to find the diameter of the circle.

Li Ye's formula is not new. The problem that he presents and its solution were solved in the earliest extant Chinese mathematical treatise, the *Nine Chapters on the Mathematical Arts*.[71] The problem and its solution presented there are as follows:

Given the short-side [of a right triangle] is eight paces, the long-side is fifteen paces, it is asked what is the diameter of the inscribed circle? Answer: Six paces.

今有句八步，股十五步，一問句中容圓徑幾何？曰六步。

Technique: The short-side is eight paces, the long-side is fifteen paces, determine the hypotenuse. The three [sides] added together are the denominator, and the short-side multiplied by the long-side then doubled is the numerator. The numerator divided by the denominator yields the diameter.

術曰：八步為句，十五步為股，為之求弦。三位并之為法，以句乘股倍之為實。實如法得徑。

This then is precisely the above formula given in the *Sea Mirror of Circle Measurement* in the first problem. The diameter of the inscribed circle does not, then, require Celestial Origin algebra for its solution. The calculation is easily made by applications of nothing more complicated than the properties of right triangles; indeed, the *Jiu zhang suan jing* had done precisely that. The *Sea Mirror of Circle Measurement*, as described above, has already given the length of the diameter of the circle in the first chapter; this length was necessary for the calculations carried out there. And by the time the Celestial Origin algebra is introduced to find it, the diameter of the circle has already been found in each of twelve preceding problems, one of which gives the *Nine Chapters of Mathematical Arts* formula.

---

[71] Bai, *Ce yuan hai jing*, 155. See *Jiuzhang suanshu* 九章算術 [Nine chapters on the mathematical arts], in *ZKJDT*.

The remainder of the *Sea Mirror of Circle Measurement* presents problem after problem to calculate the diameter of the circle—or, more rarely, lengths easily calculated once the diameter is known. Among these can be found the following problem using the Celestial Origin algebra, which has been interpreted in previous accounts as the solution of a fourth-degree polynomial:

It is asked: Alpha walks south from Caelum [*C*] 600 paces and stands there [at *A*], Beta exits the south gate and walks straight ahead, Gamma exits the east gate and walks straight ahead; the three look at each other and all fit with [i.e., are in line with the edge of] the wall [Beta is at *K*, and Gamma at *N*], and Beta and Gamma together walked 151 paces. The question and answer are the same as before.

或問：甲從乾南行六百步而立，乙出南門直行，丙出東門直行，三人相望，俱與城相直，而乙、丙共行了一百五十一步。問答同前。

Method: Square the distance Alpha walked toward the south, halve that, then multiply that by itself, and that is the constant. Double the paces walked together [by Beta and Gamma], add the distance walked to the south by Alpha, and multiply this by half the square of the distance walked by Alpha, and let this be the first term. Take the distance walked by Alpha, multiplied by the sum [of the distance walked by Beta and Gamma], as the second term. Take one-and-a-half times the distance walked to the south by Alpha as the third term. Take 2 *fen* and 5 *li* as the fourth term.

法曰：甲南行為冪，折半又以自之為實。倍共步，加甲南行，以乘半段甲行冪為從方。甲行乘共數為從廉。一個半甲南行為第二益廉。二分五厘為三乘方隅。[72]

Writing the distance walked by Alpha *AC* = *a*, the distance walked by Beta *OK* = *b*, and the distance walked by Gamma *ON* = *c*, the equation above can be translated into the following polynomial equation of the fourth degree:

$$\frac{1}{4}x^4 - \frac{3a}{2}x^3 + a(b+c)x^2 + \frac{a^2(a+2(b+c))}{2}x - \frac{a^4}{4} = 0.$$

Solving for *x* gives 360, subtracting this from the distance walked by Alpha, *AC* = 600, yields the diameter of the circle, 240.[73]

When the *Sea Mirror of Circle Measurement* is understood in its historical context—rather than appropriated to forge continuities with modern algebra in a retrospectively projected teleology of modern mathematics—an entirely different set of questions is raised. One such question is to what extent the formulas presented in the text are geometrically derived. For example, in the passage above, the text asserts that "Heaven-to-Sun and Sun-to-Heart are equal [*AK* = *KO*]. Heart-to-River and River-to-Earth are equal [*ON* = *NB*]. Sun-to-Heart and Sun-to-Mountain are equal [*KO* = *KM*]." Although from the text it may appear

---

[72] Li, *Ce yuan hai jing*, 1:800; Bai, *Ce yuan hai jing*, 366–67.

[73] Ibid., 368–72.

that the equalities are mere happenstance—$AK = 255 = KO$, $ON = 136 = NB$, $KO = 255 = KM$—they are in fact geometric properties true not just for triangles similar to this triangle, but for all right triangles constructed as in the diagram above, as can easily be verified. If we let x be the short-side of the right triangle, and y the long-side, algebraic simplifications show that

$$AK = z\frac{r}{x} = y - y\left(\frac{r}{x}\right) - r = KO,$$

$$KO = \left(y - y\frac{r}{x} - r\right) = \left(z - z\frac{r}{x} - z\frac{x - 2r}{x}\right) = KM,$$

$$NO = \frac{x(y - r)}{r} - r = z\left(\frac{r}{y}\right) = NB.$$

These equalities may prove to be important in understanding the mathematical practices that are recorded—often in a very abbreviated manner—in Chinese mathematical treatises; but such a study is beyond the scope of this book.[74]

Another question concerns the aesthetic principles embodied in the text, such as the preference for integers. For example, if we assume an aesthetic preference for integers, what conditions must x and y satisfy for z and r to also be integers? It turns out that the simplest Pythagorean triple $(3, 4, 5)$ is one solution, with $r = 1$. Moreover, there are many solutions: if z is an integer, then r is also an integer. Thus the question of the choice of the triple $(680, 320, 600)$ depends on determining the number of possible values for the sides of the largest triangle such that all the triangles have sides of integral length—it is necessary and sufficient that the lengths of all the sides of AKP, KLH, LXM, MNG, and NBQ be integers. The triangle chosen $(680, 320, 600)$ is the smallest triangle such that all the sides of all of the triangles are integers. (The greatest common divisor of all the sides of the triangles listed, or equivalently, of 64 and 135, is 1.) All operations are closed with respect to the integers—Difference, Combination, Difference-Combination, Difference-Difference, Combination-Combination, and Combination-Difference. Only rarely—in Celestial Origin calculations in the later chapters—do non-integral solutions appear.

The appendices in the commentary to the text present four other triples that also yield triangles whose sides are all integral lengths in addition to the original

---

[74] In other words, it may be possible to verify the extent to which the numerical correspondences presented in the *Sea Mirror of Circle Measurement* were the result of geometric calculations rather than fortuitous numerical equality: (1) it would be necessary to show that all the formulas are geometric; and (2) that possible formulas that were not geometric (fortuitous numerical correspondences) were not included. In any case, there is no justification offered to show that $AK = KO$. Nor is there any interest in determining whether this is true for all right triangles—as we noted, the questions posed by the *Sea Mirror of Circle Measurement* are not those of similar triangles, nor are they of demonstrating geometrical properties that must be true from other properties of the triangle. Instead, starting from the mathematical objects, we reach the mathematical fact that $NO = NB$. This knowledge is not to be extended to a class of similar triangles that were not classified together: the crafting of the triangles in fact necessitates that they not be considered the same.

text:[75] the first ratio $(600, 360, 480)$; the second ratio, $(780, 300, 720)$; the third, $(1400, 392, 1344)$; and the fourth $(3690, 810, 3600)$. Each appendix lists the lengths of the sides of all the triangles and their combinations and differences in the same detail as the first section. Although Li Rui 李銳 (1773–1817), reading the *Sea Mirror of Circle Measurement* as a treatise on the Celestial Origin algebra, suggests that these tables are for looking up specific results, most of the results from Li Ye's first set of ratios are never cited at all in the text; the commentator's additional calculations suggest that these ratios were of considerable interest in themselves.

This suggests that we might ultimately take Li Ye's preface quite seriously: the purpose of this text is literally exhausting the principles of numbers by determining all the ratios of sides of the triangles—the title announces, after all, that this text is concerned with measurements of the circle (inscribed within a triangle) rather than Celestial Origin algebra. In this context, the *Sea Mirror of Circle Measurement* represents a considerable advance in the investigation of these properties: (1) instead of one triangle, fifteen triangles are studied; (2) instead of merely investigating the relationship of the sides and the hypotenuse, the sums and differences of the sides are themselves further added together and subtracted; (3) and instead of merely finding the diameter of the circle inscribed in the triangle, this problem of measurement is solved in "all" one hundred and seventy different ways.

The social context of Li Ye's life—what can be gleaned from sources limited to biographies and one collection of his works[76]—provides at least some context for understanding this work. Li Ye's father, Li Yu, had himself withdrawn from office due to political intrigue to become a recluse; he never again took an official post. Apparently, it was in fear of possible political repercussions that his father sent Li Ye and the rest of his family to Luancheng 欒城 for safety; Li studied with Zhao Bingwen 趙秉文 and Yang Wenxian 楊文獻. In 1230 he traveled to Luoyang, passed the metropolitan examination (*ci fu ke jinshi* 詞賦 科進士), and was assigned to Gaoling 高陵 (in present-day Shaanxi). It fell to

---

[75] Bai Shangshu suggests that the commentator may have been Dai Zhen 戴震 (1724–1777). Bai, *Ce yuan hai jing*, 4.

[76] Biographical accounts of Li Ye can be found in Song Lian 宋濂 (1310–1381) et al., *Yuan shi* 元 史 [History of the Yuan dynasty], in *SKQS*, *juan* 160, 3759–60; Su Tianjue 蘇天爵 (1294–1352), *Yuanchao mingchen shilue* 元朝名臣事略 [Biographical sketches of important officials of the Yuan dynasty], in *SKQS*, *juan* 13; Feng Congwu 馮從吾 (1556–1627), *Yuan ru kao lue* 元儒考略 [Brief investigation into scholars of the Yuan dynasty], in *SKQS*, *juan* 1; Wei Yuan 魏源 (1794– 1857), *Yuanshi xin bian* 元史新編 [History of the Yuan dynasty, new compilation], in *XXSQ*, *juan* 32; Ke Shaomin 柯劭忞 (1850–1933), *Xin Yuan shi: fu kaozheng* 新元史：附考證 [New history of the Yuan dynasty, with textual research appended] (Taibei: Yiwen yinshuguan 藝文 印書館, 1956), *juan* 171; Chen Yan 陳衍 (1856–1937), *Yuan shi ji shi* 元詩紀事 [Yuan poetry chronicles], in *XXSQ*, *juan* 3; and Ruan Yuan 阮元 (1764–1849), *Chouren zhuan chu bian* 疇 人傳初編 [Biographies of men of calendrics and calculation, first volume], in *Chouren zhuan hui bian* 疇人傳彙編 [Collected editions of biographies of men of calendrics] (Taibei: Shijie shuju 世界書局, 1962), hereinafter *CRZHB*. See also Kong Guoping 孔国平, *Li Ye zhuan* 李冶 传 [Biography of Li Ye] (Shijiazhuang: Hebei jiaoyu chubanshe 河北教育出版社, 1988); idem, *Li Ye, Zhu Shijie, yu Jin-Yuan shuxue* 李冶朱世杰与金元数学 [Li Ye, Zhu Shijie, and Jin-Yuan mathematics] (Shijiazhuang: Hebei kexue jishu chubanshe 河北科学技术出版社, 2000).

Mongol invaders before he arrived; and Li Ye fled and eventually himself became a recluse, studying mathematics, literature, history, astronomy, philosophy, and medicine. It was during his flight that Li Ye encountered recluses who introduced him to the Chinese algebras (Qin Jiushao also claims to have learned algebra from a recluse);[77] and it was during this period of withdrawal that, in 1248, Li Ye wrote *Sea Mirror of Circle Measurement.*

Conclusions

This analysis of the *Sea Mirror of Circle Measurement* illustrates how little that conventional perspectives can offer toward a historical understanding of this mathematical text. On the one hand, the pejoratives often used to dismiss Chinese mathematics and science—practical, algorithmic, correlative thinking—prove irrelevant. It is a text that devotes over one hundred problems to solving in different ways—including the Celestial Origin algebra—for a single value easily found by elementary methods in the oldest Chinese mathematical text. Thus despite the artifice of solving questions about the lengths of the diameter of a city wall, one could hardly find mathematics more impractical. Yet, on the other hand, in terms of its mathematical, social, and ideological context, the *Sea Mirror of Circle Measurement* does not fit as transparently into a teleology of the solution of modern mathematical polynomial equations as has been claimed.

Cheng Dawei's *Comprehensive Source of Mathematical Methods*

Cheng Dawei's 程大位 *Comprehensive Source of Mathematical Methods* (*Suanfa tong zong* 算法統宗, 1592) was the most widely known mathematical treatise of the Ming dynasty: it was revised, republished, pirated, transmitted to Japan and Korea, and later incorporated in the imperially sponsored *Collection of Documents, Ancient and Contemporary* (*Gu jin tushu ji cheng* 古今圖書集成), one of the largest encyclopedias compiled in imperial China.[78] However, the *Comprehensive Source of Mathematical Methods* is not listed in bibliographies of Ming works. The *General Catalogue of the Complete Collection of the Four Treasuries*—the most important annotated bibliography of the Qing dynasty—finds Cheng's work vulgar and filled with mistakes.[79] And in *Biographies of Men of Calendrics and Calculation*—the first comprehensive compilation of the biographies of Chinese scholars of calendrics—the account of Cheng consists of less than 320 characters, compared with close to 10,000 characters on Xu Guangqi; of this, most is

---

[77] Libbrecht, *Chinese Mathematics in the Thirteenth Century*, 62.

[78] See "Li fa dian" 曆法典, *Gu jin tu shu ji cheng* 古今圖書集成 (Taibei: Dingwen shu ju 鼎文書局, 1977), 4:1180–1296, which incorporates the thirteen *juan* edition of *Comprehensive Source of Mathematical Methods*, comprising over half of the section devoted to mathematics (*suanfa*).

[79] *Siku quanshu zong mu* 四庫全書總目 [General catalogue of the *Complete Collection of the Four Treasuries*], ed. Yong Rong 永瑢 (1744–1790) and Ji Yun 紀昀 (1724–1805) (Beijing: Zhonghua shuju 中華書局, 1965), 1:913.

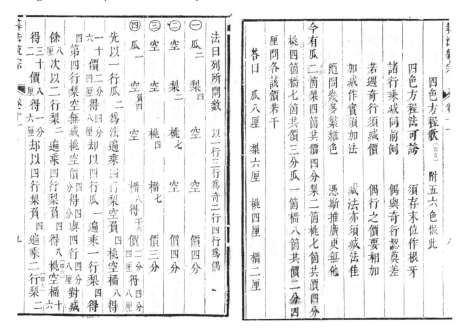

Fig. 4.5: A linear algebra problem from Cheng Dawei's *Comprehensive Source of Mathematical Methods*.

simply a copy of Cheng's bibliography in his last chapter. There are less than 40 characters on his biographical background, culminating with the following final evaluation:

> In conclusion: [Cheng] Dawei was unable to create profound mathematics. Therefore his techniques and examples are mostly erroneous. He instead arbitrarily took from all the schools.
>
> 論曰：大位算學未能深造，故其為術類多舛錯，然雜采諸家。 [80]

The low evaluation of Cheng in the *General Catalogue of the Complete Collection of the Four Treasuries* and *Biographies of Men of Calendrics and Calculation* is echoed in the few scholarly works that discuss him. The most authoritative and positive work on Cheng criticizes these negative evaluations, offering as one example the accusation that Cheng's work was "absurd in the extreme" 荒誕至極.[81] In the second section of their article, Yan and Mei assert that Cheng made several important contributions to mathematics—solving second- and third-degree polynomials with abacus calculations, introducing a surveying tool to measure distances, and his explication of Pascal's triangle.[82] In their

---

[80] Ruan, *Chouren zhuan*, 384.

[81] Yan and Mei, "Cheng Dawei ji qi shuxue zhuzuo," 52; Yan and Mei do not give the citation for this accusation.

[82] Ibid., 29–39.

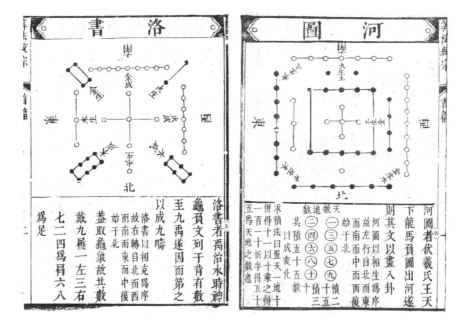

Fig. 4.6: The River Diagram and Luoshu, in Cheng's *Comprehensive Source of Mathematical Methods*.

concluding section, they assert (incorrectly) that Cheng was the greatest of the Ming mathematicians. But they then qualify that this assessment is based on the "backwardness" of mathematics during the Ming:

> The backwardness of traditional mathematics in the Ming dynasty is a historical fact. Xu Guangqi has already pointed out that there were two reasons for this backwardness, "one is Confucians who dispute [unimportant] principles and despise all evidential matters under heaven; one is the absurd techniques that state that numbers have spiritual principles." ... The important mathematical works of ancient [China] were almost completely lost, and those who researched mathematics were extremely few in number, and their level was low.[83]

Thus it is only against this background of the Ming decline in mathematics that Yan and Mei discover redeeming features in Cheng.

The *Comprehensive Source of Mathematical Methods* is, as Cheng states in his preface, a compilation of previous works:

> When I was a youth I studied [mathematics], and at twenty I traveled for business to Wu and Chu, and at the same time I visited notable masters,

---

[83] Yan and Mei, "Cheng Dawei ji qi shuxue zhuzuo," 50; quoting Xu Guangqi, "Ke *Tongwen suanzhi* xu," in *TXCH*, 5:2772–73.

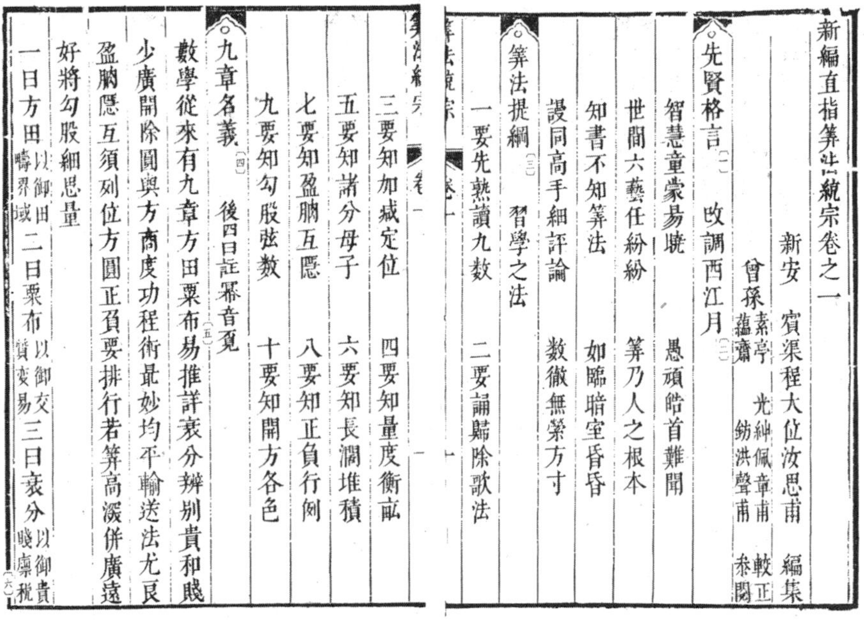

新編直指算法統宗卷之一

新安　賓渠程大位汝思甫　編集
曾孫　素亭　光紳佩章甫
　　　蘊齋　　　洪聲甫
　　　較閱　　　泰閣

先賢格言
改調西江月
智慧童蒙易曉
愚頑皓首難聞
世間六藝任紛紛
筭乃人之根本
知書不知筭法
如臨暗室昏昏
誤同高手細評論
數徹無第方寸

習學之法
一要先熟讀九數
二要誦歸除歌法

算法統宗　卷之一

九章名義
數學從來有九章
後四日定冪音覓
三要知加減定位
四要知量度衡斛
五要知諸分母子
六要知長潤堆積
七要知盈朒互隱
八要知正負行列
九要知勾股弦數
十要知開方各色

數學從來有九章方田粟布易推詳衰分辨別貴和賤
少廣開除圓與方商度功程術最妙均平輸送法尤良
盈朒互須列位方圓正負要排行若筭高深併廣遠
好將勾股細思量以御畇域
一日方田以御畇域　二日粟布以御交易　三日衰分以御貴賤

Fig. 4.7: Popular verses in Cheng's *Comprehensive Source of Mathematical Methods*.

seeking their works and meanings, investigating their methods, and after returning to the head of the Shuai River [near modern day Tunguang zhen 屯光镇, in Anhui Province] and contemplating, after twenty years, suddenly I was instantly enlightened as if something was attained. After this I consulted the methods of the various schools, adding the little that I attained, and compiled it into a volume.

予幼習是學，弱冠商遊吳楚，遍訪名師，繹其文義，審其成法，歸而覆思於率水之上，餘二十年，一旦恍然，若有所得，遂於是乎參會諸家之法，附以一得之愚，纂集成編。

The volume combines elements from different mathematical schools and different strata of society. It begins with metaphysical ruminations on the *He tu* 河圖 and *Luo shu* 洛書 diagrams and divination; it continues with elementary definitions of terminology used in abacus calculations, polynomial equations, and decimal places. It is structured following the *Nine Chapters on the Mathematical Arts*, with chapters on measuring areas (*fang tian* 方田), grains (*su mi* 粟米), proportionate distribution (*cui fen* 衰分), determining width (*shao guang* 少廣), construction consultations (*shang gong* 商功), equitable taxation (*jun shu* 均輸), excess and deficit (*ying bu zu* 盈不足), matrices (*fangscheng* 方程), and right triangles (*gou gu* 勾股). It is filled with popular jingles, ranging in subject from the importance of mathematics to calculational techniques for the abacus. It concludes with a bibliography of over 50 mathematical treatises which for Cheng comprise the tradition that he has synthesized.

The eclectic syncretism of Cheng's work, incorporating popular culture together with the traditional authoritative classic in Chinese mathematics, probably accounts for its popularity. Instead of the arcane symbols of Celestial Origin algebra, which present mathematics as an esoteric metaphysical pursuit, the *Comprehensive Source of Mathematical Methods* presents mathematics as practical, including the mathematical art known as *fangcheng* 方程.

Fangcheng in Cheng's Comprehensive Source of Mathematical Methods

Like the problems in the *Sea Mirror of Circle Measurement*, many of the problems in the *Comprehensive Source of Mathematical Methods* are practical in form only. One example of a *fangcheng* problem, Problem 3, is translated below (again, half-size characters in the original text are transcribed here in a smaller font):[84]

> Given: three inkstones, five slabs of ink, nine brushes, together cost eight *qian* one *fen*; also four inkstones, six slabs of ink, seven brushes, together cost eight *qian* nine *fen*; also five inkstones, seven slabs of ink, eight brushes, together cost one *liang* and six *fen*. How much are the inkstones, ink, and brushes?
>
> 今有硯三箇，墨五匣，筆九枝，共價八錢一分。又硯四箇，墨六匣，筆七枝，共價八錢九分。又硯五箇，墨七匣，筆八枝，共價一兩零六分。問硯、墨、筆各若干。
>
> Answer: inkstones are eight *fen* each, ink is six *fen* each, and brushes are three *fen* each.
>
> 答曰：硯每箇八分，墨每匣六分，筆每枝三分。
>
> Method: Arrange the numbers given in the problem. First take the 3 of the inkstones of the column on the right as the "divisor," and multiply the columns left and center. The two columns then have the following values:
>
> Right: Inkstones 3 as the "divisor" first multiply left and center columns; ink 5 gets 20; brushes 9 get 36; cost 8 *qian* 1 *fen* gets 3 *liang* 2 *qian* 4 *fen*.
>
> Center: Inkstones 4 gets 12; ink 6 gets 18; brushes 7 gets 21; cost 8 *qian* 9 *fen* gets 2 *liang* 6 *qian* 7 *fen*.
>
> Left: Inkstones 5 get 15; ink 7 gets 21; brushes 8 gets 24; cost 1 *liang* and 6 *fen* gets 3 *liang* 1 *qian* 8 *fen*.
>
> 法曰：列所問數。先以右行硯三為法遍乘左中。二行得數：
>
> 右：硯三為法先乘左中　墨五得二十　筆九得三十六　價八錢一分得三兩二錢四分

---

[84] The following translation is based on the modern annotated edition, Cheng Dawei 程大位 (1533–1606), *Suanfa tongzong jiaoshi* 算法统宗校释 [Comprehensive source of mathematical methods, edited with explanatory notes], ed. Mei Rongzhao 梅荣照 and Li Zhaohua 李兆华 (Hefei: Anhui jiaoyu chubanshe 安徽教育出版社, 1990). In the annotations, the *Suanfa tongzong jiaoshi* presents an outline of this problem; I have borrowed from this where appropriate. For an analysis of *fangcheng* as a mathematical practice, see the following chapter of this book.

中：硯四得一十二　　墨六得一十八　　筆七得二十一　　價八錢九分得二兩六錢七分

左：硯五得一十五　　墨七得二十一　　筆八得二十四　　價一兩令六分得三兩一錢八分

Then take the middle column inkstones 4, and multiply the right ink, brushes yielding the values ink gets 20, brushes get 36, cost gets 3 *liang* 2 *qian* 4 *fen*. Then subtract the remainder from the middle column ink 2, brushes 15, cost 5 *qian* 7 *fen*, and again place the right column. Again take the left column inkstones 5 as the "divisor," and multiply the right column ink, brushes yielding the values ink 25, brushes 45, cost 4 *liang* 0 [*qian*] 5 *fen*, and with the left column subtract the remainder ink 4, brushes 21, cost 8 *liang* 7 *fen*; and again place the left column.

卻以中行硯四遍乘右行墨筆，　得數墨得二十、筆得三十六、價得三兩二錢四分。　與中行對減餘墨二、筆十五、價五錢七分，　另列右位。又以左行硯五為法遍乘右行墨筆得數墨二十五、筆四十五、價四兩令五分與左行對減餘墨四、筆二十一、價八錢七分另列左位。

Again lay out the remainder from subtraction, with the *fen* left and right place the numbers. Take the right column ink 2 as the "divisor," and multiply the left column brushes, cost, yielding the values arranged left column:

Right: Inkstones 2; brushes 15 gets 60; cost 5 *qian* 7 *fen* gets 2 *liang* 2 *qian* 8 *fen*.

[The middle column is omitted.]

Left: Inkstones 4; brushes 21 gets 42; cost 8 *qian* 7 *fen* gets 1 *liang* 7 *qian* 4 *fen*.

再列減餘以分左右位數。以右行墨二為法，　遍乘左行筆、價得數列左位：

右：墨二　　筆十五得六十　　價五錢七分　　得二兩二錢八分

左：墨四　　筆二十一得四十二　　價八錢七分　　得一兩七錢四分

Again take the left column ink 4 as the "divisor," and multiply the right column brushes, cost to obtain the values, and place in the right column. Then take the left and right columns and subtract ink is eliminated. The remainder yields brushes 18 as the divisor. Then take the remainder of the cost to obtain the values, and subtract from the remainder 5 *qian* 4 *fen* as the dividend, and divide the divisor into the dividend, yielding the cost of brushes 3 *fen* each. Then take the cost of brushes and multiply by the latter right [column], the remainder of brushes 15, yielding 4 *qian* 5 *fen*, and subtract the cost remainder from the right column 5 *qian* 7 *fen*, leaving 1 *qian* 2 *fen*. Take the ink remainder from the right column 2 as the divisor, divide it obtaining the cost for ink 6 *fen* each. Then subtract this from the original cost 8 *qian* 1 *fen* in the right column, subtract the original brushes 9, cost 2 *qian* 7 *fen*. [Take] the original ink 5 cost 3 *qian*, remainder 2 *qian* 4 *fen* as the dividend, with the former right [column] original inkstone 3 as the divisor, divide it yielding the cost of inkstones 8 *fen* each.

復以左行墨四為法，　遍乘右行筆價得數，列右位。卻以左右對減墨盡。餘得筆一十八枝為法。又以餘價得數，相減餘五錢四分為實，以法除實，得筆價每枝三分。就以筆價乘後右，餘筆十五得四錢五分。以減右行餘價

五錢七分，餘一錢二分以右行餘墨二為法，除之得墨價每匣六分。于前右
行原價八錢一分內減原筆九價二錢七分，原墨五價三錢，餘二錢四分為實，
以前右原硯三為法，除之得硯價每箇八分。

As presented, the text itself is virtually incomprehensible, at least without a
prior understanding of mathematical practices behind it (again, *fangcheng* as a
mathematical practice will be explained in detail in the following chapter of this
book). Although formulated using inkstones (*yan* 硯), ink (*mo* 墨), and pens (*bi*
筆), these terms serve as little more than variable names to express a problem
with three conditions in three unknowns. If the problem in the text is written in
modern mathematical notation, it is equivalent to the following linear algebra
problem,

$$\begin{bmatrix} 3 & 5 & 9 & 81 \\ 4 & 6 & 7 & 89 \\ 5 & 7 & 8 & 106 \end{bmatrix}.$$

This is equivalent to the following system of three conditions in three unknowns,
again expressed in modern notation,

$$3x_1 + 5x_2 + 9x_3 = 81, \quad 4x_1 + 6x_2 + 7x_3 = 89, \quad 5x_1 + 7x_2 + 8x_3 = 106,$$

with solutions $x_1 = 8$, $x_2 = 6$, and $x_3 = 3$.

Fangcheng problems can be found in numerous extant mathematical treatises
from the Ming dynasty. The problems recorded here are not original to Cheng.
The solutions are all correct. Including Problem 3, translated above, a total of
nine *fangcheng* problems are recorded. The remaining eight are listed below,
translated into modern terminology for brevity:

PROBLEM 1.   The first problem, written as an augmented matrix, is as follows,

$$\begin{bmatrix} 3 & 2 & 114 \\ 4 & 5 & 162.5 \end{bmatrix}.$$

This is equivalent to the following system of two conditions in two unknowns,

$$3x_1 + 2x_2 = 114, \quad 4x_1 + 5x_2 = 162.5,$$

with solutions $x_1 = 35$, $x_2 = 4.5$.

PROBLEM 2.   The second problem,

$$\begin{bmatrix} 3 & 4 & 480 \\ 7 & 2 & 680 \end{bmatrix},$$

is equivalent to the following system of two conditions in two unknowns,

$$3x_1 + 4x_2 = 480, \quad 7x_1 + 2x_2 = 680,$$

with solutions $x_1 = 80$, $x_2 = 60$.

PROBLEM 4.   The fourth problem (again, Problem 3 is translated above),

$$\begin{bmatrix} 1 & 1 & 0 & 42 \\ 0 & 2 & 1 & 42 \\ 1 & 0 & 3 & 42 \end{bmatrix},$$

is equivalent to the following system of three conditions in three unknowns,

$$x_1 + x_2 = 42, \quad 2x_2 + x_3 = 42, \quad x_1 + 3x_3 = 42,$$

with solutions $x_1 = 24$, $x_2 = 18$, $x_3 = 6$.

PROBLEM 5.   The fifth,

$$\begin{bmatrix} 2 & 3 & 0 & 2040 \\ 0 & 5 & 6 & 640 \\ 3 & 0 & 7 & 2980 \end{bmatrix},$$

is equivalent to the following system of three conditions in three unknowns,

$$2x_1 + 3x_2 = 2040, \quad 5x_2 + 6x_3 = 640, \quad 3x_1 + 7x_3 = 2980,$$

with solutions $x_1 = 900$, $x_2 = 80$, $x_3 = 40$.

PROBLEM 6.   The sixth,

$$\begin{bmatrix} 4 & 3 & 0 & 750 \\ 3 & 0 & 4 & 600 \\ 0 & 5 & 6 & 810 \end{bmatrix},$$

is equivalent to the following system of three conditions in three unknowns,

$$4x_1 + 3x_2 = 750, \quad 3x_1 + 4x_3 = 600, \quad 5x_2 + 6x_3 = 810,$$

with solutions $x_1 = 120$, $x_2 = 90$, $x_3 = 60$.

PROBLEM 7.   The seventh problem,

$$\begin{bmatrix} 2 & 5 & -13 & 5 \\ 1 & -3 & 1 & 0 \\ -5 & 6 & 8 & -3 \end{bmatrix},$$

is equivalent to the following system of three conditions in three unknowns,

$$2x_1 + 5x_2 - 13x_3 = 5, \quad x_1 - 3x_2 + x_3 = 0, \quad -5x_1 + 6x_2 + 8x_3 = -3,$$

with solutions $x_1 = 6$, $x_2 = \frac{5}{2}$, $x_3 = \frac{3}{2}$.

PROBLEM 8. The next problem,

$$\begin{bmatrix} 2 & 4 & 0 & 0 & 40 \\ 0 & 2 & 7 & 0 & 40 \\ 0 & 0 & 4 & 7 & 30 \\ 1 & 0 & 0 & 8 & 24 \end{bmatrix},$$

equivalent to the following system of four conditions in four unknowns,

$$2x_1 + 4x_2 = 40, \quad 2x_2 + 7x_3 = 40, \quad 4x_3 + 7x_4 = 30, \quad x_1 + 8x_4 = 24,$$

with solutions $x_1 = \frac{8}{10}$, $x_2 = \frac{6}{10}$, $x_3 - \frac{4}{10}$, $x_4 = \frac{2}{10}$.

PROBLEM 9. The final problem,

$$\begin{bmatrix} 3 & -10 & 6 \\ -2 & 5 & 1 \end{bmatrix},$$

is equivalent to the following system of two conditions in two unknowns,

$$3x_1 - 10x_2 = 6, \quad -2x_1 + 5x_2 = 1,$$

with solutions $x_1 = 8$, $x_2 = 3$.

Cheng cannot be credited with discovering the solutions to simultaneous linear equations—these problems are similar to those found in the *Nine Chapters on the Mathematical Arts*, as will be discussed in the next chapter. Perhaps what is most important about Cheng's work is its combination of abacus calculations with the traditional problems of the *Nine Chapters*. Within the larger argument presented in this book, these problems are but one example of the mathematical practices preserved in numerous mathematical treatises commonly available during the Ming dynasty.

Quantifying Ritual: Political Cosmology, Courtly Music, and Precision Mathematics

This section will examine ritual, music, and mathematics in seventeenth-century China in its humanistic context. The focus is Zhu Zaiyu's 朱載堉 (1536–1611) *Complete Collection of Music and Pitch* (*Yuelü quanshu* 樂律全書),[85] and in particular, the mathematics presented in his *New Explanation of the Theory of*

---

[85] Zhu Zaiyu 朱載堉 (1536–1611), *Yuelü quanshu* 樂律全書 [Complete collection of music and pitch] edited by Beijing tushuguan guji chuban bianji zu 北京圖書館古籍出版編輯組 (Beijing: Shumu wenxian chubanshe 書目文獻出版社, [1988]), hereinafter *YLQS*.

Calculation (*Suanxue xin shuo* 算學新說, engraved in 1604).[86] Zhu's mathematics is perhaps some of the most sophisticated found in extant texts from the Ming Dynasty, yet he is rarely even mentioned in studies of Ming mathematics; his *New Explanation of the Theory of Calculation* has often been overlooked.[87] The research works that have been written on Zhu have been framed within the older historiography, and in particular the project of Joseph Needham's "grand titration," which was to redistribute scientific credit among civilizations; their central focus has been Zhu's work in music and in particular the asserted priority of Zhu's discovery of the equal temperament of the musical scale. In these hagiographic accounts, Zhu is portrayed as a scientist; the central focus of his work—the recovery of ancient ritual—is rarely mentioned.

This section is divided into six subsections. The first examines what has been termed "correlative thinking" and assertions that it inhibited the development of science. The second examines in detail one example of "correlative thinking"—the philosophical theories of ritual music expounded in the *Record of Music*. The third examines attempts to recover the ritual music of antiquity as a solution to Ming Dynasty crises. The fourth examines Zhu Zaiyu's proposals for the reform of astronomy and ritual music. And the fifth focuses on the mathematical solutions he offered for recovering the music of antiquity—the equal-temperament of the musical scale, calculated to twenty-five significant digits using nine abacuses. The sixth offers a summary of the conclusions. "Correlative thinking," conservative Confucian textualism, the symbolic rituals of the absolutist imperial court, and the commercial mathematics and abacus of the merchants—all have been blamed for inhibiting the development of science. This section explores how they were combined in Zhu's work.

The Problem with Correlative Thought

In previous accounts that offered grand comparisons of the sciences of civilizations so as to isolate the causes of scientific development, many explanations have been offered—institutional, social, economic, and linguistic—for the absence of Science, Modern Science, or The Scientific Revolution in China. One of the most common themes is what has been termed "correlative thinking."[88]

---

[86] Idem, *Suanxue xin shuo* 算學新說 [New explanation of the theory of calculation], in *YLQS*.

[87] Zhu's *New Explanation of the Theory of Calculation* is mentioned as an extant mathematical treatise in Mei, "Ming-Qing shuxue gailun," 1–2. However, more typical is Martzloff, *History of Chinese Mathematics*, in which Zhu is mentioned only briefly as having proposed a project to reform the calendar (p. 21).

[88] David Hall and Roger Ames provide the following explanation of what is meant by the term "correlative thinking": "Rational or logical thinking, grounded in analytic, dialectical and analogical argumentation, stresses the explanatory power of physical causation. In contrast, Chinese thinking depends upon a species of analogy which may be called 'correlative thinking.' Correlative thinking ... involves the association of image or concept-clusters related by meaningful disposition rather than physical causation. Correlative thinking is a species of spon-

For Joseph Needham, correlative thinking was one of the central differences between China and the West; one of the central examples Needham offers is Dong Zhongshu's (179?-104? BCE) theory of music and the sympathetic resonance of musical tones.[89] For later writers, "correlative thinking" and variants of Chinese metaphysics remained little more than a philosophical context that inhibited or prevented science: "as the result of a highly sophisticated metaphysics there was *always* an explanation—which of course was no explanation at all—for anything puzzling which turned up."[90] Even the most sophisticated comparative studies continue to characterize Chinese thought as essentially correlative;[91] critical studies of science continue to assert that an alleged nonseparability of "nature" and "society" is the central reason for the lack of development outside the West.[92]

---

taneous thinking grounded in informal and ad hoc analogical procedures presupposing both association and differentiation. The regulative element in this modality of thinking is shared patterns of culture and tradition rather than common assumptions about causal necessity." David L. Hall and Roger T. Ames, "Chinese Philosophy," in *Routledge Encyclopedia of Philosophy* (London: Routledge, 1998).

[89] Joseph Needham, *History of Scientific Thought*, vol. 2 of *Science and Civilisation in China* (Cambridge: Cambridge University Press, 1956). Elsewhere, Needham describes the difference as one between organic and mechanical philosophy (along with three other fundamental differences between China and the West—algebraic vs. geometrical, wave vs. particle, and practical vs. theoretical). Idem, "Poverties and Triumphs of the Chinese Scientific Tradition," in *The Grand Titration: Science and Society in East and West* (London: George Allen & Unwin, 1969), 21–23. For further accounts of "correlative thinking," see also Benjamin I. Schwartz, *The World of Thought in Ancient China* (Cambridge, MA: Belknap Press of Harvard University Press, 1985), 350–82; A. C. Graham, *Yin-Yang and the Nature of Correlative Thinking* (Singapore: Institute of East Asian Philosophies, National University of Singapore, 1986); and idem, *Disputers of the Tao: Philosophical Argument in Ancient China* (La Salle, IL: Open Court, 1989), 313–56.

[90] Mark Elvin asserts that "the consequences of this [Wang Yangming's] philosophy for Chinese science were disastrous" and continues, "given this attitude, it was unlikely that any anomaly would irritate enough for an old framework of reference to be discarded in favor of a better one." Elvin, *Pattern of the Chinese Past*, 234. Elvin's conclusion is presumably based on the analysis of shifts in conceptual schemes or paradigms articulated in Thomas S. Kuhn, *The Copernican Revolution: Planetary Astronomy in the Development of Western Thought* (Cambridge, MA: Harvard University Press, 1957) and idem, *The Structure of Scientific Revolutions*, 2nd ed. (Chicago: University of Chicago Press, 1970). Lothar von Falkenhausen, *Suspended Music: Chime-Bells in the Culture of Bronze Age China* (Berkeley: University of California Press, 1993), suggests correlative thinking formed a "numerological straightjacket" (p. 4).

[91] G. E. R. Lloyd reaffirms the general validity of the characterizations of a "cause-oriented Greek culture and a correlation-oriented Chinese one" (p. 93), while noting that the Greeks "were no strangers to correlative thinking" (p. 94) and "Chinese interest in the explanation of events is certainly highly developed in such contexts as history and medicine" (p. 109). G. E. R. Lloyd, *Adversaries and Authorities: Investigations into Ancient Greek and Chinese Science* (Cambridge: Cambridge University Press, 1996). See also G. E. R. Lloyd and Nathan Sivin, *The Way and the Word: Science and Medicine in Early China and Greece* (New Haven: Yale University Press, 2002).

[92] Although Bruno Latour does not specifically address China and correlative thinking, he argues that what differentiates the Premodern (in both the West and the non-West) from the Modern West is that "the nonseparability of natures and societies had the disadvantage of making experimentation on a large scale impossible, since every transformation of nature had to be in harmony with a social transformation, term for term, and vice versa" (p. 140). This asserted difference underwrites the three central hypotheses of his book: (1) for the Moderns,

In these accounts, having labeled Chinese thought as "correlative thinking" and thus as inhibiting science, there has been little need for further study, except for negative comparison to ideologies held to have aided the development of science. In order to better understand the historical tradition that Zhu sought to recover, and its constitutive role in Zhu Zaiyu's work, we must first seek to examine these philosophical theories.

Music and Statecraft: The Record of Music

A central text for understanding the role of music in politics and ritual in China is the *Record of Music* (*Yue ji* 樂記), the earliest extant Chinese work on the theory of music.[93] The relationship between music, ritual, and politics in China was also a topic of considerable debate in early Chinese philosophical texts. In the philosophical text *Mozi*, music is an extravagant waste of resources that burdens the subjects without providing them benefit.[94] *Zhuangzi* 莊子 contains several

---

purification makes hybrids possible; (2) for the Pre-moderns, conceiving of hybrids excluded their proliferation; and (3) the Non-modern present must "slow down, reorient, and regulate the proliferation of monsters by representing their existence officially." Bruno Latour, *We Have Never Been Modern*, trans. Catherine Porter (Cambridge, MA: Harvard University Press, 1993), 12. Latour offers no historical evidence for this claim, which is in fact strikingly similar to conventional anthropological caricatures of the non-Western and pre-modern.

[93] The *Record of Music* now exists as a section in the *Record of Rites* (*Li Ji* 禮記). According to bibliographic records, it was compiled by Liu Xiang 劉向 (77 BCE–6 CE), who collated and edited it to 23 sections (*pian* 篇); the version included in the *Record of Rites* (*Li ji* 禮記) has been further edited to eleven sections. The origin of the text is still undetermined. The edition consulted here is included in *Liji zhushu* 禮記註疏 [*Record of Rites*, with commentary and subcommentary], in *Shisanjing zhushu: fu jiaokan ji* 十三经注疏：附校勘记 [Thirteen Classics, with commentary and subcommentary, and with editorial notes appended] (Beijing: Zhonghua shuju 中华书局, 1980), hereinafter *SSJZS*. The *Record of Rites* contained in the *Thirteen Classics* has commentary by Zheng Xuan; his commentary follows the Old Text (*gu wen jing* 古文經) versions but also include material from the New Text editions (*jin wen jing* 今文經). Subcommentary was added by Kong Yingda, a scholar of the classics during the Tang Dynasty who held the titles of Erudite of the National University (*guozi boshi* 國子博士) and Chancellor of the National University (*guozi ji jiu* 國子祭酒). He was commissioned by Emperor Taizong of the Tang 唐太宗 to edit the *Correct Interpretation of the Five Classics* (*Wu jing zheng yi* 五經正義). Shen Yue 沈約 (441–513 CE) linked the *Record of Music* to the Confucian tradition, claiming that it was transmitted by Confucius to his disciple Gongsun Nizi 公孫尼子 (of the early Warring States Period). For an analysis of the compilation of the *Record of Rites*, see Loewe, *Early Chinese Texts*, s.v. For analyses of the content, see Haun Saussy, *The Problem of a Chinese Aesthetic* (Stanford: Stanford University Press, 1993), 88–105, and, in the context of the Ming Dynasty, Joseph S. C. Lam, *State Sacrifices and Music in Ming China: Orthodoxy, Creativity, and Expressiveness* (Albany, NY: State University of New York Press, 1998), 75–97.

[94] "Against Music" (*fei yue* 非樂), in *Mozi*. For analysis of the compilation of this and the following early philosophical texts, again see Loewe, *Early Chinese Texts*.

passages exhorting the rejection of ritual and music.[95] The *Analects* (*Lun yu* 論語) includes a component on aesthetics, arguing for the balanced synthesis of patterning (*wen* 文) and substance (*zhi* 質); this theory of aesthetics is then integrated into a larger ethics of the superior man (*junzi*), which includes a theory of education, proper behavior, political conduct, and morality. Archaeological evidence further verifies the important role music played in ritual ceremony and politics.[96]

The *Record of Music* adds to these early formulations a sophisticated theory of mind (*xin* 心). Music is produced as a reaction to external stimuli on the mind;[97] it is this theory of mind that establishes the necessity of the intervention of the monarch— through rites, music, government, and punishments—to regulate the minds of the subjects, thus bringing order to the state. The *Record of Music* begins with the investigation of individual sounds (*sheng* 聲) which are then combined into tones (*yin* 音),[98] and ultimately form music (*yue* 樂); parallel to music, the analysis of society begins with the investigation of the individual, which in aggregate then forms society. Thus, "Examine sound and thereby comprehend tones, examine tones and thereby comprehend music, examine music and thereby comprehend government";[99] "the principles of sound and tones (*sheng yin zhi dao* 聲音之道) are the same as those of government."[100]

The theory of individual sounds in the *Record of Music* is based on a theory of the relationship between the mind and objects (*wu* 物) outside the mind. In the nature of the human mind inheres both a state and a potential: the human mind in its original state (at birth) has no desires or feelings—this is the nature of the human mind that is derived from heaven (*tian* 天). But also in the nature of the mind is the capacity to produce feelings and desires, which result from the exposure to external objects. Through the effect of objects on the potentiality of the mind (desire), the mind is brought into activity from its original stillness, defining a new state (*xing* 形) of the mind.[101] Thus desire, feelings, and knowl-

---

[95] Zhuang Zhou 莊周 (fl. 320? BCE), *Zhuangzi ji shi* 莊子集釋 [*Zhuangzi*: Collected explanations], ed. Guo Qingfan 郭慶藩 and Wang Xiaoyu 王孝魚 (Beijing: Zhonghua shuju 中華書局, 1961), 284, 321, 341, 486.

[96] See Falkenhausen, *Suspended Music* for an analysis of the use of bells in ancestral worship (where the living sacrifice to the dead in exchange for blessings) and for weddings, entertainment, and ritual archery contests.

[97] It is this same theory of mind that is later revived as a foundation in later Confucian philosophy, and is central in the debate on the origin of desire.

[98] For a discussion of the relationship between *sheng* and *yin*, see Saussy, *Problem of a Chinese Aesthetic*, 88–105.

[99] *Liji zhushu*, 1528.

[100] Ibid., 1527.

[101] *Subcommentary*: "'Affected by objects, [the mind] is brought into activity, which is then manifested in sound' means that when the human mind has already been affected by external objects, and the mouth becomes active to express the mind, the state (*xing* 形) of the mind is expressed in sound. If the mind is affected by death or loss, and the mouth commences activity, then the form will be expressed in melancholy sound. If the mind is affected by good fortune or

edge are not innate; the form of the mind derives from its reaction to the stimuli of external reality, according to the nature the mind derives from heaven.[102]

It is through sound that this state—the reaction of the inner mind to external stimuli—is manifested in the external world. Sound, similar to the reactive nature of the mind to objects, is a reactive expression of the stimulus received by the mind. The six qualities of sound (dread, freedom from worry, coming forth, severity, solemnity, and peacefulness) reflect the six states of the mind (melancholy, joy, happiness, fury, respectfulness, adoration), which are determined by the influence of objects; the state of mind is reflected by the quality of sound, not by which particular one of the five sounds is produced. The five sounds, *gong* 宮, *shang* 商, *jue* 角, *zhi* 徵, and *yu* 羽, individually correspond to hierarchical elements of the metaphysical system: the sovereign (*jun* 君), ministers (*chen* 臣), subjects (*min* 民), affairs (*shi* 事), and objects.[103] Discord in any of these individual sounds is manifested in distortion of the sound (*gong* becomes undisciplined, *shang* becomes abnormal, *jue* becomes fretful, *zhi* evinces grief, and *yu* evinces danger), and reflects discord in the corresponding category (arrogance of the sovereign, corruption of the ministers, discontent of the subjects, overburdening of affairs, and scarcity of resources). Sound is thus individual in its production, reflecting the form of the mind of the individual; each of the five sounds represents an individual element of society.

Tones, the combination and transformation of individual sounds, correspond to society, which is the combination and transformation of individuals.[104] When sounds are manifested, they resonate with similar sounds.[105] This resonance is not enough to be tones, however: tones are the result of the combination of various sounds and their transformation into intricate patterns. As a reflection of society, the three qualities of tones (peacefulness, resentment, grief) correspond to the three qualities of the state (order, chaos, demise), and to the three qualities of government (harmony, perversity, hardship). Zheng and Wei are examples of the music of disorder; the music of the "Mulberry Grove on the River Pu" is an example of the music of the world in demise.[106] Similarly, if the five sounds are all

---

love, and the mouth commences activity, then the form will be expressed in the sound of joy." Ibid., 1527.

[102] This theory that the state of the mind is determined by the external material reality differs from earlier theories that music imitates nature, or comes from *yin* and *yang*.

[103] In the subcommentary, these are further correlated with other elements of the metaphysical system, for example earth, metal, wood, fire, and water, and varying degrees of turbidity and clarity.

[104] *Commentary*: "[The five sounds] *gong*, *shang*, *jiao*, *zhi*, and *yu* mixed together are tones (*yin*); coming forth separately they are called sound (*sheng*)." *Liji zhushu*, 1527.

[105] *Commentary*: "When the sound *gong* is played on one musical instrument, the sound *gong* will resonate on the many instruments. But this is not sufficient to be music. And so it is transformed and made intricate. The *Classic of Changes* states: 'Similar sounds resonate; similar *qi* 氣 seek each other.'" Ibid., 1527.

[106] *Commentary*: "On the upper Pu River, there is a space in the mulberry grove. The sound of a nation in ruin emanates from the river there. In times past, Emperor Zhou of Yin 殷紂 ordered his musician Yan to compose decadent music. Finished, he drowned himself in the Pu

in discord, the five elements are thus also in discord, and the demise of society is imminent.

Music is the coordination of tones and their combination with various forms of dance.[107] Music, then, is a code of behavior similar to the rites: it is part of the political system. Beasts are capable of understanding sounds only; subjects of the state are capable of understanding tones only; but the superior man can understand music.

Music, unlike sound and tones, is not reflective of the state of the individual or of the society: it is the production of a code that orders society. It is the construction of the political system, and performs a political function through the regulation of society.

The need for regulation through music is then based on this theory. The feelings of the mind are reactions to the objects that it comes in contact with; they are not constant, and not determined by human nature (which is originally still). With increasing exposure to the external reality, knowledge increases and desires multiply. After this likes and aversions are formed. But the ability of external reality to excite the mind is unlimited, and if these likes and aversions are not moderated within the inner mind, the mind will be seduced by the external, leading to the degeneration of man by objects. This degeneration results in the destruction of the principles of heaven, and the appearance of all the ills of society (deceit, rebellion, violence). This is the danger to which the theory of the *Record of Music* offers a political solution. The sovereign must prudently moderate the states of mind of the subjects, using rituals (*li*), music (*yue*), punishments (*xing*), and law (*zheng*).

Thus the monarch must intervene politically: following the example of the early sages, he is careful to moderate the minds of the subjects through controlling that which will affect them. The monarch uses the regulation of music to create harmony, rites to enforce differentiation, government to guide their will, and punishments to prevent transgressions. The rites and music are not for sensual pleasure, but to educate the subjects in the moderation of their desires and aversions, guiding them back to the correct path and creating social stability. Rites, music, government, and punishments are the ultimate unity; when all are attained, the harmonious universe is attained. The world can be governed only when the minds of the subjects are united.

Music and rites are thus complementary in governance: while the rites enforce differentiation of the vulgar and the refined, music harmonizes through common emotions. The differentiation results in mutual respect and thus preserves the existing order without conflict; the unification of music allows for closeness and thus eases resentment. The dominance of one over the other produces instability; the synthesis of rites and music is the synthesis of order and emotion. Music harmonizes heaven and earth; the rites order it. The synthesis of the differentiation

---

River. Later the musician Juan 涓 passed there. In the night he heard it and recorded it. He later performed it for Duke Ping of Jin 晉平公." *Liji zhushu*, 1528.

[107] "Juxtaposing melodies and performing [tones], and then coordinating this with shields, axes, feathers and tassels is termed music (yue 樂)." Ibid., 1527.

of rites and the harmonizing of music, thus creating an ordered universe, then becomes a natural law that provides theoretical support for the state.

Recovery of Ritual of Antiquity

During the latter part of the Ming Dynasty (1368–1644), scholar-officials sought to recover the institutions and ritual order of early China as solutions to the perceived decline of the moral order.[108] A central feature of the intellectual context of this period was the civil examination system. The received historiography— following the polemics of Chinese Westernizers after the Opium Wars—has often viewed this examination system as little more than oppressive. Work by Benjamin Elman has shown, however, that the imperially sanctioned power projected in the civil examinations was not merely a tool for reproducing political orthodoxy and social hierarchies,[109] but was also productive in enforcing competence among elite literati in Confucian moral philosophy, statecraft, literary writing, and also "natural studies." On the basis of his comprehensive analysis of extant Ming dynasty civil examinations, Elman argues that although questions on "natural studies" were less frequent than some of the other categories, these questions appeared often enough during the Ming to force between 50,000 and 100,000 candidates for each triennial administration of the provincial examinations to gain technical competence in astronomy, calendrics, and music.[110] The result at the elite level of society was, Elman argues, a scholarly reorientation that incorporated "natural studies" into Confucian "broad learning." The growth of the examination system, then, contributed to proficiency in astronomy, music, and mathematics by the scholar-officials of the period. And along with memorials on topics ranging from military policy and taxation to the moral order, solutions to the problems of calendrics and ritual music were offered in bids for imperial favor.[111]

Proposals in the 1530s for the reform of ritual music—used to reinforce the symbolic order of the state (as in Europe)[112]—have been described in some detail in Joseph Lam's study of the execution of the powerful grand secretary Xia Yan

---

[108] The best account of this perceived decline is Huang, *1587, a Year of No Significance*. For an analysis of attempts to recover ritual order, see Kai-wing Chow, *The Rise of Confucian Ritualism in Late Imperial China: Ethics, Classics, and Lineage Discourse* (Stanford, CA: Stanford University Press, 1994).

[109] Benjamin A. Elman, *A Cultural History of Civil Examinations in Late Imperial China* (Berkeley: University of California Press, 2000).

[110] Elman argues that such questions became infrequent not during the Ming but during the Qing dynasty.

[111] For an analysis of proposals in 16th-century China to reform the calendar, see Peterson, "Calendar Reform"; see also Benjamin A. Elman, "Imperial Politics and Confucian Societies in Late Imperial China: The Hanlin and Donglin Academies," *Modern China* 15 (1989): 379–418.

[112] Keith Pratt, "Art in the Service of Absolutism: Music at the Courts of Louis XIV and the Kangxi Emperor," *Seventeenth Century* 7 (1992): 83–110.

(1482–1548).[113] Xia Yan had proposed in 1530 to implement ancient sericultural ceremonials, and it was in this context that Zhang E presented solutions to the problems of ritual music: "Zhang E's proposal to measure the *huangzhong* pitch with ether [*qi*], to make the large special bell (*tezhong*), and to enforce certain musical changes were attempts to attain the perfect ritual music, which would allow communication among and coordinate human beings, natural elements, and supernatural forces, and would demonstrate the authority and power of Shizong," the Ming emperor.[114] Zhang E was banished from the court in 1536. Lam suggests that ultimately "Zhang E's efforts were, however, more ideological than practical. Given the limitations in his theories and technological data, he did not and could not resolve the musical ambiguities concerning pitch measurement, musical instruments, performance practices, and other musical matters. Zhang E's solutions were only as valid or invalid as others."[115] It was precisely this set of questions that Zhu Zaiyu addressed in his work; the nature of these debates suggests the reasons for the solutions that Zhu offered.

Zhu Zaiyu's Complete Collection of Music and Pitch

Zhu Zaiyu was the heir apparent of the sixth prince of Zheng 鄭 (Huaiqing fu 懷慶府, in present-day Henan Province), a post held by his father Zhu Houhuan 朱厚烷 (1518–1591).[116] At age 14, his father was deposed after being accused of treason by a cousin and sent to prison in Fengyang (present-day Anhui Province). Accounts describe Zhu Zaiyu as living in poverty in a simple hut outside the estate and devoting himself to learning during the period of his father's imprisonment. His work was presented to the imperial court in memorials urging reform in astronomy and music.

Secondary research works written on Zhu have focused on his studies of music,[117] and the asserted priority of his discovery of equal temperament of

---

[113] Lam, *State Sacrifices and Music*; idem, "Ritual and Musical Politics in the Court of Ming Shizong," in *Harmony and Counterpoint: Ritual Music in Chinese Context*, ed. Bell Yung, Evelyn S. Rawski, and Rubie S. Watson (Stanford, CA: Stanford University Press, 1996), 35–53. Lam's account is based primarily on Zhang Juzheng's *Da Ming Shizong Shuhuangdi Shilu*.

[114] Lam, "Ritual and Musical Politics," 52.

[115] Lam, "Ritual and Musical Politics," 52. Lam argues, too relativistically perhaps, "Third, due to the absence of definitive and objective answers, court citizens discussing state ritual and music could advance only what they psychologically and intellectually considered 'right'" (52). Here Lam cites Lucian W. Pye, *The Spirit of Chinese Politics* (Cambridge: Harvard University Press, 1968), 12–35.

[116] For biographical sources in English on Zhu, see "Chu Tsai-yü," in L. Carrington Goodrich and Chaoying Fang, eds., *Dictionary of Ming Biography, 1368–1644*, 2 vols. (New York: Columbia University Press, 1976), s.v.

[117] The most comprehensive work on Zhu is Dai, *Zhu Zaiyu*; this work contains a bibliography of secondary works on Zhu. The most important source in Western languages is Kenneth Robinson, *A Critical Study of Chu Tsai-Yü's Contribution to the Theory of Equal Temperament*

the musical scale.[118] The majority of Zhu's extant works, however, are related to ritual aspects of music, dance, and calendrics, submitted to the court as proposed solutions to the problems of ritual. These works were collected together and published in his *Complete Collection of Music and Pitch* (*Yuelü quanshu* 樂律全書),[119] which was itself presented to the court. Because of this and his status, the *Complete Collection of Music and Pitch* seems to have been widely disseminated: close to one hundred copies are recorded in lists of books presented to the editors of the *Complete Collection of the Four Treasuries* (*Siku quanshu* 四庫全書).[120] Many of the rare Ming editions listed are still extant in libraries throughout China.[121]

The treatises in Zhu's *Complete Collection of Music and Pitch* cover a variety of topics related to the recovery of the institutions and rituals of the ancients. It includes works on calendrics (examples include the *Wan nian li bei kao* 萬年曆備考, *Sheng shou wan nian li* 聖壽萬年曆, and *Lü li rong tong, yin yi* 律曆

---

in Chinese Music (Wiesbaden: Steiner, 1980). For a recent comparative perspective, see Gene J. Cho, *The Discovery of Musical Equal Temperament in China and Europe in the Sixteenth Century* (Lewiston, NY: E. Mellen Press, 2003). See also Fritz A. Kuttner, "Prince Chu Tsai-Yü's Life and Work: A Re-Evaluation of His Contribution to Equal Temperament Theory," *Ethnomusicology* 19 (1975): 163–206; Joseph Marie Amiot and Pierre-Joseph Roussier, *Mémoire sur la musique des Chinois* (Genève: Minkoff Reprint, 1973); and Hermann L. F. Helmholtz, *On the Sensations of Tone as a Physiological Basis for the Theory of Music* (Cambridge: Cambridge University Press, 2009).

[118] For the debate over priority of the discovery of equal temperament, see Robinson, *Chu Tsai-Yü's Contribution*, 2–3; for criticism of Needham and Kuttner, see Dai, *Zhu Zaiyu*, 303ff.

[119] Zhu's *Yuelü quanshu* is reprinted in the *SKQS, CJCB*, and *WYWK*. There is a rare Ming edition in the Beijing Library, reprinted as *Beijing tushuguan guji zhenben congkan* 北京圖書館古籍珍本叢刊 4, ed. Beijing tushuguan guji chuban bianji zu 北京圖書館古籍出版編輯組 (Beijing: Shumu wenxian chubanshe 書目文獻出版社, [1988?]), which is the edition I have used here. See Appendix A, "Zhu Zaiyu's *New Theory of Calculation*," on page 271 for a more complete description of editions of the *Yuelü quanshu*.

[120] See *Siku cai jin shumu* 四庫採進書目 [Bibliography of books presented for the Four Treasuries], ed. Wu Weizu 吳慰祖 (Beijing: Shangwu yinshuguan 商務印書館, 1960), originally published as *Ge sheng jin cheng shumu* 各省進呈書目 (Beijing: Shangwu yinshuguan, 1960). Among the 61 bibliographies of books presented to the *Siku* editors from various provinces and individuals (some bibliographies combine contributions from more than one individual), the following bibliographies list copies of Zhu's works: *Liang Huai yan zheng Li cheng song shumu* 兩淮鹽政李呈送書目, 46 copies of *Yue [lü quan] shu* 樂書十五種〔明朱載堉〕四十六本 (p. 57); *Zhejiang sheng di wu ci Pu shu ting song shumu* 浙江省第五次曝書亭送書目, 34 copies of *Yuelü quanshu* 樂律全書〔明鄭世子載堉著〕三十四本 (p. 115); *Shandong xunfu di er ci cheng jin shumu* 山東巡撫第二次呈進書目 lists several separate *juan* from *Yuelü quanshu*, including one copy of *Yue xue suanxue xin shuo* 樂學算學新說 (p. 150); *Henan sheng cheng song shumu* 河南省呈送書目, one copy of *Yue xue xin shuo Suanxue xin shou* 樂學新說一卷算學新說一卷 (p. 156); *Shanxi sheng cheng song shumu* 陝西省呈送書目 lists *Yuelü quanshu* 樂律全書 but does not record the numbers of copies of any of the books in the bibliography (p. 158); *Du cha yuan fu du yu shi Huang jiao chu shumu* 都察院副都御史黃交出書目, lists 17 copies of 樂律全書四十二卷〔明朱載堉著〕十七本 (p. 176); *Zhejiang cai ji yi shu zong lu jian mu* 浙江採集遺書總錄簡目 lists a 37 juan edition of the *Yuelü quanshu* (p. 245).

[121] See *Zhongguo congshu zonglu* 中國叢書綜錄 [General catalogue of Chinese collectanea], ed. Shanghai tushuguan 上海圖書館 (Shanghai: Shanghai guji chubanshe 上海古籍出版社, 1986), 1:1044–45.

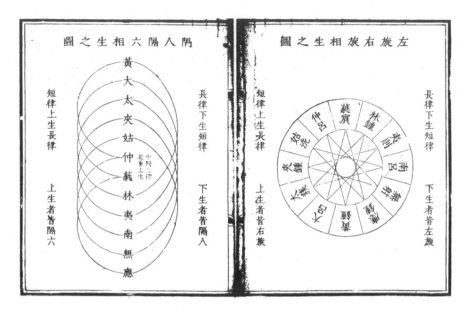

Fig. 4.8: Two diagrams showing the transformations of the tones from Zhu Zaiyu's *Complete Collection of Music and Pitch.*

融通、音義). The majority of the works are on topics related to his analysis of ritual music of antiquity: *New Explanation of Music Theory* (*Yue xue xin shuo* 樂學新說); *New Explanation of the Theory of Pitch* (*Lü xue xin shuo* 律學新說); *The Twelve Pitches: Essence and Meanings, Inner and Outer Chapters* (*Lü lü jing yi nei bian* 律呂精義內編, *Lü lü jing yi wai pian* 律呂精義外篇); *Scores for Music for Two-Tone Bells* (*Xuan gong he yue pu* 旋宮合樂譜), *Scores for Poetry and Music at Provincial Banquets* (*Xiang yin shi yue pu* 鄉飲詩樂譜), *Scores for Ancient Music and Dances with Silk* (*Cao man gu yue pu* 操縵古樂譜), *Scores for the Lesser Dances and Provincial Music of the Six Dynasties* (*Liu dai xiao wu xiang yue pu* 六代小舞鄉樂譜), and *Scores for Lesser Dances and Provincial Music* (*Xiao wu xiang yue pu* 小舞鄉樂譜). Zhu's works also include philosophical analyses of music, such as his "General Summary of Song scholar Zhu Xi's Discussion on Dance" (*Song ru Zhu Xi lun wu da lue* 宋儒朱熹論舞大略).[122]

Several examples from his illustrations will suffice to show the range of Zhu's work on the recovery of ritual. He investigated the transformations of tones (Figure 4.8) and the relation of tones to the hexagrams of the *Classic of Changes* (Figure 4.9 on the facing page). He presents scores for music (Figure 4.10 on page 122), attempts to reconstruct instruments described in the classics (Figure 4.11 on page 123), illustrations of the proper arrangements of dancers (Figure 4.12 on page 124), proper movements (Figure 4.13 on page 125), and the

---

[122] Zhu, *YLQS*, 747.

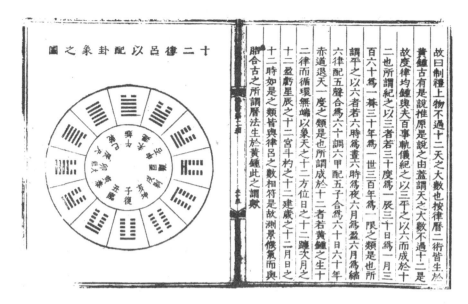

Fig. 4.9: A diagram showing the correlation between the tones and the hexagrams of the *Classic of Changes*, in Zhu's *Complete Collection of Music and Pitch*.

proper steps (Figure 4.14 on page 126). To understand his proposal for the recovery of the tones of antiquity, we must look at his mathematical work.

Precise Solutions to Problems of Ritual: The New Explanation of the Theory of Calculation

To recover the tones that harmonized heaven and earth in antiquity, Zhu proposed mathematical solutions, outlined in his *New Explanation of the Theory of Calculation* (*Suanxue xin shuo* 算學新說, engraved in 1604).[123] Zhu also published several more minor works on mathematics,[124] and an additional musical treatise.

---

[123] Zhu, *Suanxue xin shuo*. The date of engraving, Wanli 31/8/3 萬曆參拾壹年捌月初參日刻完, is from *Suanxue xin shuo*. The date has been the subject of controversy because of the debates over the priority of the discovery of equal temperament.

[124] In addition to the *Suanxue xin shuo yi juan*, extant mathematical works by Zhu Zaiyu not included in the *Yuelü quanshu* include *Zhoubi suanjing tu jie yi juan* 周髀算經圖解一卷 (Zhoubi calculational classic, illustrated and explained), in *Gujin suanxue congshu* 古今算學叢書, hereinafter *GSC*; *Yuan fang gougu tu jie yi juan* 圓方句股圖解一卷, in *GSC*; *Jialiang suanjing san juan* 嘉量算經三卷, *Jialiang suanjing san juan xu yi juan* 嘉量算經三卷序一卷; and *Jialiang suanjing san juan wenda yi juan fanlie yi juan* 嘉量算經三卷問答一卷凡例一卷, in *Xuan yin wan wei bie cang* 選印宛委別藏, *Wan wei bie cang* 宛委別藏.

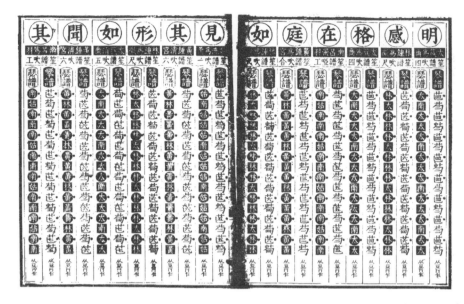

Fig. 4.10: A musical score, one of many included in Zhu's *Complete Collection of Music and Pitch*.

Zhu used geometric ratios instead of fractional proportions to divide the musical scale, in the following manner:

$$1 = 2^{\frac{0}{12}}, 2^{\frac{1}{12}}, 2^{\frac{2}{12}}, \ldots, 2^{\frac{11}{12}}, 2^{\frac{12}{12}} = 2,$$

resulting in the equal-tempered scale. Zhu then calculated the values for the equal-tempered scale to twenty-five significant digits; he also calculated the values of lengths in a base 9 numbering system for music.

Elementary Mathematics

Zhu's *New Explanation of the Theory of Calculation* presents its mathematics beginning from the most elementary definitions. Its contents include the following: a list of the numbers from one to one hundred; a list of the terms for each decimal place for "large numbers"—from 10^0 to 10^{24}, and for "small numbers" from 10^{-1} to 10^{-9} (for even smaller magnitudes, Zhu simply writes them without terms); terms for areas and for volumes; a table for looking up the nearest integral roots, which lists the squares from one to nine, ten to ninety by tens, and one hundred to nine hundred by hundreds; cubes from one to nine, ten to ninety by tens, and one hundred to nine hundred by hundreds; and a discussion on the differences between the base-10 system for measuring musical tones and the base-9 system.

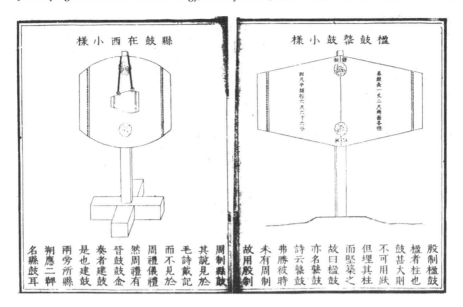

Fig. 4.11: Drums from early China, as reconstructed in Zhu's *Complete Collection of Music and Pitch.*

The River Diagram

This is followed by a short discussion of the metaphysical River Diagram (*He tu* 河 圖 diagram), which relates the five directions, the five phases (earth, metal, wood, fire, and water), the five sounds, and the five elements of the metaphysical system (see Figure 4.15 on page 127):

Five and ten reside in the center, representing earth, *gong*, and the sovereign;

Four and nine reside in the West, representing metal, *shang*, and the ministers;

Three and eight reside in the East, representing wood, *jue*, and the subjects;

Two and seven reside in the South, representing fire, *zhi*, and affairs;

One and six reside in the North, representing water, *yu*, and objects.

五與十居中央為土為宮為君。
四與九居西方為金為商為臣。
三與八居東方為木為角為民。
二與七居南方為火為徵為事。
一與六居北方為水為羽為物。

Fig. 4.12: Illustrations of dancers in Zhu's *Complete Collection of Music and Pitch*.

Calculations to 25 Significant Digits

This is followed by his precision calculations of the values for the equal-tempered system. In the received historiography, as noted above, it has been asserted that "[b]ecause the abacus could only represent a dozen or so digits in a linear array, it was useless for the most advanced algebra until it was supplemented by pen-and-paper notation."[125] One example of Zhu's mathematics is his calculation, using nine abacuses, of the square root of 200 to twenty-five significant digits, or 14.14213562373095048801689 (see Figure 4.16 on page 128).

Contrary to the simplicity of the introductory sections of the text, neither the general method nor the steps for the calculation is explained in detail. The method he uses, in modern terminology, is as follows. To find the square root of x, let r_n be the approximation of the root to the nth place. To find d_n, the value of $(n+1)^{\text{th}}$ place, we seek for the largest integer for which

$$(r_n + d_n)^2 < x.$$

Rearranging terms, this is equivalent to the following formula:

$$d_n(2r_n + d_n) < (x - r_n^2).$$

125 Sivin, "Science and Medicine in Chinese History," 172 (see note 25 on page 82 of this book).

Fig. 4.13: Poses of dancers in Zhu's *Complete Collection of Music and Pitch.*

Place	1	2	3	4	5	6	7	8	9	10	11	12
Remainder	0	0	0	0	1	0	0	7	5	9	0	0
Subtract 6 from the 6th place						−6						
Remainder						4	0	7	5	9	0	0
Subtract 24 from the 7th place						−2	4					
Remainder						1	6	7	5	9	0	0
Subtract 6 from the 8th place								−6				
Remainder						1	6	1	5	9	0	0
Subtract 24 from the 9th place								−2	4			
Remainder						1	5	9	1	9	0	0
Subtract 12 from the 10th place									−1	2		
Remainder						1	5	9	0	7	0	0
Subtract 6 from the 11th place										−6		
Remainder						1	5	9	0	6	4	0
Subtract 9 from the 12th place											−9	
Remainder						1	5	9	0	6	3	1

Table 4.2: Zhu's abacus calculation of the remainder term .0001590631.

The calculations are performed recursively on an abacus. The first term r_1 is simply the first decimal place of the root, in this case 10. Then assume that r_n is given from the previous step. All that remains is to find the largest integer d_n such that the above product $d_n(2r_n + d_n)$ is less than the remainder term: the remainder term $(x - r_n^2)$ has already been calculated from the previous step; $2r_n$ is easily calculated; in general, the term d_n^2 is small; so this essentially amounts to

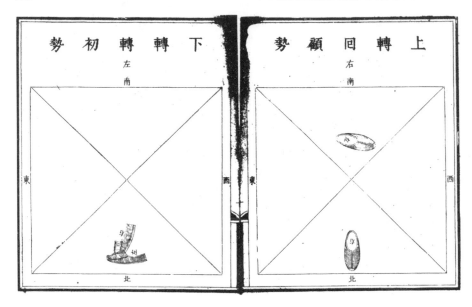

Fig. 4.14: Illustrations of dance steps in Zhu's *Complete Collection of Music and Pitch.*

Algorithm 1 Calculation of the square root of 200 to twenty-five significant digits

---

1: **procedure**
2:     $x \leftarrow 200$ ▷ We want to find the twelfth root of 200
3:     $r[1] \leftarrow 10$ ▷ Our first approximation is 10
4:     **for** $n \leftarrow 1$ **to** 25 **do**   ▷ For each n from 1 to 25
5:         $r \leftarrow r[n]$ ▷ Set r to the previous approximation $r[n]$
6:         **for** $i \leftarrow 0$ **to** 9 **do**   ▷ For each i from 0 to 9
7:             $d \leftarrow (9 - i) \times 10^{-n}$ ▷ Try digits in n^{th} decimal place
8:             **if** $d(2r + d) \leq (x - r^2)$ **then**   ▷ If d is sufficiently small
9:                 $r[n+1] \leftarrow r + d$ ▷ Add d to r to get $r[n+1]$
10:                Next n ▷ Exit for i loop to next n
11:             **end if**
12:         **end for**
13:     **end for**
14:     **return** $r[25]$ ▷ Return value calculated to 25 significant digits
15: **end procedure**

---

dividing d_n into the remainder (possibly leaving extra for d_n^2). Then subtracting $d_n(2r_n + d_n)$ from the remainder term $(x - r_n^2)$ gives the next remainder term— that is, since $r_{n+1} = r_n + d_n$,

$$(x - r_n^2) - d_n(2r_n + d_n) = x - (r_n + d_n)^2 = x - r_{n+1}^2.$$

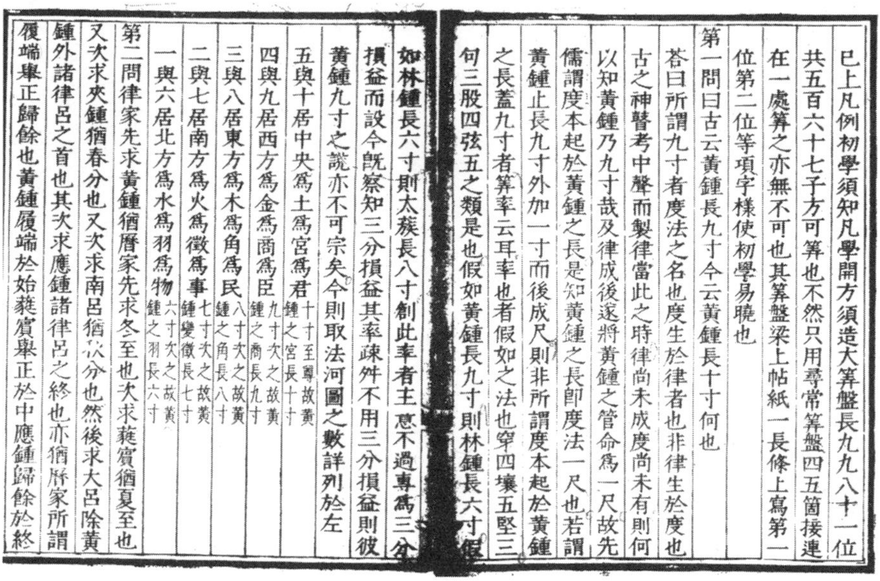

已上凡例初學須知凡學開方須造大筭盤長九九八十一位
共五百六十七子方可筭也不然只用尋常筭盤四五箇接連
在一處筭之亦無不可也其筭盤梁上帖紙一長條上寫第一
位第二位等項字樣使初學易曉也
第一問曰古云黃鍾長九寸今云黃鍾長十寸何也
荅曰所謂九寸者度法之名也度生於律者也非律生於度也
古之神瞽考中聲而製律當此之時律尚未成度尚未有則何
以知黃鍾乃九寸哉及律成後遂將黃鍾之管命爲一尺故先
儒謂度本起於黃鍾之長是知黃鍾之長卽度法一尺也若謂
黃鍾止長九寸外加一寸而後成尺則非所謂度本起於黃鍾
之長蓋九寸者筭率云耳率也者假如之法也穿四壤五堅三
句三股四弦五之類是也假如黃鍾長九寸則林鍾長六寸假
如林鍾長六寸則太蔟長八寸創此率者主意不過專爲三分
損益而設今既察知三分損益其率踈舛不用三分損益則彼
黃鍾九寸之說亦不可宗矣今則取法河圖之數詳列於左
五與十居中央爲土爲宮爲君十寸至尊故黃鍾之宮長十寸
四與九居西方爲金爲商爲臣九寸次之故黃鍾之商長九寸
三與八居東方爲木爲角爲民八寸次之故黃鍾之角長八寸
二與七居南方爲火爲徵爲事七寸次之故黃鍾變徵長七寸
一與六居北方爲水爲羽爲物六寸次之故黃鍾之羽長六寸
第二問律家先求黃鍾猶曆家先求冬至也次求蕤賓猶夏至也
又次求夾鍾猶春分也又次求南呂猶秋分也然後求大呂除黃
鍾外諸律呂之首也其次求應鍾諸律呂之終也亦猶曆家所謂
履端舉正歸餘也黃鍾履端於始蕤賓舉正於中應鍾歸餘於終

Fig. 4.15: Text corresponding to the River Diagram in Zhu's *Complete Collection of Music and Pitch.*

r_n	d_n	$2r_n + d_n$	$d_n(2r_n + d_n)$	$x - r_n^2 - d_n(2r_n + d_n)$
	10			100
10	4	24	96	4
14	.1	28.1	2.81	1.19
14.1	.04	28.24	1.1296	.0604
14.14	.002	28.282	.056564	.003836
14.142	.0001	28.2841	.00282841	.00100759
14.1421	.00003	28.28423	.0008485269	.0001590631
14.14213	.000005	28.284265	.000141421325	.000017641775
14.142135	.0000006	28.2842706	.00001697056236	.00000067121264

Table 4.3: Zhu's calculation of $\sqrt{200}$ to the first 8 significant digits (Zhu continues to 25 significant digits).

This formula can be written as shown in Algorithm 1 on the facing page. The successive values that Zhu computes are shown in Table 4.3. Zhu then calculates the following:

(1) the value for $\frac{\sqrt{2}}{2} = 2^{-\frac{1}{2}}$;

(2) he then finds the square root to calculate the value for $\sqrt{\frac{\sqrt{2}}{2}} = 2^{-\frac{1}{4}}$;

(3) finally, he takes the cube root of that to find the value for $\sqrt[3]{\sqrt{\frac{\sqrt{2}}{2}}} = 2^{-\frac{1}{12}}$.

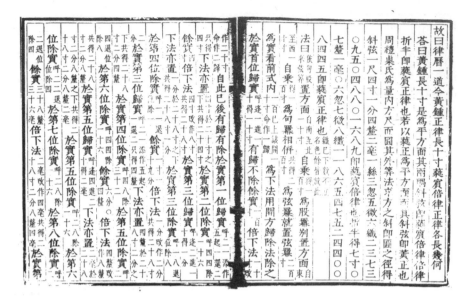

Fig. 4.16: Calculation of the square root of 200 in Zhu's *New Explanation of the Theory of Calculation.*

Conclusions—Zhu's Legitimation of Ritual

Zhu's attempt to legitimate his system of musical equal-temperament depended not just on his invocations of the classics, but also on the availability of the abacus. By placing nine abacuses together, Zhu was able to calculate the lengths of musical pitch-pipes to twenty-five significant digits. In spite of these calculations and his work in the base-nine numbering system, Zhu has rarely been mentioned in research on Ming dynasty mathematics. Zhu might seem ideal for hagiographic reconstruction as an emblem of modernity—discovering mathematical regularities in physical phenomena and calculating them with unprecedented accuracy. However, the teleologies of the secondary literature have counted the degrees of polynomial equations solved and the accuracy of the value of π as transcultural standards of mathematical progress. Zhu Zaiyu's rough approximations of π have often been cited in the historical literature on Chinese mathematics; his calculations to twenty-five significant digits have only rarely been mentioned.

Conclusions

The development of solutions for polynomial equations has seemed to provide a natural teleology, with modern mathematics as its endpoint. Increasingly higher exponents and greater numbers of variables gauged progress, giving a natural direction of growth. More important, this teleology made cross-cultural comparisons of Europe and China possible, reaffirming a golden age in the Song and early Yuan during which the glories of China outshone those of the West. Didactic histories blaming dynastic collapse on Ming intellectual decadence, together with the alleged incomprehension of earlier work on Celestial Origin algebra, the propaganda of Jesuit collaborators, and Western-impact historiography all became narrative components contributing to claims of Ming mathematical decline. These claims were ultimately self-reinforcing: the loss of treatises obviated any need for cataloguing treatises recorded in extant bibliographies; and the lack of mathematical achievements rendered unnecessary any further study.

These histories of mathematics selectively recounted benchmark achievements wrenched from mathematical context and transfigured to fit into anachronistic teleologies never available to the practitioners. In these accounts, with time collapsed, labor effaced, and applications denied, discoveries in mathematics indeed appeared miraculous. Individual genius became the only explanation; hagiographic accounts then linked genius to moral perseverance, correct ideologies, and proper attitudes. These histories (following modern ideologies) separated the context and applications from the knowledges produced; the "unreasonable effectiveness" of mathematics then became a perplexing problem in need of explanation. The ultimate result was a history that rendered the development of mathematics historically unintelligible.

The alternative approach to these histories of mathematics has often seemed to be externalistic macrohistories, studies that too often refuse to examine any of the details of the mathematics they purportedly attempt to explain. One genre has focused on social institutions and the status of mathematicians: in these, social status is viewed not as relational and the site of social contest, but rather as being conferred by society; society then becomes the culpable agent responsible for the low status of mathematicians and thus the failure to gain the benefits (known through hindsight) of mathematical development. Another genre focuses on ideologies: in these, ideological beliefs transparently produce their claimed results. Thus subscription to particular ideologies—whether Marxist objective materialism (*keguan weiwu zhuyi*), naturalistic Daoism, or interest in the "book of nature"—result in scientific development. Correlative thinking, insensitivity to anomalies, idealism, or Confucian philosophy then stunt scientific development. Ultimately, such accounts offer little more than a circular reaffirmation of the ideologies they advocate.

This chapter has attempted to demonstrate that in seeking historical explanations for the constitution of mathematical knowledge, the study of both mathematical techniques and their specific contexts is equally necessary. The analysis of the *Sea Mirror of Circle Measurement* outlined the mathematical, social, and ideological context of the remarkable Celestial Origin algebra. This context also helped make more comprehensible the loss of this mathematical practice—probably transmitted orally, without practical applications, secretively guarded by recluses—a loss that the received historiography has ironically blamed on the very Ming scholars who tried to recover its mysteries.

Cheng Dawei's *Comprehensive Source of Mathematical Methods* suggested the extent of popular interest in mathematics. Indeed, the available historical evidence shows that the late Ming was a period of unprecedented popularization of mathematics: the expansion of printing made mathematical texts more widely available; reforms of tax codes created the increased need for mathematics for surveying; commercial development increased the need for mathematical literacy and proficiency with the abacus. At the elite level of society, the growth of the civil service examinations forced competency in technical subjects on increasingly large numbers of examination candidates. Cheng Dawei's bibliography demonstrates the variety of treatises written in this period and available even outside literati circles. Thus interest in mathematics in the late Ming was scarcely revived by the translations of Western mathematics; rather, this interest in mathematics was itself the context in which the translations were conceived.

Finally, the study of Zhu Zaiyu showed that the most improbable of combinations—the political cosmology of the *Records of Music* (an example of what Needham termed "correlative thinking"), the recovery of ancient musical ritual, attempts at patronage by buttressing conservative political theories, mathematical practices obtained from recluses, and the inelegant abacus of the despised merchant class—were all constitutive elements of his work in mathematics and the sciences, work having many attributes we usually associate with scientific modernity, such as the mathematization of nature, precision calculation, and physical experiment. Zhu happened upon the mathematically tractable problem of the equal temperament of the musical scale, and provided solutions to twenty-five-digit precision.

Chapter 5
Tracing Practices Purloined by the Three Pillars

"TATTERED SANDALS," regrettably, was all that remained of Chinese mathematics, which could just as well be discarded because Western mathematics was in every respect superior—or at least so claimed Xu Guangqi 徐光啟 (1562–1633) in his preface to the *Guide to Calculation in the Unified Script* (*Tong wen suan zhi* 同文算指, 1613).[1] The *Guide to Calculation* was purported to be a translation of Western mathematics, taken from Christoph Clavius's (1538–1612) *Epitome arithmeticae practicae* (1583),[2] together with some problems from Chinese mathematics that, clearly designated as such, were included to demonstrate the alleged superiority of Western mathematics. Xu, Li Zhizao 李之藻 (1565–1630), and Yang Tingyun 杨廷筠 (1557–1627), who collectively are often referred to as the "Three Pillars" (*san da zhushi* 三大柱石) of Catholicism in Ming China, each wrote a preface for the *Guide to Calculation* denouncing contemporary Chinese mathematics while promoting the superiority of Western mathematics.

This chapter will show that the most difficult problems in their "translation" were purloined—without comparison, without analysis, without criticism, and without any mention of the source—from the very Chinese mathematical treatises Xu denigrated. In particular, the problems purloined by the Three Pillars and their collaborators, problems that were known in imperial China as *fangcheng* 方程 (sometimes translated as matrices or "rectangular arrays"), or what we

---

[1] *Tong wen suan zhi* 同文算指 [Guide to calculation in the unified script], in *Zhongguo kexue jishu dianji tonghui: Shuxue juan* 中國科學技術典籍通彙：數學卷 [Comprehensive collection of the classics of Chinese science and technology: Mathematics volumes] (Zhengzhou: Henan jiaoyu chubanshe 河南教育出版社, 1993), hereinafter *ZKJDT*. Authorship is attributed to Matteo Ricci (1552–1610) and Li Zhizao 李之藻 (1565–1630). For an analysis of the *Guide to Calculation*, see Takeda Kusuo 武田楠雄, *Dōbunzanshi no seiritsu* 同文算指の成立 [Inception of the *Guide to Calculation in the Unified Script*] (Tokyo: Iwanami Shoten 岩波書店, 1954). For a recent study of the transmission of Western Learning into China, see Ahn Daeok 安大玉, *Minmatsu Seiyō kagaku tōdenshi: "Tengaku shokan" kihen no kenkyū* 明末西洋科学東伝史：「天学初函」器編の研究 [History of the transmission of Western science to the East in the late Ming: Research on the *Qibian* of the *Tianxue chuhan*] (Tokyo: Chisen Shokan 知泉書館, 2007).

[2] Christoph Clavius, SJ (1538–1612), *Christophori Clavii Bambergensis e Societate Iesv epitome arithmetica practicae* (Romae: Ex Typographia Dominici Basae, 1583).

would now call linear algebra, are arguably the most advanced and recognizably "modern" mathematics in the *Guide to Calculation*. These problems were copied into chapter 5 of the second volume (*tong bian* 通編), without any indication that they were from contemporary Chinese treatises. These problems were then provided a new name, the title of the chapter, "Method for addition, subtraction, and multiplication of heterogeneous [elements]" (*za he jiao cheng fa* 雜和較乘法). Chinese readers of the *Guide to Calculation* could not have known that Clavius's *Epitome* contains no similar problems, but later Chinese commentators added notes remarking that similar problems could be found in Chinese works.

That nineteen *fangcheng* problems were purloined from the very Chinese mathematical texts denounced by Xu as "tattered sandals" suggests that Xu, Li, Yang, and their collaborators did not themselves believe the assertions they presented in their prefaces extolling the superiority of mathematics from "the West." Yet Xu's pronouncements have seemed so persuasive that his claims have been, at least until recently, accepted for the most part by historians—Chinese and Western alike—as fact. We should instead critically analyze the self-serving statements in their prefaces as propaganda designed to promote "Western Learning" (*Xi xue* 西學), and together with it, their own careers in the imperial bureaucracy.

More broadly, in this chapter I argue that extant mathematical treatises from this period should be viewed as epiphenomenal, preserving only fragments of mathematical practices: whereas the previous chapter of this book focused on extant mathematical treatises from the Song, Yuan, and Ming dynasties, this chapter will explore mathematical practices of imperial China by reconstructing those practices from extant mathematical treatises. More specifically, from extant mathematical treatises—which were written, compiled, circulated, and preserved by the literate elite—I reconstruct *fangcheng* mathematical practices. The earliest extant written records of these practices are recorded in Chinese mathematical treatises dating from the first century CE; these mathematical practices continued to be recorded in Chinese mathematical treatises throughout the imperial period. I argue that in general the literate elites who compiled mathematical treatises often had little more than a rudimentary understanding of the mathematical practices they recorded, especially for practices as complex as linear algebra. Using evidence preserved in extant texts, I reconstruct these practices as they were performed—with counting rods on a two-dimensional counting board. I show that fangcheng adepts were able to compute solutions to difficult problems involving *n* conditions in *n* unknowns with only an expertise with counting rods together with an understanding of a few simple two-dimensional patterns. Adepts of these practices, through repetition, would have been able to compute solutions through rote procedural memory, just as modern-day abacus adepts can compute with astonishing speed and accuracy.

A second purpose of this chapter is to question the assumption that the arrival of the Jesuits in Ming China marks the "first encounter" of China and the West. I argue that we reach very different conclusions if we trace the history of mathematical practices, instead of just focusing on the texts that preserve written records of these practices. In particular, tracing the history of *fangcheng* practices leads to the following conclusions: (1) The essentials of the methods used today in "Western" linear algebra—augmented matrices, elimination, and determinantal-style calculations—were known by the first century CE in imperial China. (2) Simple two-dimensional patterns were used to calculate the solutions on the counting board, and in particular, determinantal-style solutions to a special class of distinctive problems. (3) These practices were non-scholarly— they did not require literacy and were not transmitted by texts. (4) Some of these practices spread across the Eurasian continent and are recorded in texts from as early as the thirteenth century in Italy. In other words, these practices were not confined by the boundaries we anachronistically term "civilizations." These practices thus circulated across Eurasia long before the Jesuits traveled to China.

These conclusions can help us begin to rewrite the history of the Jesuits in China as one episode in a long history of global circulations (as opposed to a "first encounter" of two "civilizations," "China" and "the West"). It can also help us rethink one of the fundamental assumptions of the history of science—an assumption found in conventional histories of (Western) science, in Needham's *Science and Civilisation in China*, and in more recent comparative and cultural histories—that science in the premodern period was somehow "local" and trans- mitted by texts. The broader goal of this chapter then is to propose an alternative approach to the history of science, one that focuses on practices instead of texts, microhistory instead of macrohistory, and goes beyond "civilizations" toward a world history of science.

This chapter is divided into six sections. The first section uses extant written records to reconstruct *fangcheng* as a mathematical practice on the counting board, summarizing research from my recent book on *fangcheng*,[3] in order to demonstrate how adepts of this practice could quickly solve complex mathe- matical problems with n conditions in n unknowns with little more than facility with counting rods and the rote application of simple patterns on the counting board. The second section inquires into the provenance of extant written records of *fangcheng* practices. Written records of *fangcheng* practices are preserved in treatises on the mathematical arts, which were compiled by aspiring literati and presented to the imperial court as essential to ordering the empire. These literati understood only the rudiments of these practices, yet they derided methods used by adepts as arcane. The third section presents a summary of determinantal-style calculations and solutions to *fangcheng* problems, solutions seemingly so arcane that they were only rarely recorded by the literati who compiled mathematical treatises. The fourth section presents evidence that these arcane determinantal- style calculations and solutions—which are so distinctive that they can serve as "fingerprints"—circulated across the Eurasian continent, to be recorded in the works of Leonardo Pisano (c. 1170–c. 1250), more commonly known today by the name Fibonacci. The fifth section analyzes how *fangcheng* problems were pur- loined by Xu, Li, Yang, and their collaborators to support their claim that Western mathematics was in every way superior to Chinese mathematics. The concluding section outlines an alternative approach to the world history of science.

---

[3] Roger Hart, *The Chinese Roots of Linear Algebra* (Baltimore: Johns Hopkins University Press, 2011).

The *Fangcheng* Procedure as a Mathematical Practice

Today, linear algebra is one of the core courses in modern mathematics, and its main problem is the solution of systems of n linear equations in n unknowns.[4] Determinants, through what is known as Cramer's rule, offer an elegant solution for simple systems of linear equations; Gaussian elimination is the more general solution, and arguably the most important of all matrix algorithms.[5] The earliest extant records of these two approaches to solving linear equations can be found in extant mathematical treatises from imperial China.[6] This section will focus on the earliest records of the procedure we now call Gaussian elimination (determinantal calculations and solutions will be examined in the third section of this chapter).

The earliest known records of what we now recognize as Gaussian elimination are found in the *Nine Chapters on the Mathematical Arts* [*Jiuzhang suanshu* 九章算術],[7] in the eighth of those chapters, which is titled "*Fangcheng*" 方程. In that chapter, 18 problems are recorded as word problems, without diagrams. These problems range from two conditions in two unknowns to five conditions in five unknowns; there is also one problem, problem 13, with five conditions in six unknowns, which I will refer to as the "well problem."[8] The *Nine Chapters* stipulates that all 18 problems are to be solved by the *fangcheng* procedure (*fangcheng shu* 方程術), a variant of which we now call Gaussian elimination.[9] This section will illustrate how the *fangcheng* procedure, as reconstructed from written records preserved in the *Nine Chapters*, was a mathematical practice that used simple patterns on a counting board to produce solutions to what we would now call systems of linear equations with n conditions in n unknowns, in a manner similar to what we now call Gaussian elimination.

First, however, we must distinguish between written records of the *fangcheng* procedure preserved in extant Chinese mathematical treatises and the *fangcheng* procedure itself as a mathematical practice. As a mathematical practice, the *fangcheng* procedure was performed on a two-dimensional counting board with

---

[4] Gilbert Strang, *Linear Algebra and Its Applications*, 4th ed. (Belmont, CA: Thomson, Brooks/Cole, 2006).

[5] G. W. Stewart, *Matrix Algorithms* (Philadelphia: Society for Industrial and Applied Mathematics, 1998); Strang, *Linear Algebra and Its Applications*.

[6] Hart, *Chinese Roots of Linear Algebra*.

[7] *Jiuzhang suanshu* 九章算術 [Nine chapters on the mathematical arts], in *Yingyin Wenyuan ge Siku quanshu* 影印文淵閣四庫全書 [Complete collection of the Four Treasuries, photolithographic reproduction of the edition preserved at the Pavilion of Literary Erudition] (Hong Kong: Chinese University Press, 1983–1986), hereinafter *SKQS*.

[8] I thank Jean-Claude Martzloff for suggesting this name. A version of the well problem is presented below, on pages 182ff. See also Hart, *Chinese Roots of Linear Algebra*, chapter 7, "The Well Problem."

[9] Shen Kangshen, Anthony W.-C. Lun, and John N. Crossley, *The Nine Chapters on the Mathematical Art: Companion and Commentary* (New York: Oxford University Press, 1999); Karine Chemla and Guo Shuchun, *Les neuf chapitres: Le classique mathématique de la Chine ancienne et ses commentaires* (Paris: Dunod, 2004); Hart, *Chinese Roots of Linear Algebra*.

counting rods. Numbers were recorded using differing arrangements of counting rods. The Chinese numeral-rod system is not difficult to understand: the first position is for numbers 0 to 9 (0 is denoted by an empty space); 1 to 5 are denoted by the corresponding numbers of rods placed vertically, and a horizontal rod denoting 5 is placed above one to four vertical rods to denote 6 to 9. The next position to the left is for tens, which are placed horizontally. Then hundreds, and so on. A summary of the representation of numbers by counting rods is shown in Table 5.1.

Table 5.1: Chinese rod numerals for one through nine, and ten through ninety; zero is denoted by a blank space, as indicated here for numbers 10 through 90.

1	2	3	4	5	6	7	8	9
│	‖	‖‖	‖‖‖	‖‖‖‖	丅	丅丅	丅丅丅	丅丅丅丅
10	20	30	40	50	60	70	80	90
—	=	≡	≣	≣	⊥	⊥	⊥	⊥

Thus, for example, the number 721 is arranged on the counting board as $\overline{\rm II} = \rm I$.

The counting board was a powerful tool for the computation of solutions to mathematical problems ranging from addition, subtraction, multiplication, and division to the extraction of roots and the solution of systems of linear equations, namely, n conditions in n unknowns.[10] Although there are no modern practitioners, it seems reasonable to assume that counting rod adepts in imperial China had skills similar to those of modern abacus adepts. With practice, abacus calculations are fast, efficient, and virtually effortless, utilizing what is sometimes called motor learning or procedural memory, in which repetitive practice results in long-term muscle memory allowing calculations to be performed with little conscious effort.[11] Modern abacus adepts can calculate very quickly; some become so proficient that they use only procedural memory and do not require a physical abacus. It seems reasonable to assume that counting rods were similarly fast, efficient, and virtually effortless for accomplished adepts.

Expertise on a counting board, as with the abacus, did not require literacy, that is, the ability to read or write classical Chinese, the language of the elite, which was often acquired by memorizing the Confucian classics. Indeed, the

---

[10] For background on the history of Chinese mathematics, the two standard references are Jean-Claude Martzloff, *A History of Chinese Mathematics* (New York: Springer, 2006) and Li Yan and Du Shiran, *Chinese Mathematics: A Concise History*, trans. John N. Crossley and Anthony W.-C. Lun (Oxford: Clarendon Press, 1987). For background on *fangcheng*, see Hart, *Chinese Roots of Linear Algebra*; Shen, Lun, and Crossley, *Nine Chapters*. On counting board operations, see also Lay-Yong Lam and Tian-Se Ang, *Fleeting Footsteps: Tracing the Conception of Arithmetic and Algebra in Ancient China*, rev. ed. (River Edge, NJ: World Scientific, 2004).

[11] Common examples of this kind of motor learning or procedural memory include typing, playing a piano, and various sports, to name only a few.

two skills were likely for the most part mutually exclusive: it seems unlikely that many with the opportunity and status that literacy afforded would have spent the considerable time required to become expert at counting rod calculations; and for the illiterate, the use of counting rods can be thought of as a simple "language game" that provided a means for writing numbers and performing calculations in a particularly simple and intuitive manner.[12]

The only early records of *fangcheng* practices that have survived are extant mathematical treatises written in linear, one-dimensional narrative form, in classical Chinese: there are no two-dimensional diagrams in these early treatises; no other known sources, for example charts or diagrams, have survived; and in addition, as noted above, there are no modern practitioners. Early extant treatises provide few details: only brief, cryptic, and sometimes corrupt statements have been preserved. For example, this section will focus on problem 18 from "*Fangcheng*," chapter 8 of the *Nine Chapters*. The original text of the *Nine Chapters* preserves a written record of only the statement of the problem, the values found for the solution, and the instruction that the problem is to be solved by the *fangcheng* procedure together with the procedure for positive and negative numbers:[13]

Now it is given: 9 *dou*[14] of hemp, 7 *dou* of wheat, 3 *dou* of legumes, 2 *dou* of beans, and 5 *dou* of millet, [together] are valued at 140 coins; 7 *dou* of hemp, 6 *dou* of wheat, 4 *dou* of legumes, 5 *dou* of beans, and 3 *dou* of millet, [together] are valued at 128 coins; 3 *dou* of hemp, 5 *dou* of wheat, 7 *dou* of legumes, 6 *dou* of beans, and 4 *dou* of millet, [together] are valued at 116 coins; 2 *dou* of hemp, 5 *dou* of wheat, 3 *dou* of legumes, 9 *dou* of beans, and 4 *dou* of millet, [together] are valued at 112 coins; 1 *dou* of hemp, 3 *dou* of wheat, 2 *dou* of legumes, 8 *dou* of beans, and 5 *dou* of millet, [together] are valued at 95 coins. It is asked: What is the value of one *dou* [for each of the grains]? Answer: 7 coins for one *dou* of hemp; 4 coins for one *dou* of wheat; 3 coins for one *dou* of legumes; 5 coins for one *dou* of beans; 6 coins for one *dou* of millet. Procedure: Follow *fangcheng*; use the procedure for positives and negatives.

今有麻九斗、麥七斗、菽三斗、荅二斗、黍五斗，直錢一百四十 ；麻七斗、麥六斗、菽四斗、荅五斗、黍三斗，直錢一百二十八 ；麻三斗、麥五斗、菽七斗、荅六斗、黍四斗，直錢一百一十六 ；麻二斗、麥五斗、菽三斗、荅九斗、黍四斗，直錢一百一十二 ；麻一斗、麥三斗、菽二斗、荅八斗、黍五斗，直錢九十五。問一斗直幾何。答曰 ：

---

[12] On "language games," see Ludwig Wittgenstein, *Philosophical Investigations*, trans. G. E. M. Anscombe (Oxford: Blackwell, 1958). For information on counting rods, see Alexei Volkov, "Mathematics and Mathematics Education in Traditional Vietnam," in *The Oxford Handbook of the History of Mathematics*, ed. Eleanor Robson and Jacqueline A. Stedall (Oxford: Oxford University Press, 2009), 153–76.

[13] *Jiuzhang suanshu, juan* 8, 18b–19a; Chemla and Guo, *Neuf chapitres*, 649–51; Shen, Lun, and Crossley, *Nine Chapters*, 426–38.

[14] A unit of volume, equal to ten *sheng* 升. Currently, one *dou* is equal to one decaliter.

麻一斗七錢，麥一斗四錢，菽一斗三錢，荅一斗五錢，黍一斗六錢。
術曰：如方程，以正負術入之。

Although only fragmentary written records remain, these records are suffi-cient to reconstruct at least the main features of the *fangcheng* procedure as a mathematical practice. That is, although the original text of the *Nine Chapters* preserves only an outline of the *fangcheng* procedure, as applied to a simpler example, problem 1, there is enough information preserved in the *Nine Chapters* to reconstruct the procedure itself in its general form. This section will use this reconstruction of the *fangcheng* procedure to show how, in addition to basic counting rod operations, all that was required to solve complex linear algebra problems such as problem 18 was knowledge of a few simple patterns on the two-dimensional counting board.

Fangcheng problems were arranged using counting rods in a rectangular array, in a manner identical to the augmented matrix familiar from modern linear algebra, if we allow for differences in writing. The following diagram shows how problem 18 is laid out using counting rods on the counting board:

I	II	III	⫟	⫟̄
III	IIII	IIII	⊤	⫟
II	II	⫟	IIII	II
⫟	IIĪ	⊤	IIII	II
IIII	III	III	II	IIII
⊥IIII	I−II	I−⊤	I=⫟	I≡

The only difference between this and the modern mathematical notation is the orientation of the array, which corresponds to differences in writing (writing in imperial China proceeded from top to bottom, then right to left; modern English is written from left to right, then top to bottom). Written in modern mathematical notation as an augmented matrix, we have the following:

$$\begin{bmatrix} 9 & 7 & 3 & 2 & 5 & 140 \\ 7 & 6 & 4 & 5 & 3 & 128 \\ 3 & 5 & 7 & 6 & 4 & 116 \\ 2 & 5 & 3 & 9 & 4 & 112 \\ 1 & 3 & 2 & 8 & 5 & 95 \end{bmatrix} \tag{5.1}$$

This problem can also be written out as a system of n linear equations in n unknowns, which is probably the form most familiar to modern readers:

$$9x_1 + 7x_2 + 3x_3 + 2x_4 + 5x_5 = 140 \tag{5.2}$$

$$7x_1 + 6x_2 + 4x_3 + 5x_4 + 3x_5 = 128 \tag{5.3}$$

$$3x_1 + 5x_2 + 7x_3 + 6x_4 + 4x_5 = 116 \tag{5.4}$$

$$2x_1 + 5x_2 + 3x_3 + 9x_4 + 4x_5 = 112 \tag{5.5}$$

$$x_1 + 3x_2 + 2x_3 + 8x_4 + 5x_5 = 95. \tag{5.6}$$

It should be noted, however, that in modern linear algebra, the simpler augmented matrix (5.1), which corresponds to the counting board representation,
is much preferred over the more cumbersome use of x_n to denote the unknowns
(that is, in this case, x_1, x_2, x_3, x_4, x_5) in equations (5.2–5.6). In the following analysis, I will use the preferred modern notation, the augmented matrix.

The Fangcheng *Procedure as Simple Visual Patterns*

Solving *fangcheng* problems requires as a prerequisite a knowledge of counting board operations for addition, subtraction, multiplication, and division of
integers (including negative numbers). It also requires a knowledge of the corresponding operations for fractions, although fractions rarely appear. Given knowledge of these operations, the *fangcheng* procedure is perhaps best understood as
patterns on the counting board that are applied repeatedly in order to solve the
problem. Extant treatises provide specific names for only two of these patterns:
(1) Cross-multiplication (*biancheng* 徧乘), the multiplication of an entire column
of entries by one entry of another column; (2) Term-by-term subtraction (*zhi chu*
直除, which literally means "directly subtract"),[15] the term-by-term subtraction
of the entries of one column from those of another. These patterns are difficult to
explain clearly in words; they are perhaps most easily understood by observing
operations on the counting board. In addition, there are other simple patterns,
which are not provided names, that are even more difficult to explain in words.
These patterns are illustrated below.

STEP 0.   Laying out the array. Before solving problem 18, we must first represent
it as an array of numbers on the counting board (as shown previously on page
137).

STEP 1.   Cross-multiplication (*biancheng* 徧乘). Following the *fangcheng* procedure, the first step is to cross-multiply, that is, to multiply each entry in the
second column by the first entry of the first column. More specifically, the column
$(7, 6, 4, 5, 3, 128)$ is multiplied by 9, as is indicated in the following diagram:[16]

---

[15] Alternatively, Martzloff translates this as "direct reduction." Martzloff, *History of Chinese Mathematics*, 253.

[16] The counting board, as shown in this diagram, and all of the following diagrams, was used
to display the results of the calculations. In practice, the addition, multiplication, subtraction,
or division of two individual entries might have been calculated in the head or by procedural
memory; but it is also possible that recording intermediary results might have been necessary.
Written records of *fangcheng* practices do not describe these details. An additional space,
perhaps off to one side of the counting board area, might have been used as temporary registers.
For example, if two numbers being multiplied together are large (here they are not), then
using counting rods to calculate the result might have required a third register to store the
intermediary results. In this diagram, and the diagrams that follow, I have diagrammed only the
results, such as here, the multiplication of entries. I have not shown the intermediary counting
board operations involved in each individual multiplication of two entries.

Written in modern mathematical notation, this is simply the augmented matrix modified by what is called a row operation, that is, by multiplying the entire second row by the pivot in the first row, as shown in the augmented matrix below:

$$\begin{bmatrix} \boxed{9} & 7 & 3 & 2 & 5 & 140 \\ \boxed{7} & \boxed{6} & \boxed{4} & \boxed{5} & \boxed{3} & \boxed{128} \\ 3 & 5 & 7 & 6 & 4 & 116 \\ 2 & 5 & 3 & 9 & 4 & 112 \\ 1 & 3 & 2 & 8 & 5 & 95 \end{bmatrix}$$

The result of the multiplication is $9(7, 6, 4, 5, 3, 128) = (63, 54, 36, 45, 27, 1152)$, which, on the counting board, is arranged as follows:

In the augmented matrix, we write the result as follows:

$$\begin{bmatrix} 9 & 7 & 3 & 2 & 5 & 140 \\ 63 & 54 & 36 & 45 & 27 & 1152 \\ 3 & 5 & 7 & 6 & 4 & 116 \\ 2 & 5 & 3 & 9 & 4 & 112 \\ 1 & 3 & 2 & 8 & 5 & 95 \end{bmatrix}$$

STEP 2.   Cross-multiplication. The second step is another cross-multiplication, which is similar to the previous step. We multiply the entire first column by the first entry in the second column, that is, we multiply $(9, 7, 3, 2, 5, 140)$ by 7, as follows:[17]

---

[17] Again, written records of *fangcheng* practices do not fully explain the arrangement on the counting board of all the counting rods that might have been necessary at this point. In this cross-multiplication, we use the original entry in the first row of the second column, namely 7, to multiply each of the entries in the first column. I have therefore restored the original entry, 7, to show this multiplication. In the next step, the term-by-term subtraction of one column from

丨	丆	丗	丁		𝍤
川	𝍥	𝍥	𝍤 川		丁
丨	川	丁	丌 丁		丗
丅	川	丅	𝍥 川		川
𝍥	川	川	丌 丅		𝍥
𝍤 𝍥	丨 − 川	丨 − 丁	− 丨 𝍤 川		丨 𝍤

In the augmented matrix, this multiplication is represented as follows:

$$
\begin{bmatrix}
\boxed{9} & \boxed{7} & \boxed{3} & \boxed{2} & \boxed{5} & \boxed{140} \\
\boxed{7} & 54 & 36 & 45 & 27 & 1152 \\
3 & 5 & 7 & 6 & 4 & 116 \\
2 & 5 & 3 & 9 & 4 & 112 \\
1 & 3 & 2 & 8 & 5 & 95
\end{bmatrix}
$$

The result is then $7(9,7,3,2,5,140) = (63,49,21,14,35,980)$, which is displayed on the counting board as follows:

丨	川	丗	⊥ 川		⊥ 川
川	𝍥	𝍥	𝍤 川		𝍤 丅
丨	川	丁	𝍤 丁		𝍤 丨
丅	川	丅	𝍤 川		− 川
𝍥	川	川	𝍤 丁		𝍤 𝍥
𝍤 𝍥	丨 − 川	丨 − 丁	− 丨 𝍤 川		丅 𝍤

In the augmented matrix, the result is the following:

$$
\begin{bmatrix}
63 & 49 & 21 & 14 & 35 & 980 \\
63 & 54 & 36 & 45 & 27 & 1152 \\
3 & 5 & 7 & 6 & 4 & 116 \\
2 & 5 & 3 & 9 & 4 & 112 \\
1 & 3 & 2 & 8 & 5 & 95
\end{bmatrix}
$$

STEP 3. Term-by-term subtraction (*zhi chu* 直除). In the third step, we subtract, term-by-term, the entries in the first column from those in the second column, that is, $(63,54,36,45,27,1152) - (63,49,21,14,35,980)$, which, on the counting board, again follows a very simple pattern, as can be seen from the diagram below:

another, we must use the modified value for the first entry of the second column, namely 63. So it is possible that there is a register outside the counting board for preserving both values. To my knowledge, no extant written records explicitly describe these temporary registers.

I	II	III	⊥ III	⊥ III
III	IIII	IIII	≡ III	≡ III
II	II	T̄	≡ T̄	≡ I
T̄	T̄	T	≡ III	— III
IIII	III	III	≡ T̄	≡ IIII
≟ IIII	I − II	I − T	− I ≡ II	III ⊥

On the corresponding augmented matrix, this pattern is as follows:

$$
\begin{bmatrix}
63 & 49 & 21 & 14 & 35 & 980 \\
63 & 54 & 36 & 45 & 27 & 1152 \\
3 & 5 & 7 & 6 & 4 & 116 \\
2 & 5 & 3 & 9 & 4 & 112 \\
1 & 3 & 2 & 8 & 5 & 95
\end{bmatrix}
$$

The result $(63, 54, 36, 45, 27, 1152) - (63, 49, 21, 14, 35, 980) = (0, 5, 15, 31, -8, 172)$ is represented on the counting board as shown in the following diagram (here and throughout this book I use bold black to denote negative counting rods):[18]

I	II	III		T̄
III	IIII	IIII	IIII	T̄
II	II	T̄	— III	II
T̄	T̄	T	≡ I	II
IIII	III	III	T̄	IIII
≟ IIII	I − II	I − T	I ⊥ II	I ≡

Again, in modern mathematical notation, this result is precisely the augmented matrix after the first step of Gaussian elimination, where we have eliminated the first entry in the second row:

$$
\begin{bmatrix}
9 & 7 & 3 & 2 & 5 & 140 \\
0 & 5 & 15 & 31 & -8 & 172 \\
3 & 5 & 7 & 6 & 4 & 116 \\
2 & 5 & 3 & 9 & 4 & 112 \\
1 & 3 & 2 & 8 & 5 & 95
\end{bmatrix}
$$

STEP 4. Cross-multiplication. The next step is to eliminate the first entry of the third column, which is done using patterns similar to those used to eliminate the first entry of the second column. We begin by multiplying the entire third column, $(3, 5, 7, 6, 4, 116)$, by the first entry in the first column, 9:

---

[18] At this point, as shown in the diagram, the original values of the first column, $(9, 7, 3, 2, 5, 140)$, are restored. Again, extant treatises do not fully explain these details.

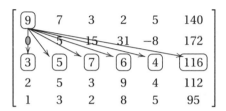

On the corresponding augmented matrix, the multiplication is as follows:

$$\begin{bmatrix} 9 & 7 & 3 & 2 & 5 & 140 \\ 0 & 5 & 15 & 31 & -8 & 172 \\ 3 & 5 & 7 & 6 & 4 & 116 \\ 2 & 5 & 3 & 9 & 4 & 112 \\ 1 & 3 & 2 & 8 & 5 & 95 \end{bmatrix}$$

This gives the result $9(3, 5, 7, 6, 4, 116) = (27, 45, 63, 54, 36, 1044)$:

In the corresponding augmented matrix, we have the following:

$$\begin{bmatrix} 9 & 7 & 3 & 2 & 5 & 140 \\ 0 & 5 & 15 & 31 & -8 & 172 \\ 27 & 45 & 63 & 54 & 36 & 1044 \\ 2 & 5 & 3 & 9 & 4 & 112 \\ 1 & 3 & 2 & 8 & 5 & 95 \end{bmatrix}$$

STEP 5. Cross-multiplication. We now multiply the entire first column by the first entry of the third column, namely, $3(9, 7, 3, 2, 5, 140)$, as shown in the following diagram:

On the corresponding augmented matrix, we have the following:

$$\begin{bmatrix} 9 & 7 & 3 & 2 & 5 & 140 \\ 0 & 5 & 15 & 31 & -8 & 172 \\ 3 & 45 & 63 & 54 & 36 & 1044 \\ 2 & 5 & 3 & 9 & 4 & 112 \\ 1 & 3 & 2 & 8 & 5 & 95 \end{bmatrix}$$

The result, $3(9,7,3,2,5,140) = (27,21,9,6,15,420)$, is displayed as follows:

On the corresponding augmented matrix, we have the following:

$$\begin{bmatrix} 27 & 21 & 9 & 6 & 15 & 420 \\ 0 & 5 & 15 & 31 & -8 & 172 \\ 27 & 45 & 63 & 54 & 36 & 1044 \\ 2 & 5 & 3 & 9 & 4 & 112 \\ 1 & 3 & 2 & 8 & 5 & 95 \end{bmatrix}$$

STEP 6. Term-by-term subtraction. The next step is to subtract the first column from the third column, namely, $(27,45,63,54,36,1044) - (27,21,9,6,15,420)$, as shown in the diagram below:

On the corresponding augmented matrix, the term-by-term subtraction is as follows:

$$\begin{bmatrix} 27 & 21 & 9 & 6 & 15 & 420 \\ 0 & 5 & 15 & 31 & -8 & 172 \\ 27 & 45 & 63 & 54 & 36 & 1044 \\ 2 & 5 & 3 & 9 & 4 & 112 \\ 1 & 3 & 2 & 8 & 5 & 95 \end{bmatrix}$$

The result, $(27, 45, 63, 54, 36, 1044) - (27, 21, 9, 6, 15, 420) = (0, 24, 54, 48, 21, 624)$, is arranged as follows:

					‾|||‾
‖	‖‖	= ‖‖	‖‖	‾|‾
‖	‖	≡ ‖‖	− ‖‖	‖
‾||‾	‾|||‾	≡ ‾||‾	=		‖
‖‖	‖	=		‾||‾	‖‖
⊥‖‖		− ‖	‾|‾ = ‖‖	|⊥‖		≡

On the corresponding augmented matrix, we have the following:

$$\begin{bmatrix} 9 & 7 & 3 & 2 & 5 & 140 \\ 0 & 5 & 15 & 31 & -8 & 172 \\ 0 & 24 & 54 & 48 & 21 & 624 \\ 2 & 5 & 3 & 9 & 4 & 112 \\ 1 & 3 & 2 & 8 & 5 & 95 \end{bmatrix}$$

STEPS 7–12.   Cross-multiplication and term-by-term subtraction. In the following steps we proceed in the same manner, repeatedly using these two simple patterns, cross-multiplication and term-by-term subtraction, until the first entries of each of the remaining columns are eliminated. In this case, on the counting board, the result is displayed as follows:

					‾|||‾
=	≡		= ‖‖	‖‖	‾|‾
− ‖‖	=		≡ ‖‖	− ‖‖	‖
⊥	⊥‾|‾	≡ ‾||‾	=		‖
≡	= ‾|‾	=		‾||‾	‖‖
‾|‾ − ‖‖	‾|‾ = ‾||‾	‾|‾ = ‖‖	|⊥‖		≡

On the augmented matrix, this corresponds to the elimination of the first entry in the remaining rows, as is familiar from Gaussian elimination:

$$\begin{bmatrix} 9 & 7 & 3 & 2 & 5 & 140 \\ 0 & 5 & 15 & 31 & -8 & 172 \\ 0 & 24 & 54 & 48 & 21 & 624 \\ 0 & 31 & 21 & 77 & 26 & 728 \\ 0 & 20 & 15 & 70 & 40 & 715 \end{bmatrix}$$

STEP 13.   Cross-multiplication. Next, using these same simple patterns, we proceed to eliminate the second entries. That is, we cross-multiply the remaining entries of the third column by the second entry of the second column, namely, $5(24, 54, 48, 21, 624)$, as shown in the following diagram:

					三
=	= 丨	= 三	〣		丅
— 〢	= 丨	≣ 三	— 〣		〢
⊥	⊥ 丅	≣ 丅	= 丨		〢
≡	= 丅	= 丨	丅		〢
丅 — 〢	丅 = 丅	丅 = 三	丨 ⊥ 〢	丨 ≡

On the corresponding augmented matrix, this is the multiplication of the third row by the pivot in the second row, as shown below:

$$
\begin{bmatrix}
9 & 7 & 3 & 2 & 5 & 140 \\
0 & 5 & 15 & 31 & -8 & 172 \\
0 & 24 & 54 & 48 & 21 & 624 \\
0 & 31 & 21 & 77 & 26 & 728 \\
0 & 20 & 15 & 70 & 40 & 715
\end{bmatrix}
$$

The result, $5(24,54,48,21,624)=(120,270,240,105,3120)$, is displayed as follows:

					三
=	= 丨	丨 =	〢		丅
— 〢	= 丨	〢 ⊥	— 〣		〢
⊥	⊥ 丅	〢 ≡	= 丨		〢
≡	= 丅	丨 〢	三		〢
丅 — 〢	丅 = 丅	≡ 丨 =	丨 ⊥ 〢	丨 ≡

On the corresponding augmented matrix, we have the following:

$$
\begin{bmatrix}
9 & 7 & 3 & 2 & 5 & 140 \\
0 & 5 & 15 & 31 & -8 & 172 \\
0 & 120 & 270 & 240 & 105 & 3120 \\
0 & 31 & 21 & 77 & 26 & 728 \\
0 & 20 & 15 & 70 & 40 & 715
\end{bmatrix}
$$

STEP 14. Cross-multiplication. Next we multiply the remaining entries of the second column by the first entry in the third column, $24(5,15,31,-8,172)$, as shown in the following diagram:

					三
=	= 丨	= 三	〣		丅
— 〢	= 丨	〢 ⊥	— 〣		〢
⊥	⊥ 丅	〢 ≡	= 丨		〢
≡	= 丅	丨 〢	三		〢
丅 — 〢	丅 = 丅	≡ 丨 =	丨 ⊥ 〢	丨 ≡

On the corresponding augmented matrix, we have the following:

$$
\begin{bmatrix}
9 & 7 & 3 & 2 & 5 & 140 \\
0 & \boxed{5} & \boxed{15} & \boxed{31} & \boxed{-8} & \boxed{172} \\
0 & \boxed{24} & 270 & 240 & 105 & 3120 \\
0 & 31 & 21 & 77 & 26 & 728 \\
0 & 20 & 15 & 70 & 40 & 715
\end{bmatrix}
$$

The result is $24(5, 15, 31, -8, 172) = (120, 360, 744, -192, 4128)$:

					$\overline{\text{III}}$
=		≡		−		−		$\overline{\text{T}}$
− ‖‖		=		‖ ⊥		‖ ⊥		‖
⊥		⊥ $\overline{\text{T}}$		‖ ≡		$\overline{\text{T}}$ ≡ ‖		‖
≡		= $\overline{\text{T}}$			‖‖			$\underline{\underline{\text{⊥}}}$ ‖		‖‖
$\overline{\text{T}}$ − ‖‖		$\overline{\text{T}}$ = $\overline{\text{II}}$		≡	=		≡	= $\overline{\text{T}}$			≡

The corresponding augmented matrix is as follows:

$$
\begin{bmatrix}
9 & 7 & 3 & 2 & 5 & 140 \\
0 & 120 & 360 & 744 & -192 & 4128 \\
0 & 120 & 270 & 240 & 105 & 3120 \\
0 & 31 & 21 & 77 & 26 & 728 \\
0 & 20 & 15 & 70 & 40 & 715
\end{bmatrix}
$$

STEP 15.   Term-by-term subtraction. We subtract the modified second column, which is now $(120, 360, 744, -192, 4128)$, from the modified third column, $(120, 270, 240, 105, 3120)$, as follows:

					$\overline{\text{III}}$
=		≡			=	→		=		$\overline{\text{T}}$
− ‖‖		=		‖ ⊥	→	‖ ⊥		‖
⊥		⊥ $\overline{\text{T}}$		‖ ≡	→	$\overline{\text{T}}$ ≡ ‖		‖
≡		= $\overline{\text{T}}$			‖‖	→		$\underline{\underline{\text{⊥}}}$ ‖		‖‖
$\overline{\text{T}}$ − ‖‖		$\overline{\text{T}}$ = $\overline{\text{II}}$		≡	=	→	≡	= $\overline{\text{T}}$			≡

On our augmented matrix, we have the following:

$$
\begin{bmatrix}
9 & 7 & 3 & 2 & 5 & 140 \\
0 & \boxed{120} & \boxed{360} & \boxed{744} & \boxed{-192} & \boxed{4128} \\
0 & \boxed{120} & \boxed{270} & \boxed{240} & \boxed{105} & \boxed{3120} \\
0 & 31 & 21 & 77 & 26 & 728 \\
0 & 20 & 15 & 70 & 40 & 715
\end{bmatrix}
$$

This then gives the result, $(120, 270, 240, 105, 3120) - (120, 360, 744, -192, 4128) = (0, -90, -504, 297, -1008)$:

					$\overline{\text{III}}$
$=$	$\equiv \mathsf{I}$			IIII	$\overline{\mathsf{I}}$
$- \text{IIII}$	$= \mathsf{I}$	$\frac{\perp}{\equiv}$		$- \text{IIII}$	III
$\frac{\perp}{\mathsf{I}}$	$\perp \overline{\mathsf{I}}$	$\text{IIIII} \ \text{IIII}$		$= \mathsf{I}$	II
$\equiv$	$= \overline{\mathsf{I}}$	$\mathsf{II} \frac{\perp}{\equiv} \overline{\mathsf{I}}$		$\overline{\text{III}}$	IIII
$\overline{\mathsf{I}} - \text{IIII}$	$\overline{\mathsf{I}} = \overline{\text{III}}$	$- \quad \overline{\text{III}}$		$\mathsf{I} \perp \mathsf{II}$	$\mathsf{I} \equiv$

On the augmented matrix, we have the following:

$$\begin{bmatrix} 9 & 7 & 3 & 2 & 5 & 140 \\ 0 & 5 & 15 & 31 & -8 & 172 \\ 0 & 0 & -90 & -504 & 297 & -1008 \\ 0 & 31 & 21 & 77 & 26 & 728 \\ 0 & 20 & 15 & 70 & 40 & 715 \end{bmatrix}$$

STEPS 16–21.　Continued cross-multiplication and term-by-term subtraction for the second column. As was the case for the first column, we continue using the same pattern of cross-multiplication and term-by-term subtraction to remove entries in the second column. After completing the eliminations, we arrive at the following:

					$\overline{\text{III}}$
				IIII	$\overline{\mathsf{I}}$
$\mathsf{II} = \text{IIII}$	$\mathsf{III} \perp$	$\frac{\perp}{\equiv}$		$- \text{IIII}$	II
$\mathsf{II} \frac{\perp}{}$	$\text{IIII} \perp \overline{\mathsf{I}}$	$\text{IIIII} \ \text{IIII}$		$= \mathsf{I}$	II
$\mathsf{III} \perp$	$\mathsf{II} \perp \overline{\text{III}}$	$\mathsf{II} \frac{\perp}{\equiv} \overline{\mathsf{I}}$		$\overline{\text{III}}$	IIII
$\mathsf{I} \equiv \text{IIII}$	$- \overline{\mathsf{I}} \frac{\perp}{\equiv} \mathsf{II}$	$- \quad \overline{\text{III}}$		$\mathsf{I} \perp \mathsf{II}$	$\mathsf{I} \equiv$

In modern mathematical notation, in the augmented matrix, we have the following:

$$\begin{bmatrix} 9 & 7 & 3 & 2 & 5 & 140 \\ 0 & 5 & 15 & 31 & -8 & 172 \\ 0 & 0 & -90 & -504 & 297 & -1008 \\ 0 & 0 & -360 & -576 & 378 & -1692 \\ 0 & 0 & -225 & -270 & 360 & 135 \end{bmatrix}$$

STEPS 22–30.　Continued cross-multiplication and term-by-term subtraction for the remaining columns. We then eliminate the entries in the remaining columns that lie above the diagonal entries, proceeding as we did for the first and second columns. That is, by simply continuing to use the same pattern of cross-multiplication and term-by-term subtraction, we eliminate the entries in the remaining columns, arriving at the following:

					川
				⦀⦀	丁
				$\perp$	— ⦀⦀	川
		— ‖ $\perp$ 丁	⦀⦀⦀ ⦀⦀	= ⎮	⎮
= = ‖ $\perp$ ⎮	丁 — 川	‖ $\perp$ 丁	川	⦀⦀⦀
⎮ = ‖ ‖ ≡ 丁	= ⎮ 丁	— 川	⎮ $\perp$ ‖	⎮ ≡

Using modern mathematical notation, in the augmented matrix, we have the following:

$$\begin{bmatrix} 9 & 7 & 3 & 2 & 5 & 140 \\ 0 & 5 & 15 & 31 & -8 & 172 \\ 0 & 0 & -90 & -504 & 297 & -1008 \\ 0 & 0 & 0 & -1296 & 729 & -2106 \\ 0 & 0 & 0 & 0 & 203391 & 1220346 \end{bmatrix}$$

This is just the triangular form familiar from modern linear algebra. We have completed the first half of the solution, namely, forward substitution. It should be noted that, as is the case with Gaussian elimination, this procedure can be applied to an array of any size.

Back Substitution as Simple Visual Patterns

If the first part of the *fangcheng* procedure is familiar from modern linear algebra, the second part, which in modern linear algebra is called "back substitution," proceeds in a counterintuitive manner, one that has hitherto not been well understood. More specifically, some modern historians of mathematics have incorrectly assimilated the *fangcheng* procedure to the modern approach to back substitution; however, several important studies of the mathematics described in the *fangcheng* procedure have noted that the approach to back substitution differs from the modern approach.[19] In *Chinese Roots of Linear Algebra*, I reconstruct this approach to demonstrate how these calculations would have been performed by following simple patterns using counting rods on a counting board. Although this approach is counterintuitive, on the counting board it is very efficient, since in general it avoids the emergence of fractions until the final step: because the counting board already uses its two dimensions to display the n conditions in n unknowns, fractions, which require at least two entries each, would considerably encumber calculations.[20] Here I provide a detailed explication of the procedure of back substitution for problem 18 using the *fangcheng* procedure on the counting board.[21] The purpose of this detailed explication of the

[19] In particular, see Shen, Lun, and Crossley, *Nine Chapters*.

[20] For more detail, see Hart, *Chinese Roots of Linear Algebra*.

[21] In *Chinese Roots of Linear Algebra*, I provide an example of this procedure for problem 1, which is a problem with three conditions in three unknowns. This problem, problem 18,

procedure of back substitution using the *fangcheng* procedure is two-fold: first, to demonstrate how simple patterns are repeatedly applied to find the solution; second, to demonstrate the complexity of these calculations. These calculations are intimidating enough that one might hesitate to compute them without the aid of some mechanical calculating device, such as counting rods; writing out these calculations (such as the so-called "brush calculations" *bi suan* 筆算 advocated in the *Guide to Calculation*) would be impracticably onerous.

STEP 31.   Cross-multiplication. First we cross-multiply.[22] This pattern of cross-multiplication will be applied repeatedly to column after column, from left to right: it is first applied to the penultimate column, then to the next column to the right, then to the next, and so on, finishing with the first column on the right. The pattern of this cross multiplication, which is quite difficult to describe in words, will gradually become clear as we follow the steps for each subsequent column. In this case, the penultimate entry in the penultimate column, 729, is multiplied by the final entry in the final column, 1220346, and the final entry in the penultimate column, −2106, is multiplied by the penultimate entry in the final column, 203391, as follows:

Using modern notation, this cross-multiplication, which has no analog in modern linear algebra, is the following:

$$
\begin{bmatrix}
9 & 7 & 3 & 2 & 5 & 140 \\
0 & 5 & 15 & 31 & -8 & 172 \\
0 & 0 & -90 & -504 & 297 & -1008 \\
0 & 0 & 0 & -1296 & \boxed{729} & \boxed{-2106} \\
0 & 0 & 0 & 0 & \boxed{203391} & \boxed{1220346}
\end{bmatrix}
$$

This gives us our result, $203391 \cdot (-2106) = -428341446$ and $1220346 \cdot 729 = 889632234$, which, on the counting board, is arranged as follows:

---

with five conditions in five unknowns, is more complex, but follows the same patterns. To my knowledge, the approach to back substitution using the *fangcheng* procedure has hitherto never been presented in this detail.

[22] This form of cross-multiplication is quite different from *biancheng*, which we saw above. To my knowledge, extant treatises do not give a name for this form of cross-multiplication.

					𝍢
				𝍠	𝍦
			⊥	− 𝍠	𝍡
		− ‖ ⊥ Т	𝍠 𝍠	≡ 𝏿	‖
=	≡ ‖ ⊥ 𝏿	𝍢 ⊥ 𝍢 ⊥ ‖ = 𝏿 = 𝍢	‖ ⊥ Т	𝍢	𝍠
𝏿 = ‖	‖ ≡ Т	𝍢 = 𝍢 ≡ 𝍢 − 𝍢 ≡ Т	− 𝍢	𝏿 ⊥ ‖	𝏿 ≡

Written using modern notation, we have the following (it should be noted that this is no longer an augmented matrix as understood in modern mathematics):[23]

$$
\begin{bmatrix}
9 & 7 & 3 & 2 & 5 & 140 \\
0 & 5 & 15 & 31 & -8 & 172 \\
0 & 0 & -90 & -504 & 297 & -1008 \\
0 & 0 & 0 & -1296 & 889632234 & -428341446 \\
0 & 0 & 0 & 0 & 203391 & 1220346
\end{bmatrix}
$$

STEP 32.   Simplification.[24] We then simplify the penultimate column. In general terms, this simplification proceeds as follows: from the final entry in the column we successively subtract each of the entries above it, until we reach the diagonal entry, that is, the first remaining nonzero entry in that column;[25] we then divide the result of the successive subtractions by this diagonal entry. In this case, we subtract the penultimate entry from the final entry, and then divide by the fourth entry, as shown in the following diagram:

					𝍢
				𝍠	𝍦
			⊥	− 𝍠	𝍡
		− ‖ ⊥ Т	𝍠 𝍠	≡ 𝏿	‖
=	≡ ‖ ⊥ 𝏿	𝍢 ⊥ 𝍢 ⊥ ‖ = 𝏿 = 𝍢	‖ ⊥ Т	𝍢	𝍠
𝏿 = ‖	‖ ≡ Т	𝍢 = 𝍢 ≡ 𝍢 − 𝍢 ≡ Т	− 𝍢	𝏿 ⊥ ‖	𝏿 ≡

Written using modern notation, we have the following:

$$
\begin{bmatrix}
9 & 7 & 3 & 2 & 5 & 140 \\
0 & 5 & 15 & 31 & -8 & 172 \\
0 & 0 & -90 & -504 & 297 & -1008 \\
0 & 0 & 0 & \boxed{-1296} & \boxed{889632234} & \boxed{-428341446} \\
0 & 0 & 0 & 0 & 203391 & 1220346
\end{bmatrix}
$$

---

[23] That is, the matrix no longer corresponds to the system of linear equations given in equations (5.2)–(5.6); in modern computer science, however, arrays of numbers are often used in this manner, and need not correspond to the coefficients of a system of linear equations.

[24] Again, to my knowledge, extant treatises do not provide a name for this form of simplification.

[25] In modern linear algebra, we would call this the pivot.

That is, we calculate $-428341446 - 889632234 = -1317973680$, and then we calculate $(-1317973680) \div (-1296) = 1016955$, with the result displayed as follows:[26]

In modern notation, the result is as follows:

$$
\begin{bmatrix}
9 & 7 & 3 & 2 & 5 & 140 \\
0 & 5 & 15 & 31 & -8 & 172 \\
0 & 0 & -90 & -504 & 297 & -1008 \\
0 & 0 & 0 & 0 & 0 & 1016955 \\
0 & 0 & 0 & 0 & 203391 & 1220346
\end{bmatrix}
$$

STEP 33. Cross-multiplication. Next we again cross multiply, extending the pattern used in the previous column. Here, we first multiply the final entry in the third column by the penultimate entry in the final column, $203391 \cdot (-1008)$. Then we multiply the penultimate entry in the third column by the final entry in the final column, $1220346 \cdot 297$. Then we multiply the fourth entry in the third column by the final entry in the fourth column, $1016955 \cdot (-504)$. These cross multiplications, displayed on the counting board, can be represented as follows, illustrating the simple visual pattern used:

Using modern notation, this can be represented as follows:

$$
\begin{bmatrix}
9 & 7 & 3 & 2 & 5 & 140 \\
0 & 5 & 15 & 31 & -8 & 172 \\
0 & 0 & -90 & \boxed{-504} & \boxed{297} & \boxed{-1008} \\
0 & 0 & 0 & 0 & 0 & \boxed{1016955} \\
0 & 0 & 0 & 0 & \boxed{203391} & \boxed{1220346}
\end{bmatrix}
$$

---

[26] I have removed the remaining entries from this column. Again, there are no descriptions specifying the arrangement of the counting board at this stage.

The result of these multiplications is then $203391 \cdot (-1008) = -205018128$ for the final entry in the third column, $1220346 \cdot 297 = 362442762$ for the penultimate entry, and $1016955 \cdot (-504) = -512545320$ for the fourth entry:

In modern notation, this is the following:

$$\begin{bmatrix} 9 & 7 & 3 & 2 & 5 & 140 \\ 0 & 5 & 15 & 31 & -8 & 172 \\ 0 & 0 & -90 & -512545320 & 362442762 & -205018128 \\ 0 & 0 & 0 & 0 & 0 & 1016955 \\ 0 & 0 & 0 & 0 & 203391 & 1220346 \end{bmatrix}$$

STEP 34. Simplification. The next step is simplification, extending the same pattern used in the previous simplification. We successively subtract from the final entry in this column the entries in the column above it, until we have reached the diagonal entry, the first nonzero entry in the column, and then we divide by the diagonal entry, as shown below:

In modern notation, this is the following:

$$\begin{bmatrix} 9 & 7 & 3 & 2 & 5 & 140 \\ 0 & 5 & 15 & 31 & -8 & 172 \\ 0 & 0 & \boxed{-90} & \boxed{-512545320} & \boxed{362442762} & \boxed{-205018128} \\ 0 & 0 & 0 & 0 & 0 & 1016955 \\ 0 & 0 & 0 & 0 & 203391 & 1220346 \end{bmatrix}$$

That is, we successively calculate $-205018128 - 362442762 = -567460890$, then $-567460890 - (-512545320) = -54915570$, and $(-54915570) \div (-90) = 610173$, giving the following result:

							𝍤
						ⅢⅢ	丅
						— ⅢⅢ	Ⅱ
						☰ Ⅰ	Ⅱ
☰	☰ Ⅱ ⊥ Ⅰ					𝍤	ⅢⅢ
Ⅰ ☰ Ⅱ	Ⅱ ☰ 丅	Ⅰ	Ⅰ ⊥ 𝍤 ☰ ⅢⅢ	⊥ Ⅰ	Ⅰ ⊥ Ⅱ	Ⅰ ⊥ Ⅱ	Ⅰ ☰

In modern notation, we have the following:

$$
\begin{bmatrix}
9 & 7 & 3 & 2 & 5 & 140 \\
0 & 5 & 15 & 31 & -8 & 172 \\
0 & 0 & 0 & 0 & 0 & 610173 \\
0 & 0 & 0 & 0 & 0 & 1016955 \\
0 & 0 & 0 & 0 & 203391 & 1220346
\end{bmatrix}
$$

STEP 35. Cross-multiplication. We again use the previous pattern for cross multiplication: we multiply the final entry in the second column by the penultimate entry in the final column, $203391 \cdot 172$; we multiply the fifth entry in the second column by the final entry in the fifth column, $1220346 \cdot (-8)$; we multiply the fourth entry in the second column by the final entry in the fourth column, $1016955 \cdot 31$; and we multiply the third entry in the second column by the final entry in the third column, $610173 \cdot 15$. These operations again follow the same simple pattern, as can easily be seen from the diagram below:

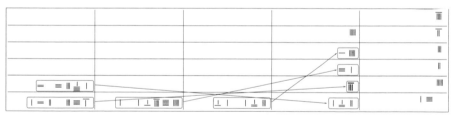

In modern notation, these cross-multiplications appear as follows:

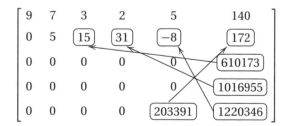

The result, recorded on the counting board, is as follows:

					〒
				⊪	〒
				〒 − ⫿⫿ = ⫿⫿⏌ ⊪	⊪
				= ⼁ 〓 ⼁ 〓 ⊤ ⊪	⼁
=	= ⫿⏌ ⼁			〒⏊⊤ = 〒⏊〒	⊪
⼁ = ⫿	⫿ 〓 ⊤	⼁	⼁⏊〒 〓 ⊪	⏊⼁ ⼁⏊⫿	= ⫿⏌〒 = ⫿ 〓 ⫿	⼁ 〓

On the augmented matrix, we have the following:

$$
\begin{bmatrix}
9 & 7 & 3 & 2 & 5 & 140 \\
0 & 5 & 9152595 & 31525605 & -9762768 & 34983252 \\
0 & 0 & 0 & 0 & 0 & 610173 \\
0 & 0 & 0 & 0 & 0 & 1016955 \\
0 & 0 & 0 & 0 & 203391 & 1220346
\end{bmatrix}
$$

STEP 36.   Simplification. Again, we extend the same pattern to simplify the second column, repeatedly subtracting from the final entry the entries above it until we reach the entry on the diagonal, namely, the second entry, 5. We then use the second entry, 5, to divide the result of the successive subtractions, which is stored in the position of the final entry. On the counting board, we have the following:

					〒
				⊪	〒
				〒 − ⫿⫿ = ⫿⫿⏌ ⊪	⊪
				= ⼁ 〓 ⼁ 〓 ⊤ ⊪	⼁
=	= ⫿⏌ ⼁			〒⏊⊤ = 〒⏊〒	⊪
⼁ = ⫿	⫿ 〓 ⊤	⼁	⼁⏊〒 〓 ⊪	⏊⼁ ⼁⏊⫿	= ⫿⏌〒 = ⫿ 〓 ⫿	⼁ 〓

On the augmented matrix, we have the following:

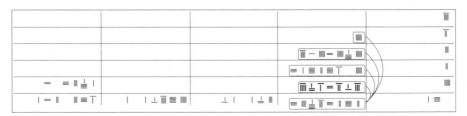

$$
\begin{bmatrix}
9 & 7 & 3 & 2 & 5 & 140 \\
0 & \boxed{5} & \boxed{9152595} & \boxed{31525605} & \boxed{-9762768} & \boxed{34983252} \\
0 & 0 & 0 & 0 & 0 & 610173 \\
0 & 0 & 0 & 0 & 0 & 1016955 \\
0 & 0 & 0 & 0 & 203391 & 1220346
\end{bmatrix}
$$

That is, we successively calculate $34983252 - (-9762768) = 44746020$, then next $44746020 - 31525605 = 13220415$, then next $13220415 - 9152595 = 4067820$, and finally $4067820 \div 5 = 813564$, which is displayed as follows:

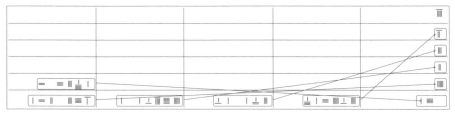

In modern notation, we have the following:

$$\begin{bmatrix} 9 & 7 & 3 & 2 & 5 & 140 \\ 0 & 0 & 0 & 0 & 0 & 813564 \\ 0 & 0 & 0 & 0 & 0 & 610173 \\ 0 & 0 & 0 & 0 & 0 & 1016955 \\ 0 & 0 & 0 & 0 & 203391 & 1220346 \end{bmatrix}$$

STEP 37.   Cross-multiplication. The final cross multiplication again simply further extends the same pattern as in the previous steps. We first multiply the final entry in the first column by the penultimate entry in the final column, $203391 \cdot 140$; we multiply the fifth entry in the first column by the final entry in the fifth column, $1220346 \cdot 5$; we multiply the fourth entry in the first column by the final entry in the fourth column, $1016955 \cdot 2$; we multiply the third entry in the first column by the final entry in the third column, $610173 \cdot 3$; and we multiply the second entry in the first column by the final entry in the second column, $813564 \cdot 7$. The pattern of this cross multiplication on the counting board is as follows:

In modern notation, these cross-multiplications follow the same simple pattern, as shown below:

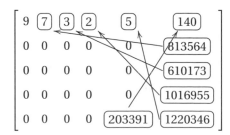

The result of the cross-multiplication, recorded on the counting board, is as follows:

In modern notation, we have the following:

$$
\begin{bmatrix}
9 & 5694948 & 1830519 & 2033910 & 6101730 & 28474740 \\
0 & 0 & 0 & 0 & 0 & 813564 \\
0 & 0 & 0 & 0 & 0 & 610173 \\
0 & 0 & 0 & 0 & 0 & 1016955 \\
0 & 0 & 0 & 0 & 203391 & 1220346
\end{bmatrix}
$$

STEP 38.   Simplification. The simplification of the first column again simply extends the same pattern used in the previous steps. From the final entry we successively subtract the entries above it until we have arrived at the entry on the diagonal, in this case the first entry, 9, which is used to divide the result.

In modern notation, we have the following:

$$
\begin{bmatrix}
\boxed{9} & \boxed{5694948} & \boxed{1830519} & \boxed{2033910} & \boxed{6101730} & \boxed{28474740} \\
0 & 0 & 0 & 0 & 0 & 813564 \\
0 & 0 & 0 & 0 & 0 & 610173 \\
0 & 0 & 0 & 0 & 0 & 1016955 \\
0 & 0 & 0 & 0 & 203391 & 1220346
\end{bmatrix}
$$

That is, we subtract the fifth entry, $28474740 - 6101730 = 22373010$; we subtract the fourth entry, $22373010 - 2033910 = 20339100$; we subtract the third entry, $20339100 - 1830519 = 18508581$; and we subtract the second entry, $18508581 - 5694948 = 12813633$. Then we then divide the result by the diagonal entry, giving $12813633 \div 9 = 1423737$:

= ⹀ ‖ ⟂ ∣					
∣ = ‖ ⹀ ‖ ≡ ⊤	∣	∣ ⟂ ⿱ ≡ ⦀	⟂ ∣	∣ ⟂ ‖	⟂ ∣ ⹀ ⦀ ⟂ ⦀ ∣ ≡ ∣ = ⊤ = ⊤

In modern notation, we have the following:

$$\begin{bmatrix} 0 & 0 & 0 & 0 & 0 & 1423737 \\ 0 & 0 & 0 & 0 & 0 & 813564 \\ 0 & 0 & 0 & 0 & 0 & 610173 \\ 0 & 0 & 0 & 0 & 0 & 1016955 \\ 0 & 0 & 0 & 0 & 203391 & 1220346 \end{bmatrix}$$

STEP 39.   Final division. The final step is division, which can be understood as simplifying fractions. We divide the final entry in each column by the penultimate entry in the final column. That is, we divide the final entry in the first column, $1423737 \div 203391 = 7$; we divide the final entry in the second column, $813564 \div 203391 = 4$; we divide the final entry in the third column, $610173 \div 203391 = 3$; we divide the final entry in the fourth column, $1016955 \div 203391 = 5$; and we divide the final entry in this fifth column, $1220346 \div 203391 = 6$. The result of this final division is the solution to the problem:

⊤	⦀	‖	⦀	⊤

In modern notation, we have the following:

$$\begin{bmatrix} 0 & 0 & 0 & 0 & 0 & 7 \\ 0 & 0 & 0 & 0 & 0 & 4 \\ 0 & 0 & 0 & 0 & 0 & 3 \\ 0 & 0 & 0 & 0 & 0 & 5 \\ 0 & 0 & 0 & 0 & 0 & 6 \end{bmatrix}$$

In sum, the *fangcheng* procedure, as described in the *Nine Chapters*, outlines what is now known to be the most powerful and general method known for solving systems of linear equations, namely, what is now called Gaussian elimination. We might summarize some of the main features of the *fangcheng* procedure as follows:

1. Prerequisites: The only prerequisites are facility with addition, subtraction, multiplication, division, and fractions on the counting board.
2. Patterns: Beyond the above prerequisites, all that is required is an understanding of simple patterns that are repeatedly applied. There are two basic patterns that are followed during the process of elimination: (i) cross-multiplication (*biancheng* 徧乘); and (ii) term-by-term subtraction (*zhi chu* 直除). There are also two additional simple patterns that are followed in the process of back substitution (neither of which is easy to describe in words, and neither of which is given a name in extant treatises): (i) cross-multiplication; and (ii) simplification. (Again, these latter two patterns can best be understood through the diagrams provided.)
3. Generality: The *fangcheng* procedure—by using these simple patterns—is completely general, and can be used (with some slight modifications) to solve any system of n conditions in n unknowns for which a solution exists.
4. Sophistication: This approach to back substitution indicates considerable sophistication—it is much more sophisticated than the intuitive approach to back substitution familiar from modern linear algebra.
5. Efficiency: These calculations, which are all rote applications of counting-rod operations and simple patterns, could likely be calculated quickly, efficiently, and with little effort.
6. Literacy: Nothing in the *fangcheng* procedure requires literacy, and in fact literacy could contribute little to mastering this practice.

Written Records of *Fangcheng* Practices

Although the *fangcheng* procedure is easily learned simply by observing two-dimensional patterns on the counting board, as noted above, it is difficult to describe using words, that is, through the medium of a one-dimensional narrative. In fact, as recorded in extant Chinese mathematical treatises, *fangcheng* practices are rendered virtually unintelligible. Extant mathematical treatises suggest that those who recorded *fangcheng* practices in writing, often to present to the imperial court, understood only the rudiments of *fangcheng* practices, and they often expressed contempt for its practitioners. The example I will present here is from the earliest extant record of *fangcheng* practices, the *Nine Chapters*. Liu Hui 劉徽 (fl. 263 CE), to whom the earliest extant commentary on the *Nine Chapters* is conventionally attributed, is usually considered the most eminent mathematician of imperial China. Yet, as I note in *Chinese Roots of Linear Algebra*, the commentary attributed to him evinces little understanding of the fact that the complexity of the *fangcheng* procedure in the *Nine Chapters* was necessary in order to avoid even more complicated calculations with fractions. In fact, the commentary expresses derision toward practitioners for their rote application of the *fangcheng* procedure:

Those who are clumsy in the essential principles vainly follow this original [*fangcheng*] procedure, some placing counting rods so numerous that they fill a carpet, seemingly so fond of complexity as to easily make mistakes. They seem to be unaware of the error [in their approach], and on the contrary, desire by the use of more [counting rods] to be highly esteemed. Therefore, of their calculations, all are ignorant of the establishment of understanding; instead, they are specialized to an extreme.

其拙于精理從按本術者，或用算而布氈，方好煩而喜誤，曾不知其非，反欲以多為貴。故其算也，莫不（同）〔闇〕于設通而專于一端。[27]

The commentary follows this accusation with several methods for solving problem 18. Here I will examine one of these methods, to show that only the rudiments of solving *fangcheng* problems were understood.

The "Old Method"

The commentary in the *Nine Chapters* presents what is asserted to be the "old method" (*jiu shu* 舊術),[28] which is sometimes assumed to be the *fangcheng* procedure.[29] The commentary, however, fails to follow the sophisticated algorithm of the *fangcheng* procedure as presented in the original text of the *Nine Chapters*. From a computational standpoint, compared with the *fangcheng* procedure, the approach taken in the commentary has several disadvantages: it does not follow an algorithm; no general method is presented; and thus it cannot be rapidly computed at each step. But most important, if the approach taken in the commentary is applied to general cases (that is, beyond the 18 contrived problems recorded in the *Nine Chapters*), fractions will often appear in back substitution, and fractions, each of which requires two dimensions on the counting board, would considerably encumber *fangcheng* calculations, which already employ the two dimensions of the counting board to record the array. In this sense, the approach presented in the commentary might at first appear to be serendipitous. I will argue that what the commentary calls the "old method" is a corrupt record of a recondite solution that attempts to minimize the number of counting rods used. What is indisputable, however, is that the commentary does not address the considerable computational advantages of the rote application of the sophisticated *fangcheng* algorithm.

---

[27] *Jiuzhang suanshu, juan* 8, 19a; Shen, Lun, and Crossley, *Nine Chapters*, 426; Chemla and Guo, *Neuf chapitres*, 650–51.

[28] *Jiu* 舊 is translated as "original" in Shen, Lun, and Crossley, *Nine Chapters*, 427. I believe the term is pejorative, so I have translated *jiu* as "old."

[29] The commentary also presents a "new method" *xin shu* 新術 and an "alternative method" *qi yi shu* 其一術. I cannot address these methods here, except to note that analysis of the "new method" and the "alternative method" yields conclusions consistent with those presented here.

The approach taken in the commentary is difficult to reconstruct, because the text here is hopelessly corrupt. Modern attempts to reconstruct this passage—so that it makes sense mathematically—require an inordinate number of emendations, transpositions, insertions, and deletions.[30] None are satisfactory. That calculations this corrupt were copied and passed on, without correction, in all extant versions of the *Nine Chapters*, including the imperially sponsored edition, is consistent with the rudimentary grasp of the *fangcheng* procedure by the literati who compiled mathematical treatises. In any case, the reconstructed versions, despite their differences, are all similar in the sense that the approach taken in the commentary does not follow any algorithm or method. As we will see, it seems that this approach is the corrupt record of a solution that requires only 77 counting rods, perhaps presented in support of the preceding criticism that the original *fangcheng* procedure used too many counting rods (185 counting rods are required to display the solution shown on pages 138–157 of this chapter).

The following is the "old method," as recorded in the commentary to the *Nine Chapters* (I have inserted the individual step numbers in brackets):

[1] First place the third column, subtract the fourth column, and subtract the third column. [2] Next,[31] place the second column, with the third column subtract from the second column, eliminating its first entry. [3] Next, place the right column and eliminate its first entry. [4] Next, with the fourth column subtract the first entry of the left column. [5] Next, with the left column, eliminate the first entry of the fourth column and the second column. [6] Next, with the fifth column subtract the first entry of the second column; the remainder can be halved. [7] Next, with the second column, eliminate the first entry of the fourth column; reduce the remainder, reduce the improper fraction to a mixed fraction, obtaining zero,[32] namely the price of millet. [8] With the divisor subtract the second column, obtaining the price of beans. [9] With the left column obtain the price of wheat; the third column is the price of hemp. [10] The right column is the price of legumes. [11] This method uses 77 counting rods.[33]

---

[30] For a comparison of several modern attempts at reconstruction, see Li Jimin 李继闵, *Jiuzhang suanshu jiaozheng* 九章算术校证 [Nine chapters on the mathematical arts, critical edition] (Xi'an: Shaanxi kexue jishu chubanshe 陕西科学技术出版社, 1993). Li notes that Guo Shuchun, one of the foremost historians of Chinese mathematics, inserts 27 characters, changes 9 characters, moves 11 characters, and deletes 1 character, changing 48 out of a total of 129 characters. Other reconstructions are similar. See also Mo Shaokui 莫绍揆, "Youguan *Jiuzhang suanshu* de yixie taolun" 有关《九章算术》的一些讨论 [Several points about the Nine chapters on the mathematical arts], *Ziran kexueshi yanjiu* 自然科学史研究 19 (2000): 97–113.

[31] More literally, *ci* 次 means "second." Here its repeated use is better translated as "next."

[32] That is, the improper fraction reduces to an integer, so the remaining numerator for the fraction is zero.

[33] The character *suan* 算 usually means to calculate. The usual interpretation is that 77 calculations are required. However, it is difficult to see how to count the number of calculations and arrive at the number 77. *Suan* 算 is also sometimes used interchangeably with *suan* 筭, which means counting rods. In this case, the evidence suggests that 77 *suan* refers to the number of counting rods, as does 124 *suan* in the following passage.

[1] 先置第三行，以減第四行，及減第三行。[2] 次置第二行，以第二行減第三行，去其頭位。[3] 次置右行去其頭位。[4] 次以第四行減左行頭位。[5] 次以左行去第四行及第二行頭位。[6] 次以第五行減第二行頭位，餘可半。[7] 次以第二行去第四行頭位，餘約之為法，實如法而一，得空即有黍價。[8] 以法減第二行，得荅價。[9] 左行得麥價，第三行麻價。[10] 右行得菽價。[11] 如此凡用七十七算。 ³⁴

Textual Reconstructions of the "Old Method"

Below I attempt to reconstruct the calculations recorded in the commentary on the counting board, interpreting the instructions in the commentary as charitably as possible, and following them as far as possible. As we shall see, it is only possible to follow the instructions to step 4. After step 4, it becomes increasingly difficult to reconcile the stated instructions with the previous steps, however they are interpreted.

STEP 1.   "First place the third column, subtract the fourth column, and subtract the third column" (lines 1 and 2 on the facing page). The first part of the first step is unambiguous—place (*zhi* 置) the third column on the counting board. We then have the following:

		III	
		IIII	
		T̄	
		T	
		III	
		I – T	

Represented as an augmented matrix, we have the following:

$$\begin{bmatrix} 3\ 5\ 7\ 6\ 4\ 116 \end{bmatrix}$$

This instruction, however, stands in contrast to the *fangcheng* procedure, as outlined in the *Nine Chapters* following problem 1, in which all of the columns are placed on the counting board at the outset, from right to left. Here we are instructed to place only one column on the counting board, and not the first column but the third. The second part of the first step—the two instructions "subtract the fourth column, and subtract the third column"—also diverges from the *fangcheng* procedure in several respects. Two columns are first operated

³⁴ *Jiuzhang suanshu, juan* 8, 20b–21a; Chemla and Guo, *Neuf chapitres*, 650–59; Shen, Lun, and Crossley, *Nine Chapters*, 527–38.

on, and only afterward are the other columns placed on the counting board. There are explicit instructions to place two of the columns in later steps: the second column is placed on the counting board in the second step (lines 2 and 3 on page 160), and the first (right) column is placed on the counting board in the third step (lines 3 and 4 on page 160). There are no explicit instructions to place the fourth and fifth columns on the counting board—presumably they are placed on the board at the time of their first operation, namely, the first step for the fourth column (lines 1 and 2 on page 160), and the fifth step for the fifth column (lines 5 to 7 on page 160). Although there is no explanation given for the delayed placement of the remaining columns, or for first operating on those columns placed on the board, it is difficult to see any other possible motive than to minimize the number of counting rods. In any case, if we place the fourth column on the counting board, the results are as follows:

	II	III		
	IIIII	IIIII		
	II	T̄		
	III	T		
	IIII	III		
	I − II	I − T		

On the augmented matrix, we have the following:

$$
\begin{bmatrix}
3\ 5\ 7\ 6\ 4\ 116 \\
2\ 5\ 3\ 9\ 4\ 112
\end{bmatrix}
$$

The two instructions that comprise the second part of the first step—"subtract the fourth column" and "subtract the third column"—are not clear. That is, the subtracting columns are given, but from which columns they are to be subtracted is only implied; there is no indication of whether the subtracting columns are subtracted once or whether they are subtracted multiple times (or equivalently, whether the subtracting columns are subtracted as they stand, or whether they are first multiplied term-by-term by constants); and the results of the operations are not given (for example, specifying which entries are eliminated). Without at least some of this information, we cannot be certain what these instructions mean. And again, these instructions diverge from the *fangcheng* procedure in several respects: in the *fangcheng* procedure, a multiple of the first column is subtracted from a multiple of the second column in order to eliminate the first entry in the second column, and then additional entries are similarly eliminated in the order familiar from modern linear algebra; in the *fangcheng* procedure, we do not subtract one column from another, and then subtract the latter from the former. Perhaps the most reasonable interpretation of the first instruction here is to subtract the fourth column, once, from the third column, $(3,5,7,6,4,116) - (2,5,3,9,4,112) = (1,0,4,-3,0,4)$. The commentary offers no explanation for this particular choice. Although it might at first appear that the purpose is to reduce the first entry in one of the columns to 1 and then use that column to eliminate the first entry in the remaining columns, the fifth column already has a 1 as its first entry. A more plausible explanation for this particular choice is that not one but two entries are eliminated, which is again consistent with the hypothesis that the motive is to minimize the number of counting rods. In any case, the result on the counting board is as follows:

		‖			
		‖‖‖			
		‖	‖‖		
		⫻	‖		
		‖‖			
			− ‖	‖‖		

On the augmented matrix, we have the following:

$$
\begin{bmatrix}
1 & 0 & 4 & -3 & 0 & 4 \\
2 & 5 & 3 & 9 & 4 & 112
\end{bmatrix}
$$

The next instruction is even more perplexing. Apparently we are to use the modified third column to subtract term-by-term from the fourth column. Perhaps the most reasonable interpretation is that we subtract the modified third column twice from the fourth column in order to eliminate the first entry of the fourth column, even though there is no mention of subtracting twice, nor is there any mention of eliminating an entry. In any case, if we subtract the third column twice from the fourth column, $(2, 5, 3, 9, 4, 112) - 2(1, 0, 4, -3, 0, 4) = (0, 5, -5, 15, 4, 104)$, the result is as follows:

		‖‖‖			
		‖‖‖‖	‖‖		
		− ‖‖	‖		
		‖			
		‖	‖‖		

On the augmented matrix, we have the following:

$$
\begin{bmatrix}
1 & 0 & 4 & -3 & 0 & 4 \\
0 & 5 & -5 & 15 & 4 & 104
\end{bmatrix}
$$

STEP 2.   "Next, place the second column, with the third column subtract from the second column, eliminating its first entry" (lines 2 and 3 on page 160). This step is clear—the subtracting column, the column subtracted from, and the result are specified. This step is consistent with our results thus far. Assuming that we are to first place the second column on the counting board before we operate with it, we have the following:

			I	⊤̄	
	ⅠⅠⅠⅠ			⊤	
	ⅠⅠⅠⅠⅠ	ⅠⅠⅠ		Ⅲ	
	— ⅠⅠⅠⅠ	ⅠⅠⅠ		ⅠⅠⅠⅠ	
	ⅠⅠⅠ			Ⅱ	
	Ⅰ ⅠⅠⅠ	ⅠⅠⅠ		Ⅰ = ⊤̄	

The corresponding augmented matrix is as follows:

$$
\begin{bmatrix}
7 & 6 & 4 & 5 & 3 & 128 \\
1 & 0 & 4 & -3 & 0 & 4 \\
0 & 5 & -5 & 15 & 4 & 104
\end{bmatrix}
$$

We then subtract the third column multiplied by 7 from the second column, namely, $(7,6,4,5,3,128) - 7(1,0,4,-3,0,4) = (0,6,-24,26,3,100)$, so that the first entry is eliminated. The result on the counting board is as follows:

			I		
	ⅠⅠⅠⅠ			⊤	
	ⅠⅠⅠⅠⅠ	ⅠⅠⅠ		= Ⅲ	
	— ⅠⅠⅠⅠ	ⅠⅠⅠ		= ⊤	
	ⅠⅠⅠ			Ⅱ	
	Ⅰ ⅠⅠⅠ	ⅠⅠⅠ		Ⅰ	

On the augmented matrix, we have the following:

$$
\begin{bmatrix}
0 & 6 & -24 & 26 & 3 & 100 \\
1 & 0 & 4 & -3 & 0 & 4 \\
0 & 5 & -5 & 15 & 4 & 104
\end{bmatrix}
$$

STEP 3.   "Next, place the right column and eliminate its first entry" (lines 3 and 4 on page 160). This instruction is at least arguably decipherable: it specifies that we are to subtract from the first (right) column, and it states the result of the subtraction; although it does not specify which column we are to use to subtract, we can reasonably infer that it is the third column, multiplied by 9. Assuming this is the case, we place the first column, $(9,7,3,2,5,140)$, on the counting board giving the following result:

			丨		⫶
	IIII			丅	丅
	IIIII	III	= III		II
	− IIII	III	= 丅		II
	III			II	IIII
	丨 III	III	丨		丨 ≡

Our augmented matrix is as follows:

$$\begin{bmatrix} 9 & 7 & 3 & 2 & 5 & 140 \\ 0 & 6 & -24 & 26 & 3 & 100 \\ 1 & 0 & 4 & -3 & 0 & 4 \\ 0 & 5 & -5 & 15 & 4 & 104 \end{bmatrix}$$

If we subtract the third column, multiplied by 9, from the first column, namely, $(9, 7, 3, 2, 5, 140) - 9(1, 0, 4, -3, 0, 4) = (0, 7, -33, 29, 5, 104)$, the result is as follows:

			丨		
	IIII			丅	丅
	IIIII	III	= III	≡ III
	− IIII	III	= 丅	= 川
	III			II	IIII
	丨 III	III	丨	丨 III

On the augmented matrix, the result is as follows:

$$\begin{bmatrix} 0 & 7 & -33 & 29 & 5 & 104 \\ 0 & 6 & -24 & 26 & 3 & 100 \\ 1 & 0 & 4 & -3 & 0 & 4 \\ 0 & 5 & -5 & 15 & 4 & 104 \end{bmatrix}$$

STEP 4.   "Next, with the fourth column subtract the first entry of the left column" (lines 4 and 5 on page 160). It is at this point that the instructions recorded in the commentary fail to make sense. If we place the fifth (left) column on the counting board, we have the following result:

	丨			丨	
	II	IIII			丅	丅
	II	IIIII	III	= III	≡ III
	丌	− IIII	III	= 丅	= 川
	IIII	III			II	IIII
⫧ IIII	丨	III	III	丨	丨 III

On the augmented matrix, we have the following:

$$\begin{bmatrix} 0 & 7 & -33 & 29 & 5 & 104 \\ 0 & 6 & -24 & 26 & 3 & 100 \\ 1 & 0 & 4 & -3 & 0 & 4 \\ 0 & 5 & -5 & 15 & 4 & 104 \\ 1 & 3 & 2 & 8 & 5 & 95 \end{bmatrix}$$

We can see that the instructions here cease to correspond to the calculations made so far: there is no first entry in the fourth column with which to subtract from the first entry of the fifth column. Other alternative interpretations of the previous steps fare no better. The following steps increasingly go awry, no matter how the previous steps are interpreted. As noted above, reconstructions of this problem all require considerable changes to the text.

Mathematical Reconstructions of the "Old Method"

From this, we can see that the elimination of entries follows no particular order. No one has provided an adequate explanation for the particular choices made. No one has provided a satisfactory reconstruction of the text—in fact the various reconstructions, and hence the final configurations on the counting board, differ perhaps as much from each other as they do from the received text.

If the text is hopelessly corrupt—far beyond any possibility of textual reconstruction—we might ask if it is perhaps possible to reconstruct the solution mathematically. That is, is there a solution that proceeds in basically the same manner and provides the answer recorded in the text, 77 counting rods? In fact, there is.[35] More specifically, the "old method" employs several techniques that are contrary to the *fangcheng* procedure, and for which the most plausible explanation would seem to be that they represent an attempt to minimize the number of counting rods: (1) delaying the placement of columns; (2) operating on columns before all of the columns have been placed; (3) eliminating entries opportunistically (as opposed to the predefined order prescribed by the *fangcheng* procedure); and (4) reducing columns by dividing all of the entries by a common divisor. If we allow (3) above—that entries can be eliminated in any order—there are approximately 14,400 variations to solving problem 18. There are more variations if we count the various possibilities in (1), (2), and (4) separately. If in solving problem 18 we allow (1)–(4) above, it turns out that 77 counting rods seems to be the minimum number necessary to display at each step the results of the calculations. That this was apparently known at the time indicates considerable expertise in the thousands of possible variations to solving this *single* problem.

---

[35] For a mathematical reconstruction showing how Problem 18 can be solved using only 77 counting rods to display the results at each step, see Roger Hart, "Tracing Practices Purloined by the 'Three Pillars,'" *The Korean Journal for the History of Science* 34 (2012): 322–28.

It also suggests that whoever recorded this solution was unable to reproduce this difficult result.

Features of the "Old Method"

We can now summarize several points about the approach presented in the commentary attributed to Liu Hui. First, the approach taken here appears to be serendipitous. On the one hand, no algorithm is presented, no method is described, and no rationale is provided for the seemingly perplexing sequence of operations. The approach taken reduces the number of counting rods employed to what appears to be the minimum number required to solve the problem, namely, 77 rods. In this sense, the "old method" presented here suggests considerable expertise with this *single* problem. On the other hand, it lacks the computational advantages of the *fangcheng* procedure, namely, the generality, speed, efficiency, and lack of effort of the rote application of counting rod operations following simple patterns, along with the virtual assurance that fractions will not emerge in the process of back substitution. The commentary itself provides considerable evidence that suggests that the writer understood only the basics of *fangcheng* practice: (1) The textual record of the "old method" preserved in the commentary is hopelessly corrupt. That these recorded steps are this vague suggests that the writer might have known the result—that problem 18 could be solved using 77 counting rods—but not the steps to achieve that result. (2) There is little criticism or discussion of the specific features of the *fangcheng* procedure. (3) In particular, there is no criticism or discussion of the counterintuitive approach to back substitution, or even recognition that it is necessary to avoid the emergence of fractions. (4) The commentary to this problem, which is over 1600 characters,[36] fails to mention many of the difficulties that would be encountered in solving general *fangcheng* problems, such as systems with no solutions or indeterminate systems. (5) The commentary reduces columns using term-by-term division, for example, "the remainder can be halved" (line 8 on page 160)—reducing columns by dividing by a common divisor of the entries results in a higher likelihood that fractions will emerge in the process of back substitution. (6) Finally, there is no mention of determinantal-style solutions to problems (this will be discussed in the following section).

Determinantal Solutions Recorded in Chinese Treatises

In addition to the more general solution using what is now called Gaussian elimination, determinantal-style calculations and solutions were also known for a

---

[36] Noted in Shen, Lun, and Crossley, *Nine Chapters*, 395.

distinctive class of *fangcheng* problems.[37] This distinctive class, in modern mathematical terms, is characterized by augmented matrix (5.8), as is explained below. Problem 13 of chapter 8 of the *Nine Chapters*, which I will refer to as the "well problem,"[38] is an exemplar for this distinctive class of *fangcheng* problems, and it is in commentaries to the *Nine Chapters* that records of determinantal calculations and solutions are preserved.

This section will demonstrate the following: (i) that determinantal calculations were used to find one of the unknowns in solving this distinctive class of problems; (ii) that this class of problems was at the time recognized as a distinct category; and (iii) that determinantal solutions were also known. It is this distinctive class of problems and solutions that can be found in Leonardo Pisano's writings, as will be shown in the section following this one.

Solutions to the Well Problem on the Counting Board

The well problem, when displayed on the counting board, is as follows:

$$\begin{bmatrix} 2 & 1 & 0 & 0 & 0 & y \\ 0 & 3 & 1 & 0 & 0 & y \\ 0 & 0 & 4 & 1 & 0 & y \\ 0 & 0 & 0 & 5 & 1 & y \\ 1 & 0 & 0 & 0 & 6 & y \end{bmatrix}. \qquad (5.7)$$

Using modern algebraic notation, we can also write the well problem as a linear system of five equations in six unknowns ($x_1, x_2, \ldots, x_5$, and y) as follows,

$$2x_1 + x_2 = y, \quad 3x_2 + x_3 = y, \quad 4x_3 + x_4 = y, \quad 5x_4 + x_5 = y, \quad 6x_5 + x_1 = y.$$

Because the well problem is a system of five conditions in six unknowns, the solutions are not unique. It is not, however, "indeterminate" in the modern sense. In the earliest received version of the *Nine Chapters*, the $(n + 1)^{\text{th}}$ unknown is simply stipulated to be 721, with no explanation provided. A later commentary by Jia Xian 賈憲 (fl. 1023–1063), preserved in Yang Hui's 楊輝 (c. 1238–c. 1298) *Nine Chapters on the Mathematical Arts, with Detailed Explanations (Xiang jie jiuzhang suanfa* 詳解九章算法, c. 1261),[39] explains how the $(n + 1)^{\text{th}}$ unknown is found. The diagonal terms are multiplied together, the remaining terms are

---

[37] Thanks are due John Crossley for suggesting the term "determinantal-style," which I will sometimes for convenience simply call "determinantal." "Determinantal-style calculations" refers to calculations made in a manner we might now call determinantal; "determinantal-style solutions" refers to solutions for all the unknowns calculated in a manner we might now call determinantal.

[38] See footnote 8 on page 134.

[39] Yang Hui, *Xiang jie jiuzhang suanfa* 詳解九章算法 [Nine chapters on the mathematical arts, with detailed explanations], in *ZKJDT*.

multiplied together (that is, the super-diagonal and the term in the corner), and then the two products are added, yielding

$$2 \cdot 3 \cdot 4 \cdot 5 \cdot 6 + 1 \cdot 1 \cdot 1 \cdot 1 \cdot 1 = 721.$$

Setting the $(n+1)^{\text{th}}$ unknown $y = 721$ gives a linear system of five conditions in five unknowns,

$$2x_1 + x_2 = 721, \ 3x_2 + x_3 = 721, \ 4x_3 + x_4 = 721, \ 5x_4 + x_5 = 721, \ 6x_5 + x_1 = 721,$$

which then has a unique solution, which can be found by Gaussian elimination,

$$x_1 = 265, \quad x_2 = 191, \quad x_3 = 148, \quad x_4 = 129, \quad x_5 = 76.$$

Jia Xian's commentary further explains that problems similar to the well problem form a category, and explains their solution. Written in modern mathematical terms, the well problem serves as an exemplar for systems of n conditions in $n + 1$ unknowns of the following distinctive form:

$$\begin{bmatrix} k_1 & l_1 & 0 & \cdots & & 0 & y \\ 0 & k_2 & l_2 & \ddots & & \vdots & y \\ \vdots & \ddots & \ddots & \ddots & & 0 & \vdots \\ 0 & \cdots & 0 & k_{n-1} & l_{n-1} & y \\ l_n & 0 & \cdots & & 0 & k_n & y \end{bmatrix}. \tag{5.8}$$

To solve problems of this distinctive class, the $(n+1)^{\text{th}}$ unknown is first assigned a value as follows. In modern terminology, the problem is transformed into n conditions in n unknowns by setting the $(n+1)^{\text{th}}$ unknown $y = \det A$, where A is the matrix of coefficients. That is,

$$y = \det A = \begin{vmatrix} k_1 & l_1 & 0 & \cdots & & 0 \\ 0 & k_2 & l_2 & \ddots & & \vdots \\ \vdots & 0 & \ddots & \ddots & & 0 \\ 0 & \vdots & \ddots & k_{n-1} & l_{n-1} \\ l_n & 0 & \cdots & 0 & k_n \end{vmatrix} = k_1 k_2 k_3 \dots k_n \pm l_1 l_2 l_3 \dots l_n,$$

where here, and throughout this chapter, $\pm$ is $+$ if n is odd and $-$ if n is even.

It should also be noted that this calculation apparently served as a "determinant" in the modern sense: the linear system of five conditions in five unknowns will have a unique solution if and only if $y = \det A \neq 0$, that is, in the well problem, if and only if the "depth of the well" (*jing shen* 井深) is not zero.

Problems of this distinctive class, exemplified by augmented matrix (5.8), also have determinantal-style solutions. For example, if, as in the well problem, we have five conditions in six unknowns, $l_i = 1$ for $1 \leq i \leq 5$, and we set the $(n+1)^{\text{th}}$ unknown $y = \det A$, the solutions for the remaining unknowns $x_1, x_2, \dots, x_5$ are as follows:

$$x_1 = (((k_2 - 1)k_3 + 1)k_4 - 1)k_5 + 1, \tag{5.9}$$

$$x_2 = (((k_3 - 1)k_4 + 1)k_5 - 1)k_1 + 1, \tag{5.10}$$

$$x_3 = (((k_4 - 1)k_5 + 1)k_1 - 1)k_2 + 1, \tag{5.11}$$

$$x_4 = (((k_5 - 1)k_1 + 1)k_2 - 1)k_3 + 1, \tag{5.12}$$

$$x_5 = (((k_1 - 1)k_2 + 1)k_3 - 1)k_4 + 1. \tag{5.13}$$

These solutions could easily have been computed on a counting board, using what might be termed a determinantal calculation, as the following diagram for the calculation of the fifth unknown, x_5, in the well problem demonstrates:

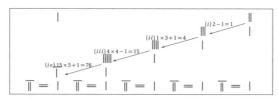

The remaining unknowns are easily calculated following similar patterns.

The earliest record of such a determinantal solution that I have found in Chinese treatises is preserved in a commentary on the well problem in Fang Zhongtong's 方中通 (1634–1698) *Numbers and Measurement, An Amplification* (*Shu du yan* 數度衍, c. 1661).[40] Fang describes, using words to denote positions in the array of numbers, a solution for the fifth unknown of the well problem. This solution is valid for a slightly more general class of problems, which, written using modern terminology as an augmented matrix, is of the following form,

$$\begin{bmatrix} k_1 & l_1 & 0 & \cdots & 0 & b_1 \\ 0 & k_2 & l_2 & \ddots & \vdots & b_2 \\ \vdots & \ddots & \ddots & \ddots & 0 & \vdots \\ 0 & \cdots & 0 & k_{n-1} & l_{n-1} & b_{n-1} \\ l_n & 0 & \cdots & 0 & k_n & b_n \end{bmatrix}. \tag{5.14}$$

The solution, as calculated by Fang, written in modern notation, is given by the following,

$$x_5 = \frac{(((k_1 b_5 - l_5 b_1)k_2 + l_5 l_1 b_2)k_3 - l_5 l_1 l_2 b_3)k_4 + l_5 l_1 l_2 l_3 b_4}{k_1 k_2 k_3 k_4 k_5 + l_5 l_1 l_2 l_3 l_4}.$$

And if, beginning with augmented matrix (5.8), we set the $(n + 1)^{\text{th}}$ unknown $y = \det A = k_1 k_2 k_3 k_4 k_5 + l_5 l_1 l_2 l_3 l_4$, we have a somewhat simpler solution for the fifth unknown,

$$x_5 = ((((k_1 - l_5)k_2 + l_5 l_1)k_3 - l_5 l_1 l_2)k_4 + l_5 l_1 l_2 l_3.$$

---

[40] Fang Zhongtong, *Shu du yan* 數度衍 [Numbers and measurement, an amplification], in *SKQS*.

Again, the remaining unknowns can be calculated in a similar manner.

The solution given by Fang, it can be shown, can be generalized for any n ($n \geq 3$). For example, given a system of n conditions in n unknowns of the more general form of augmented matrix (5.14), the value of the nth unknown is given by

$$x_n = \frac{\begin{array}{c}(\cdots((((k_1 b_n - l_n b_1)k_2 + l_n l_1 b_2)k_3 - l_n l_1 l_2 b_3)k_4 + l_n l_1 l_2 l_3 b_4)\cdots) \\ \cdot k_{n-2} \mp l_n l_1 l_2 \cdots l_{n-3} b_{n-2})k_{n-1} \pm l_n l_1 l_2 \cdots l_{n-2} b_{n-1}\end{array}}{k_1 k_2 k_3 \cdots k_{n-1} k_n \pm l_n l_1 l_2 l_3 \cdots l_{n-1}}, \quad (5.15)$$

where again $\pm$ is $+$ for n odd and $-$ for n even. A similar solution can be found for each of the n unknowns.

This solution, although it looks complicated when written in modern mathematical notation, is not difficult to compute on the counting board by following simple patterns, as illustrated below. We set the terms $l_n l_1$, $-l_n l_1 l_2$, $l_n l_1 l_2 l_3$, $\ldots$, $\mp l_n l_1 \cdots l_{n-2}$ in the empty positions in the left-hand column. The numerator given in equation (5.15) is then computed by a series of cross-multiplications, working from the outermost corners inward. In the first step, the opposite corners of the array are multiplied together, and subtracted, giving the result $k_1 b_n - l_n b_1$:

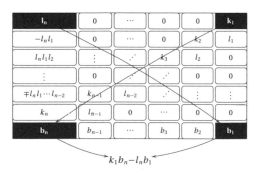

In each remaining step, we follow the same pattern, moving one position inward, cross-multiplying, and then subtracting. We continue in this manner until we reach k_{n-1}, which is the final step:

l_n	0	$\cdots$	0	0	k_1
$-l_n l_1$	0	$\cdots$	0	k_2	l_1
$l_n l_1 l_2$	$\vdots$	$\cdot\cdot$	k_3	l_2	0
$\vdots$	0	$\cdot\cdot$	$\cdot\cdot$	0	0
$\mp l_n l_1 \cdots l_{n-2}$	k_{n-1}	l_{n-2}	$\cdot\cdot$	$\vdots$	$\vdots$
k_n	l_{n-1}	0	$\cdots$	0	0
b_n	b_{n-1}	$\cdots$	b_3	b_2	b_1

$$((\cdots((k_1 b_n - l_n b_1)k_2 + l_n l_1 b_2)\cdots) \cdot k_{n-2} \mp l_n l_1 l_2 \cdots l_{n-3} b_{n-2})$$
$$\cdot k_{n-1} \pm l_n l_1 l_2 \cdots l_{n-2} b_{n-1}$$

This then gives the numerator for (5.15). Again, it is important to note that these calculations do not require literacy.

Although the earliest extant record I have found of a determinantal solution is Fang's *Numbers and Measurement* from the seventeenth century, there is important evidence that some determinantal solutions, such as those given in equations (5.9)–(5.13), were known at the time of the compilation of the *Nine Chapters*. Among the eighteen problems in "*Fangcheng*," chapter 8 of the *Nine Chapters*, there are four more problems, in addition to the well problem, that are variants of the distinctive form given by augmented matrices (5.8) and (5.14), for which the absence of entries permits relatively simple determinantal solutions, namely problems 3, 12, 14, and 15. For each of these problems, following simple patterns, it is possible to compute the values for all of the unknowns; and from the point of view of the original *fangcheng* procedure (a variant of Gaussian elimination) presented in the *Nine Chapters*, there is nothing noteworthy about any of these problems that would explain their inclusion in the text. A brief conspectus of these problems is presented in Table 5.2.

Table 5.2: From among the eighteen problems in "*Fangcheng*," chapter 8 of the *Nine Chapters*, here are four problems, in addition to the well problem given in equation (5.7), that are variants of augmented matrices (5.8) and (5.14), for which the absence of entries allows simple determinantal solutions.

Problem	Counting Board	Augmented Matrix	Notes
3		$\begin{bmatrix} 2 & 1 & 0 & 1 \\ 0 & 3 & 1 & 1 \\ 1 & 0 & 4 & 1 \end{bmatrix}$	Variant with three conditions in three unknowns.
12		$\begin{bmatrix} 1 & 1 & 0 & 40 \\ 0 & 2 & 1 & 40 \\ 1 & 0 & 3 & 40 \end{bmatrix}$	Variant with three conditions in three unknowns.
14		$\begin{bmatrix} 2 & 1 & 1 & 0 & 1 \\ 0 & 3 & 1 & 1 & 1 \\ 1 & 0 & 4 & 1 & 1 \\ 1 & 1 & 0 & 5 & 1 \end{bmatrix}$	Variant with four conditions in four unknowns.
15		$\begin{bmatrix} 2 & -1 & 0 & 1 \\ 0 & 3 & -1 & 1 \\ -1 & 0 & 4 & 1 \end{bmatrix}$	Variant with three conditions in three unknowns, with negative coefficients.

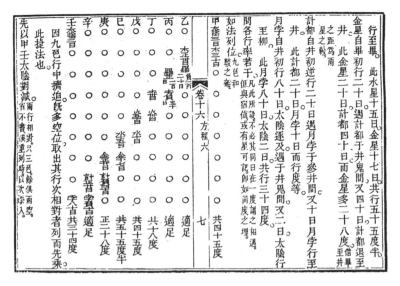

Fig. 5.1: An array of the form given in augmented matrix (5.14), representing a problem with nine conditions in nine unknowns, from Mei Wending's 梅文鼎 (1633–1721) *Fangcheng lun* 方程論 (On *fangcheng*, c. 1674, photolithographic reprint from the Mei Juecheng Chengxue tang 梅毂成承學堂 printing of the *Mei shi congshu ji yao* 梅氏叢書輯要).

Later Chinese mathematical treatises continue to record numerous problems that are variants of the distinctive form given in augmented matrices (5.8) and (5.14): often as many as 10–20% of the problems are similar. It was because of the importance of these problems in Chinese mathematical treatises that I chose an example of a *fangcheng* problem with nine conditions in nine unknowns of the form of augmented matrix (5.14), shown here in Figure 5.1, for the cover of *Chinese Roots of Linear Algebra*.

Problems Recorded in Leonardo Pisano's *Liber Abaci*

Problems of the form of augmented matrix (5.8), for which the well problem was an exemplar, are so distinctive that they can serve as a "fingerprint," a kind of unique identifier. In my research for *Chinese Roots of Linear Algebra*, I searched through thousands of matrices recorded in modern mathematical treatises,[41]

---

[41] Strang, *Linear Algebra and Its Applications*; Denis Serre, *Matrices: Theory and Applications* (New York: Springer, 2002); Fuzhen Zhang, *Matrix Theory: Basic Results and Techniques* (New York: Springer, 1999); Stewart, *Matrix Algorithms*; Roger A. Horn and Charles R. Johnson, *Topics in Matrix Analysis* (New York: Cambridge University Press, 1991); Charles G. Cullen, *Matrices and Linear Transformations*, 2nd ed. (New York: Dover, 1990); Roger A. Horn and Charles R. Johnson, *Matrix Analysis* (New York: Cambridge University Press, 1985); G. W. Stewart, *Introduction to Matrix Computations* (New York: Academic Press, 1973); Kenneth Hoffman and Ray

and found only one example of a problem that was similar (a problem with three equations in three unknowns). Because these problems are so distinctive, I then searched for similar problems in European mathematical treatises from the seventeenth and eighteenth centuries, including the extant mathematical writings of Gauss and Leibniz. Unable to find similar problems, I felt that the most that I could definitively conclude is that "[t]he essentials of the methods used today ... in modern linear algebra *were not first discovered* by Leibniz or by Gauss."[42] It turns out that I was looking for problems of the form given in augmented matrix (5.8) in seventeenth- and eighteenth-century European texts, at least four centuries after they were transmitted to Europe.

Many examples of problems equivalent to the form given in augmented matrix (5.8) can be found in the work of Leonardo Pisano (c. 1170–c. 1250), more commonly known today by the name Fibonacci. Not only are these problems themselves distinctive, the solutions to these problems are even more so. The solutions to these problems are without analog in modern mathematics, and in fact these solutions are esoteric enough that they have not been adequately analyzed in previous studies of Fibonacci's mathematics.[43] To put it another way, it was because of my familiarity with Chinese *fangcheng* problems and their solutions that the problems and solutions I found in Fibonacci made sense.

Here I will briefly examine one problem of the form of augmented matrix (5.8), "Another Problem on Five Men," and its solution (as Table 5.3 on page 179 shows, this is but one among several examples). In Fibonacci, this problem is recorded as a word problem. In modern mathematical terms, it is a system of five conditions in six unknowns, which can be written as follows:

$$x_1 + \frac{2}{3}x_2 = y, \ x_2 + \frac{4}{7}x_3 = y, \ x_3 + \frac{5}{11}x_4 = y, \ x_4 + \frac{6}{13}x_5 = y, \ x_5 + \frac{8}{19}x_1 = y. \quad (5.16)$$

If we write this as an augmented matrix, to facilitate comparison with the *fangcheng* problems, we can immediately see that it is similar in form to that given by

---

Alden Kunze, *Linear Algebra*, 2nd ed. (Englewood Cliffs, NJ: Prentice-Hall, 1971); Thomas Muir, *The Theory of Determinants in the Historical Order of Development*, 4 vols. (London: Macmillan, 1906–1923).

[42] Hart, *Chinese Roots of Linear Algebra*, p. 192, emphasis in italics added.

[43] Johannes Tropfke et al., *Geschichte der Elementarmathematik, Band 1, Arithmetik und Algebra*, 4th ed., 3 vols. (New York: Walter de Gruyter, 1980); Jacques Sesiano, "The Appearance of Negative Solutions in Mediaeval Mathematics," *Archive for History of Exact Sciences* 32 (1985): 105–50; Maryvonne Spiesser, "Problèmes linéaires dans le Compendy de la praticque des nombres de Barthélemy de Romans et Mathieu Préhoude (1471): Une approche nouvelle basée sur des sources proches du Liber abbaci de Léonard de Pise," *Historia Mathematica* 27 (2000): 362–83; John Hannah, "False Position in Leonardo of Pisa's Liber Abbaci," *Historia Mathematica* 34 (2007): 306–32; idem, "Conventions for Recreational Problems in Fibonacci's Liber Abbaci," *Archive for History of Exact Sciences* 65 (2010): 155–80.

augmented matrix (5.8):

$$\begin{bmatrix} 1 & \frac{2}{3} & 0 & 0 & 0 & y \\ 0 & 1 & \frac{4}{7} & 0 & 0 & y \\ 0 & 0 & 1 & \frac{5}{11} & 0 & y \\ 0 & 0 & 0 & 1 & \frac{6}{13} & y \\ \frac{8}{19} & 0 & 0 & 0 & 1 & y \end{bmatrix}. \tag{5.17}$$

To solve the problem, Fibonacci writes the problem out as a series of fractions arranged contiguously, as follows,

The details of the taking are written in order thus: $\frac{8}{19} \frac{6}{13} \frac{5}{11} \frac{4}{7} \frac{2}{3}$.[44]

The operations he uses to solve the problem are not those familiar from modern mathematics. First, the $(n+1)^{\text{th}}$ unknown is found as follows:

> And all of the numbers that are under the fraction are multiplied together; there will be 57057. *As the number of men is odd*, the product of the numbers that are over the fractions is *added* to this, that is, the 2 times the 4 times the 5 times the 6 [times the 8]; there will be 58977, which is had for the price of the horse.[45]

The above calculation notes that when "the number of men is odd" the products of the two sets of numbers (the numerators and the denominators) are to be added. That is, the price of the horse is given by

$$3 \cdot 7 \cdot 11 \cdot 13 \cdot 19 + 2 \cdot 4 \cdot 5 \cdot 6 \cdot 8 = 58977. \tag{5.18}$$

The problem is then solved for each unknown in succession. The calculation for the first unknown is explained in Fibonacci as follows:

> And as the first man's bezants are had, the upper number of the fraction of the takings is subtracted from the lower number of the same fraction, that is the 2 from the 3; there remains 1, and it is multiplied by the 7; there will be 7, to which you add the product of the 2 and the 4; there will be 15, which you multiply by the 11; there will be 165, from which you subtract the product of the 2 and the 4 and the 5; there remains 125, which you multiply by the 13; there will be 1625, to which you add the product of the 2 and the 4 and the 5 and the 6; there will be 1865, which is multiplied by the 19; there will be 35435, and the first man has this many.[46]

---

[44] Laurence E. Sigler, *Fibonacci's Liber Abaci: A Translation into Modern English of Leonardo Pisano's Book of Calculation* (New York: Springer, 2002), 345; *Scritti di Leonardo Pisano matematico del secolo decimoterzo. I. Il Liber abbaci di Leonardo Pisano*, ed. Baldassare Boncompagni (Roma: Tipografia delle Scienze Matematiche e Fisiche, 1857), 234.

[45] Sigler, *Fibonacci's Liber Abaci*, 345, emphasis in italics added; Boncompagni, *Liber abbaci*, 234, interpolation based on the mathematics, added.

[46] Sigler, *Fibonacci's Liber Abaci*, 345; Boncompagni, *Liber abbaci*, 234–35.

Written out using modern mathematical notation, this is the following calculation,

$$((((3-2)\cdot7+2\cdot4)\cdot11-2\cdot4\cdot5)\cdot13+2\cdot4\cdot5\cdot6)\cdot19=35435. \qquad (5.19)$$

Each of the remaining unknowns is calculated in a similar manner.

More abstractly, in modern mathematical terms, "Another Problem on Five Men" is an exemplar for the following problem with 5 conditions in 6 unknowns:

$$x_1+\frac{k_1}{l_1}x_2=y,\; x_2+\frac{k_2}{l_2}x_3=y,\; x_3+\frac{k_3}{l_3}x_4=y,\; x_4+\frac{k_4}{l_4}x_5=y,\; x_5+\frac{k_5}{l_5}x_1=y. \quad (5.20)$$

Written as an augmented matrix (again, to facilitate comparison with the *fangcheng* problems), we have the following:

$$\begin{bmatrix} 1 & \frac{k_1}{l_1} & 0 & 0 & 0 & y \\ 0 & 1 & \frac{k_2}{l_2} & 0 & 0 & y \\ 0 & 0 & 1 & \frac{k_3}{l_3} & 0 & y \\ 0 & 0 & 0 & 1 & \frac{k_4}{l_4} & y \\ \frac{k_5}{l_5} & 0 & 0 & 0 & 1 & y \end{bmatrix}. \qquad (5.21)$$

If we follow the steps given in the word problem, the first step is to write the problem as a series of fractions arranged contiguously in the following manner:

$$\frac{k_5}{l_5}\,\frac{k_4}{l_4}\,\frac{k_3}{l_3}\,\frac{k_2}{l_2}\,\frac{k_1}{l_1}. \qquad (5.22)$$

Then we calculate the $(n+1)^{\text{th}}$ unknown, y, by setting it to the value

$$k_1k_2k_3k_4k_5+l_1l_2l_3l_4l_5, \qquad (5.23)$$

which is the numerator of the determinant of the coefficient matrix,

$$\det A = \begin{vmatrix} 1 & \frac{k_1}{l_1} & 0 & 0 & 0 \\ 0 & 1 & \frac{k_2}{l_2} & 0 & 0 \\ 0 & 0 & 1 & \frac{k_3}{l_3} & 0 \\ 0 & 0 & 0 & 1 & \frac{k_4}{l_4} \\ \frac{k_5}{l_5} & 0 & 0 & 0 & 1 \end{vmatrix} = \frac{k_1k_2k_3k_4k_5+l_1l_2l_3l_4l_5}{l_1l_2l_3l_4l_5}. \qquad (5.24)$$

The solutions, again written in modern mathematical notation, are then

$$x_1=((((l_1-k_1)l_2+k_1k_2)l_3-k_1k_2k_3)l_4+k_1k_2k_3k_4)l_5, \qquad (5.25)$$
$$x_2=((((l_2-k_2)l_3+k_2k_3)l_4-k_2k_3k_4)l_5+k_2k_3k_4k_5)l_1, \qquad (5.26)$$
$$x_3=((((l_3-k_3)l_4+k_3k_4)l_5-k_3k_4k_5)l_1+k_3k_4k_5k_1)l_2, \qquad (5.27)$$
$$x_4=((((l_4-k_4)l_5+k_4k_5)l_1-k_4k_5k_1)l_2+k_4k_5k_1k_2)l_3, \qquad (5.28)$$
$$x_5=((((l_5-k_5)l_1+k_5k_1)l_2-k_5k_1k_2)l_3+k_5k_1k_2k_3)l_4. \qquad (5.29)$$

As with *fangcheng* problems, these problems were solved using simple patterns in two dimensions. Below I will reconstruct the calculations to find the first unknown.[47]

STEP 0.   In the margins of Leonardo's *Liber Abaci*, in the edition preserved by Boncompagni, there are diagrams of the fractions arranged contiguously, which can be written in general terms using modern notation as follows:

$$\frac{k_n}{l_n} \quad \frac{k_{n-1}}{l_{n-1}} \quad \cdots \quad \frac{k_3}{l_3} \quad \frac{k_2}{l_2} \quad \frac{k_1}{l_1}$$

STEP 1.   The first step is simply to begin with the lower term from the first fraction. (In this reconstruction, I have placed the calculations to the right of the contiguously arranged fractions. In the text, there is no indication of where or even whether successive results were recorded.)

STEP 2.   Second, we subtract the upper term of the first fraction from the lower term of the first fraction, yielding $l_1 - k_1$:

STEP 3.   The next step is to multiply the results of the last operation, $l_1 - k_1$, by the lower term of the second fraction, l_2:

STEP 4.   In the following step, we multiply the upper terms of the first two fractions together, giving $k_1 k_2$. In the steps that follow, we will alternate between subtracting and adding the terms $k_1 \cdots k_i$ to the previous result. In step 2 above, we subtracted k_1, so here we add $k_1 k_2$ to the previous result, $(l_1 - k_1) \cdot l_2$, giving $(l_1 - k_1) \cdot l_2 + k_1 k_2$:

---

[47] These results are a summary of research I am preparing for publication.

$$
\begin{array}{ccccccc}
k_n & k_{n-1} & & k_3 & \widehat{k_2} & \widehat{k_1} \\
\rule{1em}{0.5pt} & \rule{1em}{0.5pt} & \cdots & \rule{1em}{0.5pt} & \rule{1em}{0.5pt} & \rule{1em}{0.5pt} \\
l_n & l_{n-1} & & l_3 & l_2 & l_1
\end{array}
\qquad (l_1 - k_1)\cdot l_2 + \widehat{k_1 k_2}
$$

STEP 5.   We then multiply the results of the previous step, $((l_1 - k_1)\cdot l_2 + k_1 k_2)$, by the lower term of the third fraction, l_3:

$$
\begin{array}{ccccccc}
k_n & k_{n-1} & & k_3 & k_2 & k_1 \\
\rule{1em}{0.5pt} & \rule{1em}{0.5pt} & \cdots & \rule{1em}{0.5pt} & \rule{1em}{0.5pt} & \rule{1em}{0.5pt} \\
l_n & l_{n-1} & & \widehat{l_3} & l_2 & l_1
\end{array}
\qquad ((l_1 - k_1)\cdot l_2 + k_2 k_1)\cdot \widehat{l_3}
$$

STEP 6.   We multiply together, from right to left, the upper terms of the first three fractions, giving $k_1 k_2 k_3$. We then subtract this from the previous result $((l_1 - k_1)\cdot l_2 + k_1 k_2)\cdot l_3$, giving $((l_1 - k_1)\cdot l_2 + k_1 k_2)\cdot l_3 - k_1 k_2 k_3$:

$$
\begin{array}{ccccccc}
k_n & k_{n-1} & & \widehat{k_3} & \widehat{k_2} & \widehat{k_1} \\
\rule{1em}{0.5pt} & \rule{1em}{0.5pt} & \cdots & \rule{1em}{0.5pt} & \rule{1em}{0.5pt} & \rule{1em}{0.5pt} \\
l_n & l_{n-1} & & l_3 & l_2 & l_1
\end{array}
\qquad ((l_1 - k_1)\cdot l_2 + k_1 k_2)\cdot l_3 - \widehat{k_1 k_2 k_3}
$$

$$\vdots$$

STEP $(2n - 2)$.   We continue to follow these simple patterns. The next-to-last step is to multiply the upper terms of all but the last of the fractions, from right to left, giving $k_1 k_2 \cdots k_{n-1}$. Since we alternate between adding and subtracting these terms, we will add this term if n is odd, and subtract if n is even, as follows:

$$
\begin{array}{ccccccc}
k_n & \widehat{k_{n-1}} & \cdots & \widehat{k_3} & \widehat{k_2} & \widehat{k_1} \\
\rule{1em}{0.5pt} & \rule{1em}{0.5pt} & \cdots & \rule{1em}{0.5pt} & \rule{1em}{0.5pt} & \rule{1em}{0.5pt} \\
l_n & l_{n-1} & & l_3 & l_2 & l_1
\end{array}
\qquad (\cdots((l_1 - k_1)\cdot l_2 + k_1 k_2)\cdots)\cdot l_{n-1} \pm \widehat{k_1 k_2 \cdots k_{n-1}}
$$

STEP $(2n - 1)$.   The final step is as follows:

$$
\begin{array}{ccccccc}
k_n & k_{n-1} & & k_3 & k_2 & k_1 \\
\rule{1em}{0.5pt} & \rule{1em}{0.5pt} & \cdots & \rule{1em}{0.5pt} & \rule{1em}{0.5pt} & \rule{1em}{0.5pt} \\
\widehat{l_n} & l_{n-1} & & l_3 & l_2 & l_1
\end{array}
\qquad ((\cdots((l_1 - k_1)\cdot l_2 + k_1 k_2)\cdots)\cdot l_{n-1} \pm k_1 k_2 \cdots k_{n-1})\cdot \widehat{l_n}
$$

This is only one example of several similar problems recorded in Fibonacci's *Liber Abaci*, as shown in Table 5.3.

Table 5.3: Several examples from Fibonacci's *Liber Abaci* of problems of the form given in augmented matrix (5.8) (the last problem listed here, "On Five Men Who Bought a Horse," is solved in *Liber Abaci* by a variant of false position).

Title of problem	Problem written as an augmented matrix
"On the Purchase of a Horse by Three Men, When Each One Takes Some Bezants from the Others." Sigler, *Fibonacci's Liber Abaci*, 338–39.	$\begin{bmatrix} 1 & \frac{1}{3} & 0 & y \\ 0 & 1 & \frac{1}{4} & y \\ \frac{1}{5} & 0 & 1 & y \end{bmatrix}$
"Another Problem on Three Men According to the Abovewritten Method." Ibid., 341.	$\begin{bmatrix} 1 & \frac{2}{3} & 0 & y \\ 0 & 1 & \frac{4}{7} & y \\ \frac{5}{9} & 0 & 1 & y \end{bmatrix}$
"On the Same with Four Men." Ibid., 341–42.	$\begin{bmatrix} 1 & \frac{1}{3} & 0 & 0 & y \\ 0 & 1 & \frac{1}{4} & 0 & y \\ 0 & 0 & 1 & \frac{1}{5} & y \\ \frac{1}{6} & 0 & 0 & 1 & y \end{bmatrix}$
"On Five Men Who Bought a Horse." Ibid., 458.	$\begin{bmatrix} 1 & \frac{1}{2} & 0 & 0 & 0 & y \\ 0 & 1 & \frac{1}{3} & 0 & 0 & y \\ 0 & 0 & 1 & \frac{1}{4} & 0 & y \\ 0 & 0 & 0 & 1 & \frac{1}{5} & y \\ \frac{1}{6} & 0 & 0 & 0 & 1 & y \end{bmatrix}$

Analysis of these and other problems recorded in Leonardo's *Liber Abaci* leads to the following conclusions:

1. Problems of the distinct form of augmented matrices (5.8) and (5.14), with solutions that are quite esoteric, recorded in Chinese treatises dating from about the first century CE, are also recorded in Leonardo's *Liber Abaci*.

2. The solutions are valid for any number of unknowns, and the same simple visual patterns work for all of the unknowns.

3. Leonardo records these solutions in such detail that there can be no question as to how they were solved, and the mathematical practice can be reliably reconstructed.

4. Leonardo does not, however, provide an explanation of the actual mathematical practice, but only records the calculations.

5. These methods do not require literacy, and in fact, the translation of this two-dimensional mathematical practice into narrative renders it almost incomprehensible.

6. Calculations of this form were quite common in this period. More specifically, these calculations are similar in form to what is today known as Horner's method.[48]

7. Although these problems are preserved in Leonardo's *Liber Abaci*, to my knowledge they were not recorded in other European treatises, perhaps because the solution is so esoteric.

The *Guide to Calculation*

Having traced the history of *fangcheng* problems, we are now in a better position to analyze the *Guide to Calculation* attributed to Li and Ricci, together with Xu Guangqi's preface, in which he claims that Western mathematics is in every way superior. Most of the *Guide to Calculation* is indeed translated from Clavius's *Epitome*, and most of it consists of elementary mathematics, beginning with addition, subtraction, multiplication, and division. The most "modern" mathematics in the *Guide to Calculation* is in chapters in the second volume (*tong bian* 通編) on what we now call linear algebra. We will first examine chapter 4 of the second volume, which presents problems from Clavius's *Epitome* with 2 conditions in 2 unknowns, and then explicitly compares them to similar problems in Chinese treatises, in order to pronounce the Western mathematics superior. Next, we will examine chapter 5. There are no problems with n conditions in n unknowns ($n \geq 3$) in Clavius's *Epitome*, so *fangcheng* problems from Chinese sources were simply purloined without attribution. There is little evidence that Xu, Li, Yang, and their collaborators understood these problems: they offer no analysis, criticisms, or alternative methods. Next, we will examine Xu's preface, and in particular his claim that Western mathematics was in every way superior.

---

[48] I thank John Crossley for this important suggestion noting the similarity. For a detailed explanation of this method of finding roots, see Martzloff, *History of Chinese Mathematics*, "The Extraction of Roots," pp. 221–49. This method can be found, in various forms, in the *Nine Chapters* and later Chinese mathematical treatises. Polynomial equations, which are today written in the form

$$a_n x^n + a_{n-1} x^{n-1} + a_{n-2} x^{n-2} + \ldots + a_3 x^3 + a_2 x^2 + a_1 x + a_0,$$

were solved by calculating in the following manner,

$$(((\cdots(((a_n x + a_{n-1})x + a_{n-2})x + a_{n-3})\cdots)x + a_2)x + a_1)x + a_0.$$

In the case that $a_n = 1$, this reduces to

$$(((\cdots(((x + a_{n-1})x + a_{n-2})x + a_{n-3})\cdots)x + a_2)x + a_1)x + a_0.$$

This method is more efficient than the modern method. See Donald Ervin Knuth, *The Art of Computer Programming*, 2nd ed. (Reading, MA: Addison-Wesley, 1981), II, 467, cited in Martzloff, *History of Chinese Mathematics*, 246.

Chapter 4, "Adding and Borrowing for Mutual Proof"

The fourth chapter (*juan si* 卷四), titled "Adding and borrowing for mutual proof" (*die jie hu zheng* 疊借互徵), is a translation of chapter 23 of Clavius's *Epitome*, titled "Regula falsi duplicis positionis" (the rule of double false position). These problems, which likely circulated throughout Eurasia, are equivalent to 2 conditions in 2 unknowns, and are similar to excess-deficit problems commonly recorded in Chinese mathematical treatises of the period.

The *Guide to Calculation* employs here two main strategies to assert the superiority of Western mathematics. First, the problems from Clavius's *Epitome* are alleged to predate Chinese problems:

> Old [Chinese mathematical treatises] had a section [titled] "excess and deficit,"[49] on the whole of the same category as this; but this [the contents of this chapter of the *Guide to Calculation*] predates [the time] when [the Chinese methods of] "excess and deficit" did not yet exist.
>
> 舊有盈朒一章。大都類此。而此則於未有盈朒之先。

Second, the *Guide to Calculation* imputes superiority of methods from the *Epitome* over Chinese methods by a purported comparison. Problems from Chinese sources are appended to the end of the chapter, explicitly stating that they are copied: before the problems, "The old method 'excess and deficit' chapter is one that people are constantly studying, so [we have] appended several problems below for comparison" 舊法盈朒章人所恆習，亦附數條于後相比擬; after the problems, "the previous are 'denominator-numerator' excess-deficit" 右母子盈朒. It should be noted that there is no evidence to support either of the claims of superiority made by the Jesuits' Chinese patrons.

Chapter 5, "Method for addition, subtraction, and multiplication of heterogeneous elements"

Clavius' *Epitome* does not record any problems equivalent to *fangcheng* 方程, that is, problems equivalent to the solution of systems of n conditions in n unknowns ($n \geq 3$) in modern linear algebra. In general, such problems were rarely recorded in European treatises of the period; rather, the vast majority of written records of these problems are from Chinese treatises.

The problems in chapter 5, "Method for addition, subtraction, and multiplication of heterogeneous [elements]" [*za he jiao cheng fa* 雜和較乘法], are copied, without attribution, problem for problem, from Chinese sources. Many of these problems are copied from Cheng Dawei's 程大位 (1533–1606) *Comprehensive*

---

[49] The title used in *Jiuzhang suanshu* 九章算術 [Nine chapters on the mathematical arts], in *ZKJDT*, is "Excess and deficit" (*ying bu zu*). Wu Jing 吳敬 (fl. 1450), *Jiuzhang xiang zhu bilei suanfa da quan* 九章詳註比類算法大全 [Complete collection of the mathematical arts of the nine chapters, with detailed commentary, arranged by category], in *ZKJDT*, uses *ying bu zu*, but elsewhere uses *ying fei* 盈朏. Cheng Dawei 程大位 (1533–1606), *Suanfa tongzong* 算法统宗 [Comprehensive source of mathematical methods], in *ZKJDT*, uses "ying nü" 盈朒.

Source of Mathematical Methods (*Suanfa tongzong* 算法统宗): problem twelve is Cheng's problem three; problem fourteen is Cheng's problem five; problem sixteen is Cheng's problem seven; and problem seventeen is Cheng's problem eight. Changes are purely cosmetic: the order of sentences in the problems is changed, names of variables are changed, but none of the numbers used in the calculation are changed. All the solutions are correct, but no additional analysis or explanation is offered by the Jesuits' Chinese patrons and collaborators.

In particular, the well problem is copied, without any attribution or acknowledgment, from Chinese mathematical treatises. Again, Clavius's *Epitome* includes no similar problems. The original source of the problem is not certain: Cheng's *Comprehensive Source of Mathematical Methods* does not contain the well problem; the most likely source is Wu Jing's 吳敬 (fl. 1450) *Complete Collection of the Mathematical Arts of the Nine Chapters, with Detailed Commentary, Arranged by Category* (*Jiuzhang xiang zhu bilei suanfa da quan* 九章詳註比類算法大全).

Below is a translation of this problem, as found in the *Guide to Calculation*.

Problem: The depth of a well is unknown. Using 2 of A's ropes the water is not reached, borrowing 1 of B's ropes to supplement it, the water is reached. Using 3 of B's ropes then borrowing 1 of C's, using 4 of C's ropes then borrowing 1 of D's, using 5 of D's ropes then borrowing 1 of E's, using 6 lengths of E's ropes, then borrowing 1 of A's, then all reach the water. How much is the depth of the well? How much is [the length of] each rope?

問井不知深。用甲繩二不及泉，借乙繩一補之，及泉。用乙繩三則借丙一，用丙繩四則借丁一，用丁繩五則借戊一，用戊繩六條借甲一，乃俱及泉。其井深若干？五等繩各長若干？

The solution presented in the *Guide to Calculation* then follows. First, the depth of the well is found, in the same manner that is recorded in earlier Chinese sources, including Jia's *Detailed Notes*, Yang's *Detailed Explanations*, and Wu's *Complete Collection of Mathematical Methods*, by a determinantal calculation:

Lay out the five columns. Take the numbers of the five ropes as the "denominators" [*mu*]. Take 1, [the number of] ropes borrowed, as the "numerator" [*zi*]. First take 2 [representing the number of ropes from] A multiplied by the 3 [for the ropes from] B to obtain 6. Multiply by [4 for the ropes from] C to obtain 24. Multiply by [5 for the ropes from] D to obtain 120. Multiply by [6 for the ropes from] E to obtain 720. Add to this the "numerator" 1, and together, 721 is the "depth product" of the well.

列五行。以五繩之數為母。借繩一為子。先取甲二乘乙三得六。以乘丙得二十四。以乘丁得一百二十。以乘戊得七百二十。併入子一。共七百二十一為井深積。

The *Guide to Calculation* presents this calculation and the assignment of the result to the depth of the well without any commentary, analysis, or criticism, just as in Wu's *Complete Collection of Mathematical Methods*.

Following the determination of the depth of the well, the problem is set as an array:

Arrange in position [in an array, as shown in Figure 5.2].

列位。

Fig. 5.2: Diagram from Li and Ricci's *Guide to Calculation*.

The elimination of entries by column reductions then follows:

Then take the fifth column as "principal," and multiply [it] together with the first, second, third, and fourth [columns].

乃取五行為主，而以一二三四俱與相乘。

First take 2, [the entry for] A in the first column, as the "divisor," and multiply the fifth column term-by-term. [The entry for] A, 1, gets 2. [The entry for] E, 6, gets 12. The "product," 721, gets 1442.

先以一行甲二為法，遍乘五行 甲一得二，戊六得十二，積七百二十一得一千四百四十二。

Also, multiply the first column by 1, [the entry for] A in the fifth column, and subtract [the first column from the fifth] term-by-term. [The entry for] A, 2, has 2 subtracted from it and is eliminated. [The entry for] B, 1, gets 1, and because [the entry for] B in the fifth column is empty, set −1. [The entry for] the "product," 721, gets the original number and is subtracted from the fifth column, then the remainder is still 721.

五行甲一亦乘一行對減。 甲二得二減盡。乙一得一。因五行乙空立負一。積七百二十一得本數，以減五行，仍餘七百二十一。

Next take 3, [the entry for] B in the second column, as the "divisor," and multiply the fifth column. [The entry for] B, −1, gets −3. [The entry for] E, +12, gets 36. [The entry for] the "product," 721, gets 2163.

次以二行乙三為法乘五行。 乙負一得負三。戊正十二得三十六。積七百二十
一得二千一百六十三。

Also multiply the second column by −1, [the entry for] B in the fifth column.
[The entry] B, 3, has 3 subtracted from it and is eliminated. [The entry for] C, 1, gets 1,
and because [the entry for] C in the fifth column is empty, set −1. [The entry for] the
"product," 721, gets the original number, [to which is] added the "product" of the fifth
column, 2163, together 2884.

五行乙負一亦乘二行。 乙三得三，對減盡。丙一得一，因五行丙空，立負一。
積七百二十一得本數，併入五行積二千一百六十三，共二千八百八十四。

Next take 4, [the entry for] C in the third column, as the "divisor," and
multiply the fifth column. [The entry for] E, +36, gets 144. [The entry for] the
"product," 2884, gets 11536.

再以三行丙四為法，乘五行。 戊正三十六得一百四十四。積二千八百八十四
得一萬一千五百三十六。

Also multiply the third column by −1, [the entry for] C in the fifth column.
[The entry for] B, 4, has 4 subtracted from it and is eliminated. [The entry for] D, 1,
gets 1. Because [the entry for] D in the fifth column is empty, set −1. [The entry for]
the "product" gets the original number, and subtracted from the "product" in the fifth
column, 11536, the remainder is 10815.

五行丙負一亦乘三行。 丙四得四減盡。丁一得一，因五行丁空，立負一。積得
本數，與五行積一萬一千五百三十六對減餘一萬八百一十五。

Next take 5, [the entry for] D in the fourth column, as the "divisor," and
multiply the fifth column. [The entry for] D, −1, obtains 5. [The entry for] E, +144,
obtains 720. The "product," 10815, obtains 54075.

又以四行丁五為法乘五行。 丁負一得五。戊正一百四十四得七百二十。積一
萬八百一十五得五萬四千七十五。

Also multiply the fourth column by −1, the entry for D in the fifth column.
[The entry for] D, 5, obtains 5 and is eliminated. [The entry for] E, 1, obtains 1, added to
+720, [the entry for] E in the fifth column, together is 721. [The entry for] the "product"
obtains the original number, added to [the entry for] the "product" in the fifth column,
54075, obtaining 54796.

五行丁負一亦乘四行。 丁五得五減盡。戊一得一，併入五行戊正七百二十，共
七百二十一。積得本數併入五行積五萬四千七十五，得五萬四千七百九十六。

Then use the values obtained from the last [calculation] to find it [the
solution]. Take the "product" 54796 as the dividend, and 721 for E as the
divisor, divide it, obtaining 7 *chi* 6 *cun* for E's rope. Subtract [that] from
the total "product" in the fourth column. 721. The remainder is 645, and
divide it by 5, [the entry for] D, obtaining 129, giving 1 *zhang* 2 *chi* 9 *cun*
for D's rope. Subtract [that] from the "product" in the third column. 721,
the same for the following. The remainder is 592, and divide it by 4, [the entry
for] C, obtaining 1 *zhang* 4 *chi* 8 *cun* for C's rope. Subtract [that] from the
"product" in the second column, the remainder is 573, and divide by 3, [the

entry for] B, obtaining 1 *zhang* 9 *chi* 1 *cun* for B's rope. Subtract [that] from the "product" in the first column, the remainder is 530, and divide it by 2, [the entry for] A, obtaining 2 *zhang* 6 *chi* 5 *cun* for A's rope.

乃以最後所得求之。以積五萬四千七百九十六為實，戊七百二十一為法，除之，得戊繩七尺六寸。以減四行總積，七百二十一餘六百四十五，以丁五除之，得一百二十九，為丁繩一丈二尺九寸。以減三行積，七百二十一，後同餘五百九十二，以丙四除之，得丙繩一丈四尺八寸。亦減二行積，餘五百七十三，以乙三除，得乙繩一丈九尺一寸。以減一行積，餘五百三十，以甲二除，得甲繩二丈六尺五寸。 (*TWSZ, Tong bian, juan* 5, 18a–19b)

In addition to the well problem, there are three problems copied into the *Guide to Calculation* that are of the form of augmented matrices (5.8) and (5.14). These problems are given in Table 5.4.

Table 5.4: From among the nineteen *fangcheng* problems copied into the *Guide to Calculation*, below are three problems that are variants of augmented matrices (5.8) and (5.14), in addition to the well problem, which is problem 19 in the *Guide to Calculation*.

Problem	Chinese	Augmented Matrix	Notes
14	借牛一 ○ 正牛一 / ○ 正馬二 借馬一 / 正驢三 借驢一 ○ / 七百斤 七百斤 七百斤	$\begin{bmatrix} 1 & -1 & 0 & 700 \\ 0 & 2 & -1 & 700 \\ -1 & 0 & 3 & 700 \end{bmatrix}$	
15	硃三 ○ 硃二 / ○ 黃五 黃三 / 碌七 碌六 ○ / 價二千九百八十文 價六百四十文 價二千四十文	$\begin{bmatrix} 2 & 3 & 0 & 2040 \\ 0 & 5 & 6 & 640 \\ 3 & 0 & 7 & 2980 \end{bmatrix}$	Similar to problem 5 from Cheng's *Comprehensive Source of Mathematical Methods*.
18	丁柰一 丙 ○ 乙 ○ 甲柰二 / ○ ○ 梨二 梨四 / ○ 桃四 桃七 ○ / 榴八 榴七 ○ ○ / 二十四文 三十文 四十文 四十文	$\begin{bmatrix} 2 & 4 & 0 & 0 & 40 \\ 0 & 2 & 7 & 0 & 40 \\ 0 & 0 & 4 & 7 & 30 \\ 1 & 0 & 0 & 8 & 24 \end{bmatrix}$	Similar to problem 8 from Cheng's *Comprehensive Source of Mathematical Methods*.

The version of the well problem in the *Guide to Calculation* contains no significant additions or improvements, and is presented without additional analysis or criticisms. Most of the differences are quite minor and terminological: the fifth column is called "principal" (*zhu* 主) apparently because it is the only column that is transformed; the pivots are called "divisors" (*fa* 法); the term "set negative" (*li fu* 立負) is used for entering negative numbers. Because we do not know which mathematical treatise they used as their source, we cannot know if any of this terminology is original, but most likely it is not. For example, pivots are called "divisors" (*fa*) in Wu's *Complete Collection of Mathematical Methods*; the phrase "set negative nine" (*li fu jiu* 立負九) appears in Cheng's *Comprehensive Source of Mathematical Methods*. There is little reason to believe that the compilers of the *Guide to Calculation* understood these methods for solving *fangcheng*. The silence of the *Guide to Calculation* stands in stark contrast to Mei Wending's "On *Fangcheng*," written just half a century later, in which Mei offers 40 pages of criticism of the "well problem" alone.[50] This suggests that the compilers of the *Guide to Calculation* not only did not understand the mathematics here, they did not even notice that there was anything unusual about the method for calculating the depth of the well.

Chinese Mathematics as "Tattered Sandals"

We are now in a better position to evaluate Xu Guangqi's "Preface at the Printing of the *Guide to Calculation in the Unified Script*" (*Ke Tong wen suan zhi xu* 刻 同文算指序),[51] dated spring of 1614 (*Wanli jiayin chun yue* 萬曆甲寅春月), which illustrates the strategies he employed to assert the superiority of "Western Learning":

> The origin of numbers, could it not be at the beginning of human history? Starting with one, ending with ten, the ten fingers symbolize them and are bent to calculate them, [numbers] are of unsurpassed utility! Across the five directions and myriad countries, changes in customs are multitudinous. When it comes to calculating numbers, there are none that are not the same; that all possess ten fingers, there are none that are not the same.
>
> 數之原其與生人俱來乎？始於一，終於十，十指象之。屈而計諸，不可勝用也。五方萬國，風習千變。至于算數，無弗同者，十指之賅存，無弗同耳。

---

[50] Mei Wending 梅文鼎 (1633–1721), *Fangcheng lun* 方程論 [On *fangcheng*], in *ZKJDT*, juan 4, 40a–60a.

[51] Xu Guangqi, Xu Guangqi 徐光啟 (1562–1633), *Ke Tong wen suan zhi xu* 刻同文算指序 [Preface at the printing of the *Guide to Calculation in the Unified Script*], in *ZKJDT*.

In China, [beginning] from [the time] that the Yellow Emperor ordered Li Shou to do calculations in order to help Rong Cheng,[52] [it] was during the Zhou Dynasty (1045?–256 BCE) that [mathematics] reached its great completeness. The Duke of Zhou used it [mathematics], giving it a place in the curriculum used to choose officials, to promote the capable and virtuous [to the Imperial College],[53] to appoint them to be officials. Among the disciples of Confucius, those who mastered the Six Arts[54] were praised as having "ascended the hall and entered the chamber."[55] If mathematics were to fall to waste, the teachings of the Duke of Zhou and Confucius would fall into disorder. Some state that records and books were burned by Mr. Ying [the first emperor of China], and most of the learning of the Three Dynasties [Xia, Shang, and Zhou (ca. 2100?–256 BCE)] was not transmitted. If so, for all the early Confucians of the time of Ma [Rong (79–166)][56] and Zheng [Xuan (127–200)],[57] what was left for them to transmit? Of the *Ten Classics* [of mathematics][58] listed in the *Six Canons of the Tang*,[59] what books were left for the Erudites and disciples who studied for five years?

---

[52] Rong Cheng was the (perhaps mythical) great minister who aided the Yellow Emperor in establishing the calendar.

[53] This phrase is used in the preface to the *Nine Chapters*; it is also copied in Wu Jing's *Nine Chapters, Methods Arranged by Categories* (*Jiu zhang bi lei fa* 九章比類法, ca. 1450).

[54] The phrase *shen tong liu yi* appears in a description of Confucius' disciples in the *Records of the Grand Historian*: "Confucius taught [the classics of] Poetry, Documents, Rites, and Music, with approximately 3000 followers, those who were masters of the Six Arts [numbered] seventy-two 孔子以詩書禮樂教，弟子蓋三千焉，身通六藝者七十有二人. Sima Qian 司馬遷 (c. 145–c. 86 BCE), *Shiji ji jie* 史記集解 [*Records of the Grand Historian*, with collected explanations], in *SKQS*.

[55] The phrase *sheng tang ru shi* 升堂入室 originates from a passage in the *Analects*: "The disciples were not respectful toward Zilu. Confucius stated, 'You [Zilu], he has ascended the hall, but has not entered the chamber'" 門人不敬子路。子曰：由也升堂矣，未入於室也, using Zilu's shortcomings to indicate different levels of attainment. Later this phrase was used to indicate profound attainment. I thank Jongtae Lim for his many helpful suggestions and corrections on this and the following sentences.

[56] Ma Rong 馬融 (79–166) of the Han dynasty wrote annotations for the *Classic of Poetry*, *Classic of Changes*, *Classic of Rites*, *Classic of Documents*, *Classic of Filiality*, *Analects*, *Laozi*, *Huainanzi*, and others.

[57] Zheng Xuan 鄭玄 (127–200) of the Han dynasty studied the classics, including the *Gongyang Commentary* and the *Zuo Commentary* on the *Spring and Autumn Annals*, the *Record of Rites*, and the ancient text version of the *Classic of Documents*. He also studied the astronomical treatise *San tong li* 三統曆 and the *Nine Chapters of the Mathematical Arts*. His extant works include commentaries on the Mao commentary on the *Classic of Poetry*, the *Rites of Zhou*, *Record of Rites*, and *Decorum and Rites*; his commentaries on the *Classic of Changes*, *Spring and Autumn Annals*, and several other works were lost and exist only as reconstructed fragments.

[58] The term *Ten Classics* is also used to refer to the Confucian classics. Here Xu is referring to the ten classics of mathematics. Cheng, *Suanfa tongzong*, lists the *Ten Classics* as follows: 黃帝九章, 周髀算經, 五經算法, 海島算經, 孫子算法, 張邱建算法, 五曹算法, 緝古算法, 夏侯陽算法, 算術拾遺。

[59] A text describing government laws, regulations, and institutions in the Tang dynasty (618–906), traditionally held to have been written during the reign of the emperor Xuan Zhong (712–756). It contains a passage describing the Mathematics School, to which Xu is referring. For a

我中夏自黃帝命隸首作算，以佐容成，至周大備。周公用之，列於學
官以取士，賓興賢能，而官使之。孔門弟子身通六藝者，謂之升堂入
室，使數學可廢，則周孔之教踳矣。而或謂載籍燔於嬴氏，三代之學
多不傳，則馬、鄭諸儒先，相授何物？唐六典所列十經博士弟子，五
年而學成者，又何書也？

From the above, it can be stated only that the learning of calculation and
numbers has especially decayed over the most recent several hundreds of
years. There are two reasons for this decay: one is Confucians who dispute
[unimportant] principles and despise all evidential matters under heaven;
one is the absurd techniques that state that numbers have spiritual principles, capable of knowing what is to come and preserving what has gone,
there is nothing they cannot do. In the end, among the spiritual there is
not one that is effective; and among the evidenced there is not one that
is preserved. The great methods used by the sages from the past to order
the world and benefit [the people], once the literati were not able to obtain
them, completely following the arts and tasks of governmental affairs of
antiquity [Xia, Shang, and Zhou dynasties] became a distant [possibility].

由是言之，算數之學，特廢於近世數百年間爾。廢之緣有二：其一為
名理之儒，土苴天下之實事；其一為妖妄之術，謬言數有神理，能知
來藏往，靡所不效。卒於神者無一效，而實者亡一存。往昔聖人所以
制世利用之大法，曾不能得之士大夫間，而術業政事，盡遜於占初遠
矣。

My friend Li Zhenzhi [Zhizao] of the Bureau of Waterways and Irrigation,
a highly esteemed man of understanding, has, together with me, long
despaired because of this state [of the decline of mathematics]. Thus [we]
sought contemporary books on the techniques of calculation. [In these]
probably only one-tenth is writings from the beginning of antiquity, eight-
tenths is vulgar transmissions of recent writers, and another one-tenth is
transmissions by the early Confucians that do not betray [the works of]
the beginning of antiquity, that is all. I have once cursorily examined[60]
these vulgar transmissions, which are techniques of recluses, and most
are specious absurdities not [meriting] discussion. Even in the writings
allegedly from the beginning of antiquity and those that [allegedly] do
not betray the beginning of antiquity [of the early Confucians], there is
nothing more than just the methods, without being able to state the intent
behind establishing the methods. Moreover, again distantly thinking of the
study of the *Ten Classics* [of mathematics] in the Tang dynasty, there must
have been original, ultimately complete, and profoundly subtle meanings.
If they stopped at the contemporary transmissions, then they could be
mastered in two months—what work could require five years?

---

description of the Mathematics School, see Charles O. Hucker, *A Dictionary of Official Titles in
Imperial China* (Stanford, CA: Stanford University Press, 1985), s.v.

[60] A tentative translation for *xi mu* 戲目.

余友李水部振之，卓犖通人，生平相與慨嘆此事，行求當世算術之
書，大都古初之文十一，近代俗傳之言十八，其儒先所述作，而不倍
於古初者，亦復十一而已。俗傳者，余嘗戲目為閉關之術，多謬妄弗
論，即所謂古初之文，與其弗倍於古初者，亦僅僅具有其法，而不能
言其立法之意。益復遠想，唐學十經，必有原始通極微渺之義。若止
如今世所傳，則浹月可盡，何事乃須五年也？

Now that [I] together with [Li] accompany Mr. Ricci of the Western countries, in our spare time while discussing the Way, [we] have often touched
on principles and numbers. Since [his] discussions of the Way (*dao*) and
Principle (*li*) all return to basics and are solid (*shi*), they absolutely dispel
all theories of emptiness, profundity, illusion, and absurdity; and the studies of the numerical arts can all be traced back to the origins to recover
[the proper] tradition, the root supporting the leaves and branches, above
exhausting the nine heavens, on [all] sides completing the myriad affairs. In
the Western Countries in the academies of antiquity,[61] it also took several
years to complete the studies [of mathematics]. Even though our generation cannot see the *Ten Classics* of the Tang Dynasty, looking at the calendar
and all the affairs talked about by Mr. Ricci together with all the teachers
of the same aspirations [the Jesuits],[62] their mathematics is precise and
subtle, ten or one hundred times that compared with [the mathematics]
of the Han and Tang dynasties. Because of this [we] took seats and asked
to be benefited [by the teachings of Ricci]. Unfortunately, because of our
comings and goings, Zhenzhi and I missed each other.

既又相與從西國利先生游，論道之隙，時時及於理數。其言道言理，
既皆返本蹠實，絕去一切虛玄幻妄之說；而象數之學，亦皆溯[63]源承
流，根附葉著，上窮九天，旁該萬事，在於西國膠庠之中，亦數年而
學成者也。吾輩既不及睹[64]唐之十經，觀利公與同志諸先生所言曆法諸
事，即其數學精妙，比於漢唐之世，十百倍之，因而造席請益。惜余
與振之出入相左。

Zhenzhi came twice to live in Beijing and translated several chapters of
Ricci's mathematics. Since it was already a manuscript, I began to inquire
and [we] read it together, and discussed it together. Overall, those [Western
mathematical techniques] that were the same as the old (*jiu*)[65] [Chinese]
techniques were ones in which the old [Chinese techniques] did not reach
[the Western techniques]; those [Western mathematical techniques] that
differed from the old [Chinese] techniques were ones that the old [Chinese

---

[61] Literally, the term *jiao xiang* 膠庠 refers to academies of the Zhou dynasty; presumably Xu is
likening antiquity in Greece to the Zhou dynasty.

[62] Xu uses this circumlocution probably to avoid having to explicitly translate the "Society of
Jesus."

[63] Substituted for a variant form of the character.

[64] Substituted for variant form.

[65] The term that Xu uses here for the Chinese is "old" (*jiu* 舊) in the pejorative sense, in contrast
to the term "beginning of antiquity" (*gu chu* 古初), used in a very positive sense.

techniques] did not have. Taking the old [Chinese] techniques, [we] read them together and discussed them together. Overall, of those [Chinese techniques] that were compatible with the Western techniques, there were none that were not compatible with principle (*li*); of those [Chinese techniques] that were mistaken according to the Western ones, there were none that were not mistaken according to principle (*li*).[66] Because of this, Zhenzhi took the old techniques and considered them, discarding and selecting, using the translated Western techniques and appending [them] in parallel, printed [the manuscript], and named it *Guide to Calculation in the Unified Script*.[67] This can be called encompassing the beauty of the arts and studies, opening the path for [further] writing. Although the *Ten Classics* [of mathematics] are lost, it is just like discarding tattered sandals.[68] . . .

振之兩度居燕，譯得其算術如[69]千卷，既脫稿，余始間請而共讀之，共講之，大率與舊術同者，舊所弗及也；與舊術異者，則舊所未之有也。旋取舊術而共讀之，共講之，大率與西術合者，靡弗與理合也；與西術謬者，靡弗與理謬也。振之因取舊術，斟酌去取，用所譯西術駢附，梓之，題曰：同文算指，斯可謂網羅藝業之美，開廓著述之途，雖失十經，如棄敝屩矣。⋯

The assertions by modern historians that mathematics during the Ming dynasty (1368–1644) "had fallen into oblivion" have often been supported by little more than Xu's claim from his preface that "the learning of calculation and numbers has especially decayed over the most recent several hundreds of years." Modern historians (both Chinese and Western) have asserted that Chinese mathematics reached its zenith during the Song Dynasty (960–1279), but Xu is asserting that Chinese mathematics reached its zenith about two thousand years before his lifetime, during the Zhou Dynasty (1045?–256 BCE). Xu dismisses the entire Chinese mathematical tradition: the few works remaining from antiquity are unworthy of antiquity, the early Confucians' works are corrupt, and contemporary works are vulgar. Western Learning, Xu claims, is in every case superior to the Chinese. He offers no specifics about mathematical techniques; his claims are not based on any knowledge or evaluation of Chinese mathematics.

Despite Xu's appeals to antiquity, we should not mistake him for a textualist interested in studying early texts. There is little evidence that Xu inquired into Chinese mathematical treatises, compared with the work of his contemporaries. The itinerant merchant Cheng Dawei includes an extensive bibliography in his *Comprehensive Source of Mathematical Methods*. Jiao Hong 焦竑 (1541–1620) had written a comprehensive bibliography, *Record of Books for the Dynastic History*;[70]

---

[66] These two sets of sentences are written in parallel form, similar to an examination essay.

[67] Again, the term *tongwen* 同文 appears in the *Classic of Rites*: 車同軌，書同文. Li Zhizao seems to use it in the sense of translated into the same [Chinese] script.

[68] Literally, *jue* 屩 means cloth or straw shoes, *bi* 敝 means decrepit, shabby, dilapidated.

[69] *Ru* 如 (Wang Zhongmin's edition) should be *ruo* 若.

[70] Jiao Hong 焦竑 (1541–1620), *Guo shi jing ji zhi* 國史經籍志 [Record of books for the dynastic history] (Taibei: Taiwan shangwu yinshuguan 臺灣商務印書館, 1965).

although Xu mentions this work in a preface he later wrote for Jiao Hong, he never mentions its extensive record of mathematical treatises. Seven years after the translation of Euclid's *Elements*, Xu remains unfamiliar with even the titles of all but the most well-known Chinese mathematical texts. Furthermore, contrary to Xu's claims, the *Ten Classics* of Chinese mathematics were not lost during the Ming dynasty: many were readily available;[71] all were extant and included in the *Great Compendium of Yongle* (*Yongle dadian* 永樂大典).[72] Not only does Xu know little about Chinese mathematics, he evinces little interest in Chinese mathematical texts themselves.

Conclusions

The evidence in the *Guide to Calculation in the Unified Script* gives us more insight into the propaganda promoting Western Learning. We should not mistake Xu's claims of the decline of Chinese mathematics or his dismissal of the entire Chinese tradition for beliefs that he actually held—this was simply propaganda that he wrote to promote Western Learning. Though Xu Guangqi and his collaborators did not understand much about Chinese mathematics, they did understand how advanced parts of it were, to the extent that the most difficult problems they included in their *Guide to Calculation* were copied directly from the sources they denounced as vulgar. This parallels their borrowings from Buddhism, which they similarly denounced. In the same fashion as they had with Buddhism, they borrowed and appropriated from Chinese mathematical texts. They then transformed their copy into the original, by claiming that their copy predated the Chinese original and thus recovered meanings lost in antiquity; at the same time, they vehemently attacked the Chinese original as a corruption of their copy. We should not, of course, believe their propaganda. This chapter demonstrates, I think, that they did not believe it themselves: as they uncomprehendingly but carefully copied from the very Chinese mathematical texts that they dismissed as vulgar, they could not have believed that they were actually recovering lost meanings from the ancients.

Diffusion of Linear Algebra across Eurasia

The broader purpose of this chapter is to question the assumption that the advent of the Jesuits in late Ming China marked the introduction of Western

---

[71] For a study of extant bibliographies and an analysis of the mathematical treatises available during the Ming dynasty, see Hart, *Chinese Roots of Linear Algebra*, Appendix B, "Chinese Mathematical Treatises," pp. 213–54.

[72] This is the source from which Dai Zhen and the Qing compilers of the *Siku quanshu* "discovered" these treatises.

science into China and the "first encounter" of China and the West. To do so, we have traced the history of a distinctive class of systems of linear equations with n conditions in $n + 1$ unknowns, exemplified by augmented matrix (5.8), recorded in the *Nine Chapters* in about the first century CE. Evidence preserved in the *Nine Chapters* and in later commentaries demonstrates the following: (1) Determinantal calculations were used to set the value of the $(n+1)^{th}$ unknown; although the earliest extant Chinese record of such a determinantal calculation is a commentary dating from the twelfth century, it is likely that these calculations were used at the time of the compilation of the *Nine Chapters* in about 100 CE. (2) Determinantal solutions for all the unknowns were also known; although the first extant record of determinantal solutions is a commentary from the seventeenth century, evidence suggests that some determinantal solutions were known at the time of the compilation of the *Nine Chapters*. Essentially identical determinantal calculations and solutions are recorded in the *Liber Abaci* (1202) of Leonardo Pisano (Fibonacci). Although his problems were not displayed on a counting board, the terms were displayed as a series of fractions arranged contiguously, permitting similar calculations. First, the value of the $(n+1)^{th}$ unknown was assigned by calculating what we would in modern terms call the numerator of the determinant of the coefficient matrix. Subsequently, determinantal-style calculations resulted in the values of the remaining n unknowns.

This chapter, having demonstrated the diffusion of these problems, marks only the beginning of a broader investigation of the circulation of mathematical practices across the Eurasian continent prior to the West's "scientific revolution." In China, these mathematical practices were what we might call non-scholarly, neither based on nor transmitted primarily by texts. That is, practices such as linear algebra (*fangcheng*) were the specialty of anonymous and likely illiterate adepts; their practices were only occasionally recorded by literati who compiled treatises on the mathematical arts in their pursuit of imperial patronage; although these literati understood the basics, they did not understand (and in fact expressed disdain toward) the more esoteric calculations. What extant Chinese texts preserve is fragmentary evidence from which we can only attempt to reconstruct these practices.[73] The discoveries we now attribute to Leonardo (Fibonacci) may themselves have been the product of non-scholarly traditions.[74] The diffusion of these practices across Eurasia was likely effected by merchants, missionaries, and travelers, following the routes by which commerce, art, and religion circulated in what is increasingly understood to be the "global Middle Ages," to use Geraldine Heng's helpful term. This hypothesis may help explain the otherwise seemingly inexplicable appearance of Gaussian elimination, identical to Chinese methods, in the work of the French monk Jean Borrel (c. 1492–c. 1572). This may also help explain the almost simultaneous appearance in the seventeenth century of determinants in the works of Gottfried Wilhelm Leibniz (1646–1716), a Sinophile, and the Japanese mathematician Seki Takakazu 関孝和

---

[73] Hart, *Chinese Roots of Linear Algebra*.

[74] Jens Høyrup, "Leonardo Fibonacci and Abbaco Culture: A Proposal to Invert the Roles," *Revue d'histoire des mathèmatiques* 11 (2005): 23–56.

(1642–1708), who had studied Chinese mathematics. The considerable variety of problems, preserved in extant Chinese treatises, suggests that determinants were not suddenly discovered *ex nihilo*, but emerged from hundreds of years of explorations of increasingly complex patterns of determinantal-style computations.[75]

---

[75] Jean-Claude Martzloff, in his review of *Chinese Roots of Linear Algebra*, suggests an important possible direction for future research: "in the context of Japanese traditional mathematics, the notion of determinant developed by Seki Takakazu was prompted not exactly by the study of linear systems but rather by the problem of the elimination of unknowns between non-linear polynomial systems of equations (and the same is true in the case of Leibniz)." See Jean-Claude Martzloff, "Review of *The Chinese Roots of Linear Algebra*, by Roger Hart," *Zentralblatt MATH* (2011).

Chapter 6
Xu Guangqi, Grand Guardian

X̱U GUANGQI 徐光啟 (1562–1633) rose from presented scholar (*jin shi* 進士) to one of the more distinguished positions in the official bureaucracy of the Ming dynasty, Grand Guardian of the Heir Apparent and Grand Secretary of the Hall of Literary Profundity (*Taizi taibao, Wenyuange da xueshi* 太子太保 文淵閣大學士). I will argue in this chapter that Xu was not—as he has been anachronistically portrayed—a scientist or mathematician, but rather an official in the Ming court who indiscriminately promoted Western Learning (*Xixue* 西 學) in pursuit of imperial patronage. Yet to say that his promotion of Western Learning was indiscriminate is not to say that he was neither shrewd nor highly skilled. As we will see, he was expert in the Confucian arts: he had spent his youth memorizing the Confucian classics and then passed the civil service examinations with distinction, becoming one in perhaps ten thousand to achieve the coveted status of a presented scholar. He was then further trained at the imperial Hanlin Academy to write propaganda for the Empire, on a wide variety of issues— ranging from agriculture to military proposals and astronomy—in all of which he had little or no expertise. As Ray Huang aptly notes, "because [the Ming] empire was created to be controlled from the center by documents, field experience or lack of it made very little difference."[1]

To understand the propaganda that Xu crafted to promote Western Learning, we will examine some of his early writings.[2] We will focus here on the following

---

[1] Ray Huang, *1587, a Year of No Significance: The Ming Dynasty in Decline* (New Haven, CT: Yale University Press, 1981), 50.

[2] Ascertaining the authorship of the Chinese writings attributed to the Jesuits and their collaborators is often difficult if not impossible. The authorship of works on religion conventionally attributed to Xu is particularly controversial: see Xiaochao Wang, *Christianity and Imperial Culture: Chinese Christian Apologetics in the Seventeenth Century and Their Latin Patristic Equivalent* (Leiden: Brill, 1998); and Ad Dudink, "The Image of Xu Guangqi as Author of Christian Texts," in *Statecraft and Intellectual Renewal in Late Ming China: The Cross-Cultural Synthesis of Xu Guangqi (1562–1633)*, ed. Catherine Jami, Gregory Blue, and Peter M. Engelfriet (Leiden: Brill, 2001), 99–154. In general, for all these writings, and especially for works in Chinese attributed to the Jesuits, it would be more accurate to speak of corporate authorship, over which the Chinese patrons, and Xu in particular, exercised considerable control; authorship was then strategically assigned to legitimate these works by attributing them to particular individuals.

sources, which includes a wide range of representative writings—examinations, letters, poetry, and official documents:

1. A civil examination essay, the earliest extant writing by Xu. This is one of the most important essays he ever composed. I will argue that it bears evidence of a textual practice (sometimes vaguely described as syncretism) wherein one set of writings is explained through another. It is through this approach that Xu later argues that Western Learning recovers the lost meanings of the Chinese Confucian classics.
2. A second examination essay—a policy essay—that may be the work of Xu, or at least expresses viewpoints similar to those we see in his later writings. I will argue that this essay and the preceding examination essay offer examples through which we can better understand the textualist approach—the recovery of meanings from the writings of the ancients—employed by Xu.
3. Xu's earliest extant patronage letter, written to the famous scholar Jiao Hong 焦竑 (1541–1620), who had awarded Xu first place on his provincial examination.
4. An early work on Western Learning explaining the importance of *Tianzhu* 天主 [Lord of Heaven], prefiguring claims Xu makes in his later writings.
5. One of the few religious works attributed to Xu, an ode to Jesus [Yesu 耶穌].
6. A work that although asserted to be Xu's earliest work related to mathematics turns out to be a patronage letter with little mathematical content.
7. Xu's preface to the translation of Euclid's *Elements* into Chinese.
8. Xu's memorial in defense of Western Learning. In this memorial, Xu reiterates views similar to those presented in the earlier letter (above), claiming that Western Learning is the basis for the perfect moral order in a unified Europe that has not seen wars or changes in dynasty for thousands of years.

But before we examine these works in their historical context, we will first examine the conventional view of Xu Guangqi and the translation of Euclid's *Elements*, as presented in previous accounts. These accounts are for the most part hagiographic: Xu is portrayed as a mathematician, scientist, and polymath. Descriptions of his translation of Euclid's *Elements* into Chinese are similarly laudatory: Xu is asserted to have recognized the importance of axiomatic deduction in Euclid's *Elements* for mathematics and science in general. The next section will examine these hagiographic portrayals of Xu, and the following section presents a critical analysis of claims about the importance of axiomatic deduction in seventeenth-century China.

"China's Supreme Polymath": The History of a Hagiography

Historical accounts of the Jesuit mission in China have often been limited to what are essentially hagiographies, at least until recently. To take one notable recent example, *Science*, arguably the most prestigious peer-reviewed general science journal in the world, published a short news article on the history of science calling Xu Guangqi "China's supreme polymath," stating,

> In the early 17th century, this humanist and experimentalist helped avert starvation in China by disseminating hardier crops and devised dams and canals for irrigation and flood control. He launched a decade-long effort to improve the accuracy of the Chinese calendar by incorporating a more precise knowledge of celestial geometry. His monumental contribution was to team up with a Jesuit scholar to translate part of Euclid's *Elements*, introducing late Ming Dynasty intellectuals to new mathematical concepts—and Western thought. For his achievements, he has been compared to Leonardo da Vinci and Francis Bacon.[3]

The article recounts Xu's alleged leadership in mathematics, agriculture, defense, and calendar reform:

> The kindred spirits [Ricci and Xu] came to realize that planar geometry and other higher mathematical concepts then unknown in China were essential to progress. . . . Throughout [Xu's] life, one constant was his dedication to improving agriculture. His experiments in Shanghai with yams, then a new import from South America, led to the widespread adoption of the high-energy crop. . . . Xu also trained imperial soldiers to use a newfangled device from Europe, the cannon.[4]

This hagiographic literature has a long history, including such suggestive titles as *The Wise Man from the West*, *Matteo Ricci's Scientific Contribution to China*, and *Generation of Giants: The Story of the Jesuits in China in the Last Decades of the Ming Dynasty*,[5] material which continues to be incorporated even in more recent research.[6]

Throughout these accounts we find confident expressions of the uniform superiority of an amalgam of Western religious, philosophical, scientific, political, and moral beliefs, as exemplified by Cronin's account of how Ricci taught

---

[3] Richard Stone, "Scientists Fete China's Supreme Polymath," *Science* 318 (November 2, 2007): 733.

[4] Ibid.

[5] Vincent Cronin, *The Wise Man from the West* (London: Rupert Hart-Davis, 1955); Henri Bernard, SJ, *Matteo Ricci's Scientific Contribution to China*, trans. Edward Chalmers Werner (Westport, CT: Hyperion Press, 1973); George H. Dunne, *Generation of Giants: The Story of the Jesuits in China in the Last Decades of the Ming Dynasty* (Notre Dame, IN: University of Notre Dame Press, 1962).

[6] For example, see Michael Adas, *Machines as the Measure of Men: Science, Technology, and Ideologies of Western Dominance* (Ithaca, NY: Cornell University Press, 1989). Adas states, "The best general accounts of the Jesuits and their work in China remain Arnold Rowbotham, *Missionary and Mandarin* . . . ; and Virgile Pinot, *La Chine et la formation de l'esprit philosophique en France (1640–1740)* . . . . The recent study by Jonathan Spence, *The Memory Palace of Matteo Ricci*, provides a fine account of the background to and the first phase of the Jesuit mission" (p. 54, n. 90). Adas further notes that in his own book, the "background on Ricci and the Jesuit mission is based primarily on Henri Bernard, *Matteo Ricci's Scientific Contribution to China* . . . ; Spence, *Memory Palace*; George Harris, 'The Mission of Matteo Ricci, S.J.,' . . . ; and Rowbotham, *Missionary and Mandarin*" (p. 57, n. 101).

mathematics to Qu Taisu. Cronin portrays Chinese mathematics as "rudimentary" and "slow and clumsy"; in contrast, the "logical deduction from definitions and axioms revealed a new world to Qu." Cronin further recounts how "[e]very day during the mathematics lesson Ricci found an opportunity of mentioning Christianity"; and parallel to his conversion to Euclidean mathematics, Qu became "[c]onvinced of the truth of Christianity" and asked to be baptized. Ricci ultimately had to refuse because of Qu's moral failings in living with a concubine.[7]

But what is more important for our purposes than the hagiographic portrayal of Jesuit missionaries in these accounts is the purported evaluation of Chinese mathematics and sciences. Cronin's account is typical of the striking contrasts portrayed between Chinese mathematics, which, we are told, lacked deductive rigor, and Euclid's *Elements*. Bernard recounts that ". . . geometry was, in China, intuitive: they did not justify its theorems by abstract demonstrations, but confined themselves to numerical measures"; he contrasts this with the "wonderful 'Elements of Euclid'" in which the mind "passed in it without effort, by the smoothest descent, from one truth to another, and a whole immense doctrine unfolded itself in it with perfect regularity, without a gap and without a shock."[8]

As I previously noted, these are not conclusions derived from the analysis of sixteenth- and seventeenth-century mathematical treatises, whether Western *or* Chinese; rather, the recurring themes in these narratives derive from the uncritical incorporation of the polemics of the Jesuit missionaries into accounts that now claim for themselves historical veracity. The view of the Chinese lack of rigor in mathematics derives from similar assertions in Ricci's *Journals*,[9] such as his allegation that Chinese mathematics lacked proofs and instead proceeded through the "wildest imagination":[10]

---

[7] Cronin, *Wise Man from the West*, 100–1.

[8] Bernard, *Matteo Ricci's Scientific Contribution*, 18.

[9] Matteo Ricci, SJ, *China in the Sixteenth Century: The Journals of Matthew Ricci, 1583–1610*, trans. Louis J. Gallagher, SJ (New York: Random House, 1953), hereinafter Ricci's *Journals*, 30–32, 325–32. A note about the composition of Ricci's *Journals* is necessary to explain its unusual format. Louis J. Gallagher, SJ, in his Translator's preface, offers the following description of the sources for Trigault's original Latin version of Ricci's *Journals*: "The manuscript of his [Ricci's] Diary was brought from Macao to Rome in 1614 by Father Nicola Trigault, who translated it into Latin and published it in 1615, together with an account of Ricci's death and burial. . . . Trigault draws from Ricci sources other than the Diary, such as the Annual Mission Letters, Ricci documents [sic] to other missionaries, and personal narrative which Trigault says was omitted by Ricci because of his modesty" (ibid., xvii). Gallagher then translated Trigault's version from the Latin into English. d'Elia notes that Trigault seems to be the chief protagonist (ibid., xviii). Gallagher also notes that the original diary has been published in Italian by Father Tacchi Venturi, SJ, in 1910 under the title "The Historical Works of Matthew Ricci." The most comprehensive scholarly work on Ricci is Pasquale M. d'Elia, SJ, *Fonti Ricciane: Documenti originali concernenti Matteo Ricci e la storia delle prime relazioni tra l'Europa e la Cina (1579–1615)*, 3 vols. (Roma: Libreria dello Stato, 1942–1949).

[10] Ricci's initial evaluation is somewhat positive, stating that "The Chinese have not only made considerable progress in moral philosophy but in astronomy and in many branches of mathematics as well . . ." (Ricci, *Journals*, 30). His later evaluation of Chinese sciences becomes

[N]othing pleased the Chinese as much as the volume on the *Elements* of Euclid. This perhaps was due to the fact that no people esteem mathematics as highly as the Chinese, despite their method of teaching, in which they propose all kinds of propositions but without demonstrations. The result of such a system is that anyone is free to exercise his wildest imagination relative to mathematics, without offering a definite proof of anything. In Euclid, on the contrary, they recognized something different, namely, propositions presented in order and so definitely proven that even the most obstinate could not deny them.[11]

Ricci's assertion of the inferiority of Chinese mathematics to Euclid's *Elements* is but one facet of his claim of the absolute truths brought by the Jesuits—religious, philosophical, scientific, and moral—a claim that has been incorporated as fact into later historical accounts. These later accounts then seek plausibility for the grandiose but unevidenced claims of the efficacy of Western truths by implicitly appealing to apparent continuities with modern scientific beliefs,[12] while effacing the historical context of which these assertions were an integral part. Thus these accounts ignore similarities between Ricci's claims of the superiority of Western science and the equally confident claims he repeatedly makes for the superiority of Catholic exorcism:

There was a girl here possessed of a devil . . . . The [Chinese] exorcists were called in from the pagan temple, but the spirit only ridiculed their rites, showing no fear and causing the lights and candles to fall off the alter of the idols. . . . one of the Lay Brothers brought them a picture of Christ and told them to call upon the name of Jesus. They took all the idols off the domestic altar and the whole family began a series of instructions in Christian doctrine. From that day on, there was no evidence of the evil spirit in the house, . . . the power of God worked so readily where all the efforts of the Chinese exorcists amounted to nothing.[13]

The necessary supplement to this assertion of the universal truths of the Jesuits is the narrative of Chinese resistance—psychologized as pride, pettiness, conservatism, and selfish interests—that was required to explain the abject failure of these universal truths to obtain universal acceptance. This central theme pervades Ricci's accounts, narrated in racial terms of dullness and the inability to accept the new,[14] and again becomes directly incorporated into the hagiographic accounts as historical fact.[15]

---

much more negative with the development of his strategy of promoting the Jesuit sciences (ibid., 325ff.).

[11] Ibid., 476.

[12] For example, see ibid., 325–32.

[13] Ibid., 544–45.

[14] See ibid., passim.

[15] For an interesting example of the incorporation of Protestant missionary polemics on the racial characteristics of the Chinese, see Max Weber, *The Religion of China: Confucianism and Taoism*, trans. Hans H. Gerth (Glencoe, IL: Free Press, 1951), 231–32.

In fact, recent historical scholarship, despite the repudiation of hagiographies and considerable advances in sophistication, has remained virtually unchanged in the overall evaluation of the introduction of Western science. Perhaps the most significant transformation has been the secularization of the accounts of the Jesuits and their Chinese collaborators, which now, ironically, have become the province primarily of the history of science; even the more general historical accounts now reinterpret the Jesuits in the narrative of scientific exchange. These accounts are differentiated more by the elements of this narrative that they emphasize than by substantial differences in interpretation; it is perhaps a greater irony that Chinese accounts have been even more aggressive in affirming Western science than have the Western accounts.[16]

Mathematics during the Ming dynasty, as noted in chapter 4, has been held to have been in decline: crucial mathematical treatises and techniques from the Chinese tradition had been lost; creativity was stultified by the Neo-Confucian orthodoxy and the civil service examinations; and mathematics and science were slighted by the inwardly focused followers of Wang Yangming's 王陽明 (1472–1529) Learning of the Mind (*xin xue* 心學). This decline has been viewed as but one facet of the corruption and decay that led to the fall of the Ming Dynasty, first to domestic rebels and then to the Manchu invaders. It is against this background that the story of the introduction of the more rigorous and advanced Western science introduced by the Jesuits is narrated. The real significance of the arrival of the Jesuits in China was, we are now to believe, their introduction of Western science. The advent of the Jesuits has been interpreted by historians as the break from the stagnation or decline of the Ming to the period of importation of Western science,[17] which then stimulated a revival of Chinese science.

Because historical accounts have rarely sought to describe or compare the internal details of the sciences introduced by the Jesuits, the superiority of Jesuit sciences remains narrated through the invocation of the valorized terms "rigor," "precision," "deductive reasoning," and "science." Even criticisms of the Jesuits— usually as the precursors of imperialism—question not the superiority of Western

---

[16] For example, see the essays in *Xu Guangqi jinian lunwenji* 徐光启纪念论文集 [Collection of essays in commemoration of Xu Guangxi], ed. Zhongguo kexueyuan Zhongguo ziran kexueshi yanjiushi 中国科学院中国自然科学史研究室 (Beijing: Zhonghua shuju 中华书局, 1963); *Xu Guangqi yanjiu lunwenji* 徐光启研究论文集 [Essays on research on Xu Guangqi], ed. Xi Zezong 席泽宗 and Wu Deduo 吴德铎 (Shanghai: Xuelin chubanshe 学林出版社, 1986).

[17] Catherine Jami asserts that historians "all consider the coming of the Jesuits as a break in the history of the mathematical sciences," citing Shigeru Nakayama, "Periodization of the East Asian History of Science," *Revue de Synthese* 4 (1988): 375–79. Catherine Jami, "Western Influence and Chinese Tradition in an Eighteenth Century Chinese Mathematical Work," *Historia Mathematica* 15 (1988): 311–31. See also Li Yan and Du Shiran, *Chinese Mathematics: A Concise History*, trans. John N. Crossley and Anthony W.-C. Lun (Oxford: Clarendon Press, 1987), chapter 7; Qian Baocong 钱宝琮, *Zhongguo shuxue shi* 中国数学史 [History of Chinese mathematics] (Beijing: Kexue chubanshe 科学出版社, 1964), 234ff.; Joseph Needham, *Mathematics and the Sciences of the Heavens and the Earth*, vol. 3 of *Science and Civilisation in China* (Cambridge: Cambridge University Press, 1959), 52; Yoshio Mikami, *The Development of Mathematics in China and Japan*, 2nd ed. (repr., New York: Chelsea, 1974), 108ff.

sciences, but instead compare unfavorably the sciences introduced by the Jesuits with those of contemporary Europe.[18]

And as with the earlier hagiographies, it has been essential to the narrative of universal truths to explain their lack of universal acceptance; recent secularized accounts must now explain not the rejection of the god of Christianity, but rather the resistance to the truths of Western science. Thus the explanations for the rejection of Christianity, written in terms of conservatism, xenophobia, and pride, have become central themes in explaining the Chinese rejection of Western science.[19] This resistance of Chinese conservatives to the truths of Western science has become our central paradigm in understanding what has been termed the Western impact on China; nowhere is this account more masterfully portrayed than in Levenson's classic if psychologistic account of the Chinese historicist resistance to Western reason and value.[20]

It was then superior Western science, most accounts assert, that attracted the Chinese to the Jesuits.[21] Thus even Jacques Gernet, in perhaps the most sophisticated analysis of this period, rewrites the religious conversion of the Chinese literati in the secular story of their conversion to Western science:

> Virtually all the literate converts of the early seventeenth century were attracted to the missionaries by their scientific teaching, and they then worked alongside them to translate into Chinese the manuals that were used in Jesuit colleges. The first of them, Qu Rukui, had studied mathematics with Ricci in Shaozhou in 1590 and 1591. Xu Guangqi (1562–1633), Ricci's most famous disciple, had helped him translate the first six chapters of *The Elements of Euclid* by Clavius, a work of trigonometry, and a little treatise on Western hydraulics. For Li Zhizao (1565–1630), who had been an enthusiast of geography ever since boyhood, it had been Ricci's map of the world that had clinched it. But he also studied mathematics under Ricci and collaborated with him in the translation of a work of geometry and a treatise on arithmetic. Sun Yuanhua (1581–1632) studied mathematics with Xu Guangqi, who then introduced him to the missionaries. He took a particular interest in Western firearms. Wang Zheng (1571–1644), who had inherited from an uncle a passion and a genius for mechanics, was drawn to the missionaries when he discovered that they possessed illustrated works on mechanics and, in collaboration with Father Johann Schreck, he translated

---

[18] Qian, *Zhongguo shuxue shi*, 230; Mei Rongzhao 梅荣照, "Xu Guangqi de shuxue gongzuo" 徐光启的数学工作 [Xu Guangqi's work in mathematics], in Zhongguo ziran kexueshi yanjiushi, *Xu Guangqi jinian lunwenji*, 144.

[19] See, for example, George C. Wong, "China's Opposition to Western Science during the Late Ming and Early Ch'ing," *Isis* 54 (1963): 29–49.

[20] See Joseph R. Levenson, "'History' and 'Value': Tensions of Intellectual Choice in Modern China," in *Studies in Chinese Thought*, ed. Arthur F. Wright (Chicago: University of Chicago Press, 1953), 146–94.

[21] One important exception is Willard J. Peterson, "Why Did They Become Christians? Yang T'ing-Yun, Li Chih-Tsao, and Hsu Kuang-Ch'i," in *East Meets West: The Jesuits in China, 1582–1773*, ed. Charles E. Ronan and Bonnie B. C. Oh (Chicago: Loyola University Press, 1988), 129–52.

a treatise that he entitled *Illustrated Explanations concerning the Strange Machines of the Distant West.*[22]

Xu Guangqi, the most prominent of the Chinese said to have been attracted to Western science, is now portrayed in a secularized narrative as a scientist who introduced Western mathematics, astronomy, agriculture, and weaponry into China. Politically, he remained neutral, above the factional strife of the period, working to save the Ming Dynasty from invasion and collapse.[23] Philosophically, Xu is viewed as advocating ideas similar to modern science while relentlessly attacking contemporary superstition.[24]

In these accounts, which present Xu as attracted primarily to the rigor and precision of Western sciences, only secondarily did he become interested in Jesuit doctrines. Xu very early recognized the superiority of Western sciences, and pioneered in introducing Western techniques into a series of practical sciences, including mathematics, astronomy, and agriculture. Thus Li Yan, an eminent historian of Chinese mathematics, asserts that Xu Guangqi and Li Zhizao are representative of literati "who were anxious to strengthen the country economically and militarily" through Western science and technology.[25]

Accounts of the universality of Euclid and ultimately Western science have since been presented through the hagiographic portrayals of Xu Guangqi, adopting his polemics and apologetics as historical conclusions. Historians of Chinese

---

[22] Jacques Gernet, *China and the Christian Impact: A Conflict of Cultures,* trans. Janet Lloyd (New York: Cambridge University Press, 1985), 22. In a footnote, Gernet oddly undermines this thesis, asserting in double-negatives that China was "not without a scientific tradition," and that Jesuit sciences were "not all entirely new": "It is true that the Jesuit missionaries did introduce a few important novelties, such as the demonstration of the sphericity of the earth—a fact that had been both claimed and accepted in China in earlier periods—the introduction of new fields of mathematics, and techniques of astronomical calculation more reliable and accurate than those that had been in use in China since the fourteenth century. But these things were not all entirely new, for China was not without its own scientific tradition. Some mathematicians among the literate elite discovered analogies between the knowledge from the West and what had been taught in China ever since the Han and Sung dynasties. Quite contrary to the thesis put about in the first half of the twentieth century, when Western knowledge of the history of Chinese science was still limited, the teaching of the missionaries contained nothing of a kind to upset existing ideas. Nor did any of the Jesuits' teaching to the Chinese bear the mark of modern science or indeed convey its spirit. . . . they limited themselves to introducing into China such new knowledge as was useful to their own calculations, which was strictly of a practical and immediate nature" (pp. 251–52).

[23] See *Xu Guangqi zhuanji ziliao* 徐光啟傳記資料 [Biographic materials on Xu Guangqi], ed. Zhu Chuanyu 朱傳譽 (Taibei: Tianyi chubanshe 天一出版社, 1979); Liu Bohan 刘伯涵, "Lue lun Xu Guangqi yu Ming mo dangzheng" 略论徐光启与明末党争 [A brief discussion of Xu Guangqi and factional strife in the late Ming], in Xi and Wu, *Xu Guangqi yanjiu lunwenji,* 161; Li and Du, *Chinese Mathematics,* 193; Keizo Hashimoto, *Hsu Kuang-Ch'i and Astronomical Reform: The Process of the Chinese Acceptance of Western Astronomy, 1629–1635* (Osaka: Kansai University Press, 1988).

[24] See Mei Rongzhao 梅荣照 and Wang Yusheng 王渝生, "Xu Guangqi de shuxue sixiang" 徐光启的数学思想 [Mathematical thought of Xu Guangqi], in Xi and Wu, *Xu Guangqi yanjiu lunwenji,* 44; Qian, *Zhongguo shuxue shi,* 230, 237–38.

[25] Li and Du, *Chinese Mathematics,* 192.

mathematics have generally agreed on the fundamental importance of the translation of the *Elements* as the introduction into China of a deductive axiomatic system that China had lacked,[26] as expressed by Catherine Jami:

> In addition to being the earliest, Euclidean geometry is probably the most significant branch of Western mathematics introduced into China during this period, since there was until then no deductive system based on axioms comparable to Euclidean geometry in the Chinese mathematical tradition; its geometry was instead based on the right-angled triangle (*gougu*).[27]

And the axiomatic deductive system embodied in Euclid has been viewed as one of the keys to scientific reasoning. Needham himself assumes a special importance for "deductive Euclidean geometry," noting,

> We mentioned before that Chinese mathematical thought and practice were invariably algebraic, not geometrical. No deductive Euclidean geometry developed spontaneously in Chinese culture, and that was, no doubt, somewhat inhibitory to the advances the Chinese were able to make in optics . . . . Nevertheless it is very remarkable that the lack of Euclidean geometry did not prevent the successful realization of the great engineering inventions, including the highly complicated ones in which astronomical demonstrational and observational equipment was driven by water power through the use of elaborate gearing.[28]

More common are sentiments such as those expressed by Robert Hartwell, who, dissatisfied with Needham's apparent slighting of Euclidean deduction, invokes against Needham's encyclopedic catalogue of the scientific achievements of China an informal letter written by Albert Einstein:

> Although he tries to minimize the implications of Einstein's observation that a major impediment to the independent development of modern science in China was the absence of the formal logical system invented by the Greek philosophers, Needham completely misses the point when he

---

[26] See Qian, *Zhongguo shuxue shi*, 239; Li and Du, *Chinese Mathematics*, 194; Mei Rongzhao 梅荣照, Liu Dun 刘钝, and Wang Yusheng 王渝生, "Oujilide *Yuanben* de chuanru he dui woguo Ming Qing shuxue fazhan de yingxiang" 欧几里得《原本》的传入和对我国明清数学发展的影响 [The transmission of Euclid's *Elements* into China and its influence on the development of Ming and Qing Dynasty mathematics], in Xi and Wu, *Xu Guangqi yanjiu lunwenji*, 52–53; Ulrich Libbrecht, *Chinese Mathematics in the Thirteenth Century: The Shu-Shu Chiu-Chang of Ch'in Chiu-Shao* (Cambridge, MA: MIT Press, 1973); Needham, *Mathematics and the Sciences of the Heavens*; Jean-Claude Martzloff, *A History of Chinese Mathematics* (New York: Springer, 2006). In Chinese philosophy, Chad Hansen makes a similar assessment, arguing that although the Chinese did have some elements similar to Western thinking, "What we find in Chinese thought are counterparts of the extended uses of reasons—not of its Euclidean core." Chad Hansen, "Should the Ancient Masters Value Reason?" In *Chinese Texts and Philosophical Contexts: Essays Dedicated to Angus C. Graham*, ed. Henry Rosemont, Jr. (La Salle, IL: Open Court, 1991), 193.

[27] Jami, "Western Influence and Chinese Tradition," 312.

[28] Joseph Needham, *Science in Traditional China: A Comparative Perspective* (Cambridge, MA: Harvard University Press, 1981), 15.

attempts to refute this position by demonstrating that the lack of Euclidean geometry did not preclude the realization of many successful engineering and scientific discoveries. . . . The utility of the hypothetical-deductive method lies in the fact that it is the most efficient and reliable technique yet devised for producing fruitful theoretical statements. . . . It is possible, though highly improbable, that the transformation of the world brought about by modern science could have been achieved by technology developed by inductive procedures alone. But it would have required several millennia rather than three or four centuries.[29]

What Xu Guangqi found in the *Elements*, previous accounts assert, was this rigorous system of deductive logic, which traditional Chinese mathematics lacked.[30] For example, Li Yan asserts that

> The *Elements of Geometry* is logically precise and its mode of presentation is completely different from the traditional Chinese *Nine Chapters on the Mathematical Art*. The *Elements of Geometry* starts with a few postulates and common notions and proceeds in a deductive manner, while the *Nine Chapters on the Mathematical Art* gives three to five examples then the general method of solution is presented: it adopts the inductive form. Xu Guangqi had a relatively clear understanding of this special feature of the *Elements of Geometry*, which is quite different from the traditional mathematics of China.[31]

And, these accounts assert, interested as he was in science and practical applications—in stark contrast to his inwardly focused contemporaries—Xu extended the foundation of certainty provided by this logical system in the *Elements*. Arguing that the *Elements* was the foundation of mathematics, Xu applied mathematics to the sciences, including astronomy, water conservancy, meteorology, music, military technology, architecture, accounting, mechanical engineering, surveying, medicine, and time-keeping. This, then, is what the Chinese collaborators are claimed to have discovered in Euclid's *Elements*. For example, Liu Dun, Wang Yusheng, and Mei Rongzhao argue that it was the system of deductive logic of the *Elements* that Chinese scholars valued:

---

[29] Robert M. Hartwell, "Historical Analogism, Public Policy, and Social Science in Eleventh- and Twelfth-Century China," *American Historical Review* 76 (1971): 723. Hartwell further argues, "The appearance and development of this scientific method in the West during the sixteenth, seventeenth, and eighteenth centuries were primarily results of the conjunction of three different historical forces: first, 'the persistent use of experiment' . . . ; second, 'a widespread instinctive belief in an Order of Things' . . . ; and third, the habitual employment of the hypothetical-deductive technique of inquiry inherited from the Euclidean method of demonstration or proof." Ibid., 718. See also Derk Bodde, *Chinese Thought, Society, and Science: The Intellectual and Social Background of Science and Technology in Pre-Modern China* (Honolulu: University of Hawaii Press, 1991); Wen-yuan Qian, *The Great Inertia: Scientific Stagnation in Traditional China* (Dover, NH: Croom Helm, 1985).

[30] Mei, Liu, and Wang, "Oujilide *Yuanben* de chuanru," 52–53; Qian, *Zhongguo shuxue shi*, 239; Mei, "Xu Guangqi de shuxue gongzuo," 149.

[31] Li and Du, *Chinese Mathematics*, 194.

Instead, what made the greatest impact on (*zhendong*) the first intellectuals who encountered this system was not the geometric propositions that left one dazzled, but rather the persuasiveness and the rigorous scientific structure of the kind of deductive logic system that is embodied in the *Elements*.[32]

What these more recent historical accounts too often share with their hagiographic predecessors is the reliance on little more than the apologetics of the Jesuits and their collaborators, stripped of historical context, to function instead as historical explanation. For example, the claim of the effectiveness of science for attracting the Chinese to Christianity again derives from this same theme which pervades Ricci's own accounts, as exemplified by the following passage:

> This book [Euclid] was greatly admired by the Chinese and it had considerable effect upon the rearrangement of their calendar. For a better understanding of it, many came to Father Ricci to enroll as his pupils, and many also to Ciu Paul [Xu Guangqi], and with a teacher to direct them, they took to European scientific methods as readily as the Europeans themselves, showing a certain keenness of mind for the more subtle demonstrations.[33]

But, in contrast to earlier accounts, recent historical accounts now also incorporate the apologetics of the Chinese collaborators. For example, as noted in the previous chapters of this book, assertions of the decline of Ming mathematics rely on little more than appeals to Xu's assertion that "the studies of mathematics have fallen to waste over the last several hundred years," alleging that

> There are two reasons for this decay: one is Confucians who dispute [unimportant] principles and despise all evidential matters under heaven; one is the absurd techniques that state that numbers have spiritual principles, capable of knowing what is to come and preserving what has gone, there is nothing they cannot do. In the end, among the spiritual there is not one that is effective; and among the evidenced there is not one that is preserved.
>
> 廢之緣有二：其一為名理之儒，土苴天下之實事；其一為妖妄之術，謬言數有神理，能知來藏往，靡所不效。卒於神者無一效，而實者亡一存。[34]

---

[32] Mei, Liu, and Wang, "Oujilide *Yuanben* de chuanru," 52.

[33] Ricci, *Journals*, 477. However, contrary to the claim in Ricci's *Journals*, Xu's second preface indicated that there was little interest in the work, as noted below.

[34] Xu Guangqi, *Ke Tong wen suan zhi xu* 刻同文算指序 [Preface at the printing of the *Guide to Calculation in the Unified Script*], in *Zhongguo kexue jishu dianji tonghui: Shuxue juan* 中國科學技術典籍通彙：數學卷 [Comprehensive collection of the classics of Chinese science and technology: Mathematics volumes] (Zhengzhou: Henan jiaoyu chubanshe 河南教育出版社, 1993), hereinafter *ZKJDT*, 2772–73. This quote, with little supporting evidence, has been widely cited as conclusive evidence of the asserted decline of Ming mathematics. See Qian, *Zhongguo shuxue shi*; Qian, *Great Inertia*; Li and Du, *Chinese Mathematics*; Mei, Liu, and Wang, "Oujilide *Yuanben* de chuanru"; Mei Rongzhao 梅荣照, "Ming-Qing shuxue gailun" 明清數學概論 [Outline of Ming-Qing mathematics], in *Ming-Qing shuxueshi lunwenji* 明清數學史論

The assertions of the importance of Euclidean axiomatics as a foundation for practical sciences and logical reasoning, such as Li Yan's assertion that Xu "fully understood that the ideas of geometry develop the ability to think logically and train people to become expert in problems of finding areas and volumes,"[35] also are supported by the uncritical adoption, as historical fact, of polemics in Xu's prefaces. Quotations drawn at length from Xu's prefaces are frequently invoked to demonstrate the logical deductive system of Euclid's *Elements*; presented straightforwardly as mathematical analyses, they prove sufficient for Li to assert that Xu "concludes that the whole book's ingenuity can be described in one word, 'clarity' (*ming*): this is to point out the special characteristic of logical deduction."[36] John Henderson repeats statements in Xu's prefaces as historical conclusions, arguing that for Xu "the knowledge arrived at through the use of the geometrical method was so certain, so well-established, that it was not open to question or subject to change." He similarly borrows from Ricci in recounting that "Ricci remarked, in his own preface to the translation, that so incontrovertible were the truths deduced through Euclid's method that generations of skilled controversialists had been unable to discern any substantial error in the *Geometry*"; for Ricci and Xu, this method "yielded knowledge that was complete and unequivocal." For Xu, Henderson writes, the study of geometry was "an unsurpassed means of mental training and discipline."[37]

Axioms In Historical Context

Previous accounts that have heralded the alleged importance of the translation of Euclid's *Elements* in China rarely cite any research—historical, mathematical or philosophical—to support their laudatory assertions. These assertions have been

文集 [Collected essays on the history of Ming-Qing mathematics], ed. Mei Rongzhao 梅荣照 (Nanjing: Jiangsu jiaoyu chubanshe 江苏教育出版社, 1990), 1–20. It should be noted that Mei Rongzhao and Wang Yusheng question Xu's assessment, asserting that it is prejudiced. Mei, "Xu Guangqi de shuxue gongzuo," 42. For a detailed analysis of this passage, see pages 186–191 of this book.

[35] Li and Du, *Chinese Mathematics*, 195.

[36] Ibid.

[37] John B. Henderson, "The Assimilation of the Exact Sciences into the Ch'ing Confucian Tradition," *Journal of Asian Affairs* 5 (1980): 24–25. Henderson does not express skepticism toward these claims, much less note that several of these claims are false. It should be noted, however, that Henderson's central argument here is that the exact sciences influenced the "methodological revolution in classical studies" of the Qing. That is, Henderson's central argument is that the abandonment of tradition was detrimental—Henderson concludes that "the Confucian intellectual establishment of late-traditional China . . . did not act as such a formidable barrier to the progress of modern science in China as has commonly been thought. On the contrary, the sometimes uncritical and indiscriminate abandonment of the Qing intellectual tradition by romantic revolutionaries in the twentieth century might possibly have even inhibited the development of modern science, as well as of philosophy, literature, scholarship, and art, in twentieth-century China" (p. 29).

presented with so little concern for the mathematical and historical facts that it is difficult to know where to begin in responding to them. This section will present a brief outline of a few of the technical questions surrounding such claims.

First, we must distinguish between the "informal" axioms found in Euclid's *Elements* and modern formal axiomatic systems, which differ from informal axioms as significantly as modern logic differs from Aristotelian logic. Although the "axioms" of Euclid's *Elements* are often asserted to be self-evident, true, and the foundation of mathematics, modern axiom systems are quite different.

The concerted attempt to find formal axiomatic foundations for mathematics belongs to a specific historical period, namely, from the late nineteenth to the early twentieth century, and in particular the work of Gottlob Frege (1848–1925), David Hilbert (1862–1943), Bertrand Russell (1872–1970), and Alfred North Whitehead (1861–1947), to name only the most prominent.[38] One important inspiration for this work on the foundations of mathematics was attempts by Hilbert and by Alfred Tarski (1901–1983) to reformulate a set of formal axioms for Euclidean geometry.[39]

This attempt to provide formal axiomatic foundations for mathematics produced a series of stunning, negative technical findings that resulted in what has been termed the "loss of certainty" in mathematics.[40] Perhaps the most stunning of these findings was Kurt Gödel's "On Formally Undecidable Propositions of Principia Mathematica and Related Systems," first published in 1931, which in essence demonstrated that systems of axioms could *not* serve as the foundation of mathematics:

> [T]here exist relatively simple problems of the theory of ordinary whole numbers which cannot be decided on the basis of the axioms. This situation does not depend upon the special nature of the constructed systems,

[38] Gottlob Frege, "Begriffsschrift: A Formula Language Modeled upon that of Arithmetic, for Pure Thought," in *From Frege to Gödel: A Source Book in Mathematical Logic, 1879–1931*, ed. Jean Van Heijenoort (Cambridge, MA: Harvard University Press, 1967), 5–82; Alfred North Whitehead and Bertrand Russell, *Principia Mathematica*, 3 vols. (Cambridge: Cambridge University Press, 1910–1913). For collections of papers, see Van Heijenoort, *From Frege to Gödel*; *The Undecidable: Basic Papers on Undecidable Propositions, Unsolvable Problems and Computable Functions*, ed. Martin Davis (Mineola, NY: Dover Publications, 2004). For historical studies, see Ivor Grattan-Guinness, *The Search for Mathematical Roots, 1870–1940: Logics, Set Theories and the Foundations of Mathematics from Cantor through Russell to Gödel* (Princeton, NJ: Princeton University Press, 2000); Paolo Mancosu, ed., *From Brouwer to Hilbert: The Debate on the Foundations of Mathematics in the 1920s* (New York: Oxford University Press, 1998).

[39] David Hilbert, *The Foundations of Geometry*, trans. E. J. Townsend (La Salle, IL: Open Court, 1950); idem, *David Hilbert's Lectures on the Foundations of Geometry, 1891–1902*, ed. Michael Hallett and Ulrich Majer (Berlin: Springer, 2004); Alfred Tarski, "What is Elementary Geometry?" In *The Philosophy of Mathematics*, ed. Jaakko Hintikka (London: Oxford University Press, 1969), 164–75. See also Gottlob Frege, "On the Foundations of Geometry," in *Collected Papers on Mathematics, Logic, and Philosophy*, ed. Brian McGuinness, trans. Max Black (Oxford: B. Blackwell, 1984), 293–340; Bertrand Russell, *An Essay on the Foundations of Geometry* (Cambridge: Cambridge University Press, 2012).

[40] This phrase is from the subtitle of Morris Kline's book *Mathematics: The Loss of Certainty* (New York: Oxford University Press, 1980).

but rather holds for a very wide class of formal systems, among which are included, in particular, all of those which arise from the given systems by addition of finitely many axioms.[41]

This is only one of many findings that complicate assertions about the role of axiomatic systems in mathematics. The axioms necessary for mathematics have proven to be far from self-evident, and in fact the axioms of formal modern systems are hardly intelligible to non-specialists.[42] Self-evidence has been shown inadequate as a criteria to uniquely determine axiom systems, as shown by the development of non-Euclidean geometry that posits alternative axioms, and by the multiplicity of formal axiom systems proposed for modern mathematics.[43] Even the extent to which modern axioms are to be accepted as "true" is not always straightforward.[44] The ambivalence towards axiomatization in mathe-

---

[41] Kurt Gödel, "On Formally Undecidable Propositions of Principia Mathematica and Related Systems," reprinted in Davis, *The Undecidable*, 6, originally published as "Über formal unentscheidbare Sätze der Principia Mathematica und verwandter Systeme I," *Monatshefte für Mathematik und Physik* 38 (1931): 173–98. Cf. Michael Detlefsen, "On an Alleged Refutation of Hilbert's Program Using Gödel's First Incompleteness Theorem," *Journal of Philosophical Logic* 19 (1990): 343–77; idem, "What Does Gödel's Second Theorem Say?" *Philosophia Mathematica* 9 (2001): 37–71. For collections of papers, see either Davis, *The Undecidable*, or Van Heijenoort, *From Frege to Gödel*. For more accessible explanations, see Ernest Nagel and James Roy Newman, *Gödel's Proof*, rev. ed., ed. Douglas R. Hofstadter (New York: New York University Press, 2001); for a more popular history, see Kline, *Loss of Certainty*, 261–64.

[42] Gaisi Takeuti and Wilson M. Zaring, *Introduction to Axiomatic Set Theory*, 2nd ed. (New York: Springer-Verlag, 1982), 20–21. For example, the Axiom of Foundation, which asserts that every nonempty set must have at least one element that is not itself an element of the other elements of that set, is expressed in modern logical notation as follows:

$$\forall x(x \neq \emptyset \implies (\exists y \in x)(y \cap x = \emptyset)).$$

[43] This indeterminacy was further underscored when in 1963 the Axiom of Choice and the Generalized Continuum Hypothesis were shown to be independent of the axioms of general set theory. See Paul J. Cohen, "The Independence of the Continuum Hypothesis," *Proceedings of the National Academy of Sciences of the United States of America* 50 (1963): 1143–48; idem, "The Independence of the Continuum Hypothesis, II," *Proceedings of the National Academy of Sciences of the United States of America* 51 (1964): 105–110; idem, *Set Theory and the Continuum Hypothesis* (New York: W. A. Benjamin, 1966).

[44] Perhaps the most important example is the Axiom of Choice, which states that given a collection of nonempty sets, it is possible to choose one element from each. That is, it asserts that any (possibly uncountably infinite) collection of nonempty sets has a choice function. More formally, given $A = \{A_i : i \in I\}$, there is a function $f : I \rightarrow \cup_{i \in I} A_i$ such that for each $i \in I$, $f(i) \in A_i$. Although it seems intuitively plausible, the Axiom of Choice leads to paradoxical results: it has been shown to be equivalent to many important results, including the well-ordering principle, Zorn's lemma, Hausdorff maximal principle, and Tychonoff's theorem; although the Axiom of Choice is widely accepted, the well-ordering principle is more controversial. One particularly troubling result is the Banach-Tarski Paradox, which shows that it is mathematically possible to take one solid sphere and construct from it two identical solid spheres of the same volume as the original. More specifically, given a solid sphere ($S = \{(x, y, z) \in \mathbb{R}^3 : x^2 + y^2 + z^2 \leq 1\}$), it is possible to decompose it into finitely many pieces and reassemble those pieces to form two solid spheres, each of which is the same size as the original sphere. Stefan Banach and Alfred Tarski, "Sur la décomposition des ensembles de points en parties respectivement congruentes," *Fundamenta Mathematicae* 6 (1924): 244–77. Horst Herrlich, *Axiom of Choice* (Berlin: Springer,

matics is exemplified by the relative lack of concern about the axiomatization of infinitesimals in calculus.[45] Euclidean geometry has been cleaved from its claim to represent reality by the ascendance of non-Euclidean representations of space-time. This failure of the axiomatic method to guarantee mathematical truth has been accompanied by critiques in the philosophy of mathematics.[46]

The argument for the role of Euclidean axiomatics in the historical development of mathematics is even more tenuous. The assertion that Euclid's *Elements* develops mathematics axiomatically is not technically correct:[47] the axioms fail in the very first proposition of the first book.[48] The argument for any technical role of Euclidean axiomatics in the other sciences would also require justification. Even Ptolemy's *Almagest*, which in the introduction is written in a style similar to Eucid's *Elements*, has as its central purpose measurement, which is not permitted in Euclid's *Elements*.

In sum, rather than anachronistically reading Xu Guangqi's prefaces as presenting views from the early twentieth-century on the importance of axiomatic deduction, we must explain Xu Guangqi's appropriation of Euclid in its historical context in seventeenth-century China.

2006), notes that mathematical "disasters" occur both with and without the Axiom of Choice (see chapter 4, "Disasters without Choice," which includes examples from cardinal arithmetic, vector spaces, product spaces, and function spaces; and chapter 5, "Disasters with Choice," which includes examples from elementary analysis and paradoxical geometric decompositions). Questions about the truth of propositions that have been proven to be equivalent to the Axiom of Choice has led to the following witticism by a mathematician, in which one of these logically equivalent propositions is asserted to be true, a second is asserted to be false, and a third is asserted to be unknowable: "The Axiom of Choice is obviously true, the Well-Ordering Principle is obviously false, and who can tell about Zorn's lemma?" (Jerry Bona, quoted in Herrlich, *Axiom of Choice*, viii).

[45] Current mathematics rarely incorporates "non-standard analysis"; results appealing to non-standard analysis are sometimes noted in a manner that almost amounts to a disclaimer. For an account of Robinson's contribution to mathematics, see Joseph W. Dauben, *Abraham Robinson: The Creation of Nonstandard Analysis; A Personal and Mathematical Odyssey* (Princeton, NJ: Princeton University Press, 1995).

[46] Most prominent among these are the critiques by Ludwig Wittgenstein, who asserts that "an axiom . . . is a different part of speech" and characterizes a proof as "a custom among us, or a fact of our natural history." Ludwig Wittgenstein, *Remarks on the Foundations of Mathematics*, trans. G. E. M. Anscombe (Cambridge, MA: MIT Press, 1978), 226 and 61, respectively. See also idem, *Philosophical Investigations*, trans. G. E. M. Anscombe (Oxford: Blackwell, 1958) for his extended critique of rule-following. Important collections on the philosophy of mathematics include Paul Benacerraf and Hilary Putnam, eds., *Philosophy of Mathematics: Selected Readings* (Englewood Cliffs, NJ: Prentice-Hall, 1964); see also Paul Benacerraf and Hilary Putnam, eds., *Philosophy of Mathematics: Selected Readings*, 2nd ed. (Cambridge: Cambridge University Press, 1983), which differs significantly from the first edition.

[47] Abraham Seidenberg, "Did Euclid's Elements, Book I, Develop Geometry Axiomatically?" *Archive for History of Exact Sciences* 14 (1974): 263–95.

[48] This failure is not easy to remedy by adding another axiom: determining whether two circles intersect in zero, one, or two points requires some means of accounting for the radii of the circles and the distance between their centers; but Euclid's *Elements* does not permit measurements.

Xu's Examination Essay

To better understand Xu's writings requires that we analyze them in their historical context. I will begin here with an analysis of the earliest extant example of Xu's
writing, an examination essay for which he was awarded first place. Understanding the circumstances of Xu's passage of the Shuntian Provincial Examination in
1597 is complicated, however, by mysteries that will perhaps never be explained.
The *Chronological Biography of Xu Guangqi* states that the two Principal Readers
(*zhu kao guan* 主考官) were Quan Tianxu 全天敍 and Jiao Hong 焦竑 (1541–
1620); the Branch Examiners (*fen kao guan* 分考官) were Zhang Wudian 張五
典 (1555–1627) and others.[49] It offers the following account of Xu's passing the
examination, citing Zhang Wudian's claims to have discovered Xu's exam and
brought it to the attention of Jiao Hong, who then awarded Xu first place in the
examination:[50]

> Mr. Zhang, from a "search"[51] among the rejected examinations, came upon
> Xu's examination and recommended it to the Principal Reader. At the time,
> "only two days before the announcement of the results," Mr. Jiao "seemed
> disappointed that he had not found someone [worthy of] first place." On
> obtaining Xu's examination, "he slapped his tally and commended it, and
> after reading to the third section, he again slapped the table and shouted,
> Without doubt this is a great scholar (*ru*)[52] of considerable renown, move
> him up to first place."
>
> 張氏從落卷中 "物色" 得公卷, 薦送主考官。時, 距 "放榜前二日", 焦
> 氏 "猶以不得第一人為恨"。既得公卷, "擊節稱賞, 閱至三場, 復拍案
> 歎曰：此名世大儒無疑也, 拔置第一"。[53]

The intent of this account is straightforward, but there is something particularly
noteworthy about the last few sentences. For Jiao Hong, the examiner, was one
of the most renowned scholars of late Ming dynasty.[54] Although this account

---

[49] Liang Jiamian 梁家勉, *Xu Guangqi nianpu* 徐光啟年譜 [Chronological biography of Xu
Guangqi] (Shanghai: Shanghai guji chubanshe 上海古籍出版社, 1981), 59. Jiao Hong was not
the Principal Examiner but rather the Vice-Examiner—see Jiao Hong's memorial (below).

[50] Wang Zhongmin's account differs slightly, stating that Xu's examination was discovered by the
Principal Examining Official (zhukaoguan 主考官) Jiao Hong. See the preface in Wang Zhongmin 王重民, *Xu Guangqi ji* 徐光啟集 [Collected writings of Xu Quangqi] (Beijing: Zhonghua
shuju 中華書局, 1963), 4.

[51] The term *wu se* 物色 means to seek out. The quotation marks, from the Chinese original,
presumably indicate that Liang Jiamian, the author of *Xu Guangqi nianpu*, is quoting from
Zhang's account.

[52] The term *ru* 儒, can mean scholar, literatus, or Confucian; my choice of which one is more
suitable as a translation will depend on the context.

[53] Liang, *Xu Guangqi nianpu*, 59. Chinese punctuation in the original, quoting from Xu Ji's 徐驥
Xian Wending gong xing shu 先文定公行述.

[54] Jiao Hong 焦竑, courtesy name (*zi*) Ruohou 弱侯, sobriquets (*hao*) Yiyuan 漪園 and Danyuan
澹園, became a presented scholar (*jinshi* 進士) in 1589 and ultimately achieved the rank of

appears to assign credit to Jiao Hong for discovering Xu, its effect is to assign credit to Xu, who is discovered and termed "a great scholar" by the eminent Jiao Hong. Yet, this account seems entirely oblivious to the doubts it might unintentionally arouse during this time of rampant corruption and intense factional strife among the literati.

Xu Guangqi, Jiao Hong, and Examination Corruption

The *Official History of the Ming Dynasty* (*Ming shi* 明史) comments on the Ming civil examinations asserting, "Since fraud was commonplace, disputes were frequent" 科場弊竇既多, 議論頻數.[55] The 1597 Shuntian Provincial Examination was one such examination: Jiao Hong was indicted and demoted; afterward he never regained any but unimportant official positions. Conventional sources assert that Jiao was demoted for daring to be unorthodox. For example, the *Official History of the Ming Dynasty* contains a biography of Jiao Hong, included in the section on literati (*wenyuan* 文苑). After noting Jiao had made enemies, this accounts continues:

> In the twenty-fifth year [of the Wanli reign], [Jiao Hong] was the Principal [Examiner] at the Shuntian Provincial Examination. The writings of nine of the Provincial Graduates [successful examinees] including Cao Fan were full of sinister and absurd language. [Jiao] Hong was indicted, and demoted to Vice Magistrate of the Funing Subprefecture.
>
> 二十五年主順天鄉試, 舉子曹蕃等九人文多險誕語, 竝被劾, 謫福寧州同知.

Other sources present essentially the same picture, including the most important source on intellectual history of the period, Huang Zongxi's *Records of Ming*

Senior Compiler of the Hanlin Academy (*Hanlin yuan xiuzhuan* 翰林院修撰). His important works include the following: *Dan yuan ji* 澹園集 [Collected writings of Jiao Hong] (Beijing: Zhonghua shuju 中華書局, 1999); *Jiao shi lei lin* 焦氏類林 [Collected works of Mr. Jiao, arranged topically], in *Congshu jicheng chu bian* 叢書集成初編 [Compendium of collectanea, first edition] (Beijing: Zhonghua shuju 中华书局, 1985–1991), hereinafter *CJCB*; *Zhuangzi yi* 莊子翼 [Wings to *Zhuangzi*], in *Yingyin Wenyuan ge Siku quanshu* 影印文淵閣四庫全書 [Complete collection of the Four Treasuries, photolithographic reproduction of the edition preserved at the Pavilion of Literary Erudition] (Hong Kong: Chinese University Press, 1983–1986), hereinafter *SKQS*; and *Laozi yi* 老子翼 [Wings to *Laozi*], in *SKQS*. Secondary works on Jiao Hong include a chronological biography in Rong Zhaozu 容肇祖, *Jiao Hong ji qi sixiang* 焦竑及其思想 [Jiao Hong and his thought] (Beijing: Yanjing daxue Hafo Yanjing xueshe 燕京大學哈佛燕京學社, 1938), reprinted in *Rong Zhaozu ji* 容肇祖集 [Collected works of Rong Zhaozu] (Jinan: Qilu shushe 齊魯書社, 1989); and Edward T. Ch'ien, *Chiao Hung and the Restructuring of Neo-Confucianism in the Late Ming* (New York: Columbia University Press, 1986).

[55] *Ming shi* 明史 [Official history of the Ming dynasty], ed. Zhang Tingyu 張廷玉 (1762–1755) et al., 28 vols. (Beijing: Zhonghua shuju 中华书局, 1974), 1703.

Scholars (*Ming ru xue'an* 明儒學案).[56] The account in *Deliberations of the Dynasty* (*Guo que* 國榷) is similar but more terse;[57] most recently, Edward Ch'ien follows these accounts in suggesting that Jiao's demotion was the result of a conspiracy.[58]

According to what are probably the most reliable records for the period, the *Veritable Records of the Ming Dynasty* (*Ming shi lu* 明實錄), the charges against Jiao Hong were not limited to unorthodox writings, but included accepting bribes and influence-peddling:

> [Twenty-fifth year of the Wanli reign, eleventh month]: On the 28th day [in the sixty-day cycle] of this year, the Shuntian Provincial Examination Officials, Companion to the Heir Apparent Quan Tianxu and Senior Compiler Jiao Hong, had both been assistants [to the principal examiners].[59] At the announcement of the examination results, criticism led to pandemonium. Office of Scrutiny Minister Cao Daxian indicted[60] the successful examinees[61] Wu Yinghong, Wang Silun, Cao Fan, Zheng Fen and others, stating that [Jiao] Hong and Branch Examining Official He Chongye had accepted bribes and peddled influence.[62] [Jiao] Hong defended [in a memorial]: "[Wu] Yinghong, [Wang] Silun, and [Zheng] Fen were all selected by [Quan] Tianxu; only Cao Fan was selected by myself, because there were so few good examinations on the *Record of Rites*[63] and [Cao] Fan alone was outstanding, so I placed him first. Now [Cao] Daxian's memorial states one thousand *jin* [of silver] a toss,[64] was this thousand *jin* thrown to your humble servant? Or thrown to He Chongye?"

〔萬曆二十五年十一月〕辛卯，是年順天府考試官中允全天敘，修撰焦竑，皆陪推也。及撤棘，物議藉藉。科臣曹大咸參舉中吳應鴻、汪泗論、曹蕃、鄭莱等，謂竑與分考何崇業賄通關節，竑疏辯：應鴻、

---

[56] Huang Zongxi 黃宗羲 (1610–1695), *Huang Zongxi quanji* 黃宗羲全集 [Complete collected works of Huang Zongxi], ed. Shen Shanhong 沈善洪, 12 vols. (Hangzhou: Zhejiang guji chubanshe 浙江古籍出版社, 1985–1994), 8:83.

[57] See Tan Qian 談遷 (1594–1657), *Guo que* 國榷 [Deliberations of the dynasty], ed. Zhang Zongxiang 張宗祥, 6 vols. (Beijing: Guji chubanshe 古籍出版社, 1958), 5:4803.

[58] See Ch'ien, *Chiao Hung.*

[59] The phrase *bei tui* 陪推 refers to leaving a post to serve as an assistant to the principal examiner. Here I have followed the translation of this term used in Huang Tsung-hsi, *The Records of Ming Scholars*, ed. Julia Ching and Zhaoying Fang (Honolulu: University of Hawaii Press, 1987).

[60] Reading *can* 參 to mean indicted (*tanhe* 彈劾).

[61] Reading the phrase *ju zhong* 舉中 as similar to *juzi* 舉子 meaning "successful examinees."

[62] The phrase *hui tong guanjie* 賄通關節 also appears in an account of the examinations in Zhang et al., *Ming shi, juan* 251.

[63] The phrase *shao jia juan* 少佳卷 is a slight rewording of a phrase from Jiao Hong's memorial (below).

[64] Literally, *zhi* 擲 means per throw (as in a game).

泗論、菜皆天敘所取，惟曹蕃為臣所取，以禮記少佳卷而蕃獨勝，故
首拔之。乃大咸疏云千金一擲，不知千金投之臣乎？抑投之崇業乎？[65]

Examination corruption was rife during this period: as the *Official History of the Ming Dynasty* notes in the section on the civil service examinations, accusations of wrongdoing were frequent, and "usually the northern examinations were the worst"; among the problems were bribery and purchasing promotions, with influence-peddling [*guanjie* 關節] being the most serious problem of all.[66]

Jiao Hong's memorial responding to the charges has been preserved. As we would expect, Jiao's memorial is eloquent, authoritative, powerful, and persuasive. It is too lengthy to present in its entirety here, but the most relevant passage merits translation:

> How can it be that successful candidates Wu Yinghong, Wang Silun, Cao Fan, Zheng Fen and others passed the examinations, calling it the crime of your servant? This incredible absurdity is not true. The customary regulations of the examination [are as follows]: The Principal Examiner reads [the examination papers on] two classics, *Changes* and *Documents*; the Vice Examiner reads [the examination papers on] three classics, *Poetry*, *Spring and Autumn Annals*, and *Record of Rites*; each [examiner] does not interfere with the other. Recorded in the *Regulations of the Literati* [written by Zhang Wei—Jiao Hong's chief enemy],[67] this is most clear. Those such as Wu Yinghong, Wang Silun, and Zheng Fen all were selected by Quan Tianxu; as to whether or not there was cheating, Tianxu is responsible, your servant will not address this.
>
> 獨於舉人吳應鴻、汪泗論、曹蕃、鄭菜等中式，謂為臣罪，則尤大謬不然者。科場舊規：正考閱《易》、《書》二經，副考閱《詩》、《春秋》、《禮記》三經，各不相涉，載在《詞林典故》甚明。如吳應鴻、汪泗論、鄭菜皆正考全天敘所取也，其有無弊端，天敘任之，臣不待。
>
> Among [those named in the indictment] Cao Fan was the only one selected by your servant. *Annals* and *Record of Rites* are called the solitary classics, and good essays were from the beginning scarce, with only one hundred eighty examinees for the *Record of Rites*. Your servant edited and added evaluations of the readings, and [Cao] Fan's four [essays on the] classics and five [essays on] policy, the words and meanings alone were superior, and because of this he was placed first. Currently the failed examinations all remain, their superiority and inferiority can be investigated and leafed

---

[65] *Ming shi lu: Fu jiao kan ji* 明實錄：附校勘記 [Veritable records of the Ming dynasty, with editorial notes appended] (Nan'gang: Zhongyang yanjiuyuan lishi yuyan yanjiusuo 中央研究院歷史語言研究所, 1962–66), 5894; punctuation has been added.

[66] Zhang et al., *Ming shi*, 1705.

[67] The bibliographic treatise (*yi wen zhi* 藝文志) of the *Official History of the Ming Dynasty* lists a work by Zhang Wei 張位, *Ci lin dian gu* 詞林典故 (Zhang et al., *Ming shi*, 2394). Apparently, Jiao Hong also composed a work on administration, *Ci lin li guan biao* 詞林歷官表 (ibid., 2394).

through. [Cao] Daxian however extracted a few phrases from them, then recklessly places your servant and Branch Reader He Chongye under suspicion, to the point of the statement "One thousand *jin* per throw"; [I] don't know if the thousand *jin* [was to have been] thrown to your servant, or thrown to Chongye? In the end who is to pay for the crimes, and who is to provide evidence?

言中惟曹蕃一人為臣所取耳。《春秋》、《禮記》名為孤經，佳卷原少，《禮記》人試止百八十人，臣編加品閱，番之四經五策，詞義獨勝，是以首拔之。今落卷具在，其優劣可按覆也。大咸乃摘其數言，而遽疑臣與分校何崇業，至有「千金一擲」之語，不知千金以投之臣乎，抑投之崇業乎？果誰為過付，誰為證據乎？

Beyond Jiao's response, we know very little about the charges of corruption and counter-charges surrounding the 1597 Shuntian examination. Jiao's accuser, Cao Daxian, remained a minor enough official that not only was he not accorded a biography in the *Official History of the Ming*, but apparently did not even merit mention outside of this case; apparently none of his writings survive, including his memorial indicting Jiao Hong and the examinees. The official accused along with Jiao, He Chongye, also seems not to be mentioned elsewhere in the *Official History of the Ming Dynasty*. Other extant sources provide little additional information.[68]

What can we conclude, then, from the extant evidence? Jiao Hong was indicted on a considerable number of accusations, including passing examinations with seditious language, accepting bribes of a thousand *jin*, peddling influence, along with a number of more minor charges. In his memorial, Jiao Hong is very aggressive in attacking the character of the two officers from the Office of Scrutiny: he seeks primarily to put them into disrepute by questioning their motivations and demonstrating that they are wrong on the facts; he asserts that their charges are self-contradictory, presumably to demonstrate that they were not doing their job as censors. However, we should not exclude—just on the basis of either Jiao Hong's eloquence or his technical mastery of examination procedures—the possibility that he was guilty of at least some the charges. Indeed, Jiao Hong is strangely silent on the integrity of the examination, arguing only that if the charges are true, then Quan Tianxu must be equally guilty. The charges ultimately resulted in the demotion of Jiao Hong. Apparently the indicted examinees, including Wang Silun, were also disbarred from the examination.[69] Jiao Hong's demotion was hardly the automatic result of such an accusation—in the exemplary cases recorded in the *Official History of the Ming Dynasty*, the accusers appear to have been punished as often as those they accused.[70]

---

[68] It is possible that the *Zhang Haihong nianpu* 張海虹年譜, Huang Liji's 黃立極 *Haihong Zhang gong mubiao* 海虹張公墓表, or Quan Tianxu's 全天敘 *Shou huai xi Xu weng xu* 壽懷西徐翁序, preserved at the Beijing National Library, may contain relevant passages.

[69] Rong, *Jiao Hong*, 24.

[70] See Zhang et al., *Ming shi*, 1703ff.

What is more interesting for us here is that Xu Guangqi—who was allegedly ranked first—is never mentioned. We can be fairly certain, however, of two things: (1) from the extant patronage letters that Xu Guangqi wrote to Jiao Hong and Zhang Wudian, we know that Zhang and Jiao were indeed the readers of Xu's examination; and (2) Jiao was the Vice-Examiner in charge of three classics, *Spring and Autumn Annals*, *Record of Rites*, and *Poetry*. According to Jiao's memorial, Cao Fan was ranked first on the *Record of Rites* (or possibly *Spring and Autumn Annals*). It is likely that Xu's specialty was the *Classic of Poetry*: successful examinees often passed down within their lineage notes on the classic they studied, so as to help future generations pass the examinations; we have one such manuscript by Xu on the *Classic of Poetry*.[71]

Xu Guangqi's 1597 Shuntian Examination Essay

Since Xu was awarded first place, the examination record preserved by the Ming dynasty would have included his answers in their entirety, along with short evaluations by the examiners. But the *Record of the Shuntian Provincial Examination of 1597* (*Wanli ershiwu nain Shuntian xiangshilu* 萬曆二十五年順天鄉試錄) is apparently no longer extant: it is not among the extensive set of Ming and Qing examinations that Benjamin Elman has collected;[72] and it is not listed in any of the standard bibliographies. Only one short essay from among the several that Xu would have written has been preserved.[73]

Although the provenance of this work is complicated, there is every reason to believe that this essay is authentic. For the sake of brevity here, I will only summarize several among many reasons for believing this essay to be genuine:

1. Wang Zhongmin seems to believe that this essay is authentic, stating in a footnote: "This essay is copied from *Xu's Genealogy*. Below the title it states: 'Wanli dingyou [1597] Shuntian Provincial Examination first place Xu

---

[71] Xu Guangqi, *Shi jing chuan gao* 詩經傳稿 [Transmitted manuscript on the *Classic of Poetry*], in *Xu Guangqi zhu yi ji* 徐光啟著譯集 [Collected writings and translations of Xu Guangqi], ed. Shanghai shi wenwu baoguan weiyuanhui 上海市文物保管委員会 and Gu Tinglong 顧廷龍 (Shanghai: Shanghai guji chubanshe 上海古籍出版社, 1983), hereinafter *XGZYJ*. See also Xu Guangqi 徐光啟 (1562–1633), *Mao shi liu tie jiang yi* 毛詩六帖講意 [Notes on "six couplets" (i.e., passing exams) of the Mao *Classic of Poetry*], in *Siku quanshu cunmu congshu* 四庫全書存目叢書 [Collection of works preserved in the catalogue of the Four Treasuries] (Jinan: Qilu shushe 齊魯書社, 1995–97), hereinafter *SQCC*. The bibliography in Liang, *Xu Guangqi nianpu* also records several manuscripts from Xu's early years that are no longer extant, apparently notes Xu made on his readings of the classics of *Poetry* and *Documents*, the *Four Books*, *Conversations of Master Zhu, Arranged Topically* (*Zhuzi yu lei* 朱子語類), history and philosophers, and apparently even one treatise on mathematics.

[72] See Benjamin A. Elman, *A Cultural History of Civil Examinations in Late Imperial China* (Berkeley: University of California Press, 2000).

[73] Xu Guangqi, *Shun zhi ju shenshan zhizhong* 舜之居深山之中 [That Shun lived deep in the mountains], in *Xu Guangqi ji* 徐光啟集 [Collected works of Xu Guangqi] (Shanghai: Shanghai guji chubanshe 上海古籍出版社, 1984), hereinafter *XGQJ*.

Guangqi'; this was probably the original title in *Concise Exercises in Reading Black [Ink Examinations], One Hundred Essays*" 據《徐氏宗譜》迻錄。篇題下原題:「萬曆丁酉順天鄉試一名徐光啟」,蓋《讀墨簡練百篇》原題如此。

2. Wang also includes an excerpt written by Xu Guangqi's tenth-generation grandson, [Xu] Benceng, explaining the provenance of the exam. Xu Benceng claims to have found a black-ink essay, which then should have been in Xu Guangqi's original handwriting, with Xu's full name and place of birth, the year and place of the examination, and also the names of three generations of ancestors.

3. The format, length, and content of the examination are consistent with what we know about Ming examinations.[74]

4. It seems unlikely that this could be a forgery, especially since the unflattering content is quite contrary to later portrayals of Xu as a practical statesman.

This then is the earliest extant writing we have by Xu Guangqi.[75] This essay is his response to a question on the classics, specifically on *Mencius*, one of the *Four Books*. He would have been given the passage "That Shun lived deep in the mountains" 舜之居深山之中 and asked to write an essay interpreting the passage. Xu's response is 482 characters long: [76]

> [1. "Break open the topic" (*po ti* 破題):][77]
>
> The minds of the sacred emperors! It was only through emptiness that understanding could be attained![78]
>
> 聖帝之心,唯虛而能通也。
>
> [2. "Receiving the topic" (*cheng ti* 承題):]
>
> Thus living in the high mountains,
>
> Shun's mind was no-mind,
>
> And in no-mind there was then nothing that was not attained.

---

[74] See Elman, *Cultural History of Civil Examinations.*

[75] There is also another passage written by Xu, preserved in *XGQJ*, which Liang Jiamian dates at 1597, suggesting this could be an examination question (Liang, *Xu Guangqi nianpu*, 60). This essay has neither the format nor the literary sophistication of a formal examination essay. It appears rather to be a draft of a lecture to one of the princes, which would mean it was written much later than 1597.

[76] A note on this translation: I have presented here a more literal translation in order to preserve the parallel structure of Xu's essay. This structure was as important in the composition of the essay as were the specific concepts that he was discussing. Thus for example I have chosen "no-mind" for *wu xin* 無心 in order to preserve the parallels in the statements "Shun's mind was no-mind" and "in no-mind there was nothing that was not attained." For an explanation of the term "no-mind," see pages 226–227 of this chapter.

[77] The breaks into eight sections are my own. For a discussion of the terminology and format of the essays, see Elman, *Cultural History of Civil Examinations.*

[78] The term *tong* 通 is somewhat vague—it can mean being without any obstruction, attaining one's will, or thorough understanding. I have translated it as "attainment," and the phrase *neng tong* 能通 as "attain understanding."

The bursting forth of a great river, it is like this, is it not?

夫深山之居，

舜之心無心也，

無心斯無所不通矣。

江河之決，有是也夫？

[3. "Beginning the discussion" (*qi jiang* 起講):]

Therefore,

Mind and mind are united,

Good and good combined;

Extended to all-under-heaven,

From the start, there was nothing that could impede it.

且夫，

心與心合，

善與善同，

達之天下，

本無所滯者也。

Since the application of recognition [of objects outside the mind],

Begins in having-mind,

And from this there are the sights and sounds [of the mind];

But the sights and sounds of all-under-heaven,

From the beginning, when compared with mine,

Were not really compatible.

自知識之用，

起於有心，

於是自有其聞見；

而天下之聞見，

始與我揆，

而不相入矣。

[4. "Initial leg" (*qi gu* 起股):]

When I observe the impossibility of equaling the great Shun,

It is not in the occasions that he made use of [the doctrine of] the mean,

But rather in the days that he held fast to [the doctrine of] the mean;

It is not that he followed those who act, straining to use his clear vision and acute hearing,

But that he followed inaction, seeking the day that wisdom would burst forth and crystallize in oneness.

吾觀大舜之不可及也，

不在其用中之時，

而在其執中之日；

不從有為者窺其明目達聰之用，

而從無為者察其濬哲凝一之天。

[5. "Transition leg" (*xu gu* 續股):]

As for his residing in the high mountains, together with the trees, stones, deer and swine,

Others consider [this] to be [the reason he was] very few [in differences][79] from the [uncultured] natives [of the mountains].

I consider that which made him to be very few [in his differences from them],

Was exactly what made Shun be Shun.

即其居深山之中，

而木石鹿豕之與俱也，

人以為幾希於野人耶？

吾以為幾希於野人者，

正舜之為舜耶。

[6. "Middle leg" (*zhong gu* 中股):]

The ordinary man is unable to attain no-self, thus in place after place there is always the selfishness of having-self, but Shun then saw through his not-yet-having-a-master. At the moment of ultimate stillness, without a single fetter, [he] could thus through [this] wait for the movement of all-under-heaven.

The ordinary man is unable to attain no-intent, thus time after time there is always the chance for intent to be born, but Shun came to rest at his without-any-starting-point. Without being encompassed by objects, without forming a single intention, thus he could await those of all-under-heaven.

凡人未能無我，則在在皆有我之私，而舜方洞乎其未有主也。至靜之時，一無所係，乃可以待天下之動者也。

凡人未能無意，則時時皆意生之會，而舜方泊乎其無所起也。無物之衷，不設一意，乃可以待天下之有者也。

Without sounds that could be shared, and without anything that could be taken as sounds that he alone heard, but with the goodness of all-under-heaven, through emptiness holding to the realm of not-hearing, on his hearing the good, even though but one small fraction, yet [it] already had suffused throughout the entire body.

---

[79] Interpolated from the original passage in *Mencius*.

Without sights that could be shared, and without anything that could be taken as sights that he alone saw, but with the goodness of all-under-heaven, through silence remaining in a place of not-seeing, on his seeing the good, even though subtle and minute, yet [it] already had suffused beyond estimation.

無可共聞，亦無可執為獨聞，但以天下之善，虛涵於不聞之境，而及其聞善也，雖一隅而已融為全體矣。

無可共見，亦無可恃為獨見，但以公共之善，默存於不見之地，而及其見善也，雖幾微亦已融為不測矣。

[7. "Later leg" (*hou gu* 後股):]

Thus not hidden from me,

And not impeded in others,

Yet the feelings shared by myself and others,

Set forth from inside without an origin,

My goodness is of no-extent,

Thus, man's goodness is of no-mind,

And words and deeds all come together amidst the never-ending.

To compare this with the bursting forth of a great river,

Gushing forth without impediment,

One word or one deed of all-under-heaven,

Each is sufficient to stir the motive force of the universe?

Then in the high mountains, not recognizing, and not knowing,

As if this was to open its barriers?

蓋不蔽於我，

即不滯於人，

而我與人相感，

發於無端之內，

我善無大，

故人善無心，

而言與行俱會通於不窮之中。

譬諸江河之決，

沛然莫禦，

而天下之一言一行，

皆足以鼓其機緘者乎？

彼深山之不識不知，

尤所以開其障塞者乎？

[8. "Conclusion" (*da jie* 大結):]

If from the establishment of sights and sounds,

With goodness I set forth, and of that which arrives, there is none that is accepted;

With goodness I set out, and of that which enters, there is none that penetrates;

Then how subtle is the reaction!

若自立聞見，

我以善往，而來者莫之受矣，

吾以善出，而入者莫之通矣，

又何妙應之有哉！

Analysis of Xu's Essay

The discovery of Xu Guangqi's examination by Jiao Hong has been credited to Jiao's interest in what has been termed "practical learning" (*shixue* 實學).[80] Yet in reading Xu's examination, the problem is not just that we did not find anything practical. Under a charitable reading, Xu's essay might seem at best to be vague philosophical speculation invoking Daoist and Buddhist concepts such as "emptiness" (*xu* 虛), "no-mind" (*wuxin* 無心), "no-self" (*wuwo* 無我), and "non-action" (*wuwei* 無為). Under more careful scrutiny, Xu's essay might seem to be little more than a nonsensical and indiscriminate concatenation of Buddhist- and Daoist-inspired cant: "it was only through emptiness that understanding could be attained"; "Shun's mind was no-mind, and in no-mind there was then nothing that was not attained"; "ordinary man is unable to attain no-self . . . but Shun then saw through his not-yet-having-a-master."

To better understand this essay, and why Jiao Hong might have selected this essay for first place in the examination, we must understand more about the examination essays. The single most important and comprehensive study of the civil examinations is Benjamin Elman's *Cultural History of Civil Examinations in Late Imperial China*.[81] Although my analysis throughout this section is based on Elman's findings, I cannot summarize his work here; I will limit my analysis to the style (very briefly) and then the content (less briefly) of Xu's essay. First,

---

[80] Wang Zhongmin 王重民, "Xu yan—Xu Guangqi zhili kexue yanjiu de shiji he ta zai woguo kexue shi shang de chengjiu" 序言—徐光啟致力科學研究的事蹟和他在我國科學史上的成就 [Preface—The achievements of Xu Guangqi's devotion to scientific research and his accomplishments in the history of Chinese science], in *XGQJ*, 4. "Solid learning" (*shixue* 實學) refers to Zhu Xi's Learning of Principle; it is sometimes misinterpreted as "practical studies."

[81] Elman, *Cultural History of Civil Examinations*. Earlier studies include Andrew Plaks, "Pa-Ku Wen 八股文," in *Indiana Companion to Traditional Chinese Literature*, ed. William Nienhauser (Bloomington: Indiana University Press, 1986), 641–43; idem, "The Prose of Our Time," in *The Power of Culture: Studies in Chinese Cultural History*, ed. Willard J. Peterson, Andrew H. Plaks, and Yu Yingshi (Hong Kong: Chinese University Press, 1994), 206–17.

in terms of the style, Xu's essay seems to be extraordinarily well-written, by the highly formalized conventions of the eight-legged essay. Central to this style is the parallelism evident throughout, parallelism that operates on several levels. On one level, there is the parallelism of phrases, such as in the two couplets in "beginning the discussion": "Mind and mind are united, Good and good combined." This is found throughout the essay, with phrases repeated with only a change in one or a few characters, these changes themselves being restricted to parallel grammatical structures. The most important examples of this parallelism and its complexity are in the sixth, or "middle leg" of the essay.

Traditional Commentaries

As impressive as Xu's virtuosity with this style of writing might be, the most important aspect for us to consider is the content of this essay. For it is the content that provides a link to the textualist approaches advocated by Jiao Hong—the interpretation of one canon of texts through the citation of another canon. It is also this textualist approach that helps explain Xu's readings of Jesuit texts. We need to inquire further into the commentaries and sources that Xu was citing.[82]

Though Xu was given only a single phrase, "That Shun lived deep in the mountains," and asked to comment on it, he would have been required to know by heart the entire passage from *Mencius*, along with the major commentarial traditions. The original passage from *Mencius* states:

> Mencius stated: Shun lived amidst the high mountains, residing with rocks and trees, sporting with deer and swine; there was scarcely any difference with the natives of the high mountains. Upon hearing a good word, seeing a good deed, it was like the bursting forth of a great river—nothing could impede it.
>
> 孟子曰：舜之居深山之中，與木石居，與鹿豕遊，其所以異於深山之野人者幾希。及其聞一善言，見一善行，若決江河沛然莫之能禦也。[83]

The traditional commentaries to this passage include the commentary written by Zhao Qi 趙岐 (d. 201),[84] and the subcommentary attributed to Sun Shi 孫

---

[82] For an excellent overview, see Willard J. Peterson, "Confucian Learning in Late Ming Thought," in *The Ming Dynasty, 1368–1644*, ed. Frederick W. Mote and Denis Twitchett, 2 vols. (New York: Cambridge University Press, 1988, 1998), 708–88. The other essays in volume 2 also offer helpful overviews of various aspects of Ming culture.

[83] *Mengzi zhushu* 孟子註疏 [*Mencius*, with commentary and subcommentary], in *Shisanjing zhushu: fu jiaokan ji* 十三经注疏：附校勘记 [Thirteen Classics, with commentary and subcommentary, and with editorial notes appended] (Beijing: Zhonghua shuju 中华书局, 1980), hereinafter *SSJZS*, 2:2765.

[84] This is the earliest extant commentary on *Mencius*. For an analysis of the authenticity of the text, commentaries, and translations, see D. C. Lau, "Meng-Tzu," in *Early Chinese Texts: A Bibliographical Guide*, ed. Michael Loewe (Berkeley: Society for the Study of Early China, Institute of East Asian Studies, University of California, Berkeley, 1993), 331–35.

奭 (962–1033).[85] The commentary by Zhao follows the text very closely, adding little interpretation but explaining in more explicit terms the meaning of each character:

> During the period when Shun farmed at Lishan, he resided amongst trees and stones; deer and swine were close to people as if they sported with people. [The character] *xi* means far. At this time, how could it be possible that the distance between Shun and the natives was far? Although Shun was an outsider to the natives, he lived together with them. When he heard one good word from a person he followed it; when he saw one good deed from a person he recognized it, bursting forth without hesitation like the current of a great river, nothing could impede where it wished to go.

> 舜耕歷山之時居木石間。鹿豕近人若與人遊也。希，遠也。當此之時舜與野人相去豈遠。舜雖外與野人同其居處。聞人一善言則從之，見人一善行則識之，沛然不疑若江河之流，無能禦止其所欲行也。

Much of the commentary by Zhao Qi, then, consists of a close paraphrase of the text; although the question of Shun's difference from the natives is expanded upon, little new analysis is added.

The subcommentary attributed to Sun Shi, however, is more daring and fanciful, invoking the mythologies of spirit-dragons and the enigmatic *Classic of Changes* in interpreting this passage:

> Mencius attains the highest point of resistance.[86] *Rectified Meanings* states: This passage says that the sages hid in seclusion like the spirit-dragon,[87] both able to fly in the heavens and able to submerge into hiding, like Shun. Mencius states that at the period in the past when Yu Shun first arose farming in Lishan, [he] lived amongst the trees and stones, and it is for this reason [that the passage states that] he was close to trees and stones. Sporting with the deer and swine, it is for this reason [that the passage states that] deer and swine were close to humans. But the reason that Shun in this situation differed from the natives of the deep mountains was not that he was distant [from them], but rather that [he] had the ability, upon hearing a good word, on seeing a good deed, to follow it like the water bursting forth from a great river, abundant in power, with nothing that could impede it. Commentary: The sages hid like the spirit-dragon. *Rectified Meanings* states: this is exactly the same as the language of the "qian" hexagram in the *Classic of Changes*; Zhao's commentary cited this to explain this classic.

> 孟子至禦也。正義曰：此章言聖人潛隱若神龍，亦能飛天，亦能潛藏，同舜也。孟子言，虞舜初起於歷山耕時，居於木石之間，以其

---

[85] Zhu Xi disputed the attribution of the subcommentary to Sun Shi. Lau, "Meng-Tzu," 333.

[86] The subcommentary to each passage begins with a very brief summary (a few characters in length) of the concept that Mencius has "attained" in the passage.

[87] Because dragons were held to be supernatural, they were sometimes referred to as *shen long* 神龍.

近木石故也。與鹿豕遊，以其鹿與豕近於人也。然而舜於此，其所以有異於深山之野人，不遠但能及其聞一善言、見一善行，其從之若決江河之水，沛然其勢，莫之能禦止之也。注聖人潛隱若神龍者。《正義》曰：此盡同《易》乾卦之文也，趙注引之以解其經。

It is in the subcommentary attributed to Sun Shi that we find more concern about the central problem that Xu seems to be addressing—how was Shun different from the "natives," the uncultured peasants that farmed for a living deep in the mountains? However, we find here very little related to the answer that Xu presented—"Shun's mind was no-mind."

The most important commentaries were those by Zhu Xi 朱熹 (1130–1200), in his *Collected Commentaries on Mencius* (*Mengzi ji zhu* 孟子集注).[88] While Xu's essay seems to incorporate Daoist or Buddhist themes, at least some of these themes have textual precedents in the commentaries by Zhu Xi:[89]

> The character *xing* is [pronounced with] the falling tone. "Residing deep in the mountains" refers to the time when [Shun] farmed in Lishan. The mind of the sage reaches ultimate emptiness and ultimate clarity; in the midst of turbidity, the myriad principles are entirely complete. As soon as the senses are stirred, then the reaction is very fast, and there is nothing that is not understood. If not for the depth of the Way established by Mencius, it would not be possible to describe to this [extent].

行，去聲。居深山，謂耕歷山時也。蓋聖人之心，至虛至明，渾然之中，萬理畢具。一有感觸，則其應甚速，而無所不通，非孟子造道之深，不能形容至此也。[90]

Thus we find that Xu's first two sentences in his essay—"The minds of the sacred emperors! It was only through emptiness that understanding was reached"—which seemed to be appealing to the concept of "emptiness" from unorthodox Buddhist sources, actually refer directly to the statement in Zhu Xi's orthodox commentary: "The mind of the sage reaches ultimate emptiness and ultimate clarity." However, textual precedents for Xu's ultimate answer—"The mind of Shun was no-mind"—cannot be found in *Mencius*, the commentary by Zhao Qi, the subcommentary attributed to Sun Shi, or the commentaries by Zhu Xi.

---

[88] Zhu Xi, *Mengzi jizhu* 孟子集注 [Collected commentaries on *Mencius*], in *Sishu zhangju jizhu* 四書章句集注 [Four Books, separated into chapters and sentences, with collected commentaries] (Beijing: Zhonghua shuju 中華書局, 1983), hereinafter *SZJ*.

[89] Zhu Xi also comments on this passage elsewhere in *Conversations with Master Zhu, Arranged Topically*: It is asked: "'Shun heard a good word, saw a good deed, like the bursting forth of a great river, powerful, nothing could impede [it].' At the time when [he] had not yet seen or heard, what about phenomena (*qi xiang*)?" Answer: "Profound and clear, nothing more. His principle full and prepared, as soon as [it] was stirred, [it became] powerfully abundant, and nothing could impede [it]". Zhu Xi, *Zhuzi yu lei* 朱子語類 [Conversations with Master Zhu, arranged topically], ed. Li Jingde 黎靖德 and Wang Xingxian 王星賢, 8 vols. (Beijing: Zhonghua shuju 中華書局, 1986), 1442.

[90] Zhu, *Mengzi jizhu*, 353.

Syncretism as Philological Exegesis

To better understand Xu Guangqi's readings, and why Jiao Hong might have passed him, we must look at the writings of Jiao Hong. One example of Jiao's approach to texts is *Mr. Jiao's Writings, Continuation* (*Jiao shi bi cheng xu ji* 焦氏筆乘續集), completed in 1606 but published in parts beginning in 1580.[91] In this text, Jiao offers interpretations for selected passages from texts that have proved difficult to explicate in the Confucian commentarial tradition. For example, the first phrase that Jiao Hong chooses to explicate is from the *Analects* attributed to Confucius: "Confucius said, Silently understand it" 孔子言默而識之. The entire passage from the *Analects* is:

> The Master states: Silently understand it; study without weariness; teach others without fatigue; is it in me [to do these things] then?
>
> 子曰：默而識之，學而不厭，誨人不倦，何有於我哉？[92]

The meaning of the passage seems fairly straightforward, with the exception of the initial phrase "Silently understand it." Traditional commentaries offer little guidance on interpreting this enigmatic phrase,[93] though the theme of absence

---

[91] Jiao Hong 焦竑 (1541–1620), *Jiao shi bi sheng xu ji* 焦氏筆乘續集 [Mr. Jiao's writings, continuation], in *SQCC*. Edward Ch'ien, citing Qu Wanli, *A Catalogue of the Chinese Rare Books in the Gest Collection of the Princeton University Library* (Taibei: Yiwen yinshuguan, 1974), 289, states that this work was published first in parts in 1580; it was published in its entirety in 1606. Ch'ien, *Chiao Hung*, 324 n. 3.

[92] *Analects* 7.2; *Lunyu zhushu* 論語註疏 [*Analects*, with commentary and subcommentary], in *SSJZS*, 2:2481.

[93] The traditional commentaries present this interpretation: "Commentary: 'Zheng states: Without this practice, it is I alone who has it'" 鄭曰：無是行於我獨有之. The subcommentary adds: "[Repeats passage]. *Rectified Meanings* states: In this passage Confucius states, do not speak but write and understand it, the scholarship is solid and the heart does not grow to detest [it], teaching others without stopping to rest, others do not have this practice for me, only I have it. Thus [Confucius] states, 'is it in me [to do these things] then?'" [疏]子曰：默而識之，學而不厭，誨人不倦，何有於我哉？《正義》曰：此章仲尼言，已不言而記識之，學古而心不厭，教誨於人不有倦息，他人無是行於我，我獨有之，故曰何有於我哉？ Zhu Xi, the traditional authority, offers the following interpretation: "*Zhi* 識 is pronounced '*zhi*,' also as '*zi*.' *Zhi* means to remember. Silently remember, means to not speak but store it in the mind. It has been stated: *zhi* means to know; without speaking, the mind understands. The former interpretation is close to this. 'Is it in me [to do these things]' means 'what is it that is capable of being from me?' These three are already not the ultimate attainment of the sages. It is as if [Confucius] did not dare to take credit, thus [these] are words of humility [heaped] upon humility." 識，音志，又如字。識，記也。默識，謂不言而存諸心也。一說：識，知也，不言而心解也。前說近是。何有於我，言何者能有於我也。三者已非聖人之極至，而猶不敢當，則謙而又謙之辭也. Zhu, *Zhuzi yu lei*, 93.

of speech appears elsewhere in the *Analects*.[94] Jiao Hong offers the following reading:

> Confucius states, "Silently understand it." [This is] not silence of the mouth, [but] silence of the mind. "Silence of the mind" [means] the path of speech and thought is severed, places of the mind's activity are extinguished, but suddenly they are combined![95] It is through no-feeling that they are combined. "It is through no-feeling that they are combined" [means] something similar to "through no-speech they are combined." Therefore [the name] given it is silence. Thus [partial] study leads to weariness, [but] by studying through silent understanding, then study is without weariness. [Partial] teaching is tiring, but teaching through silent understanding, teaching is then not tiring. There is [study and teaching] without silence, thus [Confucius] states "is it in me [to do these things] then?" However, those who can truly understand through silence, even if they have [this], yet there was never a time that they were without. This is also the changelessness of having and not-having.[96]

> 孔子言默而識之，非默於口也，默於心也。默於心者，言思路斷，心行處滅，而豁然有契焉，以無情契之也。以無情契之，猶其以無言契之也，故命之曰默。夫有所學則厭，默識以為學，學不厭矣。有所誨則倦，默識以為誨，誨不倦矣。有非默也，故曰何有於我哉。雖然，真能默識者，即有亦未嘗不無，此又未易以有無論也。[97]

For Jiao Hong, this puzzling passage taken from the *Analects*—"Silently understand it"—is explained by a direct quotation from the Buddhist canon *Tripitika* (*Da zang jing* 大藏經): "The path of speech is severed, places of the mind's activity are extinguished" 言語路絕，心行處滅.[98] What interests Jiao Hong about this passage, then, is using texts from the Buddhist canon to explicate enigmatic phrases in the Confucian canon.

The following passage that Jiao Hong selects from the *Analects* follows this pattern. This time he chooses to explain the passage, "The Master stated: Do I have knowledge? [I] am without knowledge. But if a commoner asks [a question] of me—as if empty—I consider both ends until something comes out" 子曰：吾

---

[94] *Analects* 17.19: "The Master stated: I desire to be speechless. Zigong stated: If the master [Confucius] does not speak, then what can [we] the disciples record and narrate? Confucius stated: What speech has heaven? The four seasons change through [heaven], the hundred objects are born through [heaven]. What speech has heaven?" 子曰：予欲無言。子貢曰：子如不言，則小子何述焉。子曰：天何言哉，四時行焉，百物生焉，天何言哉.

[95] Reading *qi* 契 here as "combined." I thank Lim Jongtae for his many helpful suggestions, here and throughout this book.

[96] This is a very tentative translation of this enigmatic sentence. Presumably, this is another allusion to a classical source.

[97] Jiao, *Jiao shi bi sheng xu ji*.

[98] *Tripitika (Da zang jing)*. Elsewhere, in the *Ying luo jing* 瓔珞經, we find a variant of this phrase—"The way (*dao*) of speech is stopped, places of the mind's activity are extinguished" 言語道斷，心行所滅.

有知乎哉？無知也。有鄙夫問於我，空空如也，我叩其兩端而竭焉。[99] In the process of explaining this passage, Jiao Hong develops this into a discussion of Yan Hui. Then Jiao turns to Daoist texts, this time citing them explicitly by name: "Laozi states: What has not yet given a sign is easy to plan for; the brittle is easily shattered, the minute is easily scattered. Yan Hui was one who was scattered in the minute" 老子曰：其未兆易謀，其脆易破，其微易散，顏氏散之於微者也。[100] Jiao Hong's approach to commenting on these texts, then, is to explain enigmatic passages of the Confucian canon by turning to texts outside the canon, including both Buddhist and Daoist texts.

Xu's "Syncretic" Answers

Keeping in mind that the traditional commentaries were to be the basis of what Xu wrote, and Jiao Hong's unorthodox approach to philosophical exegesis, we can now return to Xu's examination. The key here is the textual sources that Xu cites in his answer. Perhaps more important than the ideas expressed, the selection of classical sources provided examinees with a venue to exhibit their erudition and mastery of the classics. Their use of sources would show what they read, what they considered important, and how they read these texts. Their use of key words would identify them as sympathetic to certain factions. In short, the examination essay provided an excellent opportunity for factional self-fashioning.

To return now to Xu's explication of this passage from *Mencius*, recall again that Xu's answer is that "Shun's mind was no-mind." "No-mind," it turns out, is indeed a term that does not appear in the traditional commentaries on this passage; in fact, this term rarely appears anywhere in the *Thirteen Classics* or the *Four Books*.[101] The term does appear frequently in texts outside the Confucian

---

[99] *Analects* 9.7. The exact meaning of this passage—and hence the translation here—is controversial.

[100] Zhu Qianzhi 朱謙之 (1899–1972), *Laozi jiao shi* 老子校釋 [*Laozi*, with emendations and explanations] (Beijing: Zhonghua shuju 中华书局, 1984), passage 64; *Lao-Tzu: Te-Tao Ching: A New Translation Based on the Recently Discovered Ma-Wang-Tui Texts*, trans. Robert G. Henricks (New York: Ballantine Books, 1989), 150. See also Chen Guying 陳鼓應, *Laozi zhu yi ji pingjie* 老子註譯及評介 [*Laozi*, annotated, translated into modern Chinese, with criticism] (Beijing: Zhonghua shuju 中華書局, 1984).

[101] The term "no-mind" (*wuxin*) appears only rarely in the *Thirteen Classics*, and when it does it is usually in commentaries, using *Laozi* to explain the classics. The term does not seem to be used in the *Four Books* 四書. The term does appear in the *Conversations of Master Zhu, Arranged Topically* (*Zhuzi yu lei* 朱子語類). However, when it does appear, it is heaven and earth that have no-mind, and the sage has mind but no-action: "[Cheng] Yichuan states: 'Heaven and earth have no-mind but become transformed; the sage has mind and inaction'" 伊川曰：『天地無心而成化，聖人有心而無為。』 Several occurrences in the *Conversations of Master Zhu* are then responses to Cheng's view on the subject. Several others relate to the interpretation of the *Classic of Changes*: Chang Ru asked: "[In the *Classic of Changes*, the hexagram] 'Heaven' is strong and [the hexagram] 'Earth' is submissive, how is it obtained that 'there is error left uncorrected'?" [Zhu Xi] stated: [The hexagrams] 'Heaven' and 'Earth,' once

canon. Although this term appears in Guo Xiang's 郭象 (d. 312 CE) introduction to *Zhuangzi* and throughout his commentaries,[102] the locus classicus for this term is a passage from the Daoist text *The Classic of the Way and Power* (*Dao de jing* 道德經):

> The sage has no-mind, taking the mind of the common people [literally, the hundred-surnames] as [his] mind. I take the good to be good, I also take the not-good to be good, thus obtaining the good.[103] I take the trustworthy to be trustworthy, I also take the not-trustworthy to be trustworthy, thus obtaining the trustworthy.
>
> 聖人無心，以百姓心為心。善者吾善之，不善者吾亦善之，得善。信者吾信之，不信者吾亦信之，得信。[104]

The term "no-mind" is particularly significant because it was a key term in the teachings of Wang Ji 王畿 (Wang Longxi 王龍溪, 1498–1583), one of the most important followers of Wang Yangming's Learning of the Mind (*xin xue* 心學).[105] Xu's use of the term in his examination essay would then represent a provocative rejection of Zhu Xi's orthodox Learning of Principle (*li xue* 理學).

We do not have Jiao Hong's written evaluations by which to confirm what he might have seen in Xu's essay to award it first place. But from the above analysis we can hypothesize about what Jiao Hong might have liked about Xu's essay. From what we know of Xu's later philosophical positions (discussed below), it turns out that this question was perfect for him. It did not ask about interpretations related to contemporary debates in Confucian philosophy. Instead, it invited the examinee to speculate about the (semi-mythical) sages of early antiquity, sages Xu revered. Xu's response is stylistically exquisite in its parallelism, and emotionally evocative in writing about the ancient sages. We have seen that in his response Xu adopts precisely the same kind of syncretic approach to the

---

qi moves in no-mind, it is not possible to err in following 'no error is left uncorrected.' The sage takes having-mind to be the master, therefore is without the mistake of 'surpassing but not arriving.' Therefore the sage is capable of admiring the development of heaven and earth; the contribution of heaven and earth awaits the sage." 長孺問：「『乾健坤順』，如何得有過不及？」曰：「乾坤者，一氣運於無心，不能無過不及之差。聖人有心以為之主，故無過不及之失。所以聖人能贊天地之化育，天地之功有待於聖人。」Zhu, *Zhuzi yu lei*, 1648.

[102] Guo Xiang 郭象 (d. 312 CE), *Zhuangzi zhu* 莊子注 [Commentary to *Zhuangzi*], in *SKQS*. The concept of no-mind (*wuxin*) also appears in Ge Hong 葛洪 (284–364), *Baopuzi nei pian* 抱樸子內篇 [Inner chapters of *Baopuzi*], in *SKQS*, and in Guan Zhong 管仲 (d. 645 BCE), *Guanzi* 管子 [Guanzi], in *SKQS*. It is used throughout commentaries and subcommentaries on *Liezi* 列子 [Liezi], in *SKQS*. See also Li Qi 李杞 (12th/13th c.), *Yong Yi xiang jie* 用易詳解 [Detailed explanations for using the *Classic of Changes*], in *SKQS*.

[103] The transmitted version of the text differs from the Mawangdui texts. Henricks translates 善者善之不善者亦善也 ... 信者信之不信者亦信之 as "Those who are good he regards as good; Those who are not good he also regards as good. ... Those who are trustworthy he trusts; those who are not trustworthy he also trusts." Henricks, *Lao-tzu: Te-tao Ching*, 120–21. I have followed the transmitted version, which is what would have been available in the Ming Dynasty.

[104] Zhu, *Laozi jiao shi*, 194.

[105] I am indebted to an anonymous reviewer for noting this important point.

reading of the classics advocated by Jiao Hong: Xu's interpretation of this passage from *Mencius* is filled with terms from texts outside the Confucian canon—"emptiness," "no-mind," "no-self," and "non-action." Xu's response, however, is more ingenious than that. His first invocation of the term "emptiness" (*xu* 虛), usually associated with Buddhism, is authorized by the orthodox commentaries in Zhu Xi's *Four Books*. We may speculate that this is precisely the kind of answer the examiner hoped to see, especially if we recall that a passage on Shun is the only passage in the *Analects* that uses the term inaction (*wuwei* 無為) usually associated with Daoism: "Among those who governed through doing nothing (*wuwei*, or inaction), there was Shun. Then what did he do? Make himself reverent and face south, that is all" 子曰：無為而治者其舜也與。夫何為哉？恭己正南面而已矣。[106]

1597 Discussion Essay: "Erudition of Objects"

Another examination essay may also be the work of Xu Guangqi.[107] This essay is preserved in a collection of examination essays titled *Huang Ming Ce Heng* 皇明策衡.[108] The essays are ordered by dates, and under the heading "1597 Shuntian Examination" (*Wanli ding you ke Shuntian*) 萬曆丁酉科順天 are listed the names of the two examiners, Jiao Hong and Quan Tianxu. There are two essays from this year: "Advisor to the Ruler" (*mo chen* 謀臣), and another essay "Erudition of Objects" (*bo wu* 博物). Both essays are unsigned. The essays that were judged worthy of inclusion in this work are not always the writing of the candidate who was awarded first place in the overall examination,[109] so even assuming that Xu was the primus, the essays included might not have been his. Still, one of these two essays answers a question about the *Classic of Poetry*, which was Xu's specialty. Furthermore, this essay expresses opinions and uses terms that are found in Xu's later writing.[110] And arguably, even if it is not Xu's writing, it can give us an idea of the sort of material the examiners passed; we might expect that Xu's essay would not have been much different.

As was the convention with policy (策 *ce*) essays, this essay is quite long—the question alone is 315 characters long (approaching the length of Xu's entire

---

[106] *Analects*, 15.4.

[107] I would like to thank Benjamin Elman, who discovered this essay, for bringing it to my attention.

[108] Mao Wei 茅維 (fl. 1605), *Huang Ming ce heng* 皇明策衡 [Weighing Ming dynasty policy examination essays], in *Siku jinhui shu congkan* 四庫禁毀书丛刊 [Collection of books banned or burned by the Four Treasuries project], ed. Wang Zhonghan 王钟翰 and Siku jin hui shu congkan bianzuan weiyuanhui 四库禁毁书丛刊编纂委员会 (Beijing: Beijing chubanshe 北京出版社, 1998), hereafter *SJSC*.

[109] Benjamin Elman, personal communication.

[110] See Xu, *Ke Tong wen suan zhi xu*, discussed below.

answer to the question on *Mencius*, 482 characters long). The first part of the question begins:

> It is asked: Knowledge about objects[111] is not something that is [considered] urgent for the superior man. Thus scholars play with theories of principle and nature, setting aside without inquiry the naming of objects. Confucius also once worried about this. Thus [Confucius] stated: "Know more about the names of birds, animals, grasses, and trees." What is valuable about birds, animals, grasses, and trees that one would want to know them? Commonly it is stated that in naming objects, nothing is more detailed than *Approaching Correctness*. . . .

> 問博物非君子所急也。然學者操理性之說，而置名物於不問。仲尼亦
> 嘗患之，故曰：「多識於鳥獸草木之名」。夫鳥獸草木何貴而欲識之
> 也？世言名物莫詳於《爾雅》。[112]

The question continues, at length, asking whether *Approaching Correctness* is compiled from the *Classic of Poetry* or written by the Duke of Zhou, what evidence there is for either view, and whether *Approaching Correctness* can be used to explain the *Classic of Poetry*.

This question develops from a lament of Confucius in the *Analects*. Although studying the *Classic of Poetry* might seem to be as far as possible from practical affairs, in the following passage, Confucius explains the importance of poetry for the study of nature and its implications for the Confucian moral order:

> The Master stated: Disciples! Why do none [of you] study poetry? Reading poetry can inspire the will, aid observation, increase fellowship, and express grievances. Nearby, it can [aid in] serving [one's] father; more distant, it can [aid in] serving [one's] superiors. [It can] also help one know the names of birds, animals, grasses, and trees.

> 子曰：「小子何莫學夫詩？詩，可以興，可以觀，可以群，可以怨。
> 邇之事父，遠之事君。多識於鳥獸草木之名。」[113]

Textual research into the *Classic of Poetry*, it seems, is supposed to provide the key to the study of the natural world.

This very leading question, posed by the examiner, then suggests to the examinee who wants to pass the examination what the answer should be; examinations were evaluated by their eloquence in giving the expected answer. The response to this question begins:

> In their studies, none of the [contemporary] Confucians (*ru*) do not take nature (*xing*) and fate (*ming*) as the target; "knowing more" [about names

---

[111] Literally, erudition about objects.

[112] Anon., "Erudition of Objects" (*bo wu* 博物), "1597 Shuntian Examination" (*Wanli ding you ke Shuntian* 萬曆丁酉科順天), in Mao, *Huang Ming ce heng*.

[113] *Analects*, 17.9. Zhu Xi, *Lunyu jizhu* 論語集注 [Collected commentaries on the *Analects*], in *SZJ, juan* 9, 4a–b. Translation adapted from Yang Bojun 楊伯峻 (1909–1992), *Lun yu yi zhu* 論語 譯注 [*Analects*, with explanations and commentary] (Beijing: Zhonghua shuju 中華書局, 1980).

of objects] then is not [their] concern. Thus currently form extinguishes substance and erudition drowns the mind, closing the profound illumination yet seeking it [illumination] through the ears and eyes. So much wasted effort! The path of language and script [of the Confucians, which one] learns as a child [results only in] confused old men, yet [they are held to be] the ones who are highly learned and specially attained. Lifted and raised! This makes [them] necessarily detest and despise everything. From this can be seen [the cause for] the falling of the Way. [They] listen only to what has been seen before, fearing [to study] even a single object; one does this, another sincerely detests [it]—thus the confusion and fractiousness of [this] vulgar learning. And the end cannot be known. [They] then split [the study of] the nature of the Way and the names of objects into two [separate fields]; the result is that current discussions are [no better than] all those who disputed "names and principle."[114] In the end, no one knows from whence to enter [these disputes], making seeking [the truth of the matter] all the more difficult. Yet the Confucians' "single [thread] that runs throughout" [all of Confucius' teachings]—from ancient to the contemporary—very strongly emphasized the study of the myriad objects. Yet, because [of the above], it has all decayed. Alas! Yet once Confucius sought a balance. Confucius stated, "My Way has a single [thread] that runs throughout it";[115] on another day [he] instructed his disciples about [the use of] the *Classic of Poetry*, also stating "know more about the names of birds, animals, grass, and trees."

夫儒者之學，則靡不以性命為鵠矣，然而非多識之患也。顧世之文
滅質，博溺心者，蔽其玄光求之耳目。揖揖焉！童習白紛於言語文字
之塗，乃高明特達者，起而矯焉，以為必厭薄一切，斯可以見道窘。
然獨聽[116]前睨[117]而懼一物之，我干也，彼誠惡，夫俗學之紛紜，而不
知其卒也。乃以岐性道名物而為二，今世之談，名理者，竟莫知其所
從入，以愈求之難。而儒者貫穿，古今極命萬物之學，亦以俱廢。嗚
呼！亦嘗折衷[118]於孔子巳乎。孔子曰「吾道一以貫之」，至他日以詩
訓小子，則又曰「多識於鳥獸草木之名」。

Although very polemical (as opposed to the more formal parallel style of the previous essay), this essay too contains many allusions from unorthodox sources. In fact, none of the phrases cited in the opening paragraph are from orthodox sources. Though the first two introductory sentences do not seem to contain allusions, the third sentence is taken directly from the chapter "Preserving [one's]

---

[114] During the Wei-Jin period, "pure discussion" (*qing tan* 清談) was divided into two factions, "discussion of the mysterious" (*xuan lun* 玄論) and "names and principles" (*ming li* 名理). Here the examinee is presumably using this term not in this specific sense but as a pejorative to refer to the generally despised "pure conversations."

[115] *Analects*, 4.15.

[116] Substituting for a variant form.

[117] Substituting for a variant.

[118] Substituted for an idiosyncratic variant.

nature" (*shan xing* 繕性) in the Daoist text *Zhuangzi* 莊子, dismissing the usefulness of knowledge, culture, and erudition: "Knowledge is insufficient to order all-under-heaven, but then to add to this culture, to increase this with erudition, substance is extinguished by culture, and erudition drowns the mind" 知而不足以定天下，然後附之以文，益之以博，文滅質，博溺心。[119] The next sentence is from the second chapter of the eclectic *Huainanzi*, asserting the futility of searching for truth with the senses: "Closing off the profound illumination but seeking knowledge through the ears and eyes" 弊其玄光而求知之于耳目。[120] The fifth sentence is a common term, which appears in *Zhuangzi*, in a parable satirizing the wasted efforts of Confucius's disciple Zigong.[121] The sixth sentence appears to be from a more obscure source—the second chapter "Wuzi" 吾子 in *Fa yan* 法言 by Yang Xiong 揚雄 (53 BCE–18 CE): "Studied from childhood, in old age confused" 童而習之，白紛如也.[122]

Though it is possible that some of the more obscure phrases might not be immediately recognized by the examiner in the haste of evaluating the examinations, undoubtedly most of the major sources, including the *Four Books*, the writings of Zhu Xi, *Zhuangzi*, *Huainanzi*, and the like were easily identified. Probably as important as the statements made in this essay is the complex scholastic aesthetic expressed. Again, we cannot know for certain whether this anonymous essay was written by Xu. But phrases and sentiments expressed here—the recovery of meanings, attacks on contemporary Confucians by likening them to the Names and Principle School (*mingli* 名理), claims of decline—are all sentiments similar to those found in Xu's later writings. And whether or not this is the writing of Xu, it gives us a more precise idea of what was meant by "solid learning" (*shixue* 實學)—which has too often anachronistically been interpreted as "practical studies" and likened to modern scientific attitudes. Dismissing contemporary debates over Confucian doctrines, the textualist writer proposes to seek truths by recovering the lost meanings of early antiquity, to explore nature through the *Classic of Poetry* and *Approaching Correctness*. It is this notion of "solid learning" that Xu and the collaborators use to promote and defend Western Learning; it is this project that justifies the Jesuit doctrines as corrections of the Buddhists.

---

[119] Zhuang Zhou 莊周 (fl. 320? BCE), *Zhuangzi ji shi* 莊子集釋 [*Zhuangzi*: Collected explanations], ed. Guo Qingfan 郭慶藩 and Wang Xiaoyu 王孝魚 (Beijing: Zhonghua shuju 中華書局, 1961), 552.

[120] Liu An 劉安 (179–122 BCE), *Huainan hong lie jie* 淮南鴻烈解 [Great illumination of Huainan, with explanations], in *SKQS, juan* 2, 25b. See also *The Huainanzi: A Guide to the Theory and Practice of Government in Early Han China*, trans. John S. Major et al. (New York: Columbia University Press, 2010).

[121] Zhuangzi, *Zhuangzi ji shi*, 433.

[122] The commentary explains the second phrase as meaning "in old age and confused" 皓首而亂; I have followed this reading in my translation. *Fa yan yi shu* 法言義疏 [*Model Words*, with meanings and commentary], ed. Wang Rongbao 汪榮寶 (1878–1933), 2 vols. (Beijing: Zhonghua shuju 中華書局, 1987). For a discussion of this text see David Knechtges, "Fa yen," in Loewe, *Early Chinese Texts*, 100–4.

Letters to Patrons

The next two extant writings authored by Xu are two letters that can be dated to the period just after Xu passed the 1597 examination (Liang's *Chronological Biography of Xu Guangqi* dates these at 1599). Xu wrote these letters to the two examiners he now, by convention, recognized as teachers—Jiao Hong and Zhang Wudian. Both letters are preserved in Wang's *Collected Writings of Xu Guangqi*. Xu's letter to Jiao Hong is probably the most complex piece of writing he ever produced, both technically and in terms of etiquette. For Xu wishes to recognize Jiao Hong, one of the greatest scholars of the Ming dynasty, as his teacher, presumably for awarding him first place on an examination for which Jiao was demoted after being accused of bribery and permitting reckless language. Many of the allusions are very obscure, and many are lost. The first half of this letter is in fact a series of allusions, one following the other:[123]

> Respectfully prostrate [before you],[124] spring returning to Hanque;[125]
>
> 伏以漢闕春迴,
>
> Obtaining wings to return to the Southern Sea;[126]
>
> 得附南溟之翮 ;
>
> The disciples of the Cheng brothers[127] [spread] afar,[128]
>
> 程門地迥,

---

[123] Xu Guangqi, *Yu Jiao laoshi* 與焦老師 [For teacher Jiao], in *XGQJ*. Parts of this letter are so obscure in meaning that the translation provided here is only tentative.

[124] The phrase *fuyi* 伏以, also written *fuwei* 伏惟, means literally to lie prone or prostrate on the ground in respect; it is conventionally used in addressing one's superiors, usually the emperor.

[125] The allusion here is not clear. *Hanque* 漢闕 may be a reference to a palace of the Han Dynasty. It is also possible that this may refer to a stone carved in the first year of the Eastern Han Dynasty (105 CE), preserved in Beijing. Finally, there is a poem "He Zhaoming taizi Zhong-shan jie jiang shi" 和昭明太子鍾山解講詩 by Lu Chui 陸倕, which states: "Hanque is close to Zhongnan" 終南鄰漢闕 (Zhongnan is one of the peaks of Qinling 秦嶺, south of the city of Xi'an in Shaanxi Province).

[126] This reference is to the opening passage of *Zhuangzi*: "In the Northern Sea there is a fish, its name is Kun. Kun is so enormous, it is not known how many thousand miles it measures. Transformed, it becomes a bird named Peng. It is not known how many thousand miles long the back of Peng is; when it rises up and flies off, its wings are like clouds covering the sky. When the sea begins to roil, this bird sets off for the Southern Sea; the Southern Sea is the Lake of Heaven" 北溟有魚, 其名曰鯤。鯤之大, 不知其幾千里也。化而為鳥, 其名為鵬。鵬之背不知幾千里也 ; 怒而飛, 其翼若垂天之雲。是鳥也, 海運將徙於南溟。南溟者, 天池也。 Zhuangzi, *Zhuangzi ji shi*; translation adapted from *Chuang Tzu: Basic Writings*, trans. Burton Watson (New York: Columbia University Press, 1964), 19.

[127] Cheng Hao 程顥 (1032–1085) and Cheng Yi 程頤 (1033–1107).

[128] The characters *di jiong* 地迥 appear in a preface "Tengwangge xu" 藤王閣序 by Wang Bo 王勃, which states "heaven is high and earth vast; [I] feel the universe is without limit" 天高地迥, 覺宇宙之無窮; *di jiong* also appears in a poem by Wang Wei 王維 which states, "The earth is vast and the ancient city is in disrepair" 地迥古城蕪.

Together suspended under the gaze of the Big Dipper.[129]

頻懸[130]北斗之瞻。

Bowing once[131] to pay respect,

拜一介以告虔,

Expressing [my] heart's[132] ardent love.[133]

攄寸衷之係戀,

In my humble view, great teacher and minister Jiao, venerated teacher, man of virtue, your honor:

恭惟大師相焦老夫子大人閣下：

Exalted[134] Keeper of the Temple Treasures,[135]

天府高華,

Great scholar of Confucian ritual,[136]

人文鴻鉅,

Burdening [yourself] with the Yin [Shang] and Zhou Dynasties' heavy burdens,

任伊周之重任,

Transmitting Confucius's and Mencius's true transmissions,[137]

傳孔孟之真傳,

---

[129] *Beidou* 北斗, the Big Dipper, occurs rarely in the *Thirteen Classics*. However, the *Analects*, 2.1, mentions Beichen 北辰 (Polaris, the North Star 北極星) in passage 2.1: "The Master stated: Using virtue to govern, one becomes like the North Star, residing in one's place, with the multitude of stars surrounding it" 子曰：為政以德，譬如北辰居其所而眾星共之. In light of the remainder of Xu's phrase, it seems that this might have been a reference to the North Star.

[130] The variant *xuan* 縣 is used here in the other extant edition of this letter, Xu Guangqi, *Yu Jiao laoshi* 與焦老師 [For teacher Jiao], in *Zengding Xu Wending gong ji* 增訂徐文定公集 [Collected works of Xu Guangqi, revised and enlarged] (Taibei: Taiwan Zhonghua shuju 臺灣中華書局, 1962), hereinafter *ZXWGJ*.

[131] Reading *yi jie* 一介 as *yi ge* 一個.

[132] The term *cun zhong* 寸衷 means *cun xin* 寸心, or *xin* 心.

[133] The term *xilian* 係戀 appears in the subcommentary to *Zhuangzi*, in the "Yang sheng zhu" chapter: 亦猶善養生者，隨變任化，與物俱遷，故吾新吾，曾無係戀，未始非我，故續而不絕者也. Zhuangzi, *Zhuangzi ji shi*, 130.

[134] Song dynasty thinker Su Shi 蘇軾, in "Xie Hanlin xueshi qi" 謝翰林學士啟, in *Dongpo's Collected Writings* 東坡集, states: 致茲朽鈍，亦踐高華.

[135] Tianfu is the name of an official post in the Zhou Dynasty, responsible for safeguarding court documents, population registers, treaties with other states, and other national treasures. Jiao Hong was in charge of the Ming bibliography project before the library burned down.

[136] This term alludes to a passage from the *Classic of Changes* (*Zhou Yi* 周易): "Observe the patterns of heaven, in order to investigate the changes in the seasons; observe the patterns of men, in order to transform all under heaven and earth" 觀乎天文以察時變，觀乎人文，以化成天下.

[137] In this line and the previous line (in parallel fashion), the character that begins the line repeats at the end of the line.

Heavenly man[138] of policy essays.[139]

策對天人,

First among all scholars.

詞林第一;

Following sun and moon,[140]

身依日月,

Historical writings without equal.

史筆無雙;

On the red steps of the palace composing elegant writings,[141]

丹陛擒詞,

Editing volumes of the *Classic of Poetry* and *Classic of Documents* without the slightest fault,

編詩書之冊而無遜;

Civil service examinations[142] and entering the lectures,[143]

金華入講,

What is not the Way of Yao [traditionally r. 2357–2256 BCE] and Shun [traditionally r. 2255 2205 BCE] is not stated.

非堯舜之道則不陳;

Of myriad horses in the open wild of Ji,[144]

冀野空萬馬之群,

Training to lead the country without fatigue.[145]

甄陶不倦。

[Yet] in the [ancient] capital Yan, it takes but three men to [create baseless rumors] of a tiger,[146]

燕市有三人之虎,

---

[138] Zhuangzi, *Zhuangzi ji shi*, 1066: 不離於宗, 謂之天人。

[139] *Cedui* 策對 or *zoudui* 奏對, memorials in response to a query by the emperor.

[140] The original source of this phrase is unclear; the phrase was later used to describe the founder of the Ming dynasty. Zhang et al., *Ming shi*, 3788.

[141] A phrase used to describe the historian Sima Biao 司馬彪 (d. ca. 306).

[142] Reading *jin hua* 金華 as *jin hua* 金花. This appears to be a reference to the civil service examinations. The phrase *jinhuatiezi* 金花帖子 refers to the written placard used in the Tang and Song dynasties to announce examinees who passed. It lists the name of the examiner and then the successful examinees; golden flowers were pasted to the placard.

[143] Jiao Hong served as a lecturer to the heir apparent.

[144] The horses of Ji (one of the nine ancient provinces, including parts of present-day Hebei, Henan, Shanxi, and Liaoning provinces) were reputed to be especially good.

[145] Yang Xiong 揚雄 *Fa yan* 法言 *Xian zhi* 先知: 甄陶天下者, 其在和乎。剛則甈, 柔則坯。

[146] Variant of the phrase "*san ren cheng hu*" 三人成虎 which summarizes the moral of an account from *Zhanguo ce* 戰國策, "Wei, er" 魏二: 夫市之無虎明矣, 然而三人言而成虎。

Facial expression luminescent pure white,[147]

神色皓如。

Calmly the single crane[148] returns to the mountain,

蕭然獨鶴以還山，

As if a solitary boat traversing the river.

宛爾孤舟之橫水。

Gray-green and red divine his entering and leaving.[149]

蒼赤卜其出處，

Could it be the same as wax shoes[150] on the Eastern Mountain?[151]

豈同蠟屐在東山？

[Could it be] that what the state depends on for safety is in the end the "Metal Coffer" opening at the Luoyang temple?[152]

社稷賴以安危，終是金縢開洛社。

This passage offers the best example of the complexity of Xu's writings. If we look beyond the excessive flattery that Xu Guangqi offers Jiao Hong, we can begin to see the brilliance of this essay, together with themes that will be repeated in Xu's later writings. First, Xu offers a series of allusions that gesture back to early antiquity, in order to deflect the accusations against Jiao. Xu compares Jiao with the Keeper of the Temple Treasures of the Zhou Dynasty (1045–256 BCE). His defense of Jiao is that Jiao's teachings go back to the ancients—the Way of Yao and Shun, the Zhou Dynasty, and the original transmissions of Confucius and Mencius. Xu implies that if Jiao was indicted for the doctrines he advocated, it is because of the corruption by contemporary Confucians of the doctrines of the ancients. In defending Jiao, Xu employs ingenious allusions. First, Xu describes Jiao Hong's demotion as the flight of the great bird Peng in the opening lines of *Zhuangzi*. The second and perhaps the most clever allusion that Xu presents is the "Metal Coffer" from the *Classic of Documents*.[153] Here Xu compares Jiao with the Duke of Zhou, whose loyalty was questioned but who was ultimately exonerated by the discovery of documents demonstrating the Duke's loyalty when the Metal Coffer was opened. Xu suggests that Jiao's loyalty, as true as that of the Duke of Zhou, awaits exoneration through the opening of the Metal Coffer. This letter perhaps best demonstrates Xu's considerable skills in writing propaganda.

---

[147] A phrase used to describe white virgin snow.

[148] This phrase possibly was drawn from a poem by He Xun.

[149] The term *chuchu* means to enter and exit; the reference to divination is not clear.

[150] From Yuan Zhen 元稹 (779–831): 謝公秋思眇天涯，蠟屐登高為菊花。

[151] From *Mencius*: "Mencius stated: Confucius climbed the Eastern Mountain, making [the state of] Lu seem small; [he] climbed Mount Tai, making all-under-heaven seem small."

[152] Li Bai 李白: 金縢若不啟，忠信誰明之。

[153] "Metal Coffer" (*Jin teng* 金縢), *juan* 12, *Shang shu zhu shu* 尚書注疏 [*Book of Documents*, with commentary and subcommentary], in *SKQS*.

We should not, of course, take this propaganda too literally, or assume that Xu actually believed the statements he makes here.

"Measurement of River Works"

Xu's earliest extant work related to anything "practical" is his "Measurement of River Works, Including Methods for Surveying Terrain" (*Liang suan he gong ji ceyuan dishi fa* 量算河工及測驗地勢法). This work is included in Xu's *Complete Book of Agricultural Policies* (*Nong zheng quan shu* 農政全書), chapter 14; Wang transcribes this version in his *Collected Writings of Xu Guangqi*. The preface is dated 1603 (Wanli 31 萬曆癸卯). Wang asserts that this work is evidence that before the translation of the *Elements*, "Xu Guangqi already had this deeply researched into the Chinese method of measurement" 徐光啟對我國傳統的 測量方法已有這樣深的研究.[154] To my knowledge, however, its contents have never been analyzed. This work is too lengthy to be translated in its entirety; I can present here only two short but representative excerpts from the work. The first passage begins:

> To measure, on a given canal, beginning from a given point and ending at a given point, all together, in actuality (*shi*), [one] should [dig] the canal how many *zhang* and *chi*?[155] Each pace is five *chi*;[156] at every twenty-five paces place a wooden border-marker, and write down the marker numbers. From the given beginning point, start [numbering] with the character A (*tian*) marker one, and finish 10 markers; then again start with the character B (*di*) marker one, and finish 10 markers, continuing to the given ending point. [Then one] must see how many numbers, how many *zhang* and *chi*.

> 一，量某河自某處起，至某處止，共實該應開河幾何丈尺？每步五 尺，每二十步立一木界樁，編定號數。自某處起天字一號，盡十號； 又起地字一號，盡十號，直編至某處止，要見若干號數，若干丈尺。

The rest of the work continues in a similar manner. The last passage begins:

> First, to measure from the opening of the canal, beginning from a given border to a given place, as with the previous method, [we have] already obtained how many *zhang* and *chi* the curved-segment hypotenuse is. Now [we] wish to know the straight hypotenuse is how many *zhang* and *chi*? The East-West straight long-side [of the right triangle] is how many *zhang* and *chi*? The North-South straight short-side [of the right triangle] is how many *zhang* and *chi*? Below the Eastern-border terrain to the Western-border is how many *zhang* and *chi*?

---

[154] Wang, *XGQJ*, 5.

[155] One *zhang* is equal to ten *chi*. At present one *chi* is one-third of a meter, and thus ten *chi* is three-and-one-third meters; Ming measurements differed.

[156] One pace (*bu* 步) was two steps (*kui* 跬).

一，量所開河某境起至某處，如前法，已得曲折弦若干丈尺，今欲知直弦幾何丈尺？東西直股幾何丈尺？南北直勾幾何丈尺？東邊地形下於西邊幾何丈尺？

In this text there is in fact little beyond the most elementary mathematics. Xu does show familiarity with the terminology and calculations of the sides of right triangles (what is now called the Pythagorean theorem), taken from the chapter on right triangles (*gougu* 句股) in popular mathematical manuals. There is little to indicate that Xu was ever involved in the construction of waterways, though he does discuss the use of the plumb line and level. This is not a description of techniques for digging canals but instead an elegant, idealized, learned essay that attempts to convince a patron of Xu's qualifications for leadership using the symbolic wealth of Xu's training in writing and the Confucian classics. The work was presented to the District Marquis Liu 劉邑侯: Wang Zhongmin has identified Liu as Liu Yikuang 劉一爌, who was the District Magistrate (*zhixian* 知縣) of Shanghai from 1602 to 1603 (Wanli 30 and 31).[157]

"Reply to a Fellow Townsman"

The earliest writings on Western Learning have little to do with anything mathematical, technical, or practical. The earliest extant writing on Western Learning that has conventionally been attributed to Xu is "Reply to a Fellow Townsman" (*Da xiang ren shu* 答鄉人書).[158] Although the letter is undated, it states, "[t]oday [I] have not yet obtained [Ricci's] writings in translation," so it was written in the early period, before the collaborators had produced their many translations. The letter presents arguments for the importance of belief in *Tianzhu* and claims for Western Learning that presage those Xu makes in his later writings. More important, these arguments also presage those that Xu makes in his memorial promoting Western Learning, which he presented in 1616 to the emperor:

> Buddhism has entered China for one-thousand eight-hundred years. The minds of the people and the way of the world, [with each passing] day cannot compare with antiquity. Accomplishments can be attained by how many people? If [we] reverently believe in *Tianzhu* [Lord of Heaven, also

---

[157] Wang, *XGQJ*, 62, citing the Kangxi 康熙 *Shanghai County Gazetteer* (*Shanghai xianzhi* 上海縣志), chapter 8, *Guan shi zhi* 官師志.

[158] *Da xiang ren shu* 答鄉人書 [Reply to a fellow townsman], in *ZXWGJ*, conventionally attributed to Xu Guangqi. Ad Dudink presents evidence that this work was originally attributed to Liu Yinchang 劉胤昌, who, like Xu, became a presented scholar (*jinshi*) in 1604. Liu, however, died in about 1618. Two later compilations, dated c. 1680 and c. 1714, also attribute this work to Liu. Dudink, "Image of Xu Guangqi," 102–104. This is particularly interesting because the content of this letter is quite similar to Xu's 1616 memorial to the emperor (see pages 245–253 of this book). This suggests that the claims presented in this letter might have been authored collaboratively, and credit assigned to Liu who, at that time, was equal to Xu in status. Then, at some point after Liu's death, and following Xu's success, attribution was assigned to Xu.

Devapati, or the Christian God],[159] it must be that within a period of several years, people will all become sages and noble gentlemen. The way of the world could then be compared to that during the [times of] Yao, Shun, and the Three Dynasties [traditionally, 2357–256 BCE], even greatly exceeding it, and the country would then for one-thousand and ten-thousand years be at length peaceful without crisis, ordered far into the future without chaos. [What I have stated above] can be deduced from principle, [it] can be tested in one area and one township [at a time].[160]

佛入中國，千八百年矣。人心世道，日不如古。成就得何許人？若崇信天主，必使數年之間，人盡為賢人君子，世道視唐虞三代且遠勝之，而國家更千萬年永安無危，長治無亂，可以理推，可以一鄉一邑試也。

What will you[161] have to follow? The Learning of Principle[162] discussed in [Matteo Ricci's] *True Meaning [of the Lord of Heaven]* raises only the broad outline. If [one] desires to completely understand the meanings (*yi*), [one] should use [Ricci's] classics, [which number] ten-thousand chapters. Today [I] have not yet obtained [his] writings in translation; on another day [they] must become the great dawn.[163] [I] am only afraid that you and I will not see [them]. If [we] are unable to deeply clarify the details, the main point is that [one] must believe that there is a Lord of Heaven [*Tianzhu*]. *Tianzhu* is just what the Confucian (*ru*) texts call the Emperor on High (*Shangdi*).[164] Once [one] believes in his existence, then all the established doctrines and admonishments cannot but be observed, all the righteousness and principles discussed cannot but be followed, just as the minister follows the ruler and the son follows the father. For which of the many places in China can this be said [to be true]?

執事將何從焉。《實義》中所論理學，止舉大概。若欲盡解其義，宜用經書萬卷。今未得遍譯，他日必當大明。恐我與執事不及見耳。若未能深明其詳，大端只宜信有天主。天主即儒書所稱上帝也。一信其有，即所立教誡，不得不守。所譚義理，不得不從。如臣從君，子從父，何中國殊方之可言乎？

---

[159] The term *tianzhu* 天主 is used both in classical Chinese sources and in Buddhist writings. In the *Records of the Grand Historian* (*Shi ji* 史記), *tianzhu* is the name of one of the eight spirits: "Of the Eight Spirits, one is named *tianzhu*, sacrificed to at Tianqi" 八神、一曰天主、祠天齊. In Buddhist writings, it is used as a transcription of Sákra Devânâm-indra and to mean Devapati, lord of the devas. Jesuits understood it to refer to the Christian God.

[160] In ancient times, *xiang* referred to 12,500 families, *yi* referred to the capital of a state.

[161] *Zhishi* is a respectful way to address the recipients of the letter.

[162] *Lixue* refers to the Cheng-Zhu Learning (or School) of Principle, the dominant school of Confucianism during the period. Here Xu describes Ricci's work as expounding these doctrines. However, Ricci (and in his later writings, Xu himself) attacked writers associated with the Learning of Principle for having corrupted the teachings of the ancients.

[163] Literally, *daming* means the sun or moon.

[164] The term *shangdi* 上帝 appears in many early Chinese texts, including the *Li ji* 禮記, *Shi jing* 詩經, *Shu jing* 書經, *Mozi* 墨子, and *Shi ji* 史記.

To give an example, a state has its master; at the great interior of the capital, [only] the stewards, minister, and attendants can see him. The commoners living by the sea or in the countryside cannot see [him]. Even though [one] cannot see [him], how is it possible not to believe that he [the ruler] exists? To not believe he exists necessarily leads to violations of the law and interferes with regulations in the extreme, just waiting to be sentenced to the palace dungeon. But if [only] after this, [after one sees the ruler, one] believes he exists and regrets one's crimes, it is too late, it is too late! The central doctrine of the teachings is all in regretting crimes and correcting mistakes. Even if one is only a moment from the end, one still can correct the old and seek the new, and avoid falling into eternal suffering! If one is of old age, the situation is already near—[one then] especially cannot but plan early. To not inquire into what can be seen to lie ahead—then there is nothing to be done. As you come to hold each other to account, truly [this] is difficult to attain, to explain to each other clearly. It is not only to save you—through you [we] can save countless people. Your pious deeds are hardly shallow!

譬如國有其主，如京師大內，宰臣侍從，方得見之。海濱草野之民不見也。雖則不見。豈可不信其有耶？不信其有，必至犯法干令。直待斷於闕下。然後信其有，悔其罪，晚矣！晚矣！教中大旨。全在悔罪改過。雖臨終一刻，尚可改舊圖新。免永遠沉淪之苦。若在高年，時勢已迫，尤不可不早計也。眼前悠不問。無可奈何。如執事來相詰難，正是難得者。相與一講明。非惟救得執事。從執事更可救無數人。執事功德，亦不淺也。

In this passage we can see several themes that appear throughout Xu Guangqi's writings on Western Learning, including the decline from the age of antiquity, increasing corruption, and appeals for the return to antiquity. The letter focuses on the failure of Buddhist doctrines to preserve the moral order of antiquity, and the increasing corruption of contemporary society. It advocates a return to antiquity, and represents Western Learning as exceeding even the beginnings of Chinese antiquity. A central focus is the moral coercion found in the doctrines of Western Learning: "Once [one] believes in [*Tianzhu*'s] existence, then all the established doctrines and admonishments cannot but be observed, all the righteousness and principles discussed cannot but be followed, just as the minister follows the ruler and the son follows the father." It is precisely this coercion to believe that Xu extols throughout his prefaces to Euclid's *Elements*.

In this letter we can also see how starkly the views of the Chinese literati differ from those of their Jesuit collaborators—again, their position of power over the Jesuits was such that they were under no compulsion to accept any more of the Jesuits' doctrines than pleased them. The letter reinterprets the writings brought by Jesuits as recovering the perfect moral order of ancient China. Missing here is the central tenet of Christianity, namely, salvation through Jesus Christ; in fact, Jesus is not even mentioned. Here, *Tianzhu* belongs to the lineage of the ancient Chinese sages: he "is just what the Confucian (*ru*) texts call the Emperor on High (*Shangdi*)." Throughout it uses the term *Tianzhu*—the object of their worship—instead of the Christian God, Deus, which is explicitly denoted by the transliteration *Dousi* 陡斯 in *True Meaning of the Lord of Heaven*. As I noted above, the term *Tianzhu* was also used by the Buddhists to mean Devapati, lord of the devas. This then explains the title *True Meaning of the Lord of Heaven*—the writings brought by the Jesuits recovered the "true meanings" from ancient China, while correcting the Buddhists' corruption of these ideas.

"Ode to the Portrait of Jesus"

There is little evidence that Xu accepted the central dogmas of Christianity, the most important of which must include the promise of personal salvation and eternal life through the divine grace of Jesus Christ, his crucifixion for the sins of mankind, and his resurrection from the dead. Xu rarely even mentions Jesus: in fact, mention of Jesus can be found only in a few of the earliest works conventionally attributed to Xu, and Xu's authorship of these works is questionable. The following is a translation of a passage possibly authored by Xu, "Ode to the Portrait of Jesus [Yesu]" (*Yesu xiang zan* 耶穌像讚).[165] Even though the poetic style provides considerable freedom to extol Jesus in an evocative ode, missing here is any mention of the central dogmas about Jesus:

> The master[166] who established heaven and earth.
>
> 立乾坤之主宰。
>
> The root ancestor[167] who commenced man and objects.
>
> 肇人物之根宗。
>
> Pushing it forward, without beginning,
>
> 推之於前無始。
>
> Leading it back, without end,

---

[165] In 1604, three years before the publication of the translation of Euclid's *Elements*, Xu wrote "Ode to Seven Virtues of Overcoming Sin" (*Ke zui qi de zhen zan* 克罪七德箴讚; the date 1604 is from Liang, *Xu Guangqi nianpu*). The *Collected Writings of Xu [Guangqi] Wending Gong* preserves several other odes, possibly written by Xu, and probably from the same period: "Ode to the Portrait of Jesus"; "Ode to the Sacred Mother" (*Shengmu xiang zan* 聖母像讚); "Ode to the Rules and Admonishments" (*Gui jie zhen zan* 規誡箴讚); "Ode to the Ten Admonishments [Commandments]" (*Shijie zhen zan* 十誡箴讚); "Ode to the Eight Aspects of True Fortune" (*Zhen fu ba duan zhen zan* 真福八端箴讚); and "Ode to the Fourteen Aspects of Compassion" (*Aijin shisi duan zhen zan* 哀矜十四端箴讚). Ad Dudink disputes attribution of the "Ode to the Portrait of Jesus" and others to Xu (Dudink, "Image of Xu Guangqi," 139–40). For a detailed analysis of religious writings attributed to Xu, see Wang, *Christianity and Imperial Culture*. Although Wang uncritically accepts the conventional view that Xu is a Christian, the evidence presented in his study demonstrates how starkly these writings differ from Christian doctrines, especially with regard to *Tianzhu* and Jesus.

[166] The term *zhuzai* appears in *Zhuangzi*, where it is criticized.

[167] This term does not appear in *Thirteen Classics* or Daoist texts.

引之於後無終。

Filling Heaven, Earth, and the Four Directions,[168] yet without a gap!

彌六合兮靡間。

Exceeding the numerous categories, yet not the same [as any of them]![169]

超庶類兮非同。

Beginning without a form[170] that could be pondered,

本無形之可擬。

The visage of [his] deceased [form], which had descended to the living,

迺降生之遺容。

Exhibiting the transformation of spirit[171] with all-encompassing love,[172]

顯神化以博愛。

Showing clearly to all his great fairness.[173]

昭勸徵[174]以大公。

[His] position is ultimate respect,[175] without any higher,

位至尊而無上。

The principle (*li*) is subtle and ingenious,[176] without exhaustion.

理微妙而莫窮。

Preface to Euclid's Elements

Xu wrote several introductions for the translation of the *Elements*. The earliest, longest, and most important of these, the "Preface at the Printing of *Jihe yuanben*" (*Ke* Jihe yuanben *xu* 刻幾何原本序, n.d.), will suggest answers to the question of what Xu Guangqi found significant in the translation.

---

[168] The term *liuhe* is from *Zhuangzi*: 六合之外，聖人存而不論.

[169] This may be a reference to the view that the Christian God is beyond all categories; in another work the Jesuits and their collaborators present Porphyry's Tree.

[170] Although the term *wuxing* appears twice in the *Thirteen Classics*—both instances in the *Record of Rites* (*Li ji* 禮記)—it is primarily a Daoist term, appearing frequently in the texts and commentaries on *Zhuangzi*, *Baopuzi*, and *Liezi*.

[171] The term *shenhua*, which appears only once in *Thirteen Classics*, appears scattered throughout the *Zhuangzi* commentaries, *Baopuzi*, and *Zhuzi yulei*.

[172] The term "encompassing love" (*bo'ai*) differs from the term "universal love" (*jian'ai* 兼愛) used throughout *Mozi*. The term "encompassing love" appears in the *Classic of Filiality* (*Xiao jing* 孝經): 是故先之以博愛.

[173] Liu Xiang 劉向 (79–8 BCE): 古有行大公者，帝堯是也.

[174] Substituted for a variant form of the character.

[175] This term appears repeatedly in *Decorum and Rituals* (*Yili* 儀禮): 天子至尊也.

[176] Term not in *Thirteen Classics*, but in *Laozi*: 微妙玄通，深不可識.

In the times of Yao and Shun,[177] from Xi and He bringing the calendar to order, to the Minister of Works, Minister of Agriculture, Minister of Forestry and Hunting, and Manager of Music, in these Five Offices, without measurement and numbers (*dushu*), no tasks could be accomplished.

唐虞之世，自羲和治曆，暨司空[178]、后稷、工虞、典樂五官者，非度數不為功。

Of the Six Arts in the *Offices of Zhou*, numbers is the first; if, when engaging in the [remaining] five arts, measurement and numbers are not used, then those too cannot attain accomplishments.

周官六藝，數與居[179]一焉，而五藝者不以度數從事，亦不得丁也。

As for Xiang and Kuang and music, Ban and Mo and weapons, can it be that there was some other secret art? They were precise in using [these] methods, that is all.

襄曠之於音，般墨之於械，豈有他謬巧哉？精于用法爾已。

Therefore in the aforementioned Three Dynasties and before, those who excelled in this endeavor prospered, possessing learning of fundamental foundations (*yuanyuan benben*) transmitted from teacher to disciple.

故嘗謂三代而上為此業者盛，有元元本本師傳[180]曹習之學。

But this was terminated in the flames of the Ancestral Dragon.[181] Since the Han Dynasty many have arbitrarily fumbled about, like blind men shooting for a bull's-eye, shooting in vain without result; some rely on imitating similarities of appearances, as if holding a feeble light[182] to illuminate an elephant, obtaining the head but losing the tail, until at the present time this Way (*dao*) has been obliterated. There is a reason this obliteration was unavoidable.[183]

而畢喪於祖龍之焰[184]。漢以來多任意揣摩，如盲人射的，虛發無效；或依儗形似，如持螢燭象，得首失尾，至於今而此道盡廢。有不得不廢者矣。

Jihe yaunben is the progenitor of measurement and numbers, that which exhausts the conditions of the square and circle, the horizontal and the vertical, completing the use of the compass, try square, level, and plumb.

幾何原本者度數之宗，所以窮方圓平直之情，盡規矩準繩之用也。

---

[177] That is, during the perfected age of antiquity of the semi-mythological sage rulers Yao and Shun, whose reign years were conventionally dated 2357–2256 BCE and 2255–2205 BCE, respectively.

[178] Substituted for a rare variant.

[179] Substituted for a rare variant.

[180] Emended to *chuan* 傳 following Wang, *Xu Guangqi ji*, 1:74; *SKQS* has a variant of *fu* 傅.

[181] The First Emperor of the Qin, who allegedly burned the Confucian classics.

[182] Literally, a firefly.

[183] More literally: "There is something for which [this] obliteration was unavoidable."

[184] Substituted for a variant.

Mr. Ricci, from the time he was a youth, in his spare time when he was not discussing the Way (*dao*), paid attention to the study of the arts. Furthermore, this endeavor is what is referred to there [in his country] as "transmitted from teacher to disciple"; his teacher was Mr. Ding [Clavius], the most renowned scholar of his time, and for this reason [Ricci] is extremely knowledgeable in its theories.

利先生從少年時，論道之暇，留意藝學。且此業在彼[185]中所謂師傅[186]曹習者，其師丁氏，又絕代名家也，以故極精其說。

Master Li [Ricci] has been in contact with me for a long period, and in the spare time between teaching and discussions, he has time after time brought up this book, and because of this I asked for all of his books on the numerical arts (*xiangshu*), to translate them into Chinese. He stated only that without having translated this book, none of the other books could be discussed. Following this, together we translated the essentials, the first six chapters.

而與不佞游久，講談餘晷，時時及之，因請其象數諸書，更以華文。獨謂此書未譯，則他書俱不可得論。遂共翻其要，約六卷。

After finishing, on returning to it again, [I found that] it begins with the obvious to penetrate the profound, from doubt it attains certainty, thereby the useless becomes useful. The foundation of myriad applications, it can truly be called the storehouse of forms of the myriad phenomena,[187] an ocean of learning for the Hundred Schools. Although it is not complete, if one takes it to study other books, one can obtain [understanding] and discuss them.

既卒業而復之，由顯入微，從疑得信，蓋不用為用，衆用所基，真可謂萬象之形囿，百家之學海。雖實未竟，然以當他書，既可得而論矣。

In my mind I said to myself: I did not expect—two thousand years after the learning of antiquity was annihilated—that I could suddenly obtain that which restores the corrupt classics and the lost meanings from the times of Yao, Shun, and the Three Dynasties. Its aid to our times certainly is not slight, and because of this—together with two or three who share my aims—I had this engraved to pass it on to others.

私心自謂：不意古學廢絕二千年後，頓獲補綴唐虞三代之闕典遺義，其神益當世，定復不小，因偕二三同志刻而傳之。

Master [Ricci] stated: "When this book is used by the Hundred Schools, soon there will be people like Xi, He, Ban, and Mo, but this seems to be the lesser use. There is a great application here: to train men's intelligence, and make it precise and accurate."

---

[185] Emended following Wang, *Xu Guangqi ji*, 1:75; *SKQS* has *bo* 波.

[186] Emended as in footnote 180 on the preceding page.

[187] It is not clear what Xu is referring to here. The phrase 萬象之形囿 occurs only once in *SKQS*, in this preface; 形囿 occurs several times only. This may derive from Buddhist terminology.

先生曰：「是書也，以當百家之用，庶幾有羲和般墨其人乎，猶其小者。有大用于此：將以習人之靈才令細而確也。」

I think that the lesser usage and greater usage are all actually up to the individual. It is like felling timber in Denglin: ridgepole, beam, rafters, or square rafters—this is up to one's choice. Of the learning of the Master [Ricci] there are probably three kinds: the greater is self-cultivation to serve heaven; the lesser is to investigate things (*gewu*) to fathom principle; one aspect of the principles of objects is also called the numerical arts (*xiangshu*). One by one all are precise, true, authoritative, and quintessential, thoroughly without anything that can be doubted.

余以為小用大用，實在其人。如鄧林伐材，棟梁榱桷，恣所取之耳。顧惟先生之學略有三種：大者修身事天，小者格物窮理，物理之一端別為象數。一一皆精實典要，洞無可疑。

Its explanations and analysis can also dispel men's doubts. So I urgently transmitted the lesser, desiring to quickly first put forth what is easily believed, so that people could understand [Mr. Ricci's] writings and wish to see his intent and principles, and so understand that his learning can be believed without doubt.

其分解擘析，亦能使人無疑。而余乃亟傳其小者，趨欲先其易信，使人繹其文想見其意埋，而知先生之學，可信不疑。

Probably it is for this that the use of this book is even greater. The manner in which this book is used by all the scholars of geometry is outlined in his preface, and will not be discussed here.

大槩如是，則是書之為用更大矣。他所說幾何諸家藉此為用，略具其自敘中，不備論。

Xu Guangqi of Wusong

吳淞徐光啟書

What deficiency in Chinese mathematics did Xu perceive that the *Elements* could supplement? What erroneous conclusions was its axiomatic rigor to correct? Xu's "Preface" provides no answer. It contains no criticism of specific authors, treatises, or mathematical techniques of the period; nor is there even any mention of specialized mathematical terminology. There is no evidence here that Xu had any specialized knowledge of mathematics before the translation. Instead, the central themes of his preface derive from his propaganda promoting Western Learning, such as the recovery of the lost meanings from ancient China. The other prefaces that Xu wrote for his translation with Matteo Ricci of Euclid's *Elements* similarly contain little mathematical content, and are in fact little more than exercises in propaganda, emphasizing that Western Learning (as interpreted and promoted by Xu) can be believed without doubt. In fact, Xu understood little of even the basics of Euclidean geometry, as shown in Appendix B, "Xu Guangqi's *Right Triangles, Meanings*" (pages 279–290 of this book).

Xu Guangqi's "West"

To better understand how Xu exploited "the West" (as he artfully depicted it) in order to promote Western Learning and advance his own career, probably the most important example is the memorial he submitted to the Wanli 萬曆 Emperor (r. 1572–1620) in 1616, titled "Memorial on Distinguishing Learning" (*Bian xue zhang shu* 辨學章疏).[188] Previous studies have interpreted this memorial following the Jesuits' own accounts of the episode—as a courageous defense of the Jesuits by Xu, their most powerful convert, against anti-Christian persecution;[189] what has been missing is an analysis from the "stranger's perspective,"[190] that is, the historical context of seventeenth-century China. The following is a translation of the first part of Xu's memorial:

> Left Secretariat of the Heir Apparent, Left Admonisher,[191] and Hanlin Academy Examining Editor, your humble servant, Xu Guangqi, cautiously submits this memorial: because the learning of the men from afar is most orthodox and your ignorant servant's knowledge [about it] is profoundly correct, I sincerely plead with [you,] the Sage of Clarity,[192] [that I may] solemnly submit this memorial, in order that the ten-thousand years of fortune may last forever, in order to bequeath ten-thousand generations with peaceful affairs.[193]

---

[188] Xu Guangqi, *Bian xue zhang shu* 辨學章疏 [Memorial on distinguishing learning], in Wang, *XGQJ*, 2:431–37. Wang also records the emperor's response to this memorial (p. 437). This response, along with the title, suggests that this is the version submitted to the emperor, so I have used this edition for this translation. There are, as Wang notes, numerous editions in circulation, some of which are titled "Draft Memorial on Distinguishing Learning" *Bian xue shu gao* 辨學疏稿, suggesting that they are drafts and not the memorial that was submitted. *Tianzhujiao dong chuan wenxian xu bian* 天主教東傳文獻續編 [Documents on the Eastern transmission of Catholicism, second collection] (Taibei: Taiwan xuesheng shuju 臺灣學生書局, 1965), hereinafter *TDCWX*, 25–26, reproduces a version printed from a Ming edition preserved at the Vatican Library, which differs from the edition in *XGQJ*. The first part of this memorial has been translated by Gail King and David Mungello in Wm. Theodore de Bary et al., eds., *Sources of Chinese Tradition*, 2nd ed., 2 vols. (New York: Columbia University Press, 1999–2000), 2:147–49; a passage has also been translated in Gernet, *China and the Christian Impact*, 110. I have consulted these works for my translation.

[189] The two most important studies of this episode are Edward Thomas Kelly, "The Anti-Christian Persecution of 1616–1617 in Nanking" (PhD diss., Columbia University, 1971) and Ad Dudink, "Opposition to Western Science and the Nanjing Persecution," in Jami et al., *Statecraft and Intellectual Renewal*, 191–224. Both studies proceed from the perspective of the Jesuits.

[190] The phrase "stranger's perspective" is from Steven Shapin and Simon Schaffer, *Leviathan and the Air-Pump: Hobbes, Boyle, and the Experimental Life* (Princeton, NJ: Princeton University Press, 1985), 13.

[191] "Responsible for giving moral and social guidance to the Heir Apparent" (Charles O. Hucker, *A Dictionary of Official Titles in Imperial China* (Stanford, CA: Stanford University Press, 1985), 516.

[192] "Sage of Clarity" is a term conventionally used to refer respectfully to the emperor.

[193] The last two phrases are written in parallel form, in a manner similar to that analyzed in Xu's 1597 examination.

左春坊左贊善兼翰林院檢討臣徐光啟謹奏：為遠人學術最正，愚臣知見甚真，懇乞聖明，表章隆重，以永萬年福祉，以貽萬世父安事。

Your servant has seen a Liaison Hostel Report [stating] the Nanjing Ministry of Rites has brought charges against Western tributary state officials Pang Diwo [Diego de Pantoja, 1571–1618, a Spanish Jesuit missionary] and others, stating: "Their teachings are flourishing; among the literati-officials there are also those who believe [in their teachings]"; it is stated, "absurdly making statements [properly reserved for] astronomical officials, scholars have also fallen into their fog." Stating "literati-officials," stating "scholars," the Ministry [of Rites] officials express concerns about those implicated by the spreading roots; among the officials of the court, left out without being explicitly named, your servant has carefully studied the way and principles of the [Western] tributary state officials, published many treatises, and thus your humble servant is among "those who believe [in their teachings]." Also, together with them I have investigated calendrics—the earlier and later memorials are all before your Majesty—thus those making statements [reserved for] "astronomical officials" also includes your servant. [If] all the [Western] tributary state officials are found to be guilty, is it possible that your servant dares to take advantage of the silence of the Ministry [of Rites] officials to unscrupulously escape [punishment]?

臣見邸報：南京禮部參西洋陪臣龐迪我等，內言：「其說浸淫，即士大夫亦有信向之者」；一云：「妄為星官之言，士人亦墮其雲霧。」曰士君子，曰士人，部臣恐根株連及，略不指名，然廷臣之中，臣嘗與諸陪臣講究道理，書多刊刻，則「信向之者」，臣也。又嘗與之考求曆法，前後疏章具在御前，則與言「星官」者亦臣也。諸陪臣果應得罪，臣豈敢幸部臣之不言以苟免乎？

In fact, your humble servant, over many years, through careful study and research together [with them], knows that all these [Western tributary state] officials are most true and most certain, not stopping to trace their thoughts until each is without doubt, truly [they] all are disciples of the worthies and sages. Furthermore, their Way is most orthodox, their discipline most rigorous, their learning most broad, their knowledge most essential, their hearts most true, their views most certain. In their country they are all the outstanding among one-thousand, the eminent among ten-thousand.

然臣累年以來，因與講究考求，知此諸臣最真最確，不止蹤[194]跡心事一無可疑，實皆聖賢之徒也。且其道甚正，其守甚嚴，其學甚博，其識甚精，其心甚真，其見甚定。在彼國中亦皆千人之英，萬人之傑。

The reason for their coming from tens of thousands of *li* is probably that their country teaches all people to take as their task self cultivation (*xiushen*) in order to serve the Sovereign on High (*Shangzhu*).[195] [When

---

[194] Substituting a variant form of the character.

[195] This term, which literally means wise and able sovereign, appears in Ban Gu 班固 (32–92 CE), *Qian Han shu* 前漢書 [History of the former Han], in *SKQS*, and is used instead of *Tianzhu* throughout this memorial. Other editions of this memorial substitute *Tianzhu*.

they] heard that the teaching of the sages and worthies of China is also self cultivation to serve Heaven (*xiu shen shi Tian*), the principles together were compatible, and because of this, despite hardships and difficulties, they crossed over peril and danger, arriving for mutual corroboration, intending to make every man do good, in order to praise the compassion of Heaven on High (*Shangtian*) toward men.

所以數萬里東來者，蓋彼國教人，皆務修身以事上主，聞中國聖賢之教，亦皆修身事天，理相符合，是以辛苦艱難，履危蹈險，來相印證，欲使人人為善，以稱上天愛人之意。

As for their teaching, its foundation is assiduously serving the Emperor on High (*Shangdi*); its essence is saving the common people; its efforts are toward loyalty [to the sovereign], filiality [to parents], compassion and kindness [to the people]; it begins with promoting good and correcting error; it furthers repentance and absolution; it offers ascending to heaven and true happiness as the high rewards of doing good; it metes out the eternal misfortune of hell as the bitter recompense for committing evil; its sermons and rules all are the ultimate of the principles of heaven and human feelings. Its methods can truly make men do good and completely eradicate evil. It is likely that this is what is referred to as the Sovereign on High's kindness to preserve and save, and the principle of rewarding good and punishing evil, clearly and distinctly, sufficient to move men's hearts, to inspire men's love and dread, by way of setting forth inner feelings.

其說以昭事上帝為宗本，以保救身靈為切要，以忠孝慈愛為工夫，以遷善改過為入門，以懺悔滌除為進修，以升天真福為作善之榮賞，以地獄永殃為作惡之苦報，一切戒訓規條，悉皆天理人情之至。其法能令人為善必真，去惡必盡，蓋所言上主生育拯救之恩，賞善罰惡之理，明白真切，足以聳動人心，使其愛信畏懼，發於繇衷故也。

Your servant has discussed the rewards and punishments of the emperors of the ancient past, the rights and wrongs of the [ancient] sages and worthies: all teach men to be good and prohibit men from evil, and are supremely detailed and utterly complete. Rewards and punishments, right and wrong, these can affect men's outward behavior, but cannot reach men's [inner] feelings. Also, as Sima Qian stated, the premature death of Yan Hui and the long life of Robber Shi cause men to doubt whether good and evil have any recompense; because of this, the more strict the standards are, the deeper cheating and lying become. When one law is established, one hundred crimes are born. Empty, with a desire to govern, despondent, without a method to govern, [rulers] then borrow the doctrines of the Buddhists for assistance. [The Buddhists] assert that the recompense for good and evil is in the afterlife, thus outward behavior converges on [inner] feelings, and Yan Hui and Robber Shi all would seem to obtain their recompense. It is stated that this is suitable for making men good and eradicating evil, but just not instantaneously. Then how is it that Buddhism has come to the East for eighteen hundred years, but the ways of the world and the minds of men still are not transformed? Thus Buddhist statements seem true but are false. Chan Buddhism develops the intent of Laozi and Zhuangzi, subtle but without use; putting in place "all the objects react to one another" and methods from miscellaneous secret Daoist texts, these are perverse and absurd without principle. They even wish to take the Buddha and put him on top of the Sovereign on High! This is counter to the intent of the ancient emperors, sages, and worthies! As a result, what do men have to follow and to depend upon?

臣嘗論古來帝王之賞罰，聖賢之是非，皆範人於善，禁人於惡，至詳極備。然賞罰是非，能及人之外行，不能及人之人情。又如司馬遷所云：顏回之夭，盜跖之壽，使人疑於善惡之無報，是以防範愈嚴，欺詐愈甚。一法立，百弊生，空有願治之心，恨無必治之術，於是假釋氏之說以輔之。其言善惡之報在於身後，則外行中情，顏回盜跖，似乎皆得其報。謂宜使人為善去惡，不旋踵。奈何佛教東來千八百年，而世道人心未能改易，則其言似是而非也。說禪宗者衍老莊之旨，幽邈而無當；行瑜迦者雜符籙之法，乖謬而無理，且欲抗佛而加於上主之上。則既與古帝王聖賢之旨悖矣，使人何所適從、何所依據乎？

Necessarily desiring to make men be as good as possible, the learning of serving Heaven transmitted by all the [Western] tributary officials can truly be used to supplement the moral influence of our sovereign, aid the arts of Confucianism, and rectify Buddhist doctrines. Thus in the West there are more than thirty neighboring kingdoms that implement this teaching. For over a thousand years, up to the present, the large and the small have compassion for one another; the superior and the inferior live together in peace; borders are without defenses; sovereigns of vassal states are from the same family; the whole country is without cheats or liars; from antiquity there has been no lasciviousness or thieving; people do not pick up objects lost on the roads; and doors are not locked at night.[196]

必欲使人盡為善，則諸陪臣所傳事天之學。真可以補益王化，左右儒術，救正佛法者也。蓋彼西洋鄰近三十餘國，奉行此教，千數百年以至于今。大小相卹，上下相安，封疆無守，邦君無姓，通國無欺謊之人。終古無淫盜之俗，路不拾遺，夜不閉關。

It is thus that the entire nation is cautious and conscientious, afraid only of loss and committing offense against the Sovereign on High. Thus his laws can in actuality cause men to be good, and publicly display [his] prominence. These kinds of civilizing customs, although these statements are made by all the tributary officials themselves, your servant has [in addition] carefully investigated their discussions, scrutinized their books, mutually

---

[196] The edition translated in Gernet, *China and the Christian Impact*, p. 110, continues as follows: "And as for disturbances and rebellions—not only are they without such affairs and without such persons—there are not even words or written characters to denote such things" 至于悖逆叛亂，非獨無其事，無其人，亦并其語言文字而無之。 This phrase is not included in the edition in Wang.

inspecting and verifying [their claims], and there is nothing preposterous about it!

然猶舉國之人，兢兢業業，惟恐失墜，獲罪於上主。則其法實能使人為善，亦旣彰明較著矣。此等教化風俗，雖諸陪臣自言；然臣審其議論，察其圖書，參互考稽，悉皆不妄。

All the schools of Buddhism and Daoism, their arts are impure, and their teachings are incomplete; for two hundred fifty years, memorials [advocating these doctrines] have not received the approval of our dynasty. If we were to believe in the Sovereign on High in the manner in which Buddha and Laozi are believed in, if we accommodated the tributary officials the way we accommodate monks and Daoists, then we would revitalize civilization and deliver governance, and in doing so must exceed the times of Yao, Shun, and the Three Dynasties. . . .

而釋道諸家，道術未純，教法未備，二百五十年來猶未能仰稱皇朝表章之盛心。若以崇奉佛老者崇奉上主，以容納僧道者容納諸陪臣，則興化致理，必出唐虞三代上矣。…

The remainder of Xu's lengthy memorial presents further arguments for the implementation of Western Learning, including specific proposals for debates at court against the Daoists and the Buddhists, plans for the translation of Western texts, and a request for the experimental implementation of his plan in one region. The memorial ends with a conventional phrase used to express modesty, in which Xu respectfully addresses the emperor as "Heaven" (*Tian*):

Your servant dares brave awesome Heaven, unable to subdue [my] terror as [I] await the arrival of your order.

臣十冒天威，不勝惶恐待命之至。

As noted above, previous studies have interpreted this memorial (together with Xu's other writings) from a perspective following that found in the Jesuits' own accounts: the Jesuits are taken to be the central historical protagonists; the contents are understood and interpreted within a European context; and statements at variance with European interpretations are explained as incomplete understandings on the part of the Chinese. More concretely, if we take as exemplary of this perspective the introduction to and translation of this memorial in Wm. Theodore de Bary's *Sources of Chinese Tradition*,[197] Xu's memorial is a selfless defense of the Jesuits against an "anti-Christian movement": Xu is

---

[197] David E. Mungello, "Chinese Responses to Early Christian Contacts," in de Bary et al., *Sources of Chinese Tradition*, 2:142–54. I have chosen this example because it is the most recent and authoritative translation of this memorial. Mungello has written extensively on the "first encounter" of China and the West: see for example David E. Mungello, *The Great Encounter of China and the West, 1500–1800*, 4th ed. (Lanham, MD: Rowman & Littlefield, 2013); idem, *Curious Land: Jesuit Accommodation and the Origins of Sinology* (Stuttgart: Steiner, 1985); and idem, *Leibniz and Confucianism: The Search for Accord* (Honolulu: University Press of Hawaii, 1977). Again, for more specialized studies of this episode, see Kelly, "Anti-Christian Persecution," and Dudink, "Opposition to Western Science."

asserted to be a convert who "was deeply committed to the new teaching and willing to risk political recrimination in defending it"; from this perspective, it then seems remarkable that "Xu's defense of this foreign teaching did not prevent him from eventually occupying one of the highest offices of the land—that of Grand Secretary." In this translation, the term "Lord of Heaven" (*Tianzhu*) is glossed in brackets unequivocally as "God," following the preferred interpretation of the Jesuits (or at least some of them); little mention is made of the fact that in the Chinese context this and other terms refer to Chinese sages, kings, and emperors. From this perspective, Xu's description of the Jesuits' doctrines appears adequate if perhaps incomplete: little mention is made of the fact that in the Chinese context, the key terms used to describe the Jesuits represent them as Confucians. In this approach, Xu's claims are uncritically presented simply as beliefs—"Xu believed Christianity to be in basic harmony with the teaching of the ancient sages of China (i.e., Confucianism)"—thus obviating any need for further inquiry into his motives in historical context. And in this approach, alibis are proffered for Xu's most outlandish claims, rationalized as a "lack of knowledge" and "naïveté": "the excerpt reveals China's lack of knowledge of European history. There is a naïveté in Xu's portrayal of European history and in his exaggeration of the positive effects of Christianity in European history." Perhaps nowhere is the presentation of the Jesuits' propaganda as historical fact more evident than in the hagiographic portrayal of Ricci himself: "Ricci was one of the most remarkable men in history—impressive in physical appearance with his blue eyes and voice like a bell, charming in manner with his facility in foreign languages and photographic memory, incisive in thinking with his ability to grasp the essentials of Chinese culture and to discern a means of entry into a sophisticated culture like that of China. . . . [F]or these and other extraordinary accomplishments, a leading European sinologue of the twentieth century has called Ricci 'the most outstanding cultural mediator between China and the West of all times.'"[198]

Interpreted in the context that this memorial was written—seventeenth-century China—in which Xu was a powerful official and the Jesuits were vulnerable supplicants for patronage and protection, the perspective is very different. Xu's memorial is presented in response to a series of memorials submitted by Shen Que 沈淮 (*jinshi* 1592), Vice Minister of the Ministry of Rites (*Libu shilang* 禮部侍郎).[199] Shen's central concern was that the Jesuits were violating Chinese laws, decrees, and rituals. Shen's first memorial begins as follows:

> Memorial presenting charges against the foreigners from afar.
>
> [This] memorial alleges that the foreigners from afar have entered the gates of the capital [Nanjing] without permission, covertly harming the civilizing influence of the sovereign. [I] sincerely plead with the brilliant understanding [of the emperor] to strictly enforce the laws and decrees, in order to make men's minds upright, and in order to safeguard customs and affairs.

---

[198] Mungello, "Chinese Responses to Early Christian Contacts," 2:142–54.

[199] For information on Shen Que, see Zhang et al., *Ming shi, juan* 280.

疏遠夷疏。奏為遠夷闌入都門，暗傷王化。懇乞聖明申嚴律令，以正
人心，以維風俗事。[200]

Shen details their violations of tributary regulations and rituals involving astronomy; many of his charges are technicalities. His memorials show how little he
understood about the Jesuits' teachings. He asserts (incorrectly) that they worship the Lord of Heaven (*Tianzhu*),[201] and counters that the Emperor is the only
Lord (*Zhu*) of Heaven (*Tian*):

> The name they have given their teaching is "Teaching of the Lord of
> Heaven." But for all under Heaven (*Tian*), inside and outside this vast
> expanse, only the Emperor on High (*Huangshang*) is the Lord (*Zhu*) whose
> [grace and virtue] covers [Heaven] and uplifts [earth], illuminating from on
> high.
>
> 自名其教曰天主教。夫普天之下，薄海內外，惟皇上為覆載炤臨之
> 主。[202]

Shen further charges that the "foreigners from afar" have appropriated the term
"Lord of Heaven" in order to place him above Heaven (that is, the Emperor):

> During the majestic Three Dynasties, [the one who] presided over the
> monarchs of the vassal states was called the Heavenly King, and [the one
> who] ruled all under Heaven was called the Son of Heaven. The present
> dynasty has investigated antiquity in order to establish its laws. At the
> closing of every edict it is stated "obey Heaven." But these foreigners craftily
> use the term "Lord of Heaven," as if to place [the Lord of Heaven] above
> [Heaven, i.e., the emperor], causing confusion among the ignorant people.
>
> 三代之隆也，臨諸侯曰天王，君天下曰天子。本朝稽古定制。每詔
> 誥之下皆曰奉天。而彼夷詭稱天主，若將駕軼其上者，然使愚民眩
> 惑。[203]

Despite his investigations, Shen apparently knows nothing about Deus (*Dousi*),
Jesus, or the Society of Jesus, and therefore he refers to them simply as "foreigners
from afar" (a term that dates back to at least the Han dynasty [206 BCE–220 CE]).
It is apparently only in the third in this series of memorials that Shen becomes
aware that the Lord of Heaven of the Jesuits is not the same as the one attested to
in Chinese texts:

> Reportedly their so-called "Lord of Heaven" was a criminal in their country.
>
> 據其所稱天主，乃是彼國罪人。[204]

[200] Shen Que, *Can yuan yi shu* 疏遠夷疏 [Memorial presenting charges against the foreigners
from afar], in *Shengchao poxie ji* 聖朝破邪集 [Collected essays on eliminating evil from this
sagely dynasty] (Beijing: Beijing chubanshe 北京出版社, 1997), hereinafter *SPJ*, 5a.

[201] It was the Chinese patrons and collaborators who worshipped *Tianzhu*, not the Jesuits; some
Jesuit missionaries considered the worship of *Tianzhu* to be heretical.

[202] Shen, *Can yuan yi shu*, 6b.

[203] Ibid., 6b–7a.

[204] Ibid., 15a.

Yet Shen still makes no mention of Jesus Christ, nor of his crucifixion, resurrection, or ascension to heaven. In sum, Shen's knowledge about the teachings of these "foreigners from afar" was limited to his misunderstanding that they worship the Lord of Heaven, who Shen alleges was not just a foreigner but also a criminal.

We are now in a better position to analyze "the West" as depicted by Xu in his memorial. It was in response to Shen's charges that Xu crafted his memorial, but Xu never even mentions the Lord of Heaven. Instead, Xu insistently misrepresents the Jesuits as Western tributary officials who have travelled from afar to serve the emperor, and as Confucians, whose teachings restore to China lost doctrines from antiquity preserved in the West, which can help the Ming dynasty return to the perfect moral order of the ancients.

More specifically, Xu exploits his prestige, power, and position to promulgate the following misrepresentations:

1. These "foreigners from afar" (*yuanyi* 遠夷), as Shen termed them, are in fact "Western tributary state officials" (*Xiyang peichen* 西洋陪臣), as Xu calls them in his opening sentences. Xu refers to them as "tributary officials" (*pei chen* 陪臣) over a dozen times in his memorial. He also calls them "men from afar" (*yuan ren* 遠人), namely, emissaries who seek the civilizing influence of the Confucian emperor and wish to serve him. Indeed, throughout the Ming dynasty, the Jesuits' Chinese patrons consistently represent them "tributary officials" (as do the Jesuits themselves), and as servants of the emperor, yearning to strengthen the empire and bearing tributary gifts. Xu does not refer to them as Jesuits, Catholics, Christians, missionaries, or monks (nor does Xu refer to himself as a Catholic or Christian).

2. These "Western state tributary officials" are Confucians. Xu appropriates the central slogan of the orthodox Cheng-Zhu Neo-Confucians—"self-cultivation [of Confucian moral character in order to] govern men" (*xiu ji zhi ren* 修己治人)—to craft his fundamental characterization of these "tributary officials": they too cultivate their moral character to serve the sovereign (*xiu shen yi shi shangzhu* 修身以事上主). Elsewhere, as we have seen, Xu characterizes them as cultivating moral character to serve Heaven (*xiu shen shi tian* 修身事天), which is how he describes the teachings of the ancient Chinese sages and worthies (*Zhongguo sheng xian zhi jiao* 中國聖賢之教). Xu also asserts these "tributary officials" from the West are followers of the ancient sages and worthies (*sheng xian zhi tu* 聖賢之徒), that is, followers of Yao, Shun, the Duke of Zhou, and Confucius. Their teachings are, Xu emphasizes, most orthodox (*zui zheng* 最正). To this description, Xu adds more Confucian keywords, including "broad learning" (*bo xue* 博学) and "orthodox way" (*zheng dao* 正道). Xu makes no mention of any specifically Christian beliefs or terms, including Deus, Jesus, or even the Lord of Heaven.

3. These "tributary officials" are loyal to the Ming emperor. The terms Xu employs in the memorial all refer to the Ming emperor, past emperors, or semi-divine figures from the Confucian classics: "Sovereign on High" (*Shangzhu* 上主, an enlightened ruler who follows the Way); "Heaven on

High" (*Shangtian* 上天, master of the fate of the myriad creatures); "Emperor on High" (*Shangdi* 上帝, lord of all under heaven); and "Heaven" (*Tian* 天, the emperor). As I noted above, Xu never mentions the Lord of Heaven, who, Shen has alleged, the Jesuits worship.

4. Western Learning can restore the Ming empire to the perfect moral order of ancient China, just as it exists in "the West," where there are no wars, rebellions, or changes in dynasty. This is in fact Xu's central point, to which he devotes most of his memorial.

The emperor's response to Xu's memorial was brief: "Understood" 御批「知道了」.[205] In sum, Xu's purpose was not to defend the Jesuits, but to refashion them for his own purposes as tributary officials charged with returning to China the ancient Confucian doctrines that had governed the countries of "the West" for two millennia. Indeed, many of Xu's claims in his memorial—and especially his most outlandish ones about Europe—would hardly have been necessary were his purpose merely to defend the Jesuits. It was for the promotion of Western Learning, and with it, his career, that Xu risked his life by presenting these false claims to the emperor, the punishment for which could have included not just his own execution but that of his extended family.

Xu's misrepresentation of "the West" in his memorial represents the culmination of a long-standing strategy of the Jesuits' Chinese patrons. Perhaps the most famous example is the "Complete Map of the Myriad Countries of the World" (*Kunyu wanguo quantu* 坤輿萬國全圖), conventionally attributed to Matteo Ricci, produced with the collaboration of Li Zhizao, and printed in 1602 at the behest of the Wanli Emperor.[206] The inscription on this map above Spain (see Fig. 6.1 on 254) depicts Europe in a manner strikingly similar to Xu's memorial:

> The continent of Europe has over thirty countries. All implement the laws of the past sovereigns. No heterodox doctrines are followed. [They] only believe in the Lord of Heaven, the Emperor on High, and the doctrines of the ancient sage kings.
>
> 此歐邏巴州。有三十餘國。皆用前王政法。一切異端不從，而獨崇奉天主、上帝、聖教。

For readers in seventeenth-century China unacquainted with the strategic appropriation of Confucian terms by the Jesuits' Chinese patrons, Europe would have appeared to be Confucian: "past sovereigns" (*qian wang* 前王) referred to former Chinese emperors and kings, especially Yao, Shun, and Yu; "heterodox doctrines" referred to teachings outside the Cheng-Zhu Neo-Confucian orthodoxy, sometimes including Wang Yangming's Learning of the Mind, Buddhism, Daoism, and various religions; "Lord of Heaven" and "Emperor on High" referred to deified monarchs attested to in the Confucian classics; and "doctrines of the ancient

---

[205] Xu, *Bian xue zhang shu*, in Wang, *XGQJ*, 2:437.

[206] For a detailed study of this map, see Qiong Zhang, "Matteo Ricci's World Maps in Late Ming Discourse of Exotica," *Horizons: Seoul Journal of Humanities* 1 (December 2010): 215–50.

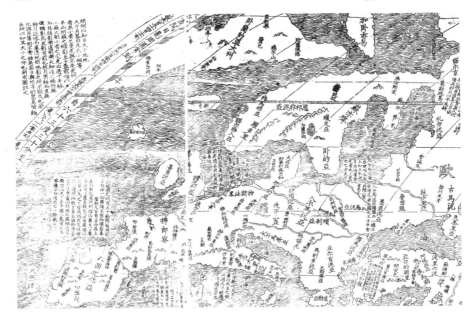

Fig. 6.1: "Complete Map of the Myriad Countries of the World" (*Kunyu wanguo quantu* 坤輿 萬國全圖), attributed to Matteo Ricci, printed in 1602. The image reproduced here is from the copy held by the James Ford Bell Library at the University of Minnesota. The four characters *Tianzhu Shangdi* 天主上帝 (Lord of Heaven, Emperor on High) have been excised from the description of Europe (see the inscription just above Spain, third column from the right, where the removal of the first four characters from the top has left a visible blank space).

sages kings" referred to the teachings of Yao, Shun, Yu, the Duke of Zhou, and Confucius.

Readers today might well question whether it was really possible for readers in seventeenth-century China to be oblivious to what would seem to us to be the obvious Christian interpretation of these documents. One might as easily ask, however, how modern scholars—Chinese and Western alike—could fail to take note of the original meanings that these terms had in China, before they were appropriated by Jesuits' Chinese patrons for their purposes. To take but one example, the sinologist Lionel Giles (1875–1958) translates this passage as follows:

> The continent of Europe contains some twenty-four states, all of which enjoy a monarchical system of government. They follow no heterodox doctrines, but are reverent adherents of the holy Christian religion, which recognizes one Supreme Deity.[207]

---

[207] Lionel Giles, "Translations from the Chinese World Map of Father Ricci," *The Geographical Journal* 52 (1918): 377. The reason for his erroneous "twenty-four" (instead of "thirty") is not clear. Lionel Giles was the son of the eminent sinologist Herbert Giles and translated several early Chinese philosophical texts, including works conventionally attributed to Sunzi, Confucius, Laozi, and Mencius.

Insistently reading Christian doctrines into his translation, Giles fails to notice the Confucian meanings these terms had in seventeenth-century China. Even the most sophisticated recent study of this map fails to mention the original meanings of these terms, translating this passage as follows:

> The continent of Europe has over thirty countries. All adopt the monarchical system of the ancient kings. They adhere to no heterodox doctrines, and all follow the holy faith of the Lord of Heaven.[208]

Toward the end of the Ming dynasty, the Jesuits and their collaborators increasingly wrote openly about Jesus,[209] and anti-Christian tracts indicate an increasing understanding of their doctrines;[210] by the early years of the Qing dynasty (1644–1911), Confucian detractors understood enough of the Jesuits' doctrines to author calumnious attacks against Jesus.[211] And at some point, these terms came to be unambiguously associated with Christianity: *Shangdi* came to be understood as referring to the Christian God; *Tianzhu* came to be understood as referring to the Catholic God; and Catholicism came to be known in China as *Tianzhujiao* 天主教. But in the early seventeenth century, when the Jesuits' teachings and the history of Europe were not yet well understood in China, Xu strategically manipulated these terms to depict "the West" as united under the perfected ancient Confucian doctrines of Yao, Shun, and the Three Dynasties.[212]

Conclusions

When Xu Guangqi's writings are understood in their historical context, the picture of him that emerges differs starkly from those presented in previous accounts. The historical context of seventeenth-century China differs considerably from that of the nineteenth and twentieth centuries during which many of these previous accounts were written: in mathematics, science, and technology, differences in development between China and the West were significantly less; externally, Western nations posed little military threat to China; and internally,

---

[208] Zhang, "Matteo Ricci's World Maps," 236.

[209] See for example Ai Rulüe 艾儒略 (Giulio Aleni, SJ, 1582–1649), *Zhifang wai ji* 職方外紀 [Records of regions outside the purview of the Bureau of Operations], in *SKQS*. For a study of the debates about terms for God, see Sangkeun Kim, *Strange Names of God: The Missionary Translation of the Divine Name and the Chinese Responses to Matteo Ricci's "Shangti" in Late Ming China, 1583–1644* (New York: P. Lang, 2004).

[210] See for example the texts collected in *SPJ* (*Shengchao poxie ji*), published in 1639.

[211] For example, see Yang Guangxian 楊光先 (1597–1669), "I Cannot Do Otherwise" (*Bu de yi* 不得已, 1665), translated in de Bary et al., *Sources of Chinese Tradition* 2:145.

[212] Is it possible that the disputes over these terms, and the historical transformation of their meaning, might have been a factor in the erasure of the four characters *Tianzhu Shangdi* 天主上帝 from the James Ford Bell Library copy of the "Complete Map of the Myriad Countries of the World"? It is not known who removed these characters or why. Again, see Fig. 6.1 on the facing page.

the Jesuits' Chinese patrons were in a position of considerable power over their Jesuit collaborators.

In historical context, Xu was not, as he has anachronistically been portrayed, a "convert," "Christian," or "Catholic," if we take as the central tenet of Christianity belief in Jesus Christ: Xu venerated *Tianzhu* [Lord of Heaven] of ancient China. Similarly, Xu was not a "mathematician," "scientist," "astronomer," or "polymath": his understanding of the sciences, like his understanding of Chinese and Western mathematics, was at best rudimentary. Xu did not seek to provide for Chinese mathematics axiomatic or logical foundations, as Frege, Hilbert, and others sought to provide for mathematics during the late nineteenth and early twentieth centuries. Xu did not strive for the synthesis of Western and Chinese mathematics sought by literati of the Qing dynasty (1644–1911), as can be seen from Xu's denigration of Chinese mathematics in toto as "tattered sandals" to be "discarded." Xu's promotion of Western Learning was not a patriotic effort to strengthen China prefiguring the "Self-Strengthening Movement" (*Ziqiang yundong* 自強運動) of the late nineteenth century: Xu promoted Western Learning indiscriminately regardless of its merits; many of Xu's claims were outlandishly false; and the Jesuits themselves were not particularly expert in mathematics, science, technology, or cannonry. Xu was not an early proponent of modernization: in fact, he sought to return to the perfection of antiquity. And Xu's mission was hardly "practical": his central aim was to promote belief in *Tianzhu* as a means to return to the perfect moral order of the legendary Chinese sage kings Yao and Shun. In sum, these anachronistic portrayals of Xu presented in previous accounts find little support beyond wishful hagiographic projections of later historical trends back onto Xu's writings wrenched from their historical context.

In historical context, Xu was a creative manipulator of Western Learning, resulting in his promotion to some of the higher official posts in the late Ming dynasty. The selections translated in this chapter, which are representative of Xu's writings, provide an outline of the central claims of his propaganda: the writings of the Jesuits returned to China ancient Confucian doctrines from the Three Dynasties; belief in *Tianzhu* [Lord of Heaven] was the key to returning to the allegedly perfected moral order of ancient China; the Jesuits were tributary officials from Western nations, "men from afar," who came to China to pay tribute the Ming emperor. That is, the Jesuits represented Xu as a convert, forging alibis and rationalizations for his failure to accept their central dogmas; and Xu, who was in a position of absolute power over the Jesuits, represented them as Confucian tributary officials who came from afar to serve the heavenly emperor of the Central Kingdom. Implementing the ancient Confucian doctrines the Jesuits had returned to China would produce results that would exceed the perfection of antiquity in the times of Yao, Shun, and the Three Dynasties, as proved by his imagined "West."

Chapter 7
Conclusions

Previous accounts of the "first encounter" of "China" and "the West" concur for the most part in their retelling of the historical events. In these accounts, China during the Ming dynasty (1368–1644), and especially during the later years of the Ming, is often held to have been in a period of profound decline—political, economic, social, military, intellectual, and scientific. Practical affairs, and in particular statecraft and science, are held to have fallen into neglect during the Ming: Ming intellectuals, it is asserted, were either consumed with the desire to gain official posts through the rote memorization of neo-Confucian doctrines so as to pass the civil service examinations, or confounded by the introspective speculation of Wang Yangming's 王陽明 (1472–1529) Learning of the Mind (*xin xue* 心學). Or they were simply deluded by what has been asserted to be vulgar Buddhist superstition. In these accounts, the Western Learning introduced by the Jesuits is held to have represented an alternative to traditional Chinese thought, one that was at the same time both new and radically different. Western Learning, these accounts assert, was recognized by a small group of concerned Chinese literati as offering solutions to the crises faced by the Ming dynasty. Working with the Jesuits, these literati translated European treatises on a broad range of subjects, including mathematics, astronomy, logic, agriculture, and military technologies, as well as Christian philosophy, morality, and the etiquette of friendship.

Xu Guangqi, the most famous among the Chinese who collaborated with the Jesuits, is often asserted to have been the greatest scientist of the period. But more than that too. These accounts assert that he dared speak openly of Chinese decline. He recognized the superiority of Western mathematics, science, technology, and agricultural techniques. Remaining above factional political fights, and at great political risk, he challenged the entrenched Ming bureaucracy, which was unconcerned with practical measures and motivated only by self-interest, to the detriment of the dynasty. Previous accounts describe how Xu strove to reform the erroneous Chinese calendar by promoting Western astronomy, how he defended the empire against the northern invaders and Japanese pirates by employing Western cannons and military techniques, and how he combatted famine by introducing Western agricultural technologies. Xu, it is claimed, dared question the orthodox clichés cherished by careerist neo-Confucians, exposed Buddhist superstitions through logical critiques, and advocated practical statecraft against fruitless metaphysical speculation.

These previous accounts, then, disagree less on the retelling of the historical events than on what conclusions are to be drawn from this alleged first encounter of China and the West. In broad terms, there have been three general approaches to the issue. Earlier works had argued that the Chinese rejected Western science and culture because of their own xenophobia, conservatism, and pride. Subsequent scholarship sought to revise this view, concluding that this first encounter was a dialogue between China and the West. Still more recent research work has countered that this was not a dialogue at all, but instead a series of unfortunate and profound misunderstandings: in one view, these misunderstandings resulted from the Jesuits' deliberate distortions of emerging modern Western science; in another view, the misunderstandings were the inevitable consequence of a radical incommensurability between China and the West. Although these three perspectives differ considerably in their ultimate conclusions about the nature of this first encounter of China and the West—whether xenophobic rejection, constructive dialogue, or unfortunate misunderstanding—what they share is an interpretive framework in which the very local, historically specific events of the cooperation between several dozen Jesuits and their Chinese collaborators have been inflated into a grand narrative about two "civilizations."

These previous accounts—despite their differences—share another important feature: they have often been based primarily on the Jesuits' own prolific writings on their mission, sometimes supplemented by the writings of the Chinese officials who collaborated with them. The result is that too often these studies have adopted, as historical conclusions, viewpoints similar to those of their sources. For example, in their own writings, the Jesuits and the Chinese literati who collaborated with them do frequently argue that China was in a state of decline—moral, political, religious, intellectual, and scientific. They do claim to employ logical reasoning, empirical observation, and clear argument to refute what they decry as the absurdities of Buddhist superstition and metaphysical speculation. They insistently emphasize the newness of Western Learning and how radically it differed from its Chinese counterparts; and they do repeatedly extol the superiority of Western Learning. The Chinese literati who collaborated with the Jesuits do argue that crucial Chinese knowledge—philosophical, metaphysical, scientific, and religious—had been lost. They do write memorials stating their concerns about the apparently intractable crises of famine, invasion, and rebellion that were crippling the Ming dynasty. And they do insistently advocate Western Learning as the solution to Ming decline.

The focus of this historical study, then, has been seventeenth-century China. The research presented here has been based on primary sources from the Ming dynasty that have largely been ignored in previous accounts, ranging from mathematical treatises to political documents, including civil service examinations and memorials to the imperial court. This study, interdisciplinary in character, combines technical analyses of mathematics and science, archival documentation, and cultural contextualization, along with critical theory and recent research on the history of science.

The approach I have taken has been microhistorical, critical, and deflationary. At the core of this book is a microhistorical analysis of the dissemination of Western Learning: instead of a grand narrative about a purported encounter of civilizations, I have analyzed these events in the historical context of seventeenth-century China. I have argued that what has been recounted in previous studies as the introduction of the axiomatic deductive proof of Euclid's *Elements* was, in historical context, propaganda to promote Western Learning presented by the Jesuits' Chinese patrons. The allegedly unquestionable certainty of their proofs—whether Euclid's *Elements*, calculations of the sizes of the crystalline spheres of the Aristotelian universe, or demonstrations using Aristotelian logic—would lead their readers inexorably to certainty in belief in *Tianzhu* 天主 [Lord of Heaven]. Their proofs were not directed at mathematical problems per se, or even addressed to mathematicians of the period, but instead used by the Jesuits' Chinese patrons to promote Western Learning and defend the Jesuits against allegations that they were a dangerous foreign cult.

The Jesuits' high-ranking Chinese patrons, and Xu Guangqi in particular, barely understood the rudiments of mathematics or science, whether Chinese or Western. Their assertions of decline were thus not the result of analyses of the state of Chinese sciences of the period, but rather one part of their claim that the meanings of the Chinese classics had been lost in the great burning of the books, almost two thousand years earlier. This lost knowledge—which had been preserved in the West and could now be restored through the writings brought to China by the Jesuits—included knowledge of Heaven, Hell, and the true meaning of the Lord of Heaven. Their claims of radical differences were made in exaggerated attempts to distinguish themselves from the Buddhists and other cults to whom their contemporaries frequently likened them. The Jesuits and their Chinese patrons borrowed much of their terminology, including the terms for heaven (*tiantang* 天堂) and hell (*diyu* 地獄), directly from the Buddhists; and in the early years of the mission the Jesuits had even donned Buddhist garb and presented themselves as monks.

Xu Guangqi and his Chinese collaborators were hardly opposed to religious and metaphysical speculation. Indeed, for them practical knowledge was far inferior to moral philosophy; their savage attacks were directed against precisely those whose religion and metaphysics seemed most similar to their own. And the extravagant claims they made for the superiority of Western Learning were not the product of any studied comparison with Chinese sciences, but rather the indiscriminate promotion of a wide variety of fields in which they possessed very little expertise, fields ranging from theories of the soul to friendship to astronomy to military affairs to mathematics. In sum, the scientific proofs and the extravagant claims of the superiority, newness, and practical efficacy of Western Learning made by these Chinese officials—elite literati with little knowledge of Chinese sciences—were in historical context merely propaganda designed to attract followers, and bids for patronage, through memorials in which they fashioned themselves as statesmen with novel solutions to late-Ming crises.

In short, I argue that Xu Guangqi was a skilled propagandist. He had spent the first half of his life memorizing passages from the canonical Four Books and Five Classics, in order to pass the civil service examinations. From this process he learned classical allusions and historical examples; he learned to write essays in a style that distinguished those who had mastered it from their contemporaries—prose that was erudite, balanced, and structurally complex. He eventually succeeded in these efforts, passing local, provincial, and metropolitan examinations in which the success rate at some levels could be as low as one in one hundred. Xu thus succeeded in becoming one in perhaps ten thousand to achieve this status. He was then trained in the Hanlin Academy, an institution established by the empire to further the training of these elite literati to write on poetry, philosophy, statecraft, and the like. But most important, he was trained to write memorials to the imperial court, documents that combined the erudition of his decades of learning, the timeless wisdom of the ancient sage kings of China, which had been preserved in the Confucian classics, and the symbolic power of the long list of prestigious official titles that had been conferred upon him.

Xu was thus trained to bring his considerable rhetorical skills to bear on a wide range of issues—water conservancy, defense and military matters, astronomy, mathematics, and agriculture—fields in which he had little if any technical familiarity. And though Xu was one in ten thousand who had succeeded on the exams, in an empire of over a hundred million, he was but one of thousands of literati who had achieved his position. What eventually distinguished him from his contemporaries was his chance encounter with the Jesuits. The writings of the Jesuits provided material for this skilled propagandist to craft beautiful, learned, compelling arguments for the virtues of Western Learning. Those writings allowed him to produce volumes of translations, laudatory comparative studies, and memorials advocating Western Learning. They allowed him to fashion himself as someone above the fights between traditional factions—for Xu was the steadfast promoter of a new faction, the cadre of Chinese who collaborated with the Jesuits. The more desperate the Ming court became, the more credence was accorded to his proposals, no matter how outlandish. Indeed, so skilled was Xu that his arguments have remained persuasive even to historians who have read him hundreds of years later.

We should not, I have argued, accept Xu's propaganda as fact. The mathematical evidence is conclusive: the Jesuits' Chinese patrons and their collaborators could not have believed in the superiority of Western Learning as they purloined linear algebra problems from the very Chinese mathematical treatises they deprecated. Previous studies that celebrate the translation of Euclid's *Elements* in China, then, are not just anachronistic, but demonstrably wrong by the anachronistic standards they themselves have employed. On the one hand, the exaggerated claims for the importance of Euclid's *Elements*—claims cobbled together from commonplaces drawn from popular histories, laudatory prefaces, and the like—do not withstand critical scrutiny, historical, mathematical, or philosophical. And on the other hand, if we were to employ such modern standards to judge mathematics in the seventeenth century, linear algebra—one of the core mathematics courses today at the high school and university level—was flourishing in the late Ming dynasty, and not in Europe.

Historicizing "China," "the West," and the "First Encounter"

This microhistorical study of the introduction of Western Learning into China has served as the starting point for a broader critical analysis of the China/West framework that previous historical studies have taken for granted. For it was from their representation as the first encounter of China and the West that these events derived their significance in the first place; it was as an encounter of two civilizations that these events were conceptualized and explained; and it was conclusions comparing China and the West that captured the imagination of a broad readership. This study demonstrates that attempting to understand these events within this China/West framework in fact renders the history of these events incomprehensible.

This is not to say that the Jesuits, the Chinese officials who collaborated with them, or even their opponents did not believe in an imagined "West." But their imagined "West" differs considerably from, for example, how it is conceived by historians in the twentieth and twenty-first centuries. For the Jesuits and their Chinese collaborators, "the West" was not fundamentally defined by "science," "democracy," "modernity," or "capitalism," all of which have since been claimed to distinguish the West from the Rest.[1] Instead, via Xu's propaganda, the West was represented as a Europe which for thousands of years had been free of war or conflict between nations, without crime, without dynastic change, without rebellion, all united under the leadership of the moral order of lost doctrines from the Zhou dynasty (1045–256 BCE). Their dispute over the true nature of "the West" was with the Buddhists, whose doctrines had also been claimed to originate in "the West," which until this time had been understood by the Chinese to mean India.

One possible remedy for the modern historian, faced with the ideological manipulation of "the West" by the Jesuits and their Chinese patrons, might be to retain these as analytic categories, while acknowledging that their understandings were mistaken. But if it is easy to see how we might criticize their mistaken representations, what is more difficult is to offer a definition of "the West" that would contribute usefully to historical analysis. The secondary historical literature is filled with attempts to do precisely this—to define "the West" and "China" by elaborating the characteristics that differentiate them. These attempts have resulted in little more than lists of the purported oppositions between the West and China, including, as I have noted, scientific versus intuitive, theoretical

---

[1] For a recent example of similar claims, see Niall Ferguson, *Civilization: The West and the Rest*, 1st American ed. (New York: Penguin Press, 2011).

versus practical, democratic versus despotic, or, more recently, adversarial versus authoritarian.

I have argued against such an approach. The problem with such characterizations is that they take but one aspect of a civilization to be the emblem for the whole, according badly with what is supposed to be the extension of that civilization over time and space. For example, claims that it is science that makes the West unique take science as representative of the civilization as a whole, on the one hand conflating quite different practices from quite different periods in order to celebrate a single thing called "science," and on the other hand dismissing the documentation of science in other civilizations as somehow not really science. Similarly, work contrasting the adversarial Greeks with the authoritarian Chinese fails to connect the macropolitics of empires to the microsocial context of scientific work, and ignores portions of the historical record in Greece wherein science was pursued under court patronage. Thus, grandiose claims of fundamental oppositions between civilizations must retreat into strained alibis of propensities, or to acknowledgements of the failure of these generalizations to justify one's asserting them. Worse, these seemingly endless lists of oppositions that are to define the West and China are often mutually contradictory: for example, one notorious attempt from the late twentieth century does not even mention science as one of the key features defining the West, vis-à-vis other civilizations, instead returning to a view closer to that of the Jesuits that saw Christianity as one of the central features.[2]

Without a way to render the concepts of "China" and "the West" precise, we cannot rely on these concepts as analytic categories to understand events of the past. The categories "China" and "the West" are simply too vague, too illusory, too crude, and too vulnerable to manipulation to be useful in historical analysis. They are instead ideological constructs too often manipulated for politicized purposes. Instead of assuming that the significance of these local events was as a first encounter of China and the West, we need to explain how the historical protagonists themselves manipulated these terms in their conflicts. In particular, we need to historically contextualize the ways in which "China" and "the West" were imagined.

Science Beyond Civilizations

More broadly, the research presented here suggests the need to rethink world history of science prior to the scientific revolution. The discovery that linear algebra is not Western in its origins is, I would argue, one of the more important discoveries in recent work in the history of Chinese science and the history of mathematics generally. It is indeed surprising that not only has the history of one of the major components in modern mathematics been unknown, but that

---

[2] Samuel P. Huntington, *The Clash of Civilizations and the Remaking of World Order* (New York: Simon & Schuster, 1996).

these developments have been incorrectly attributed to two prominent European mathematicians in whose texts these techniques could be found. And the fact that problems this specialized, with solutions as esoteric as the *fangcheng* practices I have described, would eventually spread across Eurasia—from early imperial China to Italy by the thirteenth century—suggests that the assumption that other mathematical and scientific practices were not similarly transmitted should be reconsidered. To do so, we must reconsider the relationship between scientific practices, texts, and authorship during this period. Scientific practices of this period most often did not depend on texts: the learning, teaching, and transmission of these practices did not require literacy; when these practices were recorded in texts, it was most commonly for purposes of patronage or, less frequently, displays of expertise. So we must take care to distinguish clearly between extant historical records—in this case scientific writings that have been fortuitously preserved—and the world of scientific practice. Given this, it makes little sense for historians to obligingly grant credit for scientific discoveries to those who, in their pursuit of patronage, sought to claim that credit for themselves. It makes even less sense to attribute credit to what we now anachronistically call "China" or "the West" simply on the basis of the earliest known extant text in which a practice is recorded. The idea that specific scientific practices belong to "China," "Islam," or "the West" is an assumption, and only an assumption, one which resulted in part from the twentieth-century focus on civilizations and their comparison in the history of science. It is more likely that, like the *fangcheng* practices I have studied, scientific practices had centers of activity that shifted over time from and to different parts of Eurasia.

This suggests a new direction for world history of science, one that builds upon microhistorical approaches to the history of science, and draws upon approaches adapted from literary studies, sociology, anthropology, and philosophy. In particular, this suggests an approach that incorporates the following: (1) close readings of scientific texts in their historical context; (2) critical analysis of texts as having been compiled for purposes of patronage, or displays of knowledge, or simply commercial printing; (3) recognition that texts are neither sufficient nor necessary for sciences of the period; (4) reconstructing scientific practices from the evidence preserved in texts; and (5) most important, studying possible global circulations of practices along trade routes throughout the Eurasian continent, instead of seeing sciences as belonging to civilizations. I suggest that this approach has important methodological implications that deserve further exploration: a microhistorical approach to world history focused on artisans and adepts, practices recorded in texts, and global circulations. As we increasingly come to understand that globalization did not begin in the twenty-first century, it should be the task of world history of science to trace scientific practices of the past and their global circulation.

Acknowledgments

During the period in which this book was written, I have been fortunate to have benefited from the generous support of several institutions and the encouragement of numerous colleagues; I would thus like to express my gratitude for all their help.

The Pauley Foundation at UCLA provided a four-year fellowship under which I began my graduate study of history, and, as my Ph.D. dissertation, this project. A grant from the Committee on Scholarly Communications with China, American Council of Learned Societies, provided support for a year of archival research in China at Peking University. The Department of the History of Science at Harvard University provided a year of support under which I completed the first draft of my dissertation. The following year, postdoctoral fellowships from the Fairbank Center for East Asian Studies at Harvard and the Center for Chinese Studies at the University of California, Berkeley, provided me the opportunity to complete final revisions on my dissertation, to present lectures on my work, and to participate in faculty seminars. An Andrew W. Mellon postdoctoral fellowship at Stanford University in the Program in History and Philosophy of Science gave me the opportunity to organize conferences and workshops, and to teach several colloquia that allowed me to develop considerably the ideas presented here. A fellowship from the National Endowment for the Humanities at the School of Historical Studies, Institute for Advanced Studies, Princeton, provided me support for further research, including research into linear algebra in imperial China, which developed into my first book. A year as a Visiting Assistant Professor in the Department of History with the Fishbein Center for the History of Science and the Committee on Conceptual and Historical Studies of Science at the University of Chicago provided me with an important opportunity to further explore issues related to the history of science. An ACLS/SSRC/NEH International and Area Studies Fellowship from the American Council of Learned Societies provided a year of support for my research on linear algebra in imperial China, and contributed to the reformulation of this work. A Templeton Religion Trust "Science and Religion in East Asia" Project Visiting Fellowship from the Science Culture Research Center at Seoul National University provided me the opportunity to make final revisions to my penultimate draft. In addition to the fellowship, the Templeton "Science and Religion in East Asia" Project generously provided financial support to assist the publication and indexing of this book.

Benjamin Elman, Mario Biagioli, Ted Porter, Herman Ooms, and Robert Buswell read my doctoral dissertation and offered important criticisms, suggestions, and guidance. Steve Angle, Yomi Braester, Francesca Bray, Paul Cohen, Joe Dauben, Prasenjit Duara, Sam Gilbert, Steve Harris, Catherine Jami, David Keightley, Leo Lee, Tim Lenoir, Lydia Liu, David Palumbo-Liu, Geoffrey Lloyd, Mike Mahoney, Willard Peterson, Roddey Reid, Haun Saussy, Shang Wei, Paul Smith, Richard von Glahn, and Norton Wise offered oral or written comments on one or more chapters. Amir Alexander, Peter Bol, Tim Brook, Ping-yi Chu, Paula Findlen, David Nivison, Robert Richards, Nathan Sivin, Noel Swerdlow, Shang Wei, Wang Hui, Fred Wakeman, Jr., Tom Wilson, and Wen-hsin Yeh all offered important criticisms, suggestions, help, or guidance at various stages. I have also benefited from discussions with Bridie Andrews, Jim Bono, Andrea Eberhard Bréard, Cynthia Brokaw, Karine Chemla, Mike Fischer, Charlotte Furth, Marta Hanson, Philip Huang, Ted Huters, Bob Hymes, Chris Isett, Lionel Jensen, George Krompacky, Jean-Claude Martzloff, Peter Perdue, Ed Rhoads, Benjamin Schwartz, Xin Wei Sha, Sharon Traweek, Rick Vinograd, Heinrich von Staden, Ellen Widmer, Margherita Zanasi, and Zhang Qiong. Tim Brook and Catherine Jami gave me drafts of papers on Xu Guangqi; Peter Engelfriet kindly provided me with a copy of his dissertation.

I am deeply indebted to the historians at the Institute of the History of the Natural Sciences of Academia Sinica—Guo Shuchun, He Guangshao, Wang Yusheng, and especially Liu Dun, Han Qi, and Zou Dahai—for their helpful discussions, the use of their library, and the opportunity to present and publish in Chinese an earlier version of one chapter. Without their help, and without the work in the history of mathematics by Li Yan, Qian Baocong, Yan Dunjie, and Du Shiran, this project would not have been possible.

At the University of Texas, I thank numerous colleagues who provided considerable help, especially A. G. Hopkins, Alberto Martinez, Mark Metzler, Geraldine Heng, Huaiyin Li, Gail Minault, Jorge Canizares-Esguerra, David Sena, and Avron Boretz.

I also owe special thanks to R. Bin Wong, Peter Perdue, Pamela Crossley, John Crossley, Haun Saussy, Joe Dauben, and Karine Chemla for their generous support and encouragement.

At Seoul National University, I thank Professors Kim Yung Sik 金永植, Lim Jongtae 林宗台, and the other members of the Templeton "Science and Religion in East Asia" Project for their friendship, frank discussions, patient guidance, and generous assistance. I thank in particular Lim Jongtae, who read a penultimate version of this manuscript, and graciously offered detailed written corrections, criticisms, suggestions, and improvements of the entire manuscript. Jongtae was most generous with his time during my fellowship at Seoul National University, and numerous discussions with him have done much to clarify many of the issues discussed here.

At the Johns Hopkins University Press, I thank Trevor Lipscombe, Editor-in-Chief, for his generous encouragement, enthusiasm, and support in the initial stages of publication; Jacqueline C. Wehmueller, Executive Editor, for her expert assistance in the latter stages of publication; and Juliana McCarthy, Managing Editor, for her expert help in producing this book. William Carver copyedited the manuscript with considerable care, patience, and erudition. Joseph Hunt proofread the mathematics with expertise and care. Han Qi most generously proofread the Chinese. Anne Holmes provided expert assistance in preparing the index. Any remaining shortcomings are my own responsibility.

I have learned much from the suggestions and criticisms that I have received from presentations of my work—together with the contributions by other participants and ensuing in-depth discussions—at the following conferences, workshops, and panels that I have had the privilege to organize or co-organize: "Disunity of Chinese Science" (University of Chicago, May 2002); "Rethinking Science and Civilization: The Ideologies, Disciplines, and Rhetorics of World History" (Stanford, May 1999); "Critical Studies: Writing Science" (Stanford, 1998–99); "Intersecting Areas and Disciplines: Cultural Studies of Chinese Science, Technology and Medicine" (Berkeley, February 1998); "Materializing Cultures: Science, Technology, and Medicine in Global Contexts" (Stanford, May 1998); "Empires and Cultures" (Stanford, 1997–98); "Re-Siting the Missionaries in China: Critical Analyses of Translation, Imperialism, and Historical Memory" (Association for Asian Studies, March 1998); "New Directions in the Cultural Studies of Chinese Science and Medicine" (History of Science Society, November 1997). Further details about these conferences and seminars are available through my web site: http://rhart.org.

I would also like to express my gratitude for the opportunities offered me to present lectures on my work at the following conferences, panels, and forums: "Science and Christianity in the Encounter of Confucian East Asia with the West: 1600–1800" (Templeton "Science and Religion in East Asia" Project, Seoul National University, December 15–17, 2011); "ReWired: Asian/TechnoScience/Area Studies" (University of California Humanities Research Institute Seminar in Experimental Critical Theory, August 1–10, 2011); "China in Europe, Europe in China" (University of Zürich, June 14–15, 2010); "Go-betweens, Translations, and the Circulation of Knowledge" (University College London, November 13–14, 2009); Southwest Conference on Asian Studies (University of Texas, October 16–17, 2009); "The Age of Antiquaries in Europe and China" (Bard Graduate Center, New York, March 25–27, 2004); History of Science Colloquium (UCLA, April 15, 2002); Program in History and Philosophy of Science Seminar (University of Texas at Austin, April 11, 2002); "Jesuits and the Frontiers of Science in China: New Perspectives in the History of Science and Medicine" (Annual Meeting of the American Historical Association, January 2002); "Matteo Ricci and After: Four Centuries of Cultural Interactions between China and the West" (University of Hong Kong, October 12–15, 2001); "Symposium on Chinese Science and Medicine" (Franke Institute for the Humanities, University of Chicago, May 8, 2001); "Appropriations: Between Languages, Among Nations" (Stanford University, May 11–14, 2000); "Seminars and Colloquia on Late Imperial Chinese Culture and Science," "Colloquium: Culture and Science in Late Traditional China," and "Jesuits, Textualism, and Science in China and Europe in the 17th and 18th Centuries: A Roundtable Discussion" (seminars at the Institute for Advanced Study, Princeton, 1999–2000); "Mathematical Values: Social Mores and Mathematical Practices" (History of Science Society, October 1998); "History of Mathematics: Mathematics in the Americas and the Far East, 1800–1940" (Mathematisches Forschungsinstitut, Oberwolfach, Germany, October 18–22, 1998); "Reconstructing Science and the Humanities" (UCLA, April 25–26, 1998); Chinese Historiography Studies Group (Association for Asian Studies, March 25–28, 1998); "Maritime China: Culture, Commerce, and Society" (Center for Chinese Studies, UC Berkeley, March 13–14, 1998); "Rhetorical Rambles through Epistemological Brambles" (Society for the Social Studies of Science, October 23, 1997); Research Workshop in Chinese Humanistic and Historical Studies (Center for Chinese Studies, UC Berkeley, September 19, 1997); "The Song-Yuan-Ming Transition: A Turning Point in Chinese History?" (UCLA, June 5–11, 1997); "The Jesuits: Culture, Learning, and the Arts, 1540–1773" (Boston College, May 28 to June 1, 1997); "Workshop on Chinese Science" (Center for Chinese Studies, UCLA, May 24, 1997); "Cultural Histories of Science and Medicine in Early Modern China" (Association for Asian Studies, March 14, 1997); "New Work in East Asian Humanities" (Stanford University, April 24, 1997); Freeman Center for East Asian Studies (Wesleyan University, November 7, 1996); Chinese Cultural Studies Workshop (Fairbank Center for East Asian Research, Harvard University, October 31, 1996); Colloquia in Science, Technology, and Society (MIT, October 7, 1996); Early Science Group (Department of the History of Science, Harvard, October 2, 1995); UC Graduate Student Conference (UC Berkeley, April 29, 1995); Institute for the History of the Natural Sciences (Chinese Academy of Sciences, Beijing); I wish to sincerely thank all of those who attended and offered comments and criticisms.

My life's work in scholarship would not have been possible without inspiration, instruction, guidance, and encouragement from numerous teachers and mentors along the way. Here I can only name a few: At MIT, Robert L. Halfman (ESG), Richard Dudley, and Irving E. Segal. At Stanford, in mathematics, Paul J. Cohen and Ralph S. Phillips. At UCLA, Haun Saussy, Ted Porter, and Carlo Ginzburg. At Berkeley, David N. Keightley. At Harvard, Leo Ou-fan Lee. At Stanford, Tim Lenoir and Paula Findlen. At the Institute for Advanced Study, Heinrich von Staden. At the University of Chicago, Robert Richards and Noel Swerdlow. At the University of Texas, A. G. Hopkins, who served as my mentor, and set a sterling example for me and my cohort of assistant professors. My greatest debt in writing this book is to Benjamin Elman and Mario Biagioli, who served with considerable wisdom, erudition, expertise, and patience as my teachers, dissertation co-chairs, and mentors. It is to my teachers that this book is dedicated, with special thanks to Ben and Mario.

Finally, I owe special thanks to my family for their constant love, support, patience, and faith in me: my father Walter G. Hart, Jr., my mother Charlene

C. Hart, my brother David W. Hart, my sister Carolyn J. Hart, and my beautiful daughter, Nikki Janan Hart.

Portions of this book have been published in an earlier form, and revised versions are incorporated here:

- "Translating the Untranslatable: From Copula to Incommensurable Worlds," in *Tokens of Exchange: The Problem of Translation in Global Circulations*, edited by Lydia H. Liu (Durham, N.C.: Duke University Press, 2000), 45–73. An earlier version is published as "Translating Worlds: Incommensurability and Problems of Existence in Seventeenth-Century China," *Positions: East Asia Cultures Critique* 7, no. 1 (spring 1999): 95–128.
- "Beyond Science and Civilization: A Post-Needham Critique," *East Asian Science, Technology, and Medicine* 16 (1999): 88–114. An earlier version is published as "On the Problem of Chinese Science," in *The Science Studies Reader*, edited by Mario Biagioli (New York: Routledge, 1999), 189–201.
- "Ji he yuan ben, shenwei dapao, he 'yanima': cong kexue zhishi de shehuixue de jiaodu tantao shiqi shiji Xixue zhi yinru Zhongguo" 《幾何原本》，神威大砲，和『亞尼瑪』：從科學知識的社會學的角度探討十七世紀西學之引入中國 [Euclid, Cannons and Anima: An Analysis of the Introduction of Western Studies into Seventeenth-Century China from the Perspective of the Sociology of Scientific Knowledge, in Chinese], in *Ke shi xin chuan* 科史薪傳, edited by Liu Dun 劉鈍, Han Qi 韓琦, et al. (Shenyang: Liaoning jiaoyu chubanshe, 1997), 97–111.
- "The Great Explanandum," essay review of *The Measure of Reality: Quantification and Western Society, 1250–1600*, by Alfred W. Crosby, in *American Historical Review* 105, no. 2 (April 2000).
- "Tracing Practices Purloined by the 'Three Pillars,'" *The Korean Journal for the History of Science* 34 (2012): 287–358.

Appendix A
Zhu Zaiyu's *New Theory of Calculation*

The following translation is based on the *Siku quanshu* edition of Zhu Zaiyu's *Suan xue xin shuo*.[1] In addition, I have consulted the following editions for comparison:

(1) A rare edition preserved at Harvard University;
(2) A rare edition preserved at the Beijing Library, and reprinted as *Yuelü quanshu* 樂律全書 [Complete collection of music and pitch] edited by Beijing tushuguan guji chuban bianji zu 北京圖書館古籍出版編輯組 (Beijing: Shumu wenxian chubanshe 書目文獻出版社, [1988]), hereinafter *YLQS*;[2]
(3) The *Wan you wen ku* 萬有文庫 edition.

According to the *Zhongguo congshu zong lu* 中國叢書綜錄,[3] the *Suan xue xin shuo* is included only in *Yuelü quanshu* 樂律全書 (it does not state the number

---

[1] Zhu Zaiyu 朱載堉 (1536–1611), *Suanxue xin shuo* 算學新說 [New explanation of the theory of calculation], in *Yingyin Wenyuan ge Siku quanshu* 影印文淵閣四庫全書 [Complete collection of the Four Treasuries, photolithographic reproduction of the edition preserved at the Pavilion of Literary Erudition] (Hong Kong: Chinese University Press, 1983–1986), hereinafter *SKQS*.

[2] There are no other editions of *Yuelü quanshu* in the published catalogue of the Beijing Library (*Beijing tushuguan guji shanben shumu* 北京圖書館古籍善本書目 [Beijing: Shumu wenxian chubanshe, n.d.]), and from the description it is apparently this edition that has been reprinted: 樂律全書十五種四十八卷〔明朱載堉撰 明萬曆鄭藩刻本 十九冊 十二行二十四至二十五字小字雙行同黑口四周雙邊〕; the format of the characters is not uniform. This reprint is often illegible at the columns of text near the center, apparently the result of careless photocopying.

[3] Dai Nianzu 戴念祖, *Zhu Zaiyu: Ming dai de kexue he yishu juxing* 朱載堉：明代的科學和藝術巨星 [Zhu Zaiyu: A giant of the sciences and the arts in the Ming dynasty] (Beijing: Renmin chubanshe 人民出版社, 1986) notes only the 1595–1606 edition (1595–1606 年刻版《樂律全書》, 藏北京圖書館、中國科學院及下屬自然科學史研究所圖書館), and notes that this was reprinted (萬有文庫 第一集 0735 號 樂律全書 第一至四冊 商務印書館, 1931). Du Xinfu 杜信孚, Zhou Guangpei 周光培, and Jiang Xiaoda 蔣校達, *Mingdai banke zonglu* 明代版刻綜錄 [General catalogue of books engraved during the Ming dynasty] (Yangzhou: Jiangsu Guangling guji keyinshe 江蘇廣陵古籍刻印社, 1983), lists only this one title under the publisher 鄭藩 (listing as its sources *Zhongguo congshu zonglu* 中國叢書綜錄 and *Hangzhou daxue tushuguan shanben shumu* 杭州大學圖書館善本書目, and noting a copy is preserved at Suzhou shi tushuguan 蘇州市圖書館).

of titles or chapters); all extant rare manuscript editions in China are the Ming Wanli 24 Zheng Fan edition 明萬曆二十四年鄭藩刊本. This edition is preserved in many libraries all over China,[4] suggesting that a large number of copies may have been printed and that it was widely distributed.[5]

Problem 2

The following is a translation of problem 2, the first mathematical problem in the treatise, which correctly calculates the square root of 200 to 25 significant digits to be 14.1421356237309504880168 9. Punctuation has been added.

> The second problem: Practitioners of musical tones first seek the *huang zhong*, just as the astronomer first seeks the winter solstice.[6] Second, seek the *rui bin*, just like [the astronomer seeks] the summer solstice. Then seek the *jia zhong*, just like the spring equinox. Then seek the *nan lü*, just like the autumn equinox. After this seek the *da lü* which, with the exception of the *huang zhong*, is the first of all the pitches. Next seek the *ying zhong*, which is the final of all the pitches. This is also just like what the astronomer terms the *lü duan*, *ju zheng*, and *gui yu*. *Huang zhong* and *lü duan* are at the beginning, *rui bin* and *ju zheng* are in the middle, and *ying zhong* and *gui yu* are at the end. Therefore it is said that the musical pitches and astronomy follow the same way (*yi dao* 一道). Given that the true pitch of the *huang zhong* is 10 *cun* in length, how long are the *rui bin*, the doubled pitch, and the true pitch?
>
> 第二問：律家先求黃鍾，猶曆家先求冬至也。次求蕤賓，猶夏至也。又次求夾鍾，猶春分也。又次求南呂，猶秋分也。然後求大呂，除黃鍾外，諸律呂之首也。其次求應鍾，諸律呂之終也。亦猶曆家所謂履端、舉正、歸餘也。黃鍾、履端於始，蕤賓、舉正於中，應鍾、歸餘於終。故曰律曆一道。今黃鍾正律長十寸。蕤賓、倍律、正律各長幾何。[7]

---

[4] This edition is preserved in the Beijing Library 北京, the Shoudu Library 首都, the Academia Sinica Library 科學, Qinghua University Library 清華, Shanghai Library 上海, Huadong Normal University Library 華師, Gansu Library 甘肅, Shandong University Library 山大, Nanjing Library 南京, Zhejiang Library 浙江, Hangzhou University Library 杭大, Fujian Normal University Library 福師, Henan Library 河南, Wuhan University Library 武大, Sichuan Library 四川, Heilong Jiang Library 黑龍江, Guilin Library 桂林, and the Zhongyang Minzu Xueyuan Library 民院.

[5] It should be noted that among the Ming collectanea listed in *Zhongguo congshu zonglu* that are extant, it is not unusual that several copies remain extant and that they are distributed widely over China.

[6] Kenneth Robinson notes that in the *Li Ji*, the *huang zhong* pitch pipe is blown in winter solstice. Robinson, *A Critical Study of Chu Tsai-Yü's Contribution to the Theory of Equal Temperament in Chinese Music* (Wiesbaden: Steiner, 1980), 12.

[7] There are no differences here between the *SKQS* and the Beijing tushuguan guji zhenben congkan 北京圖書館古籍珍本叢刊 editions.

Answer: The *huang zhong* is ten *cun* in length, let this be the two sides of a square. The diagonal hypotenuse joining the two corners is the doubled pitch; halving the doubled pitch yields the *rui bin* true pitch. If the true pitch of the *rui [bin]* is taken as the sides of a square, then the diagonal hypotenuse is the true pitch of the *huang [zhong]*. In the *Rites of Zhou*, the Li clan,[8] in order to measure the area of a circle circumscribed about a square, calculated the length of the diagonal, that is, the diameter of the circle.[9] This yields for the diagonal hypotenuse 1 *chi* 4 *cun* 1 *fen* 4 *li* 2 *hao* 1 *si* 3 *hu* 5 *wei* 6 *xian* 2373095048801689,[10] which is the *rui bin* doubled pitch. Halve this yields 7.071067811865475244008445, namely, the *rui bin* true pitch. In labelling the numbers following *xian*, the names of the places are not listed.[11]

答曰：黃鍾長十寸，是為平方面。其兩隅斜弦，即蕤賓倍律；倍律折半，即蕤賓正律也。若以蕤正為平方面，而其斜弦即黃正也。周禮栗氏[12]為量內方尺而圓其外，筭法求方之斜，即圓之徑。得斜弦一尺四寸一分四釐二毫一絲三忽五微六纖二三七三〇九五〇四八八〇一六八九，即蕤賓倍律也。折半得七寸〇七釐一毫〇六忽七微八纖一一八六五四七五二四四〇〇八四四五，即蕤賓正律也。纖已下數不立名色餘皆放此。

Method: Using the *gou gu* to calculate the hypotenuse. Place one side of the square from South to North, ten *cun*, multiply this by itself obtaining 100 [square] *cun*, let this be the area of the long-side [squared]. Separately place one side from East to West, ten *cun*, multiply it by itself obtaining 100 [square] *cun*, let this be the area of the short-side. Add these together obtaining 200 [square] *cun*, and let this be the area of the hypotenuse. Place the area of the hypotenuse 200 [square] *cun* as the dividend [for finding the square root]. From within the preceding table [when the dividend is] greater than 100, the factor is 10, this is called "returning one" [to the divisor]. Use the method of square-root extraction (*kai fang gui chu fa* 開方歸除法), and divide it into the dividend.[13] The first place [digit] is returned to the dividend called "when 1 is encountered, carry 10

---

[8] What is termed the "Li clan" is an administrative post in the Zhou dynasty bureaucracy, in charge of casting metals and making measuring vessels. See *Liji zhushu* 禮記註疏 [*Record of Rites*, with commentary and subcommentary], in *Shisanjing zhushu: fu jiaokan ji* 十三经注疏 : 附校勘记 [Thirteen Classics, with commentary and subcommentary, and with editorial notes appended] (Beijing: Zhonghua shuju 中华书局, 1980), hereinafter *SSJZS, juan* 42.

[9] The translation of this sentence is tentative: it is not clear exactly what mathematical technique Zhu is referring to here; his appeal to the ancients is likely spurious.

[10] This is the square root of 200, correct to 25 significant digits (including, that is, the 9 digits marked by metrological units, 1 *chi* 4 *cun* . . . 6 *xian*).

[11] Again, I have transcribed the commentary, which is written in half-size characters, in a smaller font.

[12] Substituting 栗 for its archaic form.

[13] Note that this is not ordinary division.

forward," obtaining 10 *cun*.[14] What has been "returned" [to the divisor] does not [again] divide the remainder of the dividend 100 *cun*.

法曰依句股求弦筭：置方面自南至北一十寸，自乘得一百寸，為股冪。別置方面自東至西一十寸，自乘得一百寸，為句冪。相併共得二百寸，為弦冪。就置弦冪二百寸為實，看前式內一百已上該開一十寸命作一歸為下法，用開方歸除法除之於實，首位歸實呼逢一進一十得一十寸，有歸不除餘實一百寸。

Double the lower divisor 10 *cun* is changed to 20 *cun*, called "returning" 2. Hereafter, [when] there is a "return" [to the divisor] that divides the dividend, the first place is returned to the divisor called "2 and 1 combine to 5," "start with 1, return 2," yielding only 4 *cun*.[15] Place the "lower divisor" 4 *cun* and 20 *cun* on the lower [divisor], together yielding 24 *cun*. With the second place of the dividend divide the dividend called "4 and 4, remove 16,"[16] yielding the remainder 4 *cun*.

倍下法一十寸改作二十寸命曰二歸，自此已後有歸，有除於實第一位歸實呼二十添作五起一還二只得四寸，下法亦置四寸於二十寸之下共得二十四寸，於實第二位除實呼四四除一十六，餘實四寸。

Double the "lower divisor" 4 *cun* is changed to 8 *cun*, obtaining together 28 *cun*. With the third place of the dividend return to the dividend namely, encountering 2 advance 10, obtaining 1 *fen*. Place the "lower divisor" 1 *fen* and 28 *cun* in the lower [divisor], obtaining 28 *cun* 1 *fen*. Divide the dividend with the third place of the dividend namely, 1 and 8, back one place remove 8. With the fourth place divide the dividend namely, 1 and 1, back one place, subtract 1, [yielding] the remainder of the dividend 1 *cun* 19 *fen*.

倍下法四寸改作八寸共得二十八寸，於實第三位歸實呼逢二進一十得一分，下法亦置一分於二十八寸之下共得二十八寸一分，於實第三位除實呼一八退位除八，於第四位除實呼一一退位除一，餘實一寸一十九分。

Double the "lower divisor" 1 *fen* is changed to 2 *fen*, together yielding 28 *cun* 2 *fen*. Take the third place of the dividend, and return to the dividend namely, 2 and 1 increased to make 5, take 1 and return 2, only obtaining 4 *li*. Place on the "lower dividend" 4 *li* and the lower [dividend] of 28 *cun* 2 *fen*, together obtaining 28 *cun* 2 *fen* 4 *li*. With the fourth place of the dividend, diminish the dividend namely, 4 and 8, remove 32; with the fifth place diminish the dividend namely, 2 and 4, move back one place, subtract 8; with the sixth place subtract from the dividend namely, 4 and 4, subtract 16. The remainder 6 *fen* 04 *li*.

---

[14] Here Zhu is referring to rhymes used for teaching abacus calculations; the translation is provisional. For an explanation of some of these rules of calculation, see Jean-Claude Martzloff, *A History of Chinese Mathematics* (New York: Springer, 2006), 215–21.

[15] Again, this is a rhyme for abacus calculations, as are similar phrases in the remainder of the translation.

[16] As Ulrich Libbrecht notes, *chu* 除 can be used for both subtraction and division; here it is used for subtraction. "To subtract; to divide, retaining only the remainder for the next step (in this case the division has the same result as subtracting the divisor as many times as possible." Libbrecht, *Chinese Mathematics in the Thirteenth Century: The Shu-Shu Chiu-Chang of Ch'in Chiu-Shao* (Cambridge, MA: MIT Press, 1973), 477.

倍下法一分改作二分共二十八寸二分，於實第三位歸實呼二一添作五起一還二
只得四釐，下法亦置四釐於二十八寸二分之下共得二十寸二分四釐，於實第四
位除實呼四八除三十二，於第五位除實呼二四退位除八，於第六位除實呼四
四除一十六，餘實六分〇四釐。

Double the "lower divisor" 4 *li* is changed to 8 *li*, together yielding 28 *cun* 2 *fen* 8 *li*. At the fifth place, return the remainder namely, encountering 4, carry forward 20, yielding 2 *hao*. Place the "lower divisor" 2 *hao* at the end of 28 *cun* 2 *fen* 8 *li*, together yielding 28 *cun* 2 *fen* 8 *li* 2 *hao*. At the fifth place of the remainder, subtract from the remainder namely, 2 and 8, subtract 16; at the sixth place subtract from the remainder namely, 2 and 2, carry back one place, subtract 4; at the seventh place subtract from the remainder namely, 2 and 8, subtract 16; at the eighth place, subtract from the remainder namely, 2 and 2, carry back one place, subtract 4, yielding the remainder 38 *li* 38 *hao*.[17]

倍下法四釐改作八釐共得二十八寸二分八釐，於實第五位歸實呼逢四進二十得
二毫，下法亦置二毫於二十八寸二分八釐之下共得二十八寸二分八釐二毫，於
實第五位除實呼二八除一十六，於第六位除實呼二二退位除四，於第七位
除實呼二八除一十六，於第八位除實呼二二退位除四，餘實三十八釐三十八
毫。

Double the "lower divisor" 2 *hao* is changed to 4 *hao*, together yielding 28 *cun* 2 *fen* 8 *li* 4 *hao*. At the sixth place return the remainder namely, encountering 2 move forward 10, obtaining 1 *si*. Place the "lower divisor" 1 *si* at the end of 28 *cun* 2 *fen* 8 *li* 4 *hao*, together yields 28 *cun* 2 *fen* 8 *li* 4 *hao* 1 *si*. At the sixth place of the remainder, subtract from the remainder namely, 1 and 8, move back one place, subtract 8; at the seventh place subtract from the remainder namely, 1 and 2, move back one place, and subtract 2; at the eighth place subtract from the remainder namely, 1 and 8, carry back one place, subtract 8; at the ninth place subtract from the remainder namely, 1 and 4, carry back one place, subtract 4; at the tenth place subtract from the remainder namely, 1 and 1, carry back one place, subtract 1. The remaining remainder 10 *li* 007 *hao* 59 *si*.[18]

倍下法二毫改作四毫共得二十八寸二分八釐四毫於實第六位歸實呼逢二進十得
一絲，下法亦置一絲於二十八寸二分八釐四毫之下共得二十八寸二分八釐四毫一
絲，於實第六位除實呼一八退位除八，於第七位除實呼一二退位除二，於第
八位除實呼一八退位除八，於第九位除實呼一四退位除四，於第十位除實呼
一一退位除一，餘實一十釐〇〇七毫五十九絲。

Double the "lower divisor" 1 *si* is changed to 2 *si*, together obtaining 28 *cun* 2 *fen* 8 *li* 4 *hao* 2 *si*. At the sixth place of the remainder return the remainder namely, 2 and 1 increased to 5, begin with 2 and change to 4, only obtaining 3 *hu*. Place the "lower divisor" 3 *hu* and 28 *cun* 2 *fen* 8 *li* 4 *hao* 2 *si* in the lower [divisor], together yield 28 *cun* 2 *fen* 8 *li* 4 *hao* 2 *si* 3 *hu*. At the seventh place of the remainder, subtract from the remainder namely, 3 and 8 subtract 24; at the eighth place subtract from the remainder namely, 2 and 3, go back one place, subtract 6; at the ninth place

---

[17] This appears to be a transcription error. The value should be .003836.

[18] The second 0 in 007 is excrescent.

subtract from the remainder namely, 3 and 8 subtract 24; at the tenth place subtract from the remainder namely, 3 and 4 subtract 12; at the eleventh place subtract from the remainder namely, 2 and 3, move back one place, and subtract 6; at the twelfth place subtract from the remainder namely, 3 and 3, move back one place, and subtract 9. The remaining remainder 1 *li* 59 *hao* 06 *si* 31 *hu*.

倍下法一絲改作二絲共得二十八寸二分八釐四毫二絲，於實第六位歸實呼二一添作五起二還四只得三忽，下法亦置三忽於二十八寸二分八釐四毫二絲之下共得三十八寸二分八釐四毫二絲三忽，於實第七位除實呼三八除二十四，於第八位除實呼二三退位除六，於第九位除實呼三八除二十四，於第十位除實呼三四除一十二，於第十一位除實呼二三退位除六，於第十二位除實呼三三退位除九，餘實一釐五十九毫〇六絲三十一忽。

Double the "lower divisor" 3 *hu* is changed to 6 *hu*, together obtaining 28 *cun* 2 *fen* 8 *li* 4 *hao* 2 *si* 6 *hu*. At the seventh place of the remainder return to the remainder namely, 2 and 1 increased to 5, obtaining 5 *wei*. Place the "lower divisor" 5 *wei* and 28 *cun* 2 *fen* 8 *li* 4 *hao* 2 *si* 6 *hu* in the lower [divisor], together obtaining 28 *cun* 2 *fen* 8 *li* 4 *hao* 2 *si* 6 *hu* 5 *wei*. At the eighth place of the remainder subtract from the remainder namely, 5 and 8 subtract 40; at the ninth place subtract from the remainder namely, 2 and 5 subtract 10; at the tenth place subtract from the remainder namely, 5 and 8 subtract 40; at the eleventh place subtract from the remainder namely, 5 and 4 subtract 20; at the twelfth place subtract from the remainder namely, 2 and 5 subtract 10; at the thirteenth place subtract from the remainder namely, 5 and 6 subtract 30; from the fourteenth place subtract from the remainder namely, 5 and 5 subtract 25. The remaining remainder 17 *hao* 64 *si* 17 *hu* 75 *wei*.

倍下法三忽改作六忽共得二十八寸二分八釐四毫二絲六忽，於實第七位歸實呼二一添作五得五微，下法亦置五微於二十八寸二分八釐四毫二絲六忽之下共得二十八寸二分八釐四毫二絲六忽五微，於實第八位除實呼五八除四十，於第九位除實呼二五除一十，於第十位除實呼五八除四十，於第十一位除實呼五四除二十，於第十二位除實呼二五除一十，於第十三位除實呼五六除三十，於第十四位除實呼五五除二十五，餘實一十七毫六十四絲一十七忽七十五微。

Double the "lower divisor" 5 *wei* is changed to 1 *hu* 0 *wei*, together obtaining 28 *cun* 2 *fen* 8 *li* 4 *hao* 2 *si* 7 *hu* 0 *wei* [28.284270]. Return the dividend's eighth place to the dividend namely, 20 increased [multiplied] by 5, encountering 2 advance 10 [multiplication by 5], obtaining 6. Place the "lower divisor" 6 *xian* and 28 *cun* 2 *fen* 8 *li* 4 *hao* 2 *si* 7 *hu* 0 *wei* in the lower [divisor] together yields 28 *cun* 2 *fen* 8 *li* 4 *hao* 2 *si* 7 *hu* 0 *wei* 6 *xian*. At the ninth place of the remainder, subtract from the remainder namely, 6 and 8 subtract 48; at the tenth place, subtract from the remainder namely, 2 and 6 subtract 12; at the eleventh place subtract from the remainder namely, 6 and 8 subtract 48; at the twelfth place subtract from the remainder namely, 4 and 6 subtract 24; at the thirteenth place subtract from the remainder namely, 2 and 6 subtract 12; at the fourteenth place subtract from the remainder namely, 6 and 7 subtract 42; at the fifteenth place, the "lower divisor" has zero *wei* and nothing is subtracted; at the sixteenth place subtract from the remainder namely, 6 and 6 subtract 36. The remaining remainder 67 *si* 12 *hu* 12 *wei* 64 *xian*.

倍下法五微改作一忽０微共得二十八寸二分八釐四毫二絲七忽０微，於實第八位歸實呼二十添作五逢二進一十得六纖，下法亦置六纖於二十八寸二分八釐二毫二絲七忽０微之下共得二十八寸二分八釐四毫二絲七忽０六纖，於實第九位除實呼六八除四十八，於第十位除實呼二六除一十二，於第十一位除實呼六八除四十八，於第十二位除實呼四六除二十四於第十三位除實呼二六除一十二，於第十四位除實呼六七除四十二至第十五位下法空微無除，於第十六位除實呼六六除三十六，餘實六十七絲一十二忽一十二微六十四纖。

Following this, extracting to the 25th place, the technique is the same as above, except that following *xian* the places are not named, together yielding for the diagonal hypotenuse 1 *chi* 4 *cun* 1 *fen* 4 *li* 2 *hao* 1 *si* 3 *hu* 5 *wei* 6 *xian* 2373095048801689, namely, the double pitch of the *rui bin*. Take half, i.e., obtain the true pitch of the *rui bin*, which is the same number as the true pitch of the *rui bin* extracted below.

自此已後開至二十五位其術同前，但纖已下不立名色。共得斜弦一尺四寸一分四釐二毫一絲三忽五微六纖二三七三０九五０四八八０一六八九，即蕤賓倍律也。折半即得蕤賓正律與下條開方所得蕤賓正律數同。

Appendix B
Xu Guangqi's *Right Triangles, Meanings*

This appendix presents a translation of what is arguably the most complex mathematical problem found in any of Xu Guangqi's 徐光啟 (1562–1633) writings, "Problem 7" from his *Right Triangles, Meanings* (*Gou gu yi* 勾股義).[1] It demonstrates, I will argue, how little Xu understood about mathematics, Chinese or Western. As we will see, Xu begins with the diagram from Li Ye's 李冶 (1192–1279) *Sea Mirror of Circle Measurement* (*Ce yuan hai jing* 測圓海鏡; for an analysis of the *Sea Mirror of Circle Measurement*, see pages 90–102 of this book). To this diagram, Xu adds, in an apparently haphazard manner, lines, triangles, and rectangles. Xu then provides, in an apparently equally haphazard manner, citations to Euclid's *Elements*. In the end, as this translation will show, Xu proves very little beyond his own apparent incomprehension of anything more than the most elementary propositions from Euclidean geometry. More specifically, although Xu correctly applies the propositions concerning what is now commonly known as "vertical angles" (Euclid, I. 15) and "side-angle-side" (Euclid, I. 4), his applications of other propositions are often spurious. In sum, Xu's writings on geometry provide little support for previous accounts that have anachronistically imputed to Xu an attempt to provide axiomatic foundations, rigorous logic, or deductive reasoning to Chinese mathematics.

Indeed, the stark disconnect between Euclidean geometry and Xu's writings on geometry suggests the need to fundamentally reappraise the purposes imputed to him: that Xu apparently fails to comprehend even mathematics this elementary itself requires an explanation.[2] In historical context, Xu was under no compulsion to follow the rules of Euclidean geometry, however elementary they

---

[1] On the authorship of texts attributed to the Jesuits and their collaborators, see footnote 2 on page 195 of this book. Xu's authorship of *Right Triangles, Meanings* is not contested. Based on my research, I would argue that Xu is in fact most likely the author.

[2] To adapt an example from Kripke's analysis of Wittgenstein, if 126 is given as the answer to $57 + 68$, we might suppose that an error was made in the calculation; but if 5 is given as the answer, we might instead ask whether an entirely different language game is being played. See chapter 2, "The Wittgensteinian Paradox," in Saul A. Kripke, *Wittgenstein on Rules and Private Language: An Elementary Exposition* (Cambridge, MA: Harvard University Press, 1982).

might be. Instead, as he clearly states in his prefaces, his purpose was to recover ancient "meanings" from antiquity, which he imagined to be the time of Yao, Shun, and the Three Dynasties (traditionally, 2357–2256 BCE, 2255–2205 BCE, and 2205–256 BCE, respectively). Such "meanings" were hardly limited to the words recorded in texts, and did not require textual evidence for their justification. On the contrary, for Xu, who was one of the more learned Confucians of his time, it was these "meanings"—as he conceived them—that explained ancient texts. In this sense, in his interpretations of the texts brought by the Jesuits, he was under no compulsion to accept their understandings of the doctrines they "returned" to China: there was no need for him to accept their claims about salvation through Jesus; instead, he asserted that their texts recovered lost "meanings" of an ancient Chinese *Tianzhu* [Lord of Heaven]. In a similar manner, he was under no compulsion to accept the specific propositions of Euclid as the basis of geometry; instead, Euclid's *Elements* aided him in recovering the lost "meanings" of Chinese mathematical texts.

Before presenting the translation of problem 7, I will first briefly provide some background information. The date of publication of *Right Triangles, Meanings* is not certain, with scholarly estimates ranging from about 1609 to the late 1620s: more specifically, the preface is undated; the *Si ku* editors list no date;[3] Qian Baocong states that it was written shortly after 1608, as does Wang Yusheng;[4] Liang Jiamian asserts that it was written in 1609,[5] as does Peter Engelfriet;[6] an anonymous postface (*hou ji* 後記) in *XGZYJ* also states that it was written in 1609;[7] Wang Zhongmin suggests that the date was before 1610;[8] Xu Zongze dates

---

[3] *Siku quanshu zong mu* 四庫全書總目 [General catalogue of the *Complete Collection of the Four Treasuries*], ed. Yong Rong 永瑢 (1744–1790) and Ji Yun 紀昀 (1724–1805) (Beijing: Zhonghua shuju 中華書局, 1965), 2:896.

[4] Qian Baocong 钱宝琮, *Zhongguo shuxue shi* 中国数学史 [History of Chinese mathematics] (Beijing: Kexue chubanshe 科学出版社, 1964), 239; Wang Yusheng 王渝生, *Gougu yi tiyao* 勾股義提要 [Summary of *Right Triangles, Meanings*], in *Zhongguo kexue jishu dianji tonghui: Shuxue juan* 中國科學技術典籍通彙：數學卷 [Comprehensive collection of the classics of Chinese science and technology: Mathematics volumes] (Zhengzhou: Henan jiaoyu chubanshe 河南教育出版社, 1993), hereinafter *ZKJDT*.

[5] The accompanying footnote states, however, that the date is unknown (*juan qi wei xiang* 撰期未詳), and the only basis given for the date is that the preface (*xu* 序) states that it was written after *Ce liang fa yi* 測量法義. Liang Jiamian 梁家勉, *Xu Guangqi nianpu* 徐光啟年譜 [Chronological biography of Xu Guangqi] (Shanghai: Shanghai guji chubanshe 上海古籍出版社, 1981), 92 and 93 n. 10.

[6] Peter M. Engelfriet, "Euclid in China: A Survey of the Historical Background of the First Chinese Translation of Euclid's Elements (Jihe Yuanben; Beijing, 1607), an Analysis of the Translation, and a Study of Its Influence up to 1723" (Ph.D. dissertation, Leiden, 1996), 285.

[7] *Hou ji* 後記 [Postface], in *Xu Guangqi zhu yi ji* 徐光啟著譯集 [Collected writings and translations of Xu Guangqi], ed. Shanghai shi wenwu baoguan weiyuanhui 上海市文物保管委員会 and Gu Tinglong 顧廷龍 (Shanghai: Shanghai guji chubanshe 上海古籍出版社, 1983), hereinafter *XGZYJ*, *juan* 8.

[8] Wang Zhongmin 王重民, "Xu yan—Xu Guangqi zhili kexue yanjiu de shiji he ta zai woguo kexue shi shang de chengjiu" 序言—徐光啟致力科學研究的事蹟和他在我國科學史上的成就 [Preface—The achievements of Xu Guangqi's devotion to scientific research and his accom-

it at 1617;[9] and Ding Fubao states that it was probably completed in the early years of the Chongzhen 崇禎 reign (1627–1644).[10] Textual evidence suggests that the later dates for composition may be correct: the mathematical content of *Right Triangles, Meanings* is much more sophisticated than that in Xu's other works (e.g., *Celiang yi tong* 測量異同, conventionally dated 1608); the preface also contains many more references to Chinese mathematicians than are given in his other prefaces.[11]

Right Triangles, Meanings is preserved in several collectanea: *Wenyuan ge Siku quanshu* 文淵閣四庫全書 (*SKQS*); *Xu Guangqi zhu yi ji* 徐光啟著譯集 (*XGZYJ*); *Tian xue chu han* 天學初函 (*TXCH*); *Zhi hai* 指海 (*ZH*); *Hai shan xian guan congshu* 海山仙館叢書 (*HSXGC*); *Saoye shan fang cong chao* 掃葉山房叢鈔 (*SYSF*); *Zhong-Xi suanxue congshu chu bian* 中西算學叢書初編 (*ZXSX*); *Congshu jicheng chu bian* 叢書集成初編 (*CJCB*); *Xixue da cheng shiyin ben* 西學大成石印本;[12] *Xu Wending za zhu* 徐文定雜著; the rare-edition *Zhou bi jing tian ji* 周髀井田記, preserved in the Shanghai Library;[13] and, more recently, it has been reprinted in *Zhongguo kexue jishu dianji tonghui* 中國科學技術典籍通彙 (*ZKJDT*).

This translation is based on the *SKQS* edition, which contains the fewest errors. In addition to the *SKQS* edition, the *TXCH* edition and the Shanghai Library rare edition reprinted in *XGZYJ* and in *ZKJDT* have been consulted. The main text of the *Gou gu yi* reprinted in *XGZYJ*, *ZKJDT*, and *TXCH* is apparently from the same edition.[14] The prefaces (*xu* 序), however, differ: the preface in *XGZYJ* appears to be a handwritten copy from another edition; the *TXCH* edition has what appears to be the original preface in the original type, but it is missing the last 15 characters (語絕 ... 有論) of the *XGZYJ* edition. In the translation, emendments to the *SKQS* edition and differences between editions have been noted.

plishments in the history of Chinese science], in *Xu Guangqi ji* 徐光啟集 [Collected works of Xu Guangqi] (Shanghai: Shanghai guji chubanshe 上海古籍出版社, 1984), 1:8, hereinafter *XGQJ*. However, Wang assigns no specific date to the preface.

[9] Xu Zongze 徐宗澤, *Ming-Qing jian Yesu huishi yi zhu tiyao* 明清間耶穌會士譯著提要 [Annotated bibliography of the translations and writings of the Jesuits during the Ming and Qing dynasties] (Taibei: Zhonghua shuju 中華書局, 1989), 473.

[10] Ding Fubao 丁福保, *Sibu zonglu suanfa bian* 四部總錄算法編 [General catalogue of the four sections, mathematics section] (Beijing: Wenwu chubanshe 文物出版社, 1957), 56.

[11] Xu, *Gougu yi xu* 勾股義序 [Preface to *Right Triangles, Meanings*], in *XGQJ*, 1:48. It should be noted, however, that the preface is from the *TXCH* collection, and thus may have been revised at a later date before its printing in 1635.

[12] According to Ding, *Sibu zonglu suanfa bian*, 56.

[13] According to Wang, *Gougu yi tiyao*, 25.

[14] *ZKJDT*, *XGZYJ*, and *TXCH* all reprint the same edition—they are exactly similar, down to misshapen characters (e.g., all have *bai* 百 missing the middle stroke on Problem 7, second line [*TXCH*, 3555]). The *TXCH* edition is of much poorer quality, and is illegible in places (e.g., p. 3558). The *ZKJDT* reprint states that the edition used is from the Shanghai Library (neither *XGZYJ* nor *TXCH* state the original source). The *ZKJDT* reprint may itself be a copy of the *XGZYJ* reprint: the page numbers on the first page of the table of contents are the same; Wang's preface contains phrases apparently from the postface (*hou ji* 後記).

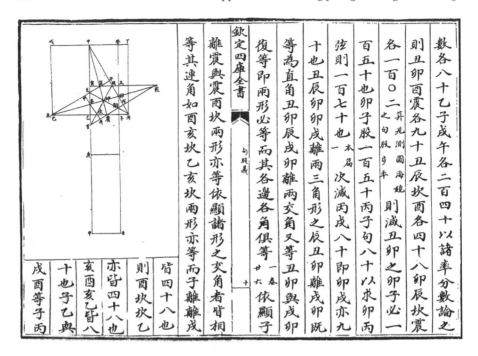

Fig. B.1: The diagram analyzed in problem 7 of Xu's *Right Angles, Meanings*.

In contrast to the considerable number of general studies on Xu Guangqi, there have been relatively few analyses of his mathematics.[15] Most evaluations of *Right Angles, Meanings* have been based directly on Xu's own claims made in his preface, rather than on any analysis of the mathematical content: for example, Qian Baocong asserts that *Right Angles, Meanings* is an attempt to make more

---

[15] General works on Xu Guangqi include: *Xu Guangqi yanjiu lunwenji* 徐光启研究论文集 [Essays on research on Xu Guangqi], ed. Xi Zezong 席泽宗 and Wu Deduo 吴德铎 (Shanghai: Xuelin chubanshe 学林出版社, 1986); *Xu Guangqi jinian lunwenji* 徐光启纪念论文集 [Collection of essays in commemoration of Xu Guangxi], ed. Zhongguo kexueyuan Zhongguo ziran kexueshi yanjiushi 中国科学院中国自然科学史研究室 (Beijing: Zhonghua shuju 中华书局, 1963); *Xu Guangqi zhuanji ziliao* 徐光啟傳記資料 [Biographic materials on Xu Guangqi] (Taibei: Tianyi chubanshe 天一出版社, 1979), hereinafter *XGZZ*; Catherine Jami, Gregory Blue, and Peter M. Engelfriet, eds., *Statecraft and Intellectual Renewal in Late Ming China: The Cross-Cultural Synthesis of Xu Guangqi (1562–1633)* (Leiden: Brill, 2001). For studies in English on Xu's translations, see Peter M. Engelfriet, "The Chinese Euclid and Its European Context," in *L'Europe en Chine: Interactions scientifiques, religieuses et culturelles aux XVIIe et XVIIIe siècles*, ed. Catherine Jami and Hubert Delahaye (Collège de France, Institut des hautes études chinoises, 1993); Engelfriet, "Euclid in China: A Survey"; idem, *Euclid in China: The Genesis of the First Chinese Translation of Euclid's Elements, Books I–VI (Jihe Yuanben, Beijing, 1607) and Its Reception up to 1723* (Boston: Brill, 1998); Horng Wann-Sheng, "The Influence of Euclid's *Elements* on Xu Guangqi and His Successors," in Jami et al., *Statecraft and Intellectual Renewal*, 380–98; Peter Engelfriet and Siu Man-Keung, "Xu Guangqi's Attempts to Integrate Western and Chinese Mathematics," in Jami et al., *Statecraft and Intellectual Renewal*, 279–310.

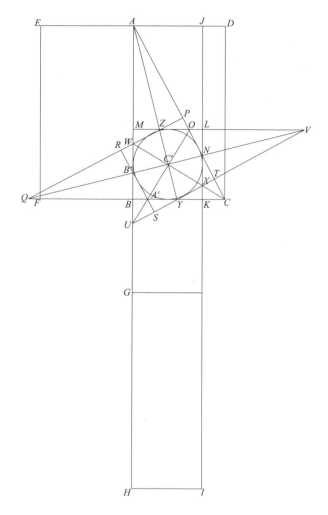

Fig. B.2: The diagram analyzed in problem 7 of Xu's *Right Angles, Meanings*. In the original, the circle inscribed in the triangle is not included (see Figure B.1 on page 282).

rigorous the Chinese theory of right triangles, demonstrating that Xu accepted the deductive logic of Euclid's *Elements*.[16] Peter Engelfriet has analyzed this work in detail, concluding, "What this bewildering 'eclectic' mélange of modes of reasoning shows above all, is that Xu's main objective in writing this treatise was to 'unearth' traditional mathematics."[17] As we will see below, Engelfriet's assessment is in fact too charitable.

---

[16] Qian, *Zhongguo shuxue shi*, 239.

[17] Engelfriet, "Euclid in China: A Survey," 295. He has also analyzed problem 7 in an unpublished manuscript. I wish to thank him for generously sharing his research.

Translation of "Problem 7"

Problem 7.

第七題

Given the shorter side and the longer side of a right triangle, find the [diameter of the] inscribed circle.

勾[18]、股，[19]求容圓。

Method: The longer side $\overline{AB}$[20] is 600, the shorter side $\overline{BC}$ is 320; find the [diameter of the] inscribed circle.[21] Multiply the shorter and longer sides together, obtaining 19200.[22] Double that, obtaining 38400,[23] and let that be the dividend. Then find the hypotenuse from the shorter and longer sides, obtaining 680 for the hypotenuse $\overline{AC}$ [Problem] one of this treatise.[24] Add the shorter side, the longer side, and the hypotenuse, and let this be the divisor. Divide into the dividend, obtaining 240 for the diameter $\overline{BK}$[25] of the inscribed circle.[26]

---

[18] *SKQS* writes *gou* 句 instead of *gou* 勾. Variations will not be noted further below.

[19] The *SKQS* edition is not punctuated. The punctuation in the *XGZYJ* edition is idiosyncratic (that is, it does not follow the usual conventions).

[20] The following substitutions have been used: for *jia* 甲 I have substituted *A*, for *yi* 乙 *B*, *bing* 丙 *C*, *ding* 丁 *D*, *wu* 戊 *E*, *ji* 己 *F*, *geng* 庚 *G*, *xing* 辛 *H*, *ren* 壬 *I*, *gui* 癸 *J*, *zi* 子 *K*, *chou* 丑 *L*, *yin* 寅 *M*, *mao* 卯 *N*, *chen* 辰 *O*, *wu* 午 *P*, *wei* 未 *Q*, *shen* 申 *R*, *you* 酉 *S*, *xu* 戌 *T*, *hai* 亥 *U*, *qian* 乾 *V*, *dui* 兌 *W*, *li* 離 *X*, *zhen* 震 *Y*, *xun* 巽 *Z*, *kan* 坎 *A'*, *gen* 艮 *B'*, *kun* 坤 *C'*. Apparently because of the similarity of *ji* 己 and *si* 巳, the latter character is not used; in transcribing the Chinese I have preserved the original; but in translating into English they are considered identical (i.e. *ji* 己 and *si* 巳 are both translated as *F*). Also, *shu* 戌, and less frequently *wu* 戊, are sometimes substituted for *xu* 戌; in the Chinese I have preserved the originals, but in the English, both are written as *T*. Note that *wu* is also an earthly stem.

[21] This is from Li Ye's 李冶 (1192–1279) *Ce yuan hai jing* 測圓海鏡 [Sea mirror of circle measurement, 1248]: the terminology (e.g., *xian he he* 弦和和) and the dimensions (600, 320, diameter 240, and *xian he he* 1600) are all the same as in the *Ce yuan hai jing*. Note that Euclid's *Elements* does not permit numerical measure of geometrical magnitudes.

[22] Here 19200 should be 192000.

[23] 38400 should be 384000.

[24] That is, the square of the hypotenuse is equal to the sum of the squares of the sides. This elementary formula was also given in most Chinese mathematical treatises.

[25] Note that $\overline{BK}$ is not the diameter of the inscribed circle but the side of a square circumscribed about the circle. Neither the circle nor its diameter is drawn in Xu's diagram. $\overline{BK}$ is the same length as the diameter, but Xu provides no proof.

[26] In this paragraph, Xu uses the *Elements* to support only one statement, namely, the well-known formula to find the hypotenuse of a right triangle, $z^2 = x^2 + y^2$ (where x and y are the two sides and z is the hypotenuse). The formula Xu uses to find the diameter of the circle, $d = 2xy/(x+y+z)$, certainly requires proof, and should be the central point in this problem. Xu provides no justification for this formula. This formula for calculating the diameter of a circle inscribed in a right triangle is the one used in the *Ce yuan hai jing* (see pages 96–98 of this book). It can also be found in the "Gou gu" chapter of one of the earliest extant texts on Chinese mathematics, the *Nine Chapters of Mathematical Arts* (*Jiu zhang suan shu* 九章算術): 術曰：八步為句，十五步為股，為之求弦。三位并之為法，以句乘股，倍之為實。實如法得徑一

法曰：甲乙股，六百。乙丙句，三百二十。求容圓，以句、股相乘[27]
得一萬九千二百，倍之，得三萬八千四白，為實。別以句、股求弦，
得甲丙弦六百八十本篇一，并勾股弦為法，除實，得容圓徑乙子二百四
十。

Justification: $\overline{AB}$ is the longer side [of the right triangle]; $\overline{BC}$ is the shorter
side. Multiply them together; this is [the area of] the rectangle $ABCD$. Dou-
ble it, and let this be the dividend; this is [the area of] the rectangle $CDEF$.[28]
Find the hypotenuse $\overline{AC}$, and combine [the hypotenuse together with] the
shorter and the longer sides, obtaining 1600. Take the line [segment] $\overline{AB}$
and extend it [to form the ray $\overline{AB}$]. Form the segment[29] $\overline{BG}$ equal to the
shorter side, and $\overline{GH}$ equal to the hypotenuse, obtaining $\overline{AH}$. This is a line
[of length equal to] the *xian he he*.[30] Take this as the divisor, and divide the
dividend, obtaining the side $\overline{HI}$ [of length] 240. This forms thc rectangle
$AHIJ$, equal to $CDEF$ Book VI, 16.[31] The side $\overline{IJ}$ intersects the shorter side
$\overline{BC}$ at K. Next, from K construct the rectangle $KLMB$.[32] Then each side [of
this rectangle] is the diameter of the inscribed circle.[33]

論曰：甲乙股、乙丙勾相乘，即甲乙丙丁直角形，倍之，為實。即丙
丁戊巳[34]直角形。求得甲丙弦，并勾、股，得一千六百。於甲乙線引長
之，截乙庚，與句等，庚辛與弦等。得甲辛，為弦和和線，以為法，
除實，得辛壬邊二百四十，即成甲辛壬癸直角形，與丙丁戊巳形等六卷

步。 See Bai Shangshu 白尚恕, *Jiu zhang suanshu zhushi* 《九章算术》注释 [*Nine Chapters
on the Mathematical Arts* with commentary and explanations] (Beijing: Kexue chubanshe 科学
出版社, 1983), 332.

[27] *SKQS* and *XGZYJ* use a common variant of *cheng* 乘.

[28] $\overline{BC}$ is 320, so $\overline{BF}$ is also 320, and thus $\overline{CF}$ is 640, but $\overline{CQ}$ is 680. Xu's diagram shows a
recognition that the points F and Q are not the same.

[29] Here *jie* 截 is used in the sense of cutting or segmenting.

[30] The term *xian he he* 弦和和, which derives from *Ce yuan hai jing*, represents the sum of the
lengths of the three sides of a right triangle. In *Ce yuan hai jing*, this term is not given an explicit
geometric representation.

[31] That is, the area of rectangle $AHIJ$ is $1600 \cdot 240 = 384000 = 640 \cdot 600$, the area of rectangle
$CDEF$. Although VI.16 is not irrelevant, Xu's appeal to it is spurious. VI.16, as translated in
Jihe yuanben 幾何原本, is as follows: 第十六題（二支）：四直線為斷比例，即首尾兩線
矩內直角形與中兩線矩內直角形等。首尾兩線與中兩線兩矩內直角形等，即四線為斷比
例。 *Jihe yuanben, juan* 6 (*XGZYJ* edition). The standard English translation of this proposition:
"If four straight lines be proportional, the rectangle contained by the extremes is equal to the
rectangle contained by the means; and, if the rectangle contained by the extremes be equal to
the rectangle contained by the means, the four straight lines will be proportional." *The Thirteen
Books of Euclid's Elements*, 2nd ed., 3 vols., ed. Sir Thomas L. Heath (New York: Dover, 1956),
2:221. To apply the first part of VI.16, Xu should first show that the sides are proportional (the
second part of VI.16 does show that since the rectangles are equal, the sides are proportional,
that is, 1600 is to 640 as 600 is to 240).

[32] Xu provides no justification. Xu does not mention IV.7, "About a given circle to circumscribe
a square."

[33] In sum, none of the statements in this paragraph are properly supported by propositions in
Euclid's *Elements*.

[34] Reading *ji* 己 for *si* 巳; *SKQS* also writes *si* 巳.

十六。而壬癸邊截乙丙句於子，次從子作子丑寅乙直角方形，即此形之
各邊皆為容圓徑。

Why is [it] called the diameter of the inscribed circle? It is so called because
inside the right triangle ABC a circle is made. Its hypotenuse $\overline{AC}$ is seg-
mented to form $\overline{NO}$ by the rectangle $KLMB$, and, along with all the sides
$\overline{BK}, \overline{KL}, \overline{LM}, \overline{MB}$, all [these] sides are tangent to the circle.[35]

曷名為容圓徑也？謂於甲乙丙三邊直角形內作一圜，其甲乙丙弦截子丑
寅乙直角方形之卯辰線，與乙子、子丑、丑寅、寅乙諸邊，皆為切圓
線也。

Then what is it that makes it evident that these five sides are all tangent
to the circle? Over the triangle[36] ABC, again make a right triangle CPQ,
overlaying (*jiao jia* 交加)[37] [the triangle ABC] on [the diagram],[38] its [side]
$\overline{PC}$ is equal to $\overline{BC}$; $\overline{QP}$ is equal to $\overline{AB}$; and $\overline{QC}$ is equal to $\overline{AC}$; then the two
triangles must be equal this can be deduced from Book I, 22.[39]

則何以顯？此五邊之皆為切圓線乎。試于甲乙丙形上復作一丙午未直
角三邊形，交加其上，其午丙[40]與乙丙等，未午與甲乙等，未丙與甲丙
等，即兩形必等一卷二十[41]二可推。

Next, based on the right angle CPQ, make the rectangle $PRST$. It is equal
to the rectangle $BKLM$. Next extend the line $\overline{TS}$ to U. This again forms
the right triangle ATU. Take A as their common angle. Overlay (*jiao jia*
交加) [this] on the triangle ABC. Then $\overline{PR}$ and $\overline{ST}$ are also [the length of]
the diameter of the inscribed circle. Next extend the two lines $\overline{UT}$ and $\overline{ML}$
to V, again forming the right triangle VMU. Let U be the common angle.
Overlaying (*jiao jia* 交加) [this] on the triangle ABC, $\overline{BK}$ and $\overline{LM}$ are also
[the length of] the diameter of the inscribed circle. Next make the line $\overline{CW}$,

---

[35] Xu does not support any of the statements in this paragraph using the *Elements*.

[36] Here Xu uses *xing* 形 for $\triangle ABC$; since $\triangle ABC$ has already been described as a right triangle
(*san bian zhi jiao xing* 三邊直角形), *xing* is translated as an abbreviated form of *san bian zhi
jiao xing* instead of the less precise term "shape."

[37] I have translated *jiao jia* 交加 as "overlaying" following the sense of the passage.

[38] The general sense of this sentence seems to be that $\triangle CQR$ is placed over $\triangle ABC$.

[39] From the point of view of Euclidean geometry, the question Xu asks is a good one: "what is
it that makes it evident that these five sides are all tangent to the circle?" Xu, however, provides
no answer. The only proposition he cites, I.22, does not answer this question. Nor does I.22
demonstrate that "the two triangles must be equal": *Jihe yuanben, juan* 1, states, 三直線，求
作三角形，其每兩線并，大于一線也。 Euclid, *Elements*, 1:292, states "Out of three straight
lines, which are equal to three given straight lines, to construct a triangle: thus it is necessary
that two of the straight lines taken together in any manner should be greater than the remaining
one." If Xu wished to justify that the two triangles are equal, he should have used proposition I.8,
which states that if two triangles have two sides equal to two sides, respectively, and the bases
are also equal, then the triangles are equal. In sum, none of the statements in this paragraph are
properly supported by Xu using the *Elements*.

[40] *XGZYJ* has *wu bing* 午丙 instead of *bing wu* 丙午.

[41] *XGZYJ* has *nian* 廿 for *er shi* 二十.

meeting the overlaid lines of each of the figures at [the points] X and W.[42]
Next make the line $\overline{AY}$, meeting the overlaid lines of each of the figures at
[the points] Z and Y. Next make the line $\overline{UO}$, meeting the overlaid lines of
each of the figures at [the points] A' and O. Next make the line $\overline{QV}$, meeting
the overlaid lines of each of the figures at [the points] B', and N. These four
lines all meet at C'. Thus $\overline{PC}$ and $\overline{BC}$ are equal.[43]

次依丙午未直角作午申酉戌直角方形，與乙子丑寅直角方形等。次于
戌酉線引之至亥，又成甲戌亥直角三邊形，以甲為同角，交加于甲乙
丙形之上。亦以午申、酉戌為容圓徑。次于亥戌、寅丑兩線引之遇于
乾，又成乾寅亥直角三邊形，以亥為同角，交加于甲乙丙形之上。亦
以乙子、丑寅為容圓徑。次作丙兌線，遇諸形之交加線于離，于兌。
次作甲震線遇諸形之交加線于巽，于震。次作亥辰線，遇諸形之交加
線于坎，于辰。次作未乾線，遇諸形之交加線于艮，于卯。而四線俱
相遇于坤，夫午丙與乙丙兩線等。

Then subtract [from $\overline{PC}$ and $\overline{BC}$ respectively] $\overline{PT}$ and $\overline{BK}$, which are equal
[in length]; then $\overline{TC}$ and $\overline{KC}$ must be equal. $\overline{CX}$ is the common line; CTX
and CKX also are equal as right angles. TXC, KXC are also both acute
angles. Then the two triangles CXT and CXK must be equal; thus all the
sides and all the angles of the two triangles must be equal Book VI, 7.[44]

而減相等之午戌[45]、乙子，即戌丙與子丙必等。丙離同線，丙戌離、丙
子離又等為直角，戌離丙、子離丙又俱小于直角。即丙離戌、丙離子
兩三角形必等，而兩形之各邊各角俱等六卷七。

Then the line $\overline{CW}$ must divide the angle ACQ into two equal parts Book I,
9.[46] And also the two sides $\overline{KX}$ and $\overline{TX}$ must be equal this essay.

則丙兌線必分甲丙未角為兩平分矣一卷九。又子離與戌離，兩邊既等本
論。

The vertical angles (*jiao jiao* 交角) KXY and TXN are also both equal Book
I, 15.[47] NTX, YKX are equal as right angles. Then each [pair] of sides and

---

[42] Roughly, the idea of this sentence is that the line $\overline{CW}$ intersects the various figures that Xu
has constructed at the points X and W.

[43] Xu provides no justification for any of the statements in this paragraph, and in particular, no
proof that "[t]hese four lines all meet at C'."

[44] *Jihe yuanben, juan* 6: 第七題：兩三角形之第一角等，而第二相當角各兩旁之邊比例等，
其第三相當角，或俱小于直角，或俱不小于直角，即兩形為等角形，而對各相似邊之角各
等。 Euclid, *Elements*, 2:206: "If two triangles have one angle equal to one angle, the sides about
other angles proportional, and the remaining angles either both less or both not less than a right
angle, the triangles will be equiangular and will have those angles equal, the sides about which
are proportional." Xu's use of proposition VI.7 here is not technically correct: the triangles are
equal, not proportional.

[45] XGZYJ has *shu* 戌. Variations will not be further noted below.

[46] *Jihe yuanben, juan* 1: 第九題：有直線角，求兩平分者。 Euclid, *Elements*, 1:264: "To bisect
a given rectilineal angle." Xu's use of I.9 here is spurious: the angles are already known to be
equal; I.9 shows how to bisect a given rectilineal angle.

[47] Here Xu's use of the *Elements* to support the equality of the vertical angles is correct. *Jihe
yuanben, juan* 1: 第十五題：凡兩直線相交作四角，每兩交角必等。 Euclid, *Elements*, 1:277:
"If two straight lines cut one another, they make the vertical angles equal to one another."

each [pair] of angles of NXT and XYK are equal, then the two triangles are also equal Book I, 26.[48] Also the two sides $\overline{KX}$ and $\overline{XT}$ are then equal. The two sides $\overline{XN}$ and $\overline{XY}$ are also equal this essay. Then the two sides $\overline{KN}$ and $\overline{TY}$ are also equal.

子離震、戌離卯兩交角又等一卷十五。卯戌離、震子離又等為直角。即卯戌離[49]、離震子之各邊各角俱等，而兩形亦等一卷廿六。又子離與離戌兩邊既等，離卯與離震兩邊又等本論，即子卯與戌震兩邊亦等。

$\overline{KL}$ and $\overline{TS}$ are sides of equal rectangles, and so must be equal. From each subtract the equal $\overline{KN}$ and $\overline{TY}$ respectively, the remaining $\overline{NL}$ and $\overline{YS}$ must be equal. The two angles LNO and $A'YS$ also are each opposite angles of the equal angles XNT and XYK, and must be equal. Again subtract the equal $\overline{KN}$ and $\overline{TY}$, the remaining $\overline{NL}$ and $\overline{YS}$ must be equal. The two angles LNO and $A'YS$ are also each the opposite angles of the equal angles XNT and XYK, and so must be equal. OLN and YSA' are also equal as right angles, then each side and each angle of NLO and YSA' are equal. Then the two triangles are also equal. Each side and each angle of ZMW and $WB'R$ are all equal, and so the two triangles are equal.[50]

子丑與戌酉各為相等之直角方形邊必等。而各減相等之子卯、戌震，其所存卯丑、震酉必等。丑卯辰、坎震酉兩角又各為離卯戌、離震子相等角之交角必等。辰丑卯、震酉坎又等為直角，即卯丑辰、震酉坎之各邊各角俱等，而兩形亦等一卷廿六。依顯午巽辰與坎艮乙之各邊各角俱等，而兩形亦等。巽寅兌與兌艮申之各邊各角俱等，而兩形亦等。

Also the lengths [lit. number] of $\overline{KC}$ and $\overline{TC}$ are both 80. $\overline{BK}$ and $\overline{TP}$ are each 240. Separate each ratio (*lü*) into [whole] numbers to discuss them. Then $\overline{LN}$ and $\overline{SY}$ are each 90. $\overline{LO}$ and $\overline{A'S}$ are each 48. $\overline{NO}$ and $\overline{A'Y}$ are each 102. For the calculations, see the *gou gu bu lü* section of *Ce yuan hai jing*. Then subtracting $\overline{NK}$ from $\overline{LN}$ must be 150. The longer side $\overline{NK}$ is 150, the shorter side $\overline{CK}$ is 80, and with these find the hypotenuse $\overline{NC}$, which is then 170 this treatise, [problem] 1. Next, subtract $\overline{CT}$ 80, namely $\overline{NT}$, which is also 90.[51]

又子丙、戌丙之數各八十。乙子、戌午各二百四十。以諸率分數論之，則丑卯、酉震各九十。丑辰、坎酉各四十八。卯辰、坎震各一百〇二算見測圓海鏡之句股步率。則減丑卯之卯子，必一百五十也。卯子股

---

[48] Here Xu's use of the *Elements* to support his statement is again correct. *Jihe yuanben, juan* 1: 第二十六題（二支）：兩三角形，有相當之兩角等，及相當之一邊等，則餘兩邊必等，餘一角亦等，其一邊，不論在兩角之內及一角之對。 Euclid, *Elements*, 1:301: "If two triangles have the two angles equal to two angles respectively, and one side equal to one side, namely, either the side adjoining the equal angles, or that subtending one of the equal angles, they will also have the remaining sides equal to the remaining sides and the remaining angle to the remaining angle."

[49] XGZYJ has *shu* 戌, and has *mao li xu* 卯離戌 instead of *mao xu li* 卯戌離.

[50] Interpreted charitably, this paragraph essentially follows Euclidean reasoning.

[51] This paragraph is based on *Ce yuan hai jing*, as noted in half-size characters.

一百五十，丙子句八十，以求卯丙弦。則一百七十也本篇一。次減丙戌八十，即卯戌亦九十也。

The [two angles] OLN and XTN of triangles LON, NTX are both right angles and are equal. The opposite angles LNO and TNX are equal, $\overline{LN}$ and $\overline{TN}$ are again equal, thus the two triangles must be equal. Then each of its sides and angles must be equal Book I, 26.[52]

丑辰卯、卯戌離兩三角形之辰丑卯、離戌卯既等為直角。丑卯辰、戊卯離兩交角又等。丑卯與戌卯復等，即兩形必等，而其各邊各角俱等一卷廿六。

Following this it is evident that the two triangles $\overline{KXY}$ and $\overline{YSA'}$ are also equal. Following this it is evident that for each triangle the opposite angles all are mutually equal.[53] The adjacent angles (*lian jiao* 連角) such as SUA', BUA' [sic][54] the two triangles are also equal. Then $\overline{KX}$, $\overline{XT}$ are all 48, then $\overline{SA'}$ and $\overline{A'B}$ are all 48. $\overline{US}$ and $\overline{UB}$ are all 80, $\overline{KB}$ and $\overline{TS}$ are equal. $\overline{KC}$ and $\overline{SU}$ are again equal. Then $\overline{BC}$ and $\overline{TU}$ must be equal.[55]

依顯子離震與震酉坎兩形亦等。依顯諸形之交角者皆相等。其連角，如酉亥坎、乙亥坎兩形亦等。而子離、離戌皆四十八也。則酉坎、坎乙亦皆四十八也。亥酉、亥乙皆八十也。子乙與戌酉等，子丙與酉亥復等，則乙丙與戌亥必等。

Then A is the common angle; ABC and ATU are also equal as right angles. Then each side and each angle of ABC and ATU are all equal, so the two triangles are equal Book I, 26.[56] $\overline{AU}$ and $\overline{AC}$ are then equal. From each subtract the equal $\overline{CT}$ and $\overline{BU}$. Also subtract the mutually equal $\overline{BM}$ and $\overline{VP}$. Then $\overline{AM}$ and $\overline{AP}$ must be equal. Then $\overline{AM}$ and $\overline{AP}$ of the two triangles AZP and AZM must be equal. $\overline{AZ}$ is the common line; APZ, AMZ are equal as right angles, so the two triangles must be equal. Then each side and each angle are all equal Book VI, 7.[57] This line, $\overline{AY}$, must divide the angle CAU into two equal parts Book I, 9.[58]

而甲為同角，甲乙丙、甲戌亥又等為直角，則甲乙丙、甲戌亥之各邊各角俱等一卷廿六。甲亥與甲丙既等，各減相等之丙戌、乙亥，又減相等之乙寅、戌午，即甲寅與甲午必等。夫甲巽午、甲巽寅兩形之甲寅、甲午既等，甲巽同線，甲午巽、甲寅巽又等為直角，

---

[52] Xu's use of I.26 here to assert the equality of the triangles is correct. On I.26, see footnote 48 on the preceding page.

[53] The meaning of Xu's statement here is not clear.

[54] This phrase is incomplete in the Chinese original. It is not clear what Xu intended.

[55] These calculations of the numeric values of the lengths of the sides are not permitted in the *Elements*.

[56] Here Xu's appeal to I.26, which asserts the equality of opposite angles, is spurious. On I.26, see footnote 48 on the preceding page.

[57] See footnote 44 on page 287.

[58] See footnote 46 on page 287.

即兩形必等。而各邊各角俱等六卷七。是甲震線必分丙甲亥角為兩平分也一卷九。 [59]

On the interior of the triangle ABC, divide the angle ACB with the line $\overline{CW}$ into two equal parts. Also divide the angle CAB into two equal parts with the line $\overline{AY}$, then they meet at C', and therefore take C' as the center, and $\overline{AB}$ as the boundary, and construct a circle. Then it must be tangent to the five sides $\overline{BK}$, $\overline{KL}$, $\overline{LM}$, $\overline{MB}$, and $\overline{NO}$. Thus it is the inscribed circle for the right triangle ABC, that is, it is the inscribed circle for each side of the square of B and L [$BKLM$] Book 4, 4. [60]

甲乙丙一形內，既以丙兌線分甲丙乙角為兩平分。又以甲震線分丙甲乙角為兩平分，而相遇于坤，則以坤為心，甲乙為界，作圓，必切乙子、子丑、丑寅、寅乙、卯辰五邊。而為甲乙丙直角三邊形之內切圓。即乙丑直角方形之各邊為容圓徑四卷四。

Further discussion: Each of the lines that divides the angles on the right triangles, each divides the angle into two equal parts. They all intersect at C'; and the circle with C' at the center is the inscribed circle for each of the figures, that is, each of the sides of the square is a diameter of the circle. [61]

展轉論之，則各大直角三邊形內之分角線，皆分本角為兩平分，皆遇于坤。而坤心圓為各形之內切圓，即兩直角方形邊，為各句股形內之容圓徑。

Another method: The long-side $\overline{AB}$ is 600, the short-side $\overline{BC}$ is 320, combining yields 920. Subtract the hypotenuse AC, which is 680, obtaining $\overline{BK}$ 240. [62]

又法曰：甲乙股，六百。乙丙句，三百二十。并得九百二十，與甲丙弦六百八十相減，亦得乙子二百四十。

Discussion: Similar to the previous discussion. The remaining short-sides of the large triangles [the diameter of the circle subtracted from the short-side of the triangle] are all 80. The differences of the combination of the short and long sides, minus the hypotenuse, are all 80. Thus the *chu fen* short-side is 240, and is the diameter of the circle. [63]

論曰：如前論，諸大句股形之分餘句[64]俱八十。諸句[65]股和與諸弦相減之較亦俱八十。則初分句二百四十為諸形之容圓徑。

---

[59] *XGZYJ* has *yi jiu juan* 一九卷.

[60] *Jihe yuanben, juan* 4: 第四題：三角形。求作形內切圓。 Euclid, *Elements*, 2:85: "In a given triangle to inscribe a circle." Although Xu cites proposition IV.4 to inscribe a circle, he does little to justify the other statements in this paragraph.

[61] Xu offers no justification for these statements.

[62] The formula $d = x + y - z$, where again x and y are the two sides and z is the hypotenuse, is also correct. Again, however, Xu provides no justification for this formula. See footnote 26 on page 285.

[63] What Xu means in this paragraph is not clear.

[64] *XGZYJ* has *gou* 句.

[65] *XGZYJ* has *gou* 句.

Appendix C
Xu Guangqi's Writings

This appendix catalogues the writings that have been attributed to Xu Guangqi 徐光啟 (1562–1633). It is based primarily on the following sources:

1. Bibliographies:

   a. Liang Jiamian 梁家勉, *Xu Guangqi nianpu* 徐光啟年譜 [Chronological biography of Xu Guangqi] (Shanghai: Shanghai guji chubanshe 上海古籍出版社, 1981), hereinafter *XGQNP*;
   b. Xu Zongze 徐宗澤, *Ming-Qing jian Yesu huishi yi zhu tiyao* 明清間耶穌會士譯著提要 [Annotated bibliography of the translations and writings of the Jesuits during the Ming and Qing dynasties] (Taibei: Zhonghua shuju 中華書局, 1989);
   c. Standard bibliographies, such as *Zhongguo congshu zonglu* 中國叢書綜錄 [General catalogue of Chinese collectanea], ed. Shanghai tushuguan 上海圖書館 (Shanghai: Shanghai guji chubanshe 上海古籍出版社, 1986).

2. Collections:

   a. *Xu Guangqi ji* 徐光啟集 [Collected works of Xu Guangqi] (Shanghai: Shanghai guji chubanshe 上海古籍出版社, 1984), hereinafter *XGQJ*;
   b. *Xu Guangqi zhu yi ji* 徐光啟著譯集 [Collected writings and translations of Xu Guangqi] edited by Shanghai shi wenwu baoguan weiyuanhui 上海市文物保管委員会 and Gu Tinglong 顧廷龍 (Shanghai: Shanghai guji chubanshe 上海古籍出版社, 1983), hereinafter *XGZYJ*;
   c. *Zengding Xu Wending gong ji* 增訂徐文定公集 [Collected works of Xu Guangqi, revised and enlarged] (Taibei: Taiwan Zhonghua shuju 臺灣中華書局, 1962), hereinafter *ZXWGJ*;
   d. Other collections, including, for example, *Ming jingshi wen bian* 明經世文編 [Collection of Ming writings on statecraft] (Beijing: Zhonghua shuju 中華書局, 1962), hereinafter *MJSWB*.

Each of these sources has its strengths and shortcomings. Liang's *Xu Guangqi nianpu* contains the most complete list of works by Xu, in chronological order.

However, Liang does not provide the collections containing these works; there is no indication which dates are Liang's conjectures, and which are in the prefaces; although Wang's *XGQJ* is listed in the bibliography, over 50 works collected there are not listed in *Xu Guangqi nianpu*. On the other hand, Wang's *XGQJ* contains no comprehensive bibliography of Xu's works; none of his longer works are included; Xu's religious works are minimized; and the writings are listed by category, often without the dates. I have also included works of Xu included in *ZXWGJ*, and other collections; and I have consulted Xu Zongze's bibliography of works in the Vatican, Xujiahui, and Paris. For future research, a particularly important resource is the *Chinese Christian Texts Database* (*CCT-Database*), an ongoing project initiated by Erik Zürcher and now continued by Ad Dudink and Nicolas Standaert, which lists both primary and secondary sources, and is available at http://www.arts.kuleuven.be/sinology/cct.[1]

For extant texts, I note the collection, e.g., "In *XGQJ*"; if a work is only listed in a bibliography, I note the bibliography in parentheses, e.g., "(*XGQNP*)"; if a bibliography states that a work is no longer extant, I include that information in parentheses, e.g., "(*XGQNP*, states no longer extant)." Full bibliographic information for the abbreviations used in this appendix for collections and bibliographies (for example, "*XGQJ*," "*XGQNP*," and "*ZXWGJ*" in the first entries below) can be found on pages 305–308 of the Bibliography.

For dates of these treatises I have followed Liang's ordering; it should be noted that in many cases Liang's dates may be speculative. Where dates are available in a preface, I have noted this in brackets. In cases where no date is given in the preface, Liang omitted the treatise, or no conventional date is known, I have tentatively assigned a date followed by a question mark, to indicate that it is my own estimate. Transliterations into pinyin are in some cases speculative.

Finally, this list should not be considered complete, and it should be noted that some of the works listed here that have been attributed to Xu were likely not authored by Xu himself. In this book I have not attempted to address the difficult problem of authorship, but I consider it likely that the memorials, prefaces, and letters attributed to Xu were most likely written by Xu himself; some of the other works, however, may have been attributed to him because of his status.[2]

Wanli Reign (1573–1620)

1597. "Shun zhi ju shenshan zhizhong" 舜之居深山之中. In *XGQJ*.

---

[1] I thank Nicolas Standaert for bringing this project to my attention.

[2] On the authorship of several texts attributed to Xu, see Ad Dudink, "The Image of Xu Guangqi as Author of Christian Texts," in *Statecraft and Intellectual Renewal in Late Ming China: The Cross-Cultural Synthesis of Xu Guangqi (1562–1633)*, ed. Catherine Jami, Gregory Blue, and Peter M. Engelfriet (Leiden: Brill, 2001), 99–154. See also footnote 2 on page 195 of this book.

[1597, see 1628]. "Zi yue: Tingsong wu you ren ye" 子曰：聽訟吾猶人也: This entry is incorrectly dated in Liang's bibliography in *XGQNP* as 1597.[3] See "Jingyan jiangyi" 經筵講義, in *XGQJ*, listed below under the date 1628.[4]

1599? "Yu Jiao laoshi" 與焦老師. In *XGQJ*, *ZXWGJ*.

1599?[5] "Yu Hai weng fuzi shu" 與海翁夫子書. In *XGQJ*, *ZXWGJ*.

1603–. "*Mao Shi* liutie jiang yi" 毛詩六帖講意. In *XGZYJ*.

Works dated 1603 or before, no longer extant

1603–. "Yuanyuan tang shi yi" 淵源堂詩藝. (*XGQNP*, states no longer extant).

1603–. "Fangrui tang shu yi" 芳蕤[6]堂書藝. (*XGQNP*, states no longer extant).

1603–. "Sishu can tong" 四書參同. (*XGQNP*, states no longer extant).

1603–. "Zi shi zhai" 子史摘. (*XGQNP*, states no longer extant).

1603–. "Fangyan zhuanzhu" 方言轉注. (*XGQNP*, states no longer extant).

1603–. "Yu lei" 語類. (*XGQNP*, states no longer extant).

1603–. "Shu shu zheng" 塾書政. (*XGQNP*, states no longer extant).

1603–. "Ershi si ze gu" 二十四則古. (*XGQNP*, states no longer extant).

1603–. "Dushu suan" 讀書算. (*XGQNP*, states no longer extant).

1603–. "Fu you" 賦囿. (*XGQNP*, states no longer extant).

1603–. "Zhi hui" 制彙. (*XGQNP*, states no longer extant).

1603–. "Shufa ji" 書法集. (*XGQNP*, states no longer extant).

1603–. "Caoshu lei" 草書類. (*XGQNP*, states no longer extant).

Works Completed before Xu Passed the Metropolitan Examination

1603 [Guimao 癸卯]. "Liangsuan hegong ji ceyan dishi fa" 量算河工及測驗地勢法. In *XGQJ*.

1604 [Wanli 32, summer solstice] "Ba *Ershi wu yan*" 跋二十五言. In *XGQJ*, *ZXWGJ*.

1604. "Ke zui qi de zhen zan" 克罪七德箴贊. In *ZXWGJ*.

---

[3] Liang lists this title and the date as 1597, but he notes that he has not seen this work. I suspect that this date is nothing more than a guess based on the assumption that this would have been an examination essay. Wang Zhongmin's *XGQJ* contains an essay titled "Jing yan jiang yi" 經筵講義, which begins with this same phrase, "Zi yue: Ting song wu you ren ye" 子曰：聽訟吾猶人也 (2:520–22). This essay is a lecture; it is not written in the style of an examination essay.

[4] Wang states that this essay was copied from *Xu Shi zong pu*, chapter four 徐氏宗譜卷四, and the original title was "Rijiang guan Xu Guangqi" 日講官徐光啟; in this case the passage was written at some time after Xu had passed the metropolitan examination in 1604. Liang lists a treatise no longer extant titled "Jing wei jiangyi" 經闈講義, which he dates at 1628; I therefore have tentatively assigned this essay the date 1628.

[5] Liang's bibliography places the date of this letter at 1604. However, in the letter, Xu suggests he has not passed the metropolitan examination: 有愧起居之未修 (XGQJ, 2:503). I have dated this tentatively as probably about the same time as his letter to Jiao Hong.

[6] Substituting for a variant form.

1604?[7] "Yesu xiang zan" 耶穌像讚. In *ZXWGJ*.

1604? "Shengmu xiang zan" 聖母像讚. In *ZXWGJ*.

1604? "Zhengdao ti gang" 正道題綱. In *ZXWGJ*.

1604? "Guijie zhen zan" 規誡箴贊. In *ZXWGJ*.

1604? "Shijie zhen zan" 十誡箴贊. In *ZXWGJ*.

1604? "Zhen fu ba duan zhen zan" 真福八端箴贊. In *ZXWGJ*.

1604? "Aijin shisi duan zhen zan" 哀矜十四端箴贊. In *ZXWGJ*.

Hanlin Academy Exercises, 1604

1604 [Wanli 32]. "Ni shang an bian yu lu shu" 擬上安邊禦虜疏. In *XGQJ*.

[1604, see 1599]. "Yu Hai weng fuzi shu" 與海翁夫子書.

1605. "Ni Han Wudi ba Tian Luntai zhao" 擬漢武帝罷田輪臺詔. In *XGQJ*.

1605. "Han Wendi zhu Bo Zhao, huo yiwei renhou zhong you shen wu; Tian Shu shao Liang yu ci, huo yiwei shan; chu ren muzi xiongdi zhijian, er shi kuan yan deshi heru dui" 漢文帝誅薄昭或以為仁厚中有神武；田叔燒梁獄詞或以為善處人母子兄弟之間二事寬嚴得失何如對. In *XGQJ*.

1605. "Zhengzhi zhonghou bian" 正直忠厚辯. In *XGQJ*.

1605. "Shengmu wan shou song" 聖母萬壽頌. In *XGQJ*.

1605. "Guo Fenyang daren song" 郭汾陽大人頌. In *XGQJ*.

1605. "Ni Dongfang Shuo Chen Taijie liu fu zou" 擬東方朔陳泰階六符奏. In *XGQJ*.

1605. "Xindu Yang Yongjia Zhang er wen zhong gong zan" 新都楊永嘉張二文忠公贊. In *XGQJ*.

1605. "Chizi zhi xin yu shengren zhi xin ruo he jie" 赤子之心與聖人之心若何解.[8] In *XGQJ*.

1605. "Ke Ziyang Zhuzi *Quanji* xu" 刻紫陽朱子全集序. In *XGQJ*.

1605. "Junchen jiao jing zhen" 君臣交儆箴. In *XGQJ*.

1605. "Wei zhi zi wo zhe dang rushi lun" 為之自我者當如是論. (*XGQNP*).

1605. "Ni huan ju dian ji chaomen gongcheng shu" 擬緩舉殿及朝門工程疏. In *XGQJ*.

1605. "Yu youren bian ya su shu" 與友人辯雅俗書. In *XGQJ*.

1605. "Ti sui han songbo tu" 題歲寒松柏圖. In *XGQJ, ZXWGJ*.

1605. "Fu de yu hu bing" 賦得玉壺冰. In *XGQJ, ZXWGJ*.

1605. "Ti Tao shi xing yun pi tu ge" 題陶士行運甓圖歌. In *XGQJ, ZXWGJ*.

1605. "Biansai kuhan yin" 邊塞苦寒吟. In *XGQJ, ZXWGJ*.

1605. "Yu ji wang Xishan" 雨霽望西山. In *XGQJ, ZXWGJ*.

1605. "Fu de cao se yao kan jin ruo wu" 賦得草色遙看近若無. In *XGQJ, ZXWGJ*.

---

[7] Liang dates "Ke zui qi de zhen zan" 克罪七德箴贊 at 1604, but offers no evidence for this date. I have tentatively dated the others in a series of odes (*zan*) at the beginning of chapter one of *ZXWGJ* with "Ke zui qi de zhen zan" at the same date, 1604. The style seems to me to be that of an earlier work.

[8] Wang Zhongmin states that this is from the Ming block-print 刻本 edition of "Jia chen ke Hanlin guan ke juan ba yi lu" 甲辰科翰林館課卷八迻錄 (511 n. 1).

1605. "Qu shuiliu shang" 曲水流觴. In *XGQJ*.

1605. "Shang yuan ting xin ying" 上苑聽新鶯. In *XGQJ*.

1605. "Nan jiao pei si you shu" 南郊陪祀有述. In *XGQJ*.

1605. "Bei jiao pei si" 北郊陪祀. In *XGQJ*.

1605. "Wen Chu bian you gan" 聞楚變有感. In *XGQJ*.

1605. "Yue *Songshi* jianmen Zheng Xia shangshu min tu you gan" 閱宋史監門鄭俠上疏民圖有感. In *XGQJ*.

1605. "Jiu ri lian fang ju" 九日憐芳菊. In *XGQJ*.

1605. "Jiachen Hanlin guan ke" 甲辰翰林館課. (*XGQNP*).

1605. "Kao yi ji jie" 考一記解. (*XGQNP*).

1605. "Ji li gu che tu jie" 記里鼓車圖解. (*XGQNP*).

1605.[9] "Cao he yi" 漕河議. In *XGQJ, ZXWGJ*.

1605. "Tong cao lei bian " 通漕類編. (*XGQNP*).

1605. "Tong cao kao ping" 通漕考評. (*XGQNP*).

1605. "Cao he ping zheng" 漕河評正. (*XGQNP*).

1605. "Chuzhi zong lu chahe bian xiang yi" 處置宗祿查核邊餉議. In *XGQJ, ZXWGJ*.

Works after the Hanlin Academy

1605.[10] "*Shan hai yu di tu* jing jie" 山海輿地圖經解. Also titled "Ti *Wanguo er yuan tu* xu " 題萬國二圜圖序.[11] In *XGQJ*.

1607 [Wanli 34/08/10]. "Jia shu, yi" 家書一. In *XGQJ*.

1607. *Jihe yuanben* 幾何原本. In SKQS, others. "*Jihe yuanben* xu" 幾何原本序, in *ZXWGJ*.

1607? [n.d.]. "Ke *Jihe yuanben* xu" 刻幾何原本序. In *XGQJ; XGZYJ*.

1607. *Celiang fa yi* 測量法義. In *SKQS*.

1607. "Xuan lian lun" 選練論. (*XGQNP*).

1607 [Wanli 35/01 or 02?].[12] "Jia shu, er" 家書二. In *XGQJ*.

1608. "Ti *Celiang fa yi*" 題測量法義. In *XGQJ, ZXWGJ*.

1608. *Celiang yi tong* 測量異同一卷. In *CJCB, XGZYJ, SKQS, TXCH*.

1608. *Ganshu shu* 甘藷疏. (*XGQNP*).

1608 [n.d.]. "*Ganshu shu* xu" 甘藷疏序. In *XGQJ*.

1608. "Fu gongduan Quan zuoshi" 復宮端全座師. (*XGQNP*).

1609. "Yu zi ru xiansheng xiang zan" 俞子如先生像贊. In *XGQJ, ZXWGJ*.

1609. "Wujing shu" 蕪菁疏. (*XGQNP*).

1609. *Gougu yi* 勾股義. In *SKQS*. "*Gougu yi* xu" 勾股義序, in *XGQJ, ZXWGJ*.

1611. "Jiao shi *Danyuan xuji* xu" 焦氏澹園續集序. In *XGQJ, ZXWGJ*.

1611. "Ba *Jihe yuanben*" 跋幾何原本. (*XGQNP*).

---

[9] Wang Zhongmin dates this at 1607 (1:36).

[10] Wang Zhongmin dates this at 1604 or 1605 (1:18).

[11] Liang states these are the same title, 232.

[12] Wang, *XGQJ*, 2:482.

1611 [Xinhai, fall 辛亥秋月]. *Jian ping yi shuo* 簡平儀說. In *CJCB*, *XGZYJ*, *SKQS*,
    TXCH. "*Jian ping yi shuo* xu" 簡平儀說序, in *XGQJ*, *ZXWGJ*.

1611. "Ping hun tu shuo" 平渾圖說. (*XGQNP*).

1611. "Rigui tu shuo" 日晷圖說. (*XGQNP*).

1611. "Yegui tu shuo" 夜晷圖說. (*XGQNP*).

1612 [Wanli 39, spring?].[13] "Jia shu, san" 家書三. In *XGQJ*.

1612 [Wanli renzi, spring 萬曆壬子春月]. *Taixi shui fa* 泰西水法, transcribed (*bi
    shu* 筆述) by Xu. In *XGZYJ*, *SKQS*, *TXCH*. "*Taixi shui fa* xu" 泰西水法序. In
    XGQJ, *ZXWGJ*, *XGZYJ*, *SKQS*, *TXCH*.

1612. "Nong yi za shu" 農遺雜疏. (*XGQNP*).

1612. "Zhi qinjia" 致親家. (*XGQNP*).

1612 [Wanli 40?].[14] "Jia shu, si" 家書四. In *XGQJ*.

1613 [Wanli 41?].[15] "Jia shu, wu" 家書五. In *XGQJ*.

1614. "Ke *Tong wen suan zhi* xu" 刻同文算指序. In *XGQJ*, *ZXWGJ*.

1614 [Wanli 41]. "Jia shu, liu" 家書六. In *XGQJ*.

1614 [Wanli 41/08?].[16] "Jia shu, qi" 家書七. In *XGQJ*.

1614. "Yi ken ming" 宜墾命. (*XGQNP*).

1614. "Bei geng lu" 北耕錄. (*XGQNP*).

1615. "Bi wang" 闢妄. (*XGQNP*). Extant. Published by Shanghai Tushanwan yin-
    shuguan 上海土山灣印書館發行, 1931.

1615. "Zou zi ou bian" 諏諮偶編. (*XGQNP*).

1615. "Ni fu zhu chuang shuo" 擬復竹窗說. (*XGQNP*).

1615. "Yi fang kao" 醫方考. (*XGQNP*).

1615. "Zhong mianhua fa" 種棉花法. (*XGQNP*).

1616 [Wanli 44/07]. "Bian xue zhang shu" 辨學章疏. In *XGQJ*, *ZXWGJ*, *TDCWX*.[17]

1617. "Fen yong guize" 糞壅規則. (*XGQNP*).

1618. "Haifang yu shuo" 海防迂說. In *XGQJ*, *ZXWGJ*.

1618. "Haifang kao ping" 海防考評. (*XGQNP*).

1618. "Fu Jiao zuoshi" 復焦座師. (*XGQNP*).

1618 [Wu wu 午戊]. "Fu Lü Yixuan zhongcheng" 復呂益軒中丞. In *XGQJ*.

1618 [Wu wu 午戊]. "Fu Qian yourong" 復錢游戎. In *XGQJ*.

1618 [Wanli 46]. "Fu taishi Jiao zuoshi" 復太史焦座師. In *XGQJ*.

1619 [Ji wei 己未]. "Fu Qian yourong" 復錢游戎. In *XGQJ*.

1619. "Fu Zhuang yourong" 復莊游戎. (*XGQNP*).

1619. "Jia shu" 家書. (*XGQNP*).

1619 [Wanli 47/03/20]. "Fu chen mo yi yi tian xiong qiu shu" 敷陳末議以殄兇酋
    疏. In *XGQJ*.

1619 [Wanli 47/04/05]. "Bing fei xuan lian jue nan zhan shou shu" 兵非選練決難
    戰守疏. In *XGQJ*.

---

[13] Wang, XGQJ, 2:483.

[14] Wang, XGQJ, 2:484.

[15] Wang, XGQJ, 2:484.

[16] Wang, XGQJ, 2:489.

[17] The version in *TDCWX* differs from the version in *XGQJ*.

1619 [Jiwei 己未 04]. "Fu Wang Xiaolian" 復王孝廉. In *XGQJ*.

1619 [Jiwei 己未]. "Fu Xiong Zhigang jing lüe" 復熊芝岡經略. In *XGQJ*.

1619 [Wanli 47/06/28]. "Liao zuo dian wei yi shen shu" 遼左阽危已甚疏. In *XGQJ*.

1619 [Wanli 47/09/15]. "Gong cheng xin ming jin chen ji qie shi zhi shu" 恭承新命謹陳急切事直疏. In *XGQJ*.

1619 [Wanli 47/09/25]. "Bing shi bai bu xiangying shu" 兵事百不相應疏. In *XGQJ*.

1619. "Xuan lian bai zi kuo" 選練百字括. (*XGQNP*).

1619. "Xuan lian tiao ge" 選練條格. (*XGQNP*).

1619. "Lian yi tiao ge" 練藝條格. (*XGQNP*).

1619. "Shu wu tiao ge" 束伍條格. (*XGQNP*).

1619. "Xing ming tiao ge" 形名條格. (*XGQNP*).

1619 [Ji wei 己未]. "Fu Yuan xianshi weiyu" 復袁憲使位宇. In *XGQJ*.

1619 [Wanli 47/10/05]. "Shishi ji po ji qun shu" 時事極迫極窘疏. In *XGQJ*.

1619 [Ji wei 己未]. "Fu Huang xianfu gucheng xiansheng" 復黃憲副穀城先生. In *XGQJ*.

1619 [Wanli 47]. "Fu taishi Jiao zuoshi" 復太史焦座師. In *XGQJ*.

1619. "Bing ji yaolüe" 兵機要略. (*XGQNP*).

1619. "Huo gong yaolüe" 火攻要略. (*XGQNP*).

1619. "Lu qing di yi" 虜情第一. (*XGQNP*).

1619.[18] "Da zheng di er" 大征第二. In *XGQJ*, *ZXWGJ*.

1619.[19] "Qi sheng di san" 器勝第三. Also titled "Qi sheng ce" 器勝策, in *XGQJ*, *ZXWGJ*.

1619. "Fu rong di si" 服戎第四. Also titled "Fu rong ce" 服戎策, in *XGQJ*, *ZXWGJ*.

1619. "Bian bei di wu" 邊備第五. (*XGQNP*, states not seen).

1619. "Jin lü di liu" 禁旅第六. (*XGQNP*, states not seen).

1619. "Yong ren di qi" 用人第七. (*XGQNP*, states not seen).

1619. "Cai ji di ba" 財計第八. (*XGQNP*, states not seen).

1619. "Ying tian di jiu" 營田第九. (*XGQNP*, states not seen).

1620 [Wanli 47/11/19]. "Pouxi shili reng qi ba chi shu" 剖析事理仍祈罷斥疏. In *XGQJ*.

1620 [Wanli 48/04/01]. "Dong shi jingji lianxi fangyu shu" 東事警急練習防禦疏. In *XGQJ*.

1620. "Yu Mao Zhisheng" 與茅止生. (*XGQNP*).

1620. "Yu Lu Boshun" 與鹿伯順. (*XGQNP*). *XGQJ* contains 致鹿善繼簡.

---

[18] Liang's bibliography lists the title "Da zheng di er" 大征第二, noting that another title is "Da zheng ce" 大征策 and assigns the date 1619.

[19] Similarly, Liang's bibliography lists "Qi sheng di san" 器勝第三, noting that another title is "Qi sheng ce" 器勝策 and giving 1619 as the date. This suggests that "Da zheng ce" and "Qi sheng ce" were a series of essays, possibly written at the Hanlin Academy.

Taichang Reign (1620)

1620 [Taichang 01/08/20]. "Tong yu shi yi shu" 統馭事宜疏. In *XGQJ*.

1620 [Taichang 01/10/16]. "Xun li yi zhou shi chen shi shi bing qing shu" 巡歷已周實陳事勢兵情疏. In *XGQJ*.

1620 [Taichang 01/11/10]. "Zhuo chu min bing shi yi shu" 酌處民兵事宜疏. In *XGQJ*.

1620 [Tiachang 01/11/15]. "Xun li kong ci shu" 巡歷控辭疏. In *XGQJ*.

1621 [Taichang 01/12/11]. "Jian bing jiang jun gou ji qi xu shu" 簡兵將竣遘疾乞休疏. In *XGQJ*.

Tianqi Reign (1621–1627)

1621 [Tianqi 01/01/21]. "Jian bing shi jun shu" 簡兵事竣疏. In *XGQJ*.

1621 [Tianqi 01 天啟元年]. Li Zhizao 李之藻, "Zhi sheng wu xu xi chong qi chi su qu shu" 制勝務須西銃乞敕[20]速取疏, in *ZXWGJ*.[21]

1621 [Tianqi 01/05/01]. "Cui Jingrong deng zou wei sheng zhi wu xu xi chong jing shu gou mu shi mo shu" 崔景榮等奏為勝制務須西銃敬述購募始末疏.[22] In *ZXWGJ*.

1621 [Tianqi 01/02/25]. "Xie Huang shang shu" 謝皇賞疏. In *XGQJ*.

1621 [Tianqi 01/02/27]. "Jin chen ren nei shi li shu" 謹陳任內事理疏. In *XGQJ*.

1621 [Xinyou 辛酉 03]. "Yu Li Wocun Taipu" 與李我存太僕. In *XGQJ*.

1621 [Xinyou 05] "Yu Li Wocun Taipu" 與李我存太僕. In *XGQJ*.

1621. "Zhi qin jia" 致親家. (*XGQNP*).

1621 [Tianqi 01/04/01]. "Zhi mou tong nian shu" 致某同年書. In *XGQJ*.

1621 [Tianqi 01/04/26]. "Jin shen yi de yi bao wan quan shu" 謹申一得以保萬全疏. In *XGQJ*.

1621. [Xinyou]. "Yu Dasitu Li Mengbai" 與大司徒李孟白. In *XGQJ*.

1621. [Xinyou]. "Fu Dasima Zhang zuoshi" 復大司馬張座師. In *XGQJ*.

1621 [Tianqi 01/05/09]. "Shen ming chu yi lu cheng yuan shu shu" 申明初意錄呈原疏疏. In *XGQJ*.

1621 [Tianqi 01/05/09]. "Tai chong shi yi shu" 臺銃事宜疏. In *XGQJ*.

1621 [Tianqi 01/05/12]. "Yang cheng en ming liang li zhi nan shu" 仰承恩命量力知難疏. In *XGQJ*.

1621 [Tianqi 01/05/15]. "Fu guan fei fen shu" 服官非分疏. In *XGQJ*.

1621 [Xinyou 07]. "Yu Zhou Ziyi ji jian" 與周子儀給諫. In *XGQJ*.

---

[20] Substituted for a variant.

[21] *ZXWGJ* includes this memorial written by Li Zhizao; it is not clear if this memorial is included because the editor believed that Xu contributed to its composition.

[22] This memorial states that it is written by "Cui Jingrong and others." Near the end, the memorial notes in support that "[Xu Guangqi's] methods are also transmitted from the West" 其法亦自西洋傳來. It is not clear whether Xu Guangqi was one of the authors. See the preceding footnote, on the memorial authored by Li Zhizao included in this collection.

1621 [Xinyou 08]. "Yu Zhou Ziyi ji jian" 與周子儀給諫. In *XGQJ*.

1621 [Xinyou]. "Fu lin xian yin zhu ge dan ming" 復臨縣尹諸葛澹明. In *XGQJ*.

1621 [Xinyou]. "Yu Hu ji reng bi bu" 與胡季仍比部. In *XGQJ*.

1621 [Xinyou 07]. "Yu Yang Qiyuan jing zhao" 與楊淇園京兆. In *XGQJ*.

1621 [Xinyou 08]. "Yu Wang Taimeng Dasikong" 與王泰蒙大司空. In *XGQJ*.

1621. "Yu Mao Zhisheng" 與茅止生. (*XGQNP*).

1621 [Tianqi 01/07]. "Lüe chen tai chong shi yi bing shen yu jian shu" 略陳臺銃
事宜并申愚見疏. In *XGQJ*.

1621. "Yangming xiansheng pi *Wu jing* xu" 陽明先生批《武經》序.[23] In *XGQJ*,
ZXWGJ.

1622 [Renxu 壬戌]. "Yu Li Wocun Taipu" 與李我存太僕. In *XGQJ*.

1622 [Renxu]. "Yu Wu Shengbai Fangbo" 與吳生白方伯. In *XGQJ*.

1623. "Duan wei zou cao" 端闈奏草. (*XGQNP*).

1624 [Jiazi 甲子]. "Fu Zhou Wuyi xue xian" 復周無逸學憲. In *XGQJ*.

1624. "Zhang Quan zhuan" 張銓傳. (*XGQNP*).

1624 [Jiazi]. "Fu Zhang Shen zhi si li" 復張深之司隸. In *XGQJ*.

1624. "Ling yan li shao" 靈言蠡勺二卷. In *TXCH*.

1624 [Jiazi]. "Yu Lü gong yuan bi bu" 與呂公原比部. In *XGQJ*.

1624. "Xian zu shi lüe" 先祖事略. In *XGQJ*, *ZXWGJ*.

1624. "Xian zu bi shi lüe" 先祖妣事略. In *XGQJ*, *ZXWGJ*.

1624. "Xian kao shi lüe" 先考事略. In *XGQJ*, *ZXWGJ*.

1624. "Xian bi shi lüe" 先妣事略. In *XGQJ*, *ZXWGJ*.

1624. "Wu fu ren shi lüe" 吳夫人事略. (*XGQNP*).

1625 [Yichou 乙丑]. "Yu Wang Wujin Duanyin" 與王無近端尹. In *XGQJ*.

1625. "Shi zhi zhai gao xu" 適志齋稿序. In *XGQJ*.

1625. "Shu bian" 疏辨. (*XGQNP*).

1625 [Yichou]. "Yu Li Junxu zhushi" 與李君敘柱史. In *XGQJ*.

1627 [Dingmao]. "Fu Su Borun zhushi" 復蘇伯潤柱史. In *XGQJ*.

1627. *Nong shu* 農書. (*XGQNP*).

1627. *Pao yan* 庖言. (*XGQNP*).

Chongzhen Reign (1628–1644)

1628 [Chongzhen 01/09/02]. "Jing chen jiangyan shiyi yi bi shengxue zhengshi
shu" 敬陳講筵事宜以裨聖學政事疏. In *XGQJ*, *ZXWGJ*.

1628. "Jing wei jiang yi" 經闈講義. In *XGQJ*.

1629 [Chongzhen 01]. "Zi chen bu zhi qi si ba chi shu" 自陳不職乞賜罷斥疏. In
XGQJ, *ZXWGJ*.

1629 [Chongzhen 01]. "Zai li xue cheng bian ming yuan wu shu" 再瀝血誠辨明
冤誣疏. In *XGQJ*, *ZXWGJ*.

1629. "Xian shi qiang bing shu" 先事強兵疏. (*XGQNP*, states no longer extant).

---

[23] Wang Zhongming states that this preface is copied from 武經七書, preserved in the Dongbei
Library.

1629. "Chong xiu Tianjin Weixue gong ji" 重修天津衛學宮記. In *XGQJ*.

1629 [Chongzhen 02/05/03]. "Nei ge ti fu qin tian jian tui suan ri shi qian hou ke shu bu dui shu" 內閣題覆欽天監推算日食前後刻數不對疏. In *ZXWGJ*.

1629 [Chongzhen 02/05/10]. "Li bu wei rishi ke shu bu dui qing chi bu xiugai shu" 禮部為日食刻數不對請敕部修改疏. In *XGQJ, ZXWGJ*.

1629 [Chongzhen 02/07/11]. "Li bu wei feng zhi xiugai lifa kai lie shi yi qi cai shu" 禮部為奉旨修改曆法開列事宜乞裁疏. In *XGQJ, ZXWGJ*.

1629 [Chongzhen 02/07/21]. "Li bu ti qing xiugai lifa chi shu guan fang shu" 禮部題請修改曆法敕書關防疏. In *XGQJ*. Titled "Li bu ti wei feng zhi xiu gai li fa kai lie shi yi qi cai shu" 禮部題為奉旨修改曆法開列事宜乞裁疏, in *ZXWGJ*.

1629 [Chongzhen 02/07/26]. "Tiao yi lifa xiu zheng sui cha shu" 條議曆法修正歲差疏. In *XGQJ, ZXWGJ*.

1629. "Bao gao ce hou rishi shu" 報告測候日食疏. (*XGQNP*, states no longer extant).

1629 [Chongzhen 02/09/13]. "Yu du ling gai xiu li fa" 諭督領改修曆法. In *ZXWGJ*.

1629 [Chongzhen 02/09/23]. "Feng zhi xiugai lifa kai lie shi yi qi cai shu" 奉旨修改曆法開列事宜乞裁疏. In *XGQJ, ZXWGJ*.

1629 [Chongzhen 02/11]. *Shou cheng tiao yi* 守城條議. In *XGQJ*.

1630 [Chongzhen 02/12/09]. "Zai chen yi de yi pi miao sheng shu" 再陳一得以裨廟勝疏. In *XGQJ, ZXWGJ*.

1630 [Chongzhen 02/12]. "Kong chen ying chong shi yi shu" 控陳迎銃事宜疏. In *XGQJ, ZXWGJ*.

1630 [Chongzhen 02/12/22]. "Po lu zhi ce shen jin shen yi shu" 破虜之策甚近甚易疏. In *XGQJ, ZXWGJ*.

n.d. "Ji Chongzhen er nian shiyi yue chu si ri ping tai zhao dui shi" 記崇禎二年十一月初四日平臺召對事. In *XGQJ, ZXWGJ*.

n.d. "Ji Chongzhen er nian shiyi yue ershi ba ri ping tai zhao dui shi" 記崇禎二年十一月二十八日平臺召對事. In *XGQJ, ZXWGJ*.

1630 [Chongzhen 03/01/02]. "Chou lu zhan dong chou miu yi ji jin shu chu yan yi bei zhan shou shu" 醜虜暫東綢繆宜亟謹述初[24]言以備戰守疏. In *XGQJ, ZXWGJ*.

1630 [Chongzhen 03/01/22]. "Xi yang shen qi ji jian qi yi yi jin qi yong shu" 西洋神器既見其益宜盡其用疏. In *XGQJ, ZXWGJ*.

1630 [Chongzhen 03/02/11]. "Gong bao jiao yan riqi shu" 恭報教演日期疏. In *XGQJ, ZXWGJ*.

1630 [Chongzhen 03/03]. "Yao ju shi huo shu" 藥局失火疏. In *XGQJ*.

1630 [Chongzhen 03/04/02]. "Zhen chen zou qiu zhi chong jin ju shi zhang shu" 鎮臣驟求製銃謹據職掌疏. In *XGQJ, ZXWGJ*.

1630 [Chongzhen 03]. "Wen feng fen ji zhi xian chu rao shu" 聞風憤激直獻芻蕘疏. In *XGQJ, ZXWGJ*.

1630 [Chongzhen 03/05]. "Yi bing bu zhao hui" 移兵部照會. In *XGQJ*.

---

[24] ZXWGJ has *shu* 數.

1630 [Chongzhen 03/05/16]. "Xiugai lifa qing fang yong Tang Ruowang, Luo Yagu
shu" 修改曆法請訪用湯若望羅雅谷疏. In *XGQJ*. Titled "Xiugai lifa qing fang
yong Tang Ruowang shu" 修改曆法請訪用湯若望疏, in *ZXWGJ*.

1630 [Chongzhen 03/06/09]. "Qin feng ming zhi tiao hua dun tian shu" 欽奉明
旨條畫屯田疏 ["Ken tian di yi" 墾田第一, "Yong shui di er" 用水第二, "Chu
huang di san" 除蝗第三, "Jin si yan di si" 禁私鹽第四, "Shai yan di wu" 曬鹽
第五]. In *XGQJ*, *ZXWGJ*.

1630. "Yan dun yan shi yi shu" 言屯鹽事宜疏. (*XGQNP*).

1630. "Feng zhi tiao hua dun yan shu" 奉旨條畫屯鹽疏. (*XGQNP*).

1630. "Xiugai lifa shu" 修改曆法疏. In *ZXWGJ*.

1630 [Chongzhen 03]. "Qin feng ming zhi jin chen yu jian shu" 欽奉明旨謹陳愚
見疏. In *XGQJ*, *ZXWGJ*.

1630 [Chongzhen 03/07/02?]. "Xiugai lifa yuan chen Luo Yagu dao jing shu" 修改
曆法遠臣羅雅谷到京疏. In *XGQJ*.

1630 [Chongzhen 03/09]. "Qin feng sheng zhi fu zou shu" 欽奉聖旨復奏疏. In
XGQJ, *ZXWGJ*.

1630. "Ti wei yue shi shi yi shu" 題為月食事宜疏. (*XGQNP*).

1630 [Chongzhen 03/09/20]. "Feng ming xiu li yin shi zhan chuo jin lüe chen shi
xu yi ming zhi shou shu" 奉命修曆因事暫輟謹略陳事緒以明職守疏 [*XGQJ*
contains 修曆因事暫輟略陳事緒疏]. (*XGQNP*).

1630 [Chongzhen 03/09/20]. "Tuisuan yueshi qi fu fangwei bing ju tuxiang shu"
推算月食起復方位並具圖象疏. In *XGQJ*.

1630. "Bao gao quan ce yueshi shu" 報告觀測月食疏. (*XGQNP*).

1630. "Shen cha leng shou zhong lifa yi yi jian" 審查冷守中曆法議意見.
(*XGQNP*).

1630. "Wei yi shi liang zhi qian hou hu yi jin ju shi zhi chen ken kan ming gui yi yi
quan zhong jie yi zhao xin shi shi shu" 為一事兩旨前後互異謹據實直陳懇勘
明歸一以勸忠節以昭信史事疏. (*XGQNP*).

1630. "Feng zhi hui zou shu" 奉旨回奏疏. In *ZXWGJ*.

1630 [Chongzhen 03/10/17]. "Yueshi hui zou shu" 月食回奏疏. In *XGQJ*.

1630 [Chongzhen 03/11]. "Zi libu zhuan zi du cha yuan wen" 咨禮部轉咨都察院
文. In *XGQJ*.

1630 [Chongzhen 03/11/24]. "Ce hou yueshi feng zhi hui zou shu" 測候月食奉旨
回奏疏. In *XGQJ*.

1631 [Chongzhen 03/12/02]. "Yin bing zai shen qian qing yi wan da dian shu" 因
病再申前請以完大典疏. In *XGQJ*, *ZXWGJ*.

1631 [Chongzhen 03/12/03]. "Yueshi qi fu fangwei ju tu cheng lan shu" 月食起復
方位具圖呈覽疏. In *XGQJ*.

1631. "Yu tui yueshi shu" 預推月食疏. (*XGQNP*).

1631 [Chongzhen 04]. "Fang Xiaoru yi feng ci shu" 方孝儒裔奉祠疏. In *XGQJ*.

1631[Chongzhen 04/01/28]. "Li shu zong mu biao" 曆書總目表. In *XGQJ*. Titled
"Zou cheng li shu zong mu biao" 奏呈曆書總目表 in *ZXWGJ*.

1631 [Chongzhen 04/01/28]. "Feng zhi gong jin li shu shu" 奉旨恭進曆書疏. In
XGQJ, *ZXWGJ*.

1631 [Chongzhen 04/03/09]. "Zun li yin nian ken qi xu zhi shu" 遵例引年懇乞休
致疏. In *XGQJ, ZXWGJ*.

1631. "Yu Zhou Mingyu" 與周明璵. (*XGQNP*).

1631. "Bao gao ce yan yueshi shu" 報告測驗月食疏. (*XGQNP*).

1631. "Shen cha Wei Wenkui suo zhu li yuan yi shu yi jian" 審查魏文魁所著曆元
一書意見. (*XGQNP*).

1631. "Xue li xiao bian" 學曆小辨. (*XGQNP*).

1631 [Chongzhen 04/04/16]. "Yueshi tuisuan li cha shu" 月食推算里差疏. In
XGQJ.

1631 [Chongzhen 04/06/11]. "Yueshi xian qi jin cheng qi fu fangwei bing ju tu
xiang shu" 月食先期進呈起復方位並具圖象疏. In *XGQJ*.

1631 [Chongzhen 04/08/01]. "Feng zhi xu jin li shu shu" 奉旨續進曆書疏. In
XGQJ, ZXWGJ.

1631 [Chongzhen 04/09/08]. "Rishi fen shu fei duo lüe chen yi ju yi daihou yan
shu" 日食分數非多略陳義據以待候驗疏. In *XGQJ, ZXWGJ*.

1631. "Yu tui rishi shu" 預推日食疏. (*XGQNP*).

1631. "Bao gao ce yan rishi shu" 報告測驗日食疏. (*XGQNP*).

1631 [Chongzhen 04/10/15]. "Chu bu de bu zhan zhi shi yi qiu bi zhan bi sheng
zhi ce shu" 處不得不戰之勢宜求必戰必勝之策疏. In *XGQJ, ZXWGJ*.

1631 [Chongzhen 04/10/02]. "Rishi yong yiqi ceyan shu" 日食用儀器測驗疏. In
XGQJ. Titled "Ri shi shu" 日食疏 in *ZXWGJ*.

1631 [Chongzhen 04/10/21]. "Qin feng ming zhi fu chen yu jian shu" 欽奉明旨敷
陳愚見疏. In *XGQJ, ZXWGJ*.

1631 [Chongzhen 04/10/17]. "Yueshi hui zou shu" 月食回奏疏. In *XGQJ*.

1631 [Chongzhen 04/11r/06]. "Yueshi yi fa tuibu ju tu cheng lan shu" 月食依法
推步具圖呈覽疏. In *XGQJ, ZXWGJ*.

1631. "Qin wei yue yi shi shu" 秦為月食事疏. (*XGQNP*).

1631. "Bing shi huo wen" 兵事或問. (*XGQNP*).

1631. "Bing shi shu" 兵事疏. (*XGQNP*).

1631. "Liu han hui ji" 六函彙輯. (*XGQNP*).

1631. "Shang lüe xia lüe" 上略下略. (*XGQNP*).

1632. "Bao gao ce yan yue yi shu" 報告測驗月食疏. (*XGQNP*).

1632 [Chongzhen 05/03/17].[25] "Yueshi shu" 月食疏. In *XGQJ, ZXWGJ*.

1632. [Chongzhen 05/04/04] "Feng zhi gong jin di san ci li shu shu" 奉旨恭進第
三次曆書疏. In *XGQJ, ZXWGJ*.

1632. "Yu tui yueshi shu" 預推月食疏. (*XGQNP*).

1632. "Nan gong qin cao" 南宮秦草. (*XGQNP*).

1632 [Chongzhen 05/04/29]. "Wei yueshi ju tu cheng lan qi ceyan shixing shu" 為
月食具圖呈覽乞測驗施行疏. In *XGQJ, ZXWGJ*.

1632. [Chongzhen 05/09/12] "Yueshi qi zhao qian deng tai shi yan shu" 月食乞照
前登臺實驗疏. In *XGQJ, ZXWGJ*.

---

[25] The dates copied in *XGQJ* and *ZXWGJ* transpose two characters, *san* 三 and *qi* 七: Chongzhen
05/03/17 (former) and Chongzhen 05/07/13 (latter).

1632 [Chongzhen 05/09/15]. "Feng zhi ce hou yueshi yun qi yin bi wu ping ceyan shu" 奉旨測候月食雲氣隱蔽無憑測驗疏. In *XGQJ*.[26]

1632 [Chongzhen 05/09/15]. "Feng zhi ce hou yueshi wu ping ceyan shu" 奉旨測候月食無憑測驗疏. In *ZXWGJ*.

1632. [Chongzhen 05/10/11] "Yueshi xian hou ge fa bu tong yuanyou ji ceyan er fa shu" 月食先後各法不同緣由及測驗二法疏. In *XGQJ*, *ZXWGJ*.

1632. [Chongzhen 05/10/11] "Xiu li que yuan jin shen qian qing yi jun da dian shu" 修曆缺員謹申前請以竣大典疏. In *XGQJ*, *ZXWGJ*.

1632. "Yu Zhou Mingyu" 與周明璵. (*XGQNP*).

1632. "Jingjiao tang bei ji" 景教堂碑記. In *XGQJ*.

1633 [Chongzhen 05/12]. "Wei Huang san zi ni ming shu" 為皇三子擬名疏. In *XGQJ*.

1633. "Ken yu jia diao li shu" 懇予假調理疏. (*XGQNP*).

1633 [Chongzhen 06/02]. "Ken qi sheng en yu jia diao li shu" 懇乞聖恩予假調理疏. In *XGQJ*.

1633 [Chongzhen 06/02]. "Ji shi shao jian ru zhi ban shi shu" 疾勢少減入直辦事疏. In *XGQJ*.

1633. [Chongzhen 06/07] "Shuai bing shi shen ken si chi shu" 衰病實深懇賜斥疏. In *XGQJ*, *ZXWGJ*.

1633. [Chongzhen 06/07/25] "Kao ke wu neng qi yun ci mian shu" 考課無能乞允辭免疏. In *ZXWGJ*.

1633. [Chongzhen 06/07] "Gong xie tian en shu" 恭謝天恩疏. In *XGQJ*, *ZXWGJ*.

1633 [Chongzhen 06/07/16]. "Gong cheng ming ming ru zhi ban shi shu" 恭承明命入直辦事疏. In *XGQJ*.

1633 [Chongzhen 06/07/29]. "Ru zhi ban shi shu" 入直辦事疏. In *XGQJ*.

1633 [Chongzhen 06/09]. "Ce feng guifei li cheng ban si xie en shu" 冊封貴妃禮成頒賜謝恩疏. In *XGQJ*.

1633 [Chongzhen 06/09]. "Gong xie ban si shu" 恭謝頒賜疏. In *XGQJ*.

1633 [Chongzhen 06/09/13]. "Gong shi xie zhuan jin feng guifei ce yin ban si xie en shu" 恭視寫篆進封貴妃冊印頒賜謝恩疏. In *XGQJ*.

1633 [Chongzhen 06/09/29]. "Yueshi yi xin xiu jiao shi li tui bu bing ju tu xiang cheng lan shu" 月食依新修交食曆推步並具圖像呈覽疏. In *XGQJ*.

1633. "Zhang Haihong xiansheng *Wenji* xu" 張海虹先生文集序. (*XGQNP*).

1633. "Chidao nan bei liang zong xing tu xu" 赤道南北兩總星圖敘. In *XGQJ*.

1633. "Gong xie ban si shu" 恭謝頒賜疏. (*XGQNP*).

1633. "Yu tui yueshi shu" 預推月食疏. (*XGQNP*).

1633. [Chongzhen 06/09/29] "Lifa xiu zheng gao cheng shu qi shan zhi you dai qing yi Li Tianjing ren Li ju shu" 曆法修正告成書器繕治有待請以李天經任曆局疏. In *XGQJ*, *ZXWGJ*. (*XGQNP*).

1633. "Lun fei zou cao" 綸扉奏草. (*XGQNP*, states no longer extant).

1633. "Qing tai zou cao" 清臺奏草. (*XGQNP*, states no longer extant).

1633. *Nong ji* 農輯. (*XGQNP*, states no longer extant).

1633. "Jia shu" 家書. (*XGQNP*, states no longer extant).

---

[26] The memorial in *XGQJ* is shorter and differs from the memorial of the same date in *ZXWGJ*.

1633. [Chongzhen 06/10/06] "Zhi li yi you cheng mo ken qi en xu shu" 治曆已有成模懇祈恩敘疏. In *XGQJ*, *ZXWGJ*.

1633 [Chongzhen 06/10/07]. "Jin zhuo chi yin kai bao qian liang shu" 進繳敕印開報錢糧疏. In *XGQJ*, *ZXWGJ*.

1633. *Wenji* 文集. (*XGQNP*, states no longer extant).

1633. *Xu yi* 序議. (*XGQNP*, states no longer extant).

1633. *Shu du* 書牘. (*XGQNP*, states no longer extant).

1633. *Shi pian* 詩篇. (*XGQNP*, states no longer extant).

1635. *Chongzhen li shu* 崇禎曆書. Extant. Republished by Shanghai guji chuban-she 上海古籍出版社 in 2009.

Bibliography

Chinese Collectanea

BTSC Baifu tang suanxue congshu 白芙堂算學叢書 [White Lotus Hall collectanea of books on calculation]. Edited by Ding Quzhong 丁取忠 (fl. 1805–1875). Reprint. Shanghai: Shaowen shuju 紹文書局, 1897.

CFZXS Cehaishan fang Zhong Xi suanxue congke chu bian 測海山房中西算學叢刻初編 [First compilation of the collectanea of Chinese and Western books on calculation engraved by the Cehai Mountain House]. Compiled by "Master of the Cehai Mountain House" (*Cehaishan fang zhuren* 測海山房主人). 1896.

CJCB Congshu jicheng chu bian 叢書集成初編 [Compendium of collectanea, first edition]. Edited by Wang Yunwu 王雲五 (1888–1979). 4000 vols. Reprint of 3467 vols. published by Shanghai shangwu yinshuguan 商務印書館 in 1935, together with 533 previously unpublished vols. Beijing: Zhonghua shuju 中华书局, 1985–1991.

CJXB Congshu jicheng xu bian 叢書集成續編 [Compendium of collectanea, continuation]. Shanghai: Shanghai shudian 上海書店, 1994.

CRZHB Chouren zhuan hui bian 疇人傳彙編 [Collected editions of biographies of men of calendrics]. Edited by Ruan Yuan 阮元 (1764–1849) et al. Reprint. Taibei: Shijie shuju 世界書局, 1962.

CZLS Chongzhen li shu 崇禎曆書 [Chongzhen book of calendrics]. Attributed to Xu Guangqi 徐光啟 (1562–1633). Reprint. Haikou: Hainan chubanshe 海南出版社, 2000.

GSC Gujin suanxue congshu 古今算學叢書 [Collection of works on mathematics, ancient and contemporary]. 1898. Compiled by Liu Duo 劉鐸 (fl. 1898). Rare book. At the Institute for the History of Natural Sciences Library, Beijing, PRC.

GTJ Gujin tushu jicheng 古今圖書集成 [Collection of documents, ancient and contemporary]. 1725. Imperially sponsored. Compiled by Jiang Tingxi 蔣廷

錫, Chen Menglei 陳夢雷 et al. Reprint, photolithographic reproduction of palace edition. Beijing: Zhonghua shuju 中華書局, 1934.

HSXGC Haishanxian guan congshu 海山仙館叢書 [Collectanea of Haishanxian Building]. Compiled by Pan Shicheng 潘仕成 (1804–1873). Reprint. Taibei: Yiwen yinshuguan 藝文印書館, 1967.

JFCS Jifu congshu 畿輔叢書 [Capital collectanea]. Compiled by Wang Hao 王灝. Taibei: Yiwen yinshuguan 藝文印書館, 1966.

LSQS Li suan quanshu 歷算全書 [Complete book of astronomy and mathematics]. 1720. By Mei Wending 梅文鼎 (1633–1721). Reprint. In *SKQS*.

MJSWB Ming jingshi wen bian 明經世文編 [Collection of Ming writings on statecraft]. Compiled by Chen Zilong 陳子龍 (1608–1647). Reprint. Beijing: Zhonghua shuju 中華書局, 1962.

NJZJ Nan jing zha ji 南菁札記 [Nanjing reading notes]. Compiled by Fu Liang 溥良. Jiangyin shi shu 江陰使署, 1894.

QSZSX Qiushi zhai suanxue si zhong 求是齋算學四種 [Four books on calculation from the Truth-seeking Studio]. By Zhang Chuzhong 張楚鍾. Beijing: Beijing tushuguan chubanshe 北京圖書館出版社.

SBBY Sibu beiyao 四部備要 [Complete essentials of the four categories]. Reprint. Shanghai: Zhonghua shuju 中華書局, 1920–1936.

SBCK Sibu congkan 四部叢刊 [Publication of the four categories]. Shanghai: Shangwu yinshuguan 商務印書館, 1922–.

SJSC Wang Zhonghan 王钟翰 and Siku jin hui shu congkan bianzuan weiyuanhui 四庫禁毀书丛刊编纂委员会, eds. *Siku jinhui shu congkan* 四庫禁毀书丛刊 [Collection of books banned or burned by the Four Treasuries project]. 311 vols. Beijing: Beijing chubanshe 北京出版社, 1998.

SJSS Suan jing shi shu 算經十書 [Mathematical classics, ten books]. 1774. Edited by Dai Zhen 戴震 (1724–1777). Reprint. *Wan you wen ku* 萬有文庫. Shanghai: Shangwu yinshuguan 商務印書館, 1930.

SKQS Yingyin Wenyuan ge Siku quanshu 影印文淵閣四庫全書 [Complete collection of the Four Treasuries, photolithographic reproduction of the edition preserved at the Pavilion of Literary Erudition]. 1773–1782. Edited by Ji Yun 紀昀 (1724–1805) and Lu Xixiong 陸錫熊. 3461 titles in 1500 vols. Reprinted from the collection of the Guoli gugong bowuyuan 國立故宮博物院, Taipei: Taiwan shangwu yinshuguan 臺灣商務印書館. Digitized full text, searchable edition. Hong Kong: Chinese University Press, 1983–1986.

SKSJ Song ke suan jing liu zhong: fu yi zhong 宋刻算经六种：附一种 [Six mathematical classics, with one appended, engraved during the Song Dynasty]. 1213. Reprint. Beijing: Wenwu chubanshe 文物出版社, 1981.

SPJ Xu Changzhi 徐昌治 (fl. 1639). *Shengchao poxie ji* 聖朝破邪集 [Collected essays on eliminating evil from this sagely dynasty]. Beijing: Beijing chubanshe 北京出版社, 1997.

SQCC Siku quanshu cunmu congshu 四庫全書存目叢書 [Collection of works preserved in the catalogue of the Four Treasuries]. 4508 titles. Jinan: Qilu shushe 齊魯書社, 1995–97.

SQCCB Siku quanshu cunmu congshu bu bian 四庫全書存目叢書補編 [Collection of works preserved in the catalogue of the Four Treasuries, supplemental series]. 219 titles in 100 vols. Jinan: Qilu shushe 齊魯書社, 2001.

SSJZS Shisanjing zhushu: fu jiaokan ji 十三经注疏：附校勘记 [Thirteen Classics, with commentary and subcommentary, and with editorial notes appended]. Edited by Ruan Yuan 阮元 (1764–1849). Beijing: Zhonghua shuju 中华书局, 1980.

SZJ Zhu Xi 朱熹 (1130–1200). *Sishu zhangju jizhu* 四書章句集注 [Four Books, separated into chapters and sentences, with collected commentaries]. Beijing: Zhonghua shuju 中華書局, 1983.

TDCW Tianzhujiao dong chuan wenxian 天主教東傳文獻 [Documents on the Eastern transmission of Catholicism]. Attributed to Matteo Ricci (1552–1610) et al. Reprint. Zhongguo shixue congshu 中國史學叢書 24. Taibei: Taiwan xuesheng shuju 臺灣學生書局, 1965.

TDCWS Tianzhujiao dong chuan wenxian san bian 天主教東傳文獻三編 [Documents on the Eastern transmission of Catholicism, third collection]. Attributed to Giulio Aleni (1582–1649) et al. Reprint. Zhongguo shixue congshu xu bian 中國史學叢書續編 21. Taibei: Taiwan xuesheng shuju 臺灣學生書局, 1972.

TDCWX Tianzhujiao dong chuan wenxian xu bian 天主教東傳文獻續編 [Documents on the Eastern transmission of Catholicism, second collection]. Attributed to Xu Guangqi 徐光啟 (1562–1633). Reprint. Zhongguo shixue congshu 中國史學叢書 40. Taibei: Taiwan xuesheng shuju 臺灣學生書局, 1965.

TJW Tang Jingchuan wenji 唐荊川文集 [Collected works of Tang Shunzhi]. Tang Shunzhi 唐順之 (1507–1560). Reprint. Shanghai: Shangwu yinshuguan 商務印書館, 1922.

TJXW Tang Jingchuan xiansheng wenji 唐荊川先生文集 [Collected works of Tang Shunzhi]. Tang Shunzhi 唐順之 (1507–1560). Reprint. Shanghai shudian 上海書店, 1994.

TXCH Tianxue chu han 天學初函 [Learning of heaven, first series]. Attributed to Li Zhizao 李之藻 (1565–1630). Reprint. Taibei: Taiwan xuesheng shuju 臺灣學生書局, 1965.

WWBC Wanwei biecang 宛委別藏 [Secondary collection from Wanwei]. Compiled by Ruan Yuan 阮元 (1764–1849). Shanghai: Shangwu yinshuguan 商務印書館, 1935.

WYWK Wanyou wenku 萬有文庫 [Universal library]. Edited by Wang Yunwu 王雲五 (1888–1979). 1710 titles in 4000 volumes. Shanghai: Shangwu yinshuguan 商務印書館, 1929–1937.

XGQJ Xu Guangqi ji 徐光啟集 [Collected works of Xu Guangqi]. Edited by Wang Zhongmin 王重民. Shanghai: Shanghai guji chubanshe 上海古籍出版社, 1984.

XGZYJ Shanghai shi wenwu baoguan weiyuanhui 上海市文物保管委員会 and Gu Tinglong 顧廷龍, eds. *Xu Guangqi zhu yi ji* 徐光啟著譯集 [Collected writ-

ings and translations of Xu Guangqi]. Shanghai: Shanghai guji chubanshe 上
海古籍出版社, 1983.

XGZZ Xu Guangqi zhuanji ziliao 徐光啟傳記資料 [Biographic materials on Xu
Guangqi]. Compiled by Zhu Chuanyu 朱傳譽. Taibei: Tianyi chubanshe 天
一出版社, 1979.

XLTCS Xuanlan tang congshu 玄覽堂叢書 [Collectanea of Xuanlan Hall]. Taipei:
Guoli zhongyang tushuguan 國立中央圖書館, 1981.

XQSL Xian Qing shi liao 先清史料 [Historical materials of the early Qing]. Edited
by Li Shutian 李澍田 and Diao Shuren 刁书仁. Changchun: Jilin wenshi
chubanshe 吉林文史出版社, 1990.

XSQT Xuxiu Siku quanshu tiyao 續修四庫全書提要 [Annotated bibliography of
the continuation of the complete collection of the Four Treasuries]. Edited
by Ji Yun 紀昀 (1724–1805) and Wang Yunwu 王雲五 (1888–1979). Taibei:
Taiwan shangwu yinshuguan 臺灣商務印書館, 1972.

XXSQ Xu xiu Siku quanshu 續修四庫全書 [Continuation of the complete collection of the Four Treasuries]. 5213 titles dating from 1735 to 1911. Shanghai:
Shanghai guji chubanshe 上海古籍出版社, 1995–2002.

YLDD Yongle da dian 永樂大典 [Great encyclopedia of the Yongle reign]. 1403–
1407. Compiled by Xie Jin 解縉 (1369–1415) et al. Reprint of extant fragments.
10 vols. Beijing: Zhonghua shuju 中华书局, 1986.

YLDMQ Yesuhui Luomu Dang'anguan Ming-Qing Tianzhujiao wenxian 耶穌會
羅馬檔案館明清天主教文獻 "Chinese Christian texts from the Roman
Archives of the Society of Jesus." Edited by Nicolas Standaert (Zhong Ming-
dan 鐘鳴旦) and Adrian Dudink (Du Dingke 杜鼎克). Reprint. Taibei: Taibei
Li shi xueshe 臺北利氏學社, 2002.

YLQS Zhu Zaiyu 朱載堉 (1536–1611). *Yuelü quanshu* 樂律全書 [Complete collection of music and pitch]. Edited by Beijing tushuguan guji chuban bianji zu
北京圖書館古籍出版編輯組. Beijing tushuguan guji zhenben congkan 北
京圖書館古籍珍本叢刊 4. Beijing: Shumu wenxian chubanshe 書目文獻出
版社, [1988].

ZBZZC Zhibuzu zhai congshu 知不足齋叢書 [Collection of the Zhibuzu Studio].
Compiled by Bao Tingbo 鮑廷博 (1728–1814). Reprint. Taibei: Xingzhong
shuju 興中書局, 1964.

ZH Zhi hai 指海 [Pointing to the ocean]. Edited by Qian Xizuo 錢熙祚, Qian
Peirang 錢培讓, and Qian Peijie 錢培杰. Reprint. Taibei: Yiwen yinshuguan
藝文印書館, 1967.

ZKJDT Zhongguo kexue jishu dianji tonghui: Shuxue juan 中國科學技術典籍
通彙：數學卷 [Comprehensive collection of the classics of Chinese science
and technology: Mathematics volumes]. Edited by Guo Shuchun 郭书春. 5
vols. Zhengzhou: Henan jiaoyu chubanshe 河南教育出版社, 1993.

ZXWGJ Zengding Xu Wending gong ji 增訂徐文定公集 [Collected works of Xu
Guangqi, revised and enlarged]. Taibei: Taiwan Zhonghua shuju 臺灣中華書
局, 1962.

Primary Sources in Chinese

Ai Rulüe 艾儒略 (Giulio Aleni, SJ, 1582–1649). *Zhifang wai ji* 職方外紀 [Records of regions outside the purview of the Bureau of Operations]. In *SKQS*.

———. *Da xi Xitai Li xiansheng xingji* 大西西泰利先生行蹟 [Life of Master Ricci of the great West]. Taibei: Taibei Li shi xueshe 臺北利氏學社, 2002.

Ban Gu 班固 (32–92 CE). *Qian Han shu* 前漢書 [History of the former Han]. Commentary by Yan Shigu 顏師古 (581–645). In *SKQS*.

Chen Yan 陳衍 (1856–1937). *Yuan shi ji shi* 元詩紀事 [Yuan poetry chronicles]. In *XXSQ*.

Cheng Dawei 程大位 (1533–1606). *Suanfa tongzong* 算法统宗 [Comprehensive source of mathematical methods]. In *ZKJDT*.

———. *Suanfa tongzong jiaoshi* 算法统宗校释 [Comprehensive source of mathematical methods, edited with explanatory notes]. Edited by Mei Rongzhao 梅荣照 and Li Zhaohua 李兆华. Hefei: Anhui jiaoyu chubanshe 安徽教育出版社, 1990.

Dong Qichang 董其昌 (1555–1636). *Xuanshang zhai shumu* 玄賞齋書目 [Bibliography of the Studio of Occult Rewards]. Beijing: Guoli Beiping tushuguan 國立北平圖書館, 1932–1937.

Fang Zhongtong 方中通 (1634–1698). *Shu du yan* 數度衍 [Numbers and measurement, an amplification]. Reprint. In *SKQS*.

Feng Congwu 馮從吾 (1556–1627). *Yuan ru kao lue* 元儒考略 [Brief investigation into scholars of the Yuan dynasty]. In *SKQS*.

Fu Weilin 傅維鱗 (d. 1667). *Ming shu* 明書 [Book of the Ming]. In *SQCC*.

Ge Hong 葛洪 (284–364). *Baopuzi nei pian* 抱樸子內篇 [Inner chapters of *Baopuzi*]. In *SKQS*.

Gu Yingxiang 顧應祥 (1483–1565). *Ce yuan hai jing fen lei shi shu* 測圓海鏡分類釋術 [Sea mirror of circle measurement, arranged by categories, with explanations of the methods]. Reprint. In *SKQS*.

Guan Zhong 管仲 (d. 645 BCE). *Guanzi* 管子 [Guanzi]. In *SKQS*.

Guo Tingxun 過廷訓 (fl. 1604). *Ben chao fen sheng renwu kao* 本朝分省人物考 [Investigation of men of this dynasty, by province]. Reprint. Shanghai guji chubanshe 上海古籍出版社, 2002.

Guo Xiang 郭象 (d. 312 CE). *Zhuangzi zhu* 莊子注 [Commentary to *Zhuangzi*]. In *SKQS*.

Hua Shifang 華世芳 (1854–1905). *Jindai chouren zhu shu ji* 近代疇人著述記 [Record of the writings of modern men of calendrics and calculation]. In *CRZHB*.

Huang Yuji 黃虞稷 (1629–1691). *Qianqing tang shumu* 千頃堂書目 [Bibliography of the Thousand Qing Hall]. In *CJXB*.

Huang Zhongjun 黃鐘駿. *Chouren zhuan si bian* 疇人傳四編 [Biographies of men of calendrics and calculation, fourth volume]. In *CRZHB*.

Huang Zongxi 黃宗羲 (1610–1695). *Huang Zongxi quanji* 黃宗羲全集 [Complete collected works of Huang Zongxi]. Edited by Shen Shanhong 沈善洪. 12 vols. Hangzhou: Zhejiang guji chubanshe 浙江古籍出版社, 1985–1994.

Jiao Hong 焦竑 (1541–1620). *Dan yuan ji* 澹園集 [Collected writings of Jiao Hong]. Reprint. Beijing: Zhonghua shuju 中華書局, 1999.

———. *Guo shi jing ji zhi* 國史經籍志 [Record of books for the dynastic history]. Reprint. Taibei: Taiwan shangwu yinshuguan 臺灣商務印書館, 1965.

———. *Jiao shi bi sheng* 焦氏筆乘 [Mr. Jiao's writings]. Reprint. Shanghai: Shanghai guji chubanshe 上海古籍出版社, 1986.

———. *Jiao shi bi sheng xu ji* 焦氏筆乘續集 [Mr. Jiao's writings, continuation]. In *SQCC*.

———. *Jiao shi lei lin* 焦氏類林 [Collected works of Mr. Jiao, arranged topically]. In *CJCB*.

———. *Laozi yi* 老子翼 [Wings to *Laozi*]. In *SKQS*.

———. *Zhuangzi yi* 莊子翼 [Wings to *Zhuangzi*]. In *SKQS*.

Jiguge ying Song chaoben Jiuzhang suan jing, wu juan 汲古閣影宋抄本九章算經五卷 [Nine chapters on the mathematical arts, five chapters, Song Dynasty hand-copied manuscript, reprinted by the Drawing-from-the-Ancients Pavilion]. Reprint, *Tianlulinlang congshu, di yi ji* 天祿琳琅叢書, 第一集 [13–14]. Beijing: Gugong bowuyuan 故宮博物院, 1932.

Jihe yuanben 幾何原本 [Euclid's *Elements*]. Orally translated (*koyi* 口譯) by Li Madou 利瑪竇 (Matteo Ricci, SJ, 1552–1610), transcribed (*bishou* 筆受) by Xu Guangqi 徐光啟 (1562–1633). In *SKQS*.

Jingu tang shumu 近古堂書目 [Bibliography of the Hall of Approaching the Ancients]. In *CJXB*.

Jiuzhang suanshu 九章算術 [Nine chapters on the mathematical arts]. Commentary by Liu Hui 劉徽 (fl. 263 CE), subcommentary by Li Chunfeng 李淳風 (602–670). Wuyingdian juzhen ban congshu 武英殿聚珍版叢書 edition (1774). Reprint. In *ZKJDT*.

Jiuzhang suanshu 九章算術 [Nine chapters on the mathematical arts]. Commentary by Liu Hui 劉徽 (fl. 263 CE), subcommentary by Li Chunfeng 李淳風 (602–670). In *SKQS*.

Li Dongyang 李東陽 (1447–1516) et al. *Daming huidian* 大明會典 [Record of government of the Ming dynasty]. Reprint. Taipei: Dongnan shubaoshe 東南書報社, 1963.

Li Dupei 李篤培 (1575–1631). *Zhong Xi shuxue tu shuo* 中西數學圖說 [Chinese and Western mathematics, explained with diagrams]. Rare book. At the Institute for the History of Natural Sciences Library, Beijing, PRC.

Li Huang 李潢 (?–1812). *Jiuzhang suanshu xi cao tu shuo jiu juan* 九章算術細草圖說九卷 [Nine chapters on the mathematical arts, with detailed explanations and explanatory diagrams]. 1812. Reprint. In *ZKJDT*.

Li Madou 利瑪竇 (Matteo Ricci, SJ, 1552–1610). *Li Madou Zhongwen zhu yi ji* 利瑪竇中文著譯集 [Collected Chinese writings and translations of Matteo Ricci]. Edited by Zhu Weizheng 朱維錚, Deng Zhifeng 鄧志峰, et al. Hong Kong: Xianggang chengshi daxue chubanshe 香港城市大學出版社, 2001.

Li Qi 李杞 (12th/13th c.) *Yong Yi xiang jie* 用易詳解 [Detailed explanations for using the *Classic of Changes*]. In *SKQS*.

Li Ye 李冶 (1192–1279). *Ce yuan hai jing* 測圓海鏡 [Sea mirror of circle measurement]. In *ZKJDT*.

Liezi 列子 [Liezi]. Attributed to Lie Yukou 列禦寇 (c. 4th cent. BCE). In *SKQS*.

Liji zhushu 禮記註疏 [*Record of Rites*, with commentary and subcommentary]. Commentary by Zheng Xuan 鄭玄 (127–200 CE), and subcommentary by Kong Yingda 孔穎達 (547–648). In *SSJZS*.

Ling yan li shao 靈言蠡勺 [A preliminary discussion of anima]. Orally transmitted (*ko shi* 口譯) by Bi Fangji 畢方濟 (Franciscus Sambiasi, SJ, 1582–1649) and recorded (*bi lu* 筆錄) by Xu Guangqi 徐光啟 (1562–1633). In *TXCH*.

Ling yan li shao yin 靈言蠡勺引 [Preface to *A Preliminary Discussion of Anima*]. Attributed to Bi Fangji 畢方濟 (Franciscus Sambiasi, SJ, 1582–1649). In *TXCH*.

Liu An 劉安 (179–122 BCE). *Huainan hong lie jie* 淮南鴻烈解 [Great illumination of Huainan, with explanations]. Commentary by Gao You 高誘 (fl. 205–212). In *SKQS*.

Lunyu zhushu 論語註疏 [*Analects*, with commentary and subcommentary]. Attributed to Kongzi 孔子 (Confucius, 551–479 BCE), with commentary by He Yan 何晏 (?–249), and subcommentary by Xing Bing 邢昺 (932–1010). In *SSJZS*.

Luo Shilin 羅士琳 (1774?–1853). *Chouren zhuan xubian* 疇人傳續編 [Biographies of men of calendrics and calculation, second volume]. In *CRZHB*.

Mao Wei 茅維 (fl. 1605). *Huang Ming ce heng* 皇明策衡 [Weighing Ming dynasty policy examination essays]. In *SJSC*.

Mei Wending 梅文鼎 (1633–1721). *Meishi congshu jiyao* 梅氏叢書輯要 [Essential anthology of Mei Wending]. Reprint. Taibei: Yiwen yinshuguan 藝文印書館, 1971.

———. *Fangcheng lun* 方程論 [On *fangcheng*]. Photolithographic reprint from the Mei Juecheng Chengxuetang 梅瑴成承學堂 printing of the *Mei shi congshu ji yao* 梅氏叢書輯要. In *ZKJDT*.

Mengzi zhushu 孟子註疏 [*Mencius*, with commentary and subcommentary]. Commentary by Zhao Qi 趙岐 (108–201 CE), and subcommentary attributed to Sun Shi 孫奭 (962–1033). In *SSJZS*.

Ming shi lu: Fu jiao kan ji 明實錄：附校勘記 [Veritable records of the Ming dynasty, with editorial notes appended]. Compiled by Zhongyang yanjiuyuan lishi yuyan yanjiusuo 中央研究院歷史語言研究所. Nan'gang: Zhongyang yanjiuyuan lishi yuyan yanjiusuo 中央研究院歷史語言研究所, 1962–66.

Ouyang Xiu 歐陽修 (1007–1072) and Song Qi 宋祁 (998–1061). *Xin Tang shu* 新唐書 [New history of the Tang dynasty]. 20 vols. 225 *juan*. Beijing: Zhonghua shuju 中华书局, 1975.

Peng Sunyi 彭孫貽 (1615–1673). *Shanzhong wenjian lu* 山中聞見錄 [Record of things seen and heard amidst the mountains]. In *XQSL*.

Qian Pu 錢溥 (*jinshi* 1439). *Bige shumu* 秘閣書目 [Bibliography of the Imperial Pavilion]. In *SQCCB*.

Qin Jiushao 秦九韶 (13th c.) *Shu shu jiu zhang* 數書九章 [Mathematical treatise in nine sections]. Reprint. In *ZKJDT*.

Ruan Yuan 阮元 (1764–1849). *Chouren zhuan chu bian* 疇人傳初編 [Biographies of men of calendrics and calculation, first volume]. In *CRZHB*.

Shang shu zhu shu 尚書注疏 [*Book of Documents*, with commentary and sub-commentary]. Traditionally attributed to Kongzi 孔子 (Confucius, 551–479 BCE), commentary by Kong Yingda 孔穎達 (547–648), pronunciation and meanings by Lu Deming 陸德明 (556–627). In *SKQS*.

Shanghai shi wenwu baoguan weiyuanhui 上海市文物保管委員会 and Gu Tinglong 顧廷龍, eds. *Hou ji* 後記 [Postface]. In *XGZYJ*.

Shen Que 沈㴶 (*jinshi* 1592). *Can yuan yi shu* 㕥遠夷疏 [Memorial presenting charges against the foreigners from afar]. In *SPJ*.

Sima Qian 司馬遷 (c. 145–c. 86 BCE). *Shiji ji jie* 史記集解 [*Records of the Grand Historian*, with collected explanations]. Commentary by Pei Yin 裴駰 (fl. 438). In *SKQS*.

———. *Shiji hui zhu kaozheng* 史記會注考證 [*Records of the Grand Historian*, with collected commentaries and textual research]. Edited by Takigawa Kametarō 瀧川資言 (1865–1946). Beijing: Wenxue guji kanxing she 文學古籍刊行社, 1955.

Song Lian 宋濂 (1310–1381) et al. *Yuan shi* 元史 [History of the Yuan dynasty]. In *SKQS*.

Su Tianjue 蘇天爵 (1294–1352). *Yuanchao mingchen shilue* 元朝名臣事略 [Biographical sketches of important officials of the Yuan dynasty]. In *SKQS*.

Sun Nengchuan 孫能傳 (*juren* 1582) and Zhang Xuan 張萱. *Neige cangshu mulu* 內閣藏書目錄 [Catalogue of books of the Grand Secretariat]. In *XXSQ*.

Tan Qian 談遷 (1594–1657). *Guo que* 國榷 [Deliberations of the dynasty]. Edited by Zhang Zongxiang 張宗祥. 6 vols. 104 *juan*. Beijing: Guji chubanshe 古籍出版社, 1958.

Tang Shunzhi 唐順之 (1507–1560). *Gougu cewang lun* 句股測望論 [On surveying with right triangles]. In *TJW*.

———. *Gougu deng liu lun* 句股等六論 [On right triangles]. In *TJXW*.

Tianzhu shi yi 天主實義 [True meaning of the lord of heaven]. Attributed to Li Madou 利瑪竇 (Matteo Ricci, SJ, 1552–1610). In *TXCH*.

Tianzhu shi yi yin 天主實義引 [Preface to *True Meaning of the Lord of Heaven*]. Attributed to Li Madou 利瑪竇 (Matteo Ricci, SJ, 1552–1610). In *TXCH*.

Tong wen suan zhi 同文算指 [Guide to calculation in the unified script]. Attributed to Li Madou 利瑪竇 (Matteo Ricci, SJ, 1552–1610) and Li Zhizao 李之藻 (1565–1630). In *ZKJDT*.

Wang Chang 王昶 (1725–1806). *Ming ci zong* 明詞綜 [Compendium of Ming *ci*]. Taibei: Taiwan Zhonghua shuju 臺灣中華書局, 1965.

Wang Hongxu 王鴻緒 (1645–1723). *Ming shi gao* 明史稿 [Draft history of the Ming dynasty]. Reprint. Taibei: Wenhai chubanshe 文海出版社, 1962.

Wang Yunwu 王雲五 (1888–1979), Yong Rong 永瑢 (1744–1790), and Ji Yun 紀昀 (1724–1805), eds. *Siku quanshu zong mu tiyao ji Siku wei shou shumu, jinhui shumu* 四庫全書總目提要及四庫未收書目・禁燬書目 [General catalogue with annotations of the *Complete Collection of the Four Treasuries*, with catalogue of books not included in the Four Treasuries, and catalogue of prohibited and burned books]. Reprint. Taibei: Taiwan shangwu yinshuguan 臺灣商務印書館, 1971.

Wei Yuan 魏源 (1794–1857). *Yuanshi xin bian* 元史新編 [History of the Yuan dynasty, new compilation]. In *XXSQ*.

Wu Jing 吳敬 (fl. 1450). *Jiuzhang xiang zhu bilei suanfa da quan* 九章詳註比類算法大全 [Complete collection of the mathematical arts of the nine chapters, with detailed commentary, arranged by category]. Engraved in 1450. In *ZKJDT*.

Xu Guangqi 徐光啟 (1562–1633). *Bian xue zhang shu* 辨學章疏 [Memorial on distinguishing learning]. In *TDCWX*.

———. *Da xiang ren shu* 答鄉人書 [Reply to a fellow townsman]. In *ZXWGJ*.

———. *Gougu yi xu* 勾股義序 [Preface to *Right Triangles, Meanings*]. In *XGQJ*.

———. *Jia shu* 家書 [Family letters]. In *XGQJ*.

———. *Ke Tong wen suan zhi xu* 刻同文算指序 [Preface at the printing of the *Guide to Calculation in the Unified Script*]. In *ZKJDT*.

———. *Mao shi liu tie jiang yi* 毛詩六帖講意 [Notes on "six couplets" (i.e., passing exams) of the Mao *Classic of Poetry*]. In *SQCC*.

———. *Shi jing chuan gao* 詩經傳稿 [Transmitted manuscript on the *Classic of Poetry*]. In *XGZYJ*.

———. *Shun zhi ju shenshan zhizhong* 舜之居深山之中 [That Shun lived deep in the mountains]. In *XGQJ*.

———. *Yu Jiao laoshi* 與焦老師 [For teacher Jiao]. In *XGQJ*.

———. *Yu Jiao laoshi* 與焦老師 [For teacher Jiao]. In *ZXWGJ*.

Yang Hui 楊輝 (c. 1238–c. 1298). *Xiang jie jiuzhang suanfa* 詳解九章算法 [Nine chapters on the mathematical arts, with detailed explanations]. Extant 5 juan, Yijiatang congshu 宜稼堂叢書 edition. In *ZKJDT*.

Yang Shiqi 楊士奇 (1365–1444) and Wang Chengxiang 王呈祥, eds. *Wenyuan ge shumu* 文淵閣書目 [Bibliography of the Hall of Literary Depth]. In *CJCB*.

Yong Rong 永瑢 (1744–1790) and Ji Yun 紀昀 (1724–1805), eds. *Siku quanshu zong mu* 四庫全書總目 [General catalogue of the *Complete Collection of the Four Treasuries*]. Reprint. Beijing: Zhonghua shuju 中華書局, 1965.

Yuzhi shu li jing yun 御製數理精蘊 [The emperor's collected essential principles of mathematics]. Attributed to the Kangxi 康熙 Emperor (r. 1662–1722). In *ZKJDT*.

Zhang Quijian suan jing 張邱建算經 [Zhang Qiujian's mathematical classic]. In *ZKJDT*.

Zhang Tingyu 張廷玉 (1762–1755) et al., eds. *Ming shi* 明史 [Official history of the Ming dynasty]. 28 vols. 332 *juan*. Beijing: Zhonghua shuju 中华书局, 1974.

Zhao Qimei 趙琦美 (1563–1624). *Maiwang guan shumu* 脈望館書目 [Bibliography of the Maiwang Hall]. In *CJXB*.

Zhouyi zhu shu 周易注疏 [*Zhou Changes*, with commentary and subcommentary]. Commentary by Wang Bi 王弼 (226–249 CE), pronunciation and meanings by Lu Deming 陸德明 (556–627), subcommentary by Kong Yingda 孔穎達 (547–648). In *SKQS*.

Zhu Kebao 諸可寶 (1845–1903). *Chouren zhuan san bian* 疇人傳三編 [Collected biographies of men of calendrics and calculation, third volume]. In *CRZHB*.

Zhu Shijie 朱世傑 (1249–1314). *Si yuan yu jian* 四元玉鑑 [Jade mirror of the four origins]. In *ZKJDT*.

Zhu Xi 朱熹 (1130–1200). *Da xue zhang ju* 大學章句 [*Great Learning*, separated into chapters and sentences]. In *SZJ*.

———. *Lunyu jizhu* 論語集注 [Collected commentaries on the *Analects*]. In *SZJ*.

———. *Mengzi jizhu* 孟子集注 [Collected commentaries on *Mencius*]. In *SZJ*.

———. *Zhuzi yu lei* 朱子語類 [Conversations with Master Zhu, arranged topically]. Edited by Li Jingde 黎靖德 and Wang Xingxian 王星賢. 8 vols. 140 juan. Beijing: Zhonghua shuju 中華書局, 1986.

Zhu Yizun 朱彝尊 (1629–1709). *Ming shi zong* 明詩綜 [Compendium of Ming poetry]. In *SKQS*.

Zhu Zaiyu 朱載堉 (1536–1611). *Suanxue xin shuo* 算學新說 [New explanation of the theory of calculation]. In *YLQS*.

———. *Suanxue xin shuo* 算學新說 [New explanation of the theory of calculation]. In *SKQS*.

Zhuang Zhou 莊周 (fl. 320? BCE). *Zhuangzi ji shi* 莊子集釋 [*Zhuangzi*: Collected explanations]. Edited by Guo Qingfan 郭慶藩 and Wang Xiaoyu 王孝魚. Beijing: Zhonghua shuju 中華書局, 1961.

Secondary Sources in East Asian Languages

Ahn Daeok 安大玉. *Minmatsu Seiyō kagaku tōdenshi: "Tengaku shokan" kihen no kenkyū* 明末西洋科学東伝史：「天学初函」器編の研究 [History of the transmission of Western science to the East in the late Ming: Research on the *Qibian* of the *Tianxue chuhan*]. Tokyo: Chisen Shokan 知泉書館, 2007.

Bai Shangshu 白尚恕. *Jiu zhang suanshu zhushi* 《九章算术》注释 [*Nine Chapters on the Mathematical Arts* with commentary and explanations]. Beijing: Kexue chubanshe 科学出版社, 1983.

———. *Ce yuan hai jing jin yi* 測圓海鏡今譯 [*Sea mirror of circle measurement*, translated into modern Chinese]. Shandong: Shandong jiaoyu chubanshe 山东教育出版社, 1985.

———. *Zhongguo shuxueshi yanjiu: Bai Shangshu wenji* 中国数学史研究：白尚恕文集 [Research on the history of Chinese mathematics: Collected essays of Bai Shangshu]. Beijing: Beijing shifan daxue chubanshe 北京师范大学出版社, 2008.

Beijing daxue tushuguan xuexi 北京大学图书馆学系 and Wuhan daxue tushuguan xuexi 武汉大学图书馆学系, eds. *Tushuguan guji bianmu* 图书馆古籍编目 [Cataloging ancient Chinese books]. Beijing: Zhonghua shuju 中华书局, 1985.

Chen Guying 陳鼓應. *Laozi zhu yi ji pingjie* 老子註譯及評介 [*Laozi*, annotated, translated into modern Chinese, with criticism]. Beijing: Zhonghua shuju 中華書局, 1984.

Chen Tian 陳田 (1849–1921). *Ming shi ji shi* 明詩紀事 [Ming poetry chronicles]. Reprint. Shanghai: Shanghai guji chubanshe 上海古籍出版社, 1993.

Chen Weiping 陈卫平 and Li Chunyong 李春勇. *Xu Guangqi ping zhuan* 徐光启评传 [Critical biography of Xu Guangqi]. Nanjing: Nanjing daxue chubanshe 南京大学出版社, 2006.

Chu Xiaobo 初晓波. *Cong Huayi dao wanguo de xiansheng: Xu Guangqi duiwai guannian yanjiu* 从华夷到万国的先声：徐光启对外观念研究 [Herald of the shift from the Chinese/barbarian dichotomy to multinationalism: Xu Guangqi's conceptions of foreign affairs analyzed]. Beijing: Beijing daxue chubanshe 北京大学出版社, 2008.

Dai Nianzu 戴念祖. *Zhu Zaiyu: Ming dai de kexue he yishu juxing* 朱載堉：明代的科學和藝術巨星 [Zhu Zaiyu: A giant of the sciences and the arts in the Ming dynasty]. Beijing: Renmin chubanshe 人民出版社, 1986.

Ding Fubao 丁福保. *Sibu zonglu suanfa bian* 四部總錄算法編 [General catalogue of the four sections, mathematics section]. Beijing: Wenwu chubanshe 文物出版社, 1957.

Du Shiran 杜石然. "Shilun Song-Yuan shiqi Zhongguo he Yisilan guojia jian de shuxue jiaoliu" 試論宋元時期中國和伊斯蘭國家間的數學交流 [Preliminary essay on mathematical exchanges between the Islamic countries and China during the Song and Yuan Dynasties]. In Qian et al., *Song Yuan shuxueshi lunwenji*.

Du Xinfu 杜信孚, Zhou Guangpei 周光培, and Jiang Xiaoda 蔣校達. *Mingdai banke zonglu* 明代版刻綜錄 [General catalogue of books engraved during the Ming dynasty]. Yangzhou: Jiangsu Guangling guji keyinshe 江蘇廣陵古籍刻印社, 1983.

Fang Hao 方豪. *Xu Guangqi* 徐光啓 [Xu Guangqi]. Chongqing: Shengli chubanshe 勝利出版社, 1944.

———. *Zhongguo tianzhujiao shi luncong: jia ji* 中國天主教史論叢：甲集 [Collected works on the history of Catholicism in China, first set]. Shanghai: Shangwu yinshuguan 商務印書館, 1947.

———. *Ming mo Qing chu Tianzhujiao bi fu rujia xueshuo zhi yanjiu* 明末清初天主教比附儒家學說之研究 [Catholicism at the end of the Ming and beginning of the Qing compared, with research on theories of Confucianism appended]. Taibei: Guoli Taiwan daxue 國立臺灣大學, 1962.

———. *Li Zhizao yanjiu* 李之藻研究 [Research on Li Zhizao]. Taibei: Taiwan shangwu yinshuguan 臺灣商務印書館, 1966.

———. *Zhongguo Tianzhujiao shi renwu zhuan* 中國天主教史人物傳 [Biographies of persons in the history of Chinese Catholicism]. Beijing: Zhonghua shuju 中華書局, 1988.

Fu Pu 傅溥. *Zhongguo shuxue fazhanshi* 中國數學發展史 [History of the development of Chinese mathematics]. Taibei: Zhongyang wenwu gongyingshe 中央文物供應社, 1982.

Guo Shirong 郭世荣. *Zhongguo shuxue dianji zai Chaoxian bandao de liuchuan yu yingxiang* 中国数学典籍在朝鲜半岛的流传与影响 [The transmission to and influence on the Korean peninsula of ancient Chinese mathematical texts]. Jinan: Shandong jiaoyu chubanshe 山东教育出版社, 2009.

Guo Shuchun 郭书春. *Jiu zhang suanshu huijiaoben* 九章算術匯校本 [Critical edition of the *Nine Chapters on Mathematical Arts*]. Shenyang: Liaoning jiaoyu chubanshe 辽宁教育出版社, 1990.

———, ed. *Zhongguo kexue jishu dianji tonghui: Shuxue juan* 中國科學技術典籍通彙：數學卷 [Comprehensive compilation of the Chinese technological and scientific classics: Mathematics]. Zhengzhou: Henan jiaoyu chubanshe 河南教育出版社, 1993.

Guo Shuchun 郭书春 and Liu Dun 刘钝, eds. *Suan jing shi shu* 算經十書 [Mathematical classics, ten books]. Shenyang: Liaoning jiaoyu chubanshe 辽宁教育出版社, 1998.

He Aisheng 何艾生 and Liang Chengrui 梁成瑞. "Jihe yuanben zai Zhongguo de chuanbo" 幾何原本及其在中國的傳播 [The *Elements* and its diffusion in China]. *Zhongguo kejishi* 中國科技史 5, no. 3 (1984): 32–42.

Hong Wansheng 洪萬生. "Gudai Zongguo de jihexue" 古代中國的幾何學 [The geometry of ancient China]. *Kexue yuekan* 科學月刊 [Taibei] 12, no. 8 (1981): 22–30.

———. "Zhongshi zhengming de shidai: Wei, Jin, Nanbeichao de keji" 重試証明的時代魏晉南北朝的科技 [The period that emphasized proofs: the science and technology of the Wei, Jin, and Southern and Northern Dynasties]. In Hong and Liu, *Gewu yu chengqi*, 105–64.

———, ed. *Cong Li Yuese chufa—shuxue shi, kexue shi wenji* 從李約瑟出發—數學史科學史文集 [Joseph Needham as a point of departure—collection of works on the history of mathematics and science]. Taibei: Jiuzhang chubanshe 九章出版社, 1999.

Hong Wansheng 洪萬生 and Liu Dai 劉岱, eds. *Gewu yu chengqi* 格物與成器 [Investigating things and becoming useful]. Taibei: Lianjing chuban shiye gongsi 聯經出版事業工司, 1982.

Hong Wansheng 洪萬生 et al. *Tantian sanyou* 談天三友 [Three friends (Jiao Xun 焦循, Wang Lai 汪萊, and Li Rui 李銳)]. Taibei: Mingwen shuju 明文書局, 1993.

Huang Yinong 黃一農. *Liang tou she: Ming mo Qing chu de di yi dai Tianzhu jiaotu* 兩頭蛇：明末清初的第一代天主教徒 [Two-headed snakes: The first generation of Catholics in the late Ming and early Qing]. Xinzhu: Guoli qinghua daxue chubanshe 國立清華大學出版社, 2005.

Ke Shaomin 柯劭忞 (1850–1933). *Xin Yuan shi: fu kaozheng* 新元史：附考證 [New history of the Yuan dynasty, with textual research appended]. Taibei: Yiwen yinshuguan 藝文印書館, 1956.

Kong Guoping 孔国平. *Li Ye zhuan* 李冶传 [Biography of Li Ye]. Shijiazhuang: Hebei jiaoyu chubanshe 河北教育出版社, 1988.

———. "Li Ye" 李冶 [Li Ye]. In *Zhongguo gudai kexuejia zhuanji* 中国古代科学家传记 [Biographies of scientists in ancient China], edited by Du Shiran 杜石然. Beijing: Kexue chubanshe 科学出版社, 1992–1993.

———. *Li Ye, Zhu Shijie, yu Jin-Yuan shuxue* 李冶朱世杰与金元数学 [Li Ye, Zhu Shijie, and Jin-Yuan mathematics]. Zhongguo shuxueshi daxi 中国数学史大系. Shijiazhuang: Hebei kexue jishu chubanshe 河北科学技术出版社, 2000.

Lao Hansheng 劳汉生, ed. *Zhusuan yu shiyong suanshu* 珠算与实用算术 [Calculations with the abacus and practical mathematics]. Zhongguo shuxueshi daxi 中国数学史大系. Shijiazhuang: Hebei kexue jishu chubanshe 河北科学技术出版社, 2000.

Li Di 李迪. *Zhongguo shuxue shi jianbian* 中國數學史簡編 [Concise history of Chinese mathematics]. Shenyang: Liaoning renmin chubanshe 辽宁人民出版社, 1984.

———, ed. *Ming mo dao Qing zhong qi* 明末到清中期 [Late Ming to mid-Qing]. Vol. 7 of *Zhongguo shuxueshi daxi*, 中国数学史大系 [Compendium of the history of Chinese mathematics]. Beijing: Beijing shifan daxue chubanshe 北京师范大学出版社, 2000.

———, ed. *Zhongguo suanxue shumu huibian* 中国算学书目汇编 [Compilation of bibliographies of Chinese mathematical treatises]. Zhongguo shuxueshi daxi 中国数学史大系. Supplemental volume 2. Beijing: Beijing shifan daxue chubanshe 北京师范大学出版社, 2000.

———. *Zhongguo shuxue tongshi* 中国数学通史 [Comprehensive history of Chinese mathematics]. Nanjing: Jiangsu jiaoyu chubanshe 江苏教育出版社, 2004.

Li Di 李迪 and Guo Shirong 郭世荣. *Mei Wending: Qing dai zhuming tianwen shuxue jia* 梅文鼎：清代著名天文数学家 [Mei Wending: Famous astronomer and mathematician of the Qing dynasty]. Shanghai: Shanghai kexue jishu wenxian chubanshe 上海科学技术文献出版社, 1988.

Li Jimin 李继闵. *Jiuzhang suanshu jiaozheng* 九章算术校证 [Nine chapters on the mathematical arts, critical edition]. Xi'an: Shaanxi kexue jishu chubanshe 陕西科学技术出版社, 1993.

———. *Suanfa de yuanliu: Dongfang gudian shuxue de tezheng* 算法的源流：东方古典数学的特征 [The origins of mathematics: Distinctive characteristics of classical Eastern mathematics]. Edited by Qu Anjing 曲安京 and Jin Yingji 金英姬. Beijing: Kexue chubanshe 科学出版社, 2007.

Li Yan 李儼. "Zhongguo jinguqi zhi suanxue" 中國近古期之算學 [Chinese mathematics from the Song to Qing]. *Xueyi* 學藝 9, nos. 4–5 (1928): 1–28.

———. *Zhongguo suanxue shi* 中國算學史 [History of Chinese mathematics]. Shanghai: Shangwu yinshuguan 商務印書館, 1937.

———. "Zengxiu Mingdai suanxue shuzhi" 增修明代算學書志 [Additions and revisions to the bibliography of mathematical works of the Ming Dynasty]. In *Zhongguo suanxue shi luncong*.

———. *Zhongguo suanxue shi luncong* 中國算學史論叢 [Collected essays on the history of Chinese mathematics]. Taibei: Zhengzhong shuju 正中書局, 1954.

Li Yan 李儼. "Jindai Zhongsuan zhushu ji" 近代中算著述記 [Notes on compilations of Chinese mathematics in the modern period]. In *Zhong suan shi lun
cong.*

———. "Ming dai suanxue shu zhi" 明代算學書志 [Bibliography of mathematical
works of the Ming Dynasty]. In *Zhong suan shi lun cong,* 2: 86–102.

———. "Zhong suan ru Riben zhi jingguo" 中算入日本之經過 [The process of
the introduction of Chinese mathematics into Japan]. In *Zhong suan shi lun
cong,* 5:168–86.

———. *Zhong suan shi lun cong* 中算史論叢 [Collected essays on the history of
Chinese mathematics]. 5 vols. Beijing: Zhongguo kexueyuan 中國科學院,
1954–55.

———. "Zhongsuanjia zhi Pythagoras dingli yanjiu" 中算家之 Pythagoras 定理
研究 [Research on the Pythagorean theorem of Chinese mathematicians]. In
Zhong suan shi lun cong.

———. *Zhongguo gudai shuxue shiliao* 中國古代數學史料 [Historical materials
on mathematics in ancient China]. Shanghai: Shanghai kexue jishu chubanshe 上海科学技术出版社, 1956.

———. *Zhongguo shuxue dagang* 中國數學大綱 [An outline of Chinese mathe-
matics]. Beijing: Kexue chubanshe 科学出版社, 1958.

———. "Woguo diyiben weijifenxue de yiben *Dai wei ji shi ji* chuban yibai
zhounian" 我國的一本微積分學的譯本代微積拾級出版一周年 [The
hundredth-year anniversary of the publication of China's first translation of
calculus, *Dai wei ji shi ji*]. *Kexueshi jikan* 科學史季刊 3 (1960): 59–64.

Li Yan 李儼 and Li Di 李迪. *Zhongguo shuxue shi lunwen mulu, guonei zhi bu*
中國數學史論文目錄國內之部 [A bibliography of essays on the history
of Chinese mathematics published in China]. [Hohhot]: Neimenggu shifan
xueyuan keyanchu 内蒙古师范学院科研处, 1980.

Li Yan 李儼 and Qian Baocong 钱宝琮. *Li Yan Qian Baocong kexueshi quanji*
李俨钱宝琮科学史全集 [Complete collection of works by Li Yan and
Qian Baocong on the history of science]. 10 vols. Shenyang: Liaoning jiaoyu
chubanshe 辽宁教育出版社, 1998.

Li Zhaohua 李兆华 and Guo Shuchun 郭书春, eds. *Zhongguo kexue jishu shi:
Shuxue juan* 中国科学技术史：数学卷 [History of Chinese science and
technology: Mathematics volume]. Beijing: Kexue chubanshe 科学出版社,
2010.

Liang Jiamian 梁家勉. *Xu Guangqi nianpu* 徐光啟年譜 [Chronological biography
of Xu Guangqi]. Shanghai: Shanghai guji chubanshe 上海古籍出版社, 1981.

Liang Qichao 梁啟超 (1873–1929). *Zhongguo jin sanbai nian xueshushi* 中國近
三百年學術史 [History of Chinese intellectual thought during the last 300
years]. Yangzhou: Jiangsu Guangling guji keyinshe 江蘇廣陵古籍刻印社,
1990.

Liu Bohan 刘伯涵. "Lue lun Xu Guangqi yu Ming mo dangzheng" 略论徐光启
与明末党争 [A brief discussion of Xu Guangqi and factional strife in the late
Ming]. In Xi and Wu, *Xu Guangqi yanjiu lunwenji.*

Liu Dun 刘钝 and Wang Yangzong 王扬宗, eds. *Zhongguo kexue yu kexue geming: Li Yuese nanti ji qi xiangguan wenti yanjiu lunzhu xuan* 中国科学与科学革命：李约瑟难题及其相关问题研究论著选 [Chinese science and scientific revolution: Selected writings on the Needham problem and related questions]. Shenyang: Liaoning jiaoyu chubanshe 辽宁教育出版社, 2002.

Mei Rongzhao 梅荣照. "Xu Guangqi de shuxue gongzuo" 徐光启的数学工作 [Xu Guangqi's work in mathematics]. In Zhongguo ziran kexueshi yanjiushi, *Xu Guangqi jinian lunwenji*.

———. "Ming-Qing shuxue gailun" 明清數學概論 [Outline of Ming-Qing mathematics]. In *Ming-Qing shuxueshi lunwenji*, 1–20.

———, ed. *Ming-Qing shuxueshi lunwenji* 明清數學史論文集 [Collected essays on the history of Ming-Qing mathematics]. Nanjing: Jiangsu jiaoyu chubanshe 江苏教育出版社, 1990.

Mei Rongzhao 梅荣照, Liu Dun 刘钝, and Wang Yusheng 王渝生. "Oujilide *Yuanben* de chuanru he dui woguo Ming Qing shuxue fazhan de yingxiang" 欧几里得《原本》的传入和对我国明清数学发展的影响 [The transmission of Euclid's *Elements* into China and its influence on the development of Ming and Qing Dynasty mathematics]. In Xi and Wu, *Xu Guangqi yanjiu lunwenji*.

Mei Rongzhao 梅荣照 and Wang Yusheng 王渝生. "Xu Guangqi de shuxue sixiang" 徐光启的数学思想 [Mathematical thought of Xu Guangqi]. In Xi and Wu, *Xu Guangqi yanjiu lunwenji*, 37–44.

Mo Shaokui 莫绍揆. "Youguan *Jiuzhang suanshu* de yixie taolun" 有关《九章算术》的一些讨论 [Several points about the Nine chapters on the mathematical arts]. *Ziran kexueshi yanjiu* 自然科学史研究 19, no. 2 (2000): 97–113.

Qian Baocong 钱宝琮. *Suan jing shi shu* 算經十書 [Mathematical classics, ten books]. Beijing: Zhonghua shuju 中華書局, 1963.

———. *Zhongguo shuxue shi* 中国数学史 [History of Chinese mathematics]. Beijing: Kexue chubanshe 科学出版社, 1964.

———. "Zeng cheng kaifang fa de lishi fa zhan" 增乘開方法的歷史發展 [Historical development of the *zeng cheng* method for root extraction]. In Qian et al., *Song Yuan shuxueshi lunwenji*, 36–59.

———. *Qian Baocong kexueshi lunwen xuanji* 钱宝琮科学史论文选集 [Selected essays by Qian Baocong on the history of science]. Edited by Zhongguo kexueyuan ziran kexueshi yanjiusuo 中国科学院自然科学史研究所. Beijing: Kexue chubanshe 科学出版社, 1983.

———. "Song-Yuan shiqi shuxue yu daoxue de guanxi" 宋元时期数学与道学的关系 [Relationship between mathematics and Song-Yuan Learning of the Way]. In *Qian Baocong kexueshi lunwen xuanji*, 225–40.

Qian Baocong 钱宝琮 et al., eds. *Song Yuan shuxueshi lunwenji* 宋元數學史論文集 [Collected essays on the history of mathematics during the Song and Yuan Dynasties]. Beijing: Kexue chubanshe 科学出版社, 1966.

Qian Mu 錢穆. *Zhongguo jin sanbai nain xueshushi* 中國近三百年學術史 [Chinese intellectual history of the last 300 years]. Taibei: Taiwan shangwu yinshuguan 臺灣商務印書館, 1957.

Rong Zhaozu 容肇祖. *Jiao Hong ji qi sixiang* 焦竑及其思想 [Jiao Hong and his thought]. Beijing: Yanjing daxue Hafo Yanjing xueshe 燕京大學哈佛燕京學社, 1938.

———. *Rong Zhaozu ji* 容肇祖集 [Collected works of Rong Zhaozu]. Jinan: Qilu shushe 齊魯書社, 1989.

Shanghai tushuguan 上海圖書館, ed. *Zhongguo congshu zonglu* 中國叢書綜錄 [General catalogue of Chinese collectanea]. Shanghai: Shanghai guji chubanshe 上海古籍出版社, 1986.

Shen Kangshen 沈康身, ed. *Zaoqi shuxue wenxian* 早期数学文献 [Early mathematical treatises]. Zhongguo shuxueshi daxi 中国数学史大系. Supplemental volume 1. Beijing: Beijing shifan daxue chubanshe 北京师范大学出版社, 2004.

Shi Tingyong 施廷镛, ed. *Zhongguo congshu zonglu xu bian* 中国丛书综录续编 [Continuation of the general catalogue of Chinese collectanea]. Beijing: Beijing tushuguan chubanshe 北京图书馆出版社, 2003.

Song Haojie 宋浩杰, ed. *Zhongxi wenhua huitong di yi ren: Xu Guangqi xueshu yantaohui lunwenji* 中西文化会通第一人：徐光启学术研讨会论文集 "The first person that connected Western and Chinese culture together: Essays for the learning proseminar on Xu Guangqi." Shanghai: Shanghai guji chubanshe 上海古籍出版社, 2006.

Sun Shangyang 孙尚扬 and Tang Yijie 汤一介. *Li Madou yu Xu Guangqi* 利玛窦与徐光启 [Matteo Ricci and Xu Guangqi]. Beijing: Xinhua chubanshe 新华出版社, 1993.

Sun Yuxiu 孫毓修, ed. *Ge sheng jin cheng shumu* 各省進呈書目 [Bibliography of works presented from all the provinces]. Shanghai: Shangwu yinshuguan 商務印書館, 1916–1921.

Takeda Kusuo 武田楠雄. *Dōbunzanshi no seiritsu* 同文算指の成立 [Inception of the *Guide to Calculation in the Unified Script*]. Tokyo: Iwanami Shoten 岩波書店, 1954.

Tian Miao 田淼. *Zhongguo shuxue de Xihua licheng* 中国数学的西化历程 "The Westernization of mathematics in China." Jinan: Shandong jiaoyu chubanshe 山东教育出版社, 2005.

Wang Fukang 王福康. "Jin sanbai nian Xu Guangqi yanjiu zhuzuo lunwen mulu" 近三百年徐光啟研究著作：論文目錄 [A bibliography of the research works and essays on Xu Guangqi in the last three hundred years]. In Zhongguo ziran kexueshi yanjiushi, *Xu Guangqi jinian lunwenji*.

Wang Ping 王萍. *Xifang lisuanxue zhi chuanru* 西方曆算學之輸入 [The introduction of Western astronomical and mathematical sciences into China]. Zhongyang yanjiuyuan jindai shi yanjiusuo zhuankan 中央研究院近代史研究所專刊 17. Taibei: Zhongyang yanjiuyuan jindaishi yanjiusuo 中央研究院近代史研究所, 1966.

Wang Qingjian 王青建. *Kexue yizhu xianshi: Xu Guangqi* 科学译著先师：徐光啓 [Early master of translation and authorship of science: Xu Guangqi]. Beijing: Kexue chubanshe 科学出版社, 2000.

Wang Rongbao 汪榮寶 (1878–1933), ed. *Fa yan yi shu* 法言義疏 [*Model Words*, with meanings and commentary*]. 2 vols. Reprinted with punctuation and emendations by Chen Zhongfu 陳仲夫. Beijing: Zhonghua shuju 中華書局, 1987.

Wang Shounan 王壽南. *Xu Guangqi* 徐光啓 [Xu Guangqi]. Taibei: Taiwan shangwu yinshuguan 臺灣商務印書館, 2007.

Wang Xinzhi 王欣之. *Ming dai da kexuejia Xu Guangqi* 明代大科学家徐光启 [Xu Guangqi, great scientist of the Ming dynasty]. Shanghai: Shanghai renmin chubanshe 上海人民出版社, 1985.

Wang Yusheng 王渝生. *Gougu yi tiyao* 勾股義提要 [Summary of *Right Triangles, Meanings*]. In *ZKJDT*.

———. *Zhongguo suanxue shi* 中国算学史 [History of Chinese mathematics]. Shanghai: Shanghai renmin chubanshe 上海人民出版社, 2006.

Wang Zhongmin 王重民. *Xu Guangqi ji* 徐光啟集 [Collected writings of Xu Quangqi]. Beijing: Zhonghua shuju 中華書局, 1963.

———. *Xu Guangqi* 徐光啟 [Xu Quangqi]. Shanghai: Shanghai renmin chubanshe 上海人民出版社, 1981.

———. "Xu yan—Xu Guangqi zhili kexue yanjiu de shiji he ta zai woguo kexue shi shang de chengjiu" 序言—徐光啟致力科學研究的事蹟和他在我國科學史上的成就 [Preface—The achievements of Xu Guangqi's devotion to scientific research and his accomplishments in the history of Chinese science]. In *XGQJ*.

Wu Weizu 吳慰祖, ed. *Siku cai jin shumu* 四庫採進書目 [Bibliography of books presented for the Four Treasuries]. Beijing: Shangwu yinshuguan 商務印書館, 1960. Originally published as Sun Yuxiu 孫毓修, ed., *Ge sheng jin cheng shumu* 各省進呈書目 [Bibliography of works presented from all the provinces] (Shanghai: Shangwu yinshuguan 商務印書館, 1916–1921).

Wu Wen-tsün 吳文俊, ed. *Zhongguo shuxueshi lunwenji* 中国数学史论文集 [Collected essays on the history of Chinese mathematics]. 4 vols. Jinan: Shandong jiaoyu chubanshe 山东教育出版社, 1985–1996.

———, ed. *Zhongguo shuxueshi daxi* 中国数学史大系 [Compendium of the history of Chinese mathematics]. Beijing: Beijing shifan daxue chubanshe 北京师范大学出版社, 1998–.

Wu Xin 吳馨 (1873–1919), ed. *Shanghai xianzhi* 上海縣志 [Records of Shanghai county]. Taibei: Chengwen chubanshe 成文出版社, 1935.

Xi Zezong 席泽宗 and Wu Deduo 吴德铎, eds. *Xu Guangqi yanjiu lunwenji* 徐光启研究论文集 [Essays on research on Xu Guangqi]. Shanghai: Xuelin chubanshe 学林出版社, 1986.

Xu hui qu wenhua ju 徐汇区文化局, ed. *Xu Guangqi yu "Jihe yuanben": "Jinian Xu Guangqi ji 'Jihe yuanben' fanyi chuban si bai zhounian guoji xueshu yantaohui" lunwen ji* 徐光启与《几何原本》：『纪念徐光启暨《几何原本》翻译出版四百周年国际学术研讨会』论文集 [Xu Guangqi and Euclid's *Elements*: Collected essays from the international academic conference commemorating Xu Guangqi and the four-hundredth anniversary of the publication of the translation of Euclid's *Elements*]. Shanghai: Shanghai jiaotong daxue chubanshe 上海交通大学出版社, 2011.

Xu Zongze 徐宗澤. *Ming-Qing jian Yesu huishi yi zhu tiyao* 明清間耶穌會士譯著提要 [Annotated bibliography of the translations and writings of the Jesuits during the Ming and Qing dynasties]. Taibei: Zhonghua shuju 中華書局, 1989.

Yan Dunjie 严敦杰. "Oujilide de *Jihe yaunben* Yuandai shuru Zhongguo shuo" 歐几里得幾何原本元代輸入中國說 [A theory that the *Elements* of Euclid entered China in the Yuan Dynasty]. *Dongfang zazhi* 東方雜志 13, no. 39 (1943): 59–61.

———. "Jihe bushi 'Geo' de yiyin" 几何不是 Geo 的译音 ["Jihe" is not a transliteration of "geo"]. *Shuxue tongbao* 教学通报 11 (1959): 31.

———. "Ming Qing zhi ji Xifang chuanru Zhongguo zhi lisuan jilu" 明清之際西方傳入中國之曆算記錄 [Record of the transmission of calendrics and mathematics into China during the Ming-Qing period]. In Mei, *Ming-Qing shuxueshi lunwenji*, 114–81.

Yan Dunjie 严敦杰 and Mei Rongzhao 梅荣照. "Cheng Dawei ji qi shuxue zhuzuo" 程大位及其數學著作 [Cheng Dawei and his mathematical works]. In Mei, *Ming-Qing shuxueshi lunwenji*, 26–52.

Yang Bojun 楊伯峻 (1909–1992). *Lun yu yi zhu* 論語譯注 [*Analects*, with explanations and commentary]. Beijing: Zhonghua shuju 中華書局, 1980.

Yang Haiqing 陽海清 and Jiang Xiaoda 蔣校達, eds. *Zhongguo congshu zonglu buzheng* 中國叢書綜錄補正 [Supplement to the general catalogue of Chinese collectanea]. Yangzhou: Jiangsu Guangling guji keyinshe 江蘇廣陵古籍刻印社, 1984.

Yu Shiyi 余石屹. *Han yi ying lilun duben* 汉译英理论读本 [Theoretical reader on translating Chinese into English]. Beijing: Kexue chubanshe 科学出版社, 2008.

Zhang Baichun 张柏春, Tian Miao 田淼, Matthias Schemmel, Jürgen Renn, and Peter Damerow. *Chuanbo yu huitong:* Qi qi tu shuo *yanjiu yu jiaozhu* 传播与会通：《奇器图说》研究与校注 [Transmission and integration: *Qi qi tu shuo* (Marvelous devices, with illustrations and explanations), research and emendations]. 2 vols. Nanjing: Jiangsu kexue jishu chubanshe 江苏科学技术出版社, 2008.

Zhongguo kexueyuan Zhongguo ziran kexueshi yanjiushi 中国科学院中国自然科学史研究室, ed. *Xu Guangqi jinian lunwenji* 徐光启纪念论文集 [Collection of essays in commemoration of Xu Guangxi]. Beijing: Zhonghua shuju 中华书局, 1963.

Zhu Chuanyu 朱傳譽, ed. *Xu Guangqi zhuanji ziliao* 徐光啟傳記資料 [Biographic materials on Xu Guangqi]. Taibei: Tianyi chubanshe 天一出版社, 1979.

Zhu Qianzhi 朱謙之 (1899–1972). *Laozi jiao shi* 老子校釋 [*Laozi*, with emendations and explanations]. Beijing: Zhonghua shuju 中华书局, 1984.

Zou Dahai 邹大海. *Zhongguo shuxue de xingqi yu xian-Qin shuxue* 中国数学的兴起与先秦数学 [The emergence of Chinese mathematics and pre-Qin mathematics]. Zhongguo shuxueshi daxi 中国数学史大系. Shijiazhuang: Hebei kexue jishu chubanshe 河北科学技术出版社, 2001.

Western Language Sources

Abu-Lughod, Janet L. *Before European Hegemony: The World System A.D. 1250–1350*. New York: Oxford University Press, 1989.

Adas, Michael. *Machines as the Measure of Men: Science, Technology, and Ideologies of Western Dominance*. Ithaca, NY: Cornell University Press, 1989.

Alexander, Amir R. *Geometrical Landscapes: The Voyages of Discovery and the Transformation of Mathematical Practice*. Writing Science 1. Stanford, CA: Stanford University Press, 2002.

Althusser, Louis. *Essays on Ideology*. London: Verso, 1976.

———. *Philosophy and the Spontaneous Philosophy of the Scientists & Other Essays*. Edited by Gregory Elliott. Translated by Ben Brewster. London: Verso, 1990.

Amiot, Joseph Marie, and Pierre-Joseph Roussier. *Mémoire sur la musique des Chinois*. Genève: Minkoff Reprint, 1973. Originally published: Paris: Nyon, 1779.

Anderson, Benedict R. *The Spectre of Comparisons: Nationalism, Southeast Asia, and the World*. London: Verso, 1998.

———. "Reimagining Asia." *Kyoto Journal* 45 (2000): 22–27.

———. "Western Nationalism and Eastern Nationalism: Is There a Difference That Matters?" *New Left Review* 9 (May-June 2001): 31–42.

———. *Imagined Communities: Reflections on the Origin and Spread of Nationalism*. Rev. ed. London: Verso, 2006. First ed. published in 1983.

———. "Introduction." In Mullaney, *Coming to Terms with the Nation*, xv–xx.

Anderson, Perry. *In the Tracks of Historical Materialism*. Chicago: University of Chicago Press, 1984.

———. "Sinomania." *London Review of Books* 32, no. 2 (2010): 3–6.

Andrews, Bridie, and Andrew Cunningham. *Western Medicine as Contested Knowledge*. Studies in Imperialism. New York: Manchester University Press, 1997.

Appadurai, Arjun. *Modernity at Large: Cultural Dimensions of Globalization*. Public Worlds 1. Minneapolis, MN: University of Minnesota Press, 1996.

Asad, Talal. "The Concept of Cultural Translation in British Social Anthropology." In Clifford and Marcus, *Writing Culture*.

———. *Genealogies of Religion: Discipline and Reasons of Power in Christianity and Islam*. Baltimore: Johns Hopkins University Press, 1993.

Ashmore, Malcolm. *The Reflexive Thesis: Wrighting Sociology of Scientific Knowledge*. Chicago: University of Chicago Press, 1989.

Aspray, William, and Philip Kitcher, eds. *History and Philosophy of Modern Mathematics*. Minnesota Studies in the Philosophy of Science 11. Minneapolis: University of Minnesota Press, 1988.

Austin, J. L. *How to Do Things with Words*. 2nd ed. William James Lectures, 1955. Cambridge, MA: Harvard University Press, 1975.

———. *Philosophical Papers*. 3rd ed. Edited by J. O. Urmson and G. J. Warnock. Oxford: Oxford University Press, 1979.

Bachelard, Gaston. *The Formation of the Scientific Mind: A Contribution to a Psychoanalysis of Objective Knowledge*. Translated by Mary McAllester Jones. Philosophy of Science. Manchester: Clinamen, 2002. Originally published as *La formation de l'esprit scientifique: Contribution à une psychanalyse de la connaissance objective* (Paris: J. Vrin, 1938).

Baker, Donald L. "Jesuit Science through Korean Eyes." *Journal of Korean Studies* 4 (1982–1983): 207–39.

Bakhtin, M. M. *The Dialogic Imagination: Four Essays*. Translated by Caryl Emerson and Michael Holquist. University of Texas Press Slavic Series 1. Austin: University of Texas Press, 1981. Originally published as *Voprosy literatury i estetiki* (Moscow: Khudojestvennaja literatura, 1975).

———. *Speech Genres and Other Late Essays*. Edited by Caryl Emerson and Michael Holquist. Translated by Vern W. McGee. University of Texas Press Slavic Series 8. Austin: University of Texas Press, 1986. Originally published as *Estetika slovesnogo tvorchestva* (Moscow: Iskusstvo, 1979).

Bala, Arun. *The Dialogue of Civilizations in the Birth of Modern Science*. New York: Palgrave Macmillan, 2006.

———, ed. *Asia, Europe, and the Emergence of Modern Science: Knowledge Crossing Boundaries*. New York, NY: Palgrave Macmillan, 2012.

Banach, Stefan, and Alfred Tarski. "Sur la décomposition des ensembles de points en parties respectivement congruentes." *Fundamenta Mathematicae* 6 (1924): 244–77.

Barthes, Roland. *The Rustle of Language*. Translated by Richard Howard. New York: Hill and Wang, 1986. Originally published as *Le bruissement de la langue* (Paris: Seuil, 1984).

Batchelor, Robert. "On the Movement of Porcelains: Rethinking the Birth of Consumer Society as Interactions of Exchange Networks, 1600–1750." In *Consuming Cultures, Global Perspectives: Historical Trajectories, Transnational Exchanges*, edited by John Brewer and Frank Trentmann, 95–122. Oxford: Berg, 2006.

Baudrillard, Jean. *Symbolic Exchange and Death*. Translated by Iain Hamilton Grant. Theory, Culture & Society. Thousand Oaks, CA: Sage Publications, 1993. Originally published as *L'échange symbolique et la mort* (Paris: Gallimard, 1976).

Bays, Daniel H. *A New History of Christianity in China*. Blackwell Guides to Global Christianity. Malden, MA: Wiley-Blackwell, 2012.

Benacerraf, Paul, and Hilary Putnam, eds. *Philosophy of Mathematics: Selected Readings.* Prentice-Hall Philosophy Series. Englewood Cliffs, NJ: Prentice-Hall, 1964.

———, eds. *Philosophy of Mathematics: Selected Readings.* 2nd ed. Cambridge: Cambridge University Press, 1983.

Benite, Zvi Ben-Dor. *The Dao of Muhammad: A Cultural History of Muslims in Late Imperial China.* Harvard East Asian Monographs 248. Cambridge, MA: Harvard University Asia Center, 2005.

Benjamin, Walter. *Illuminations.* Edited by Hannah Arendt. Translated by Harry Zohn. New York: Harcourt, Brace & World, 1968.

Benoit, Paul, Karine Chemla, and Jim Ritter, eds. *Histoire de fractions, fractions d'histoire.* Science Networks: Historical Studies 10. Berlin: Birkhäuser, 1992.

Benveniste, Emile. *Problems in General Linguistics.* Translated by Mary Elizabeth Meek. Coral Gables, FL: University of Miami Press, 1971. Originally published as *Problèmes de linguistique générale* (Paris: Gallimard, 1966).

Ben-Zaken, Avner. *Cross-Cultural Scientific Exchanges in the Eastern Mediterranean, 1560–1660.* Baltimore: Johns Hopkins University Press, 2010.

Berggren, J. L. *Episodes in the Mathematics of Medieval Islam.* New York: Springer, 2003. First edition published 1986.

Berling, Judith A. *The Syncretic Religion of Lin Chao-En.* IASWR Series. New York: Columbia University Press, 1980.

Bernal, Martin. *Black Athena Writes Back: Martin Bernal Responds to His Critics.* Edited by David Chioni Moore. Durham, NC: Duke University Press, 2001.

Bernard, Henri, SJ. *Le père Matthieu Ricci et la société chinoise de son temps (1552–1610).* 2 vols. Tientsin: Hautes études, 1937.

———. "Les adaptations chinoises d'ouvrages européens: Bibliographie chronologique depuis la venue des Portugais à Canton jusqu'à la mission française de Pékin (1514–1688)." *Monumenta Serica* 10 (1945): 1–57, 309–88.

———. *Matteo Ricci's Scientific Contribution to China.* Translated by Edward Chalmers Werner. Westport, CT: Hyperion Press, 1973. Originally published as *L'apport scientifique du père Matthieu Ricci à la Chine* (Beijing: H. Vetch, 1935).

Bhabha, Homi K., ed. *Nation and Narration.* New York: Routledge, 1990.

———. *The Location of Culture.* New York: Routledge, 1994.

Biagioli, Mario. "The Social Status of Italian Mathematicians: 1450–1600." *History of Science* 17 (1989): 41–95.

———. "The Anthropology of Incommensurability." *Studies in History and Philosophy of Science* 21, no. 2 (1990): 183–209.

———. *Galileo, Courtier: The Practice of Science in the Culture of Absolutism.* Science and Its Conceptual Foundations. Chicago: University of Chicago Press, 1993.

———, ed. *The Science Studies Reader.* New York: Routledge, 1999.

———. "Stress in the Book of Nature: The Supplemental Logic of Galileo's Realism." *MLN* 118, no. 3, German Issue (April 2003): 557–85.

Biagioli, Mario. *Galileo's Instruments of Credit: Telescopes, Images, Secrecy.* Chicago: University of Chicago Press, 2006.

———. "Postdisciplinary Liaisons: Science Studies and the Humanities." *Critical Inquiry* 35 (2009): 816–33.

Biagioli, Mario, and Peter L. Galison, eds. *Scientific Authorship: Credit and Intellectual Property in Science.* New York: Routledge, 2003.

Biderman, Shlomo, and Ben-Ami Scharfstein, eds. *Rationality in Question: On Eastern and Western Views of Rationality.* Leiden: Brill, 1989.

Billeter, Jean François. *Li Zhi, philosophe maudit (1527–1602): Contribution à une sociologie du mandarinat chinois de la fin des Ming.* Travaux de droit, d'économie, de sociologie et de sciences politiques 116. Genève: Droz, 1979.

Bishop, John. "Some Limitations of Chinese Fiction." *Far Eastern Quarterly* 15 (1956): 239–47.

Blair, Ann. *The Theater of Nature: Jean Bodin and Renaissance Science.* Princeton, NJ: Princeton University Press, 1997.

Bloch, Marc. "The Advent and Triumph of the Watermill." In *Land and Work in Mediaeval Europe: Selected Papers,* translated by J. E. Anderson, 136–68. Berkeley: University of California Press, 1967.

Bloom, Alfred H. *The Linguistic Shaping of Thought: A Study in the Impact of Language on Thinking in China and the West.* Hillsdale, NJ: Lawrence Erlbaum, 1981.

Bloor, David. *Wittgenstein: A Social Theory of Knowledge.* Contemporary Social Theory: Theoretical Traditions in the Social Sciences. London: Macmillan Press, 1983.

———. *Knowledge and Social Imagery.* 2nd ed. Chicago: University of Chicago Press, 1991. First ed. published in 1976.

Bloor, David, and Barry Barnes. "Relativism, Rationalism and the Sociology of Knowledge." In Hollis and Lukes, *Rationality and Relativism,* 21–47.

Blue, Gregory. "Xu Guangqi in the West: Early Jesuit Sources and the Construction of an Identity." In Jami et al., *Statecraft and Intellectual Renewal,* 19–71.

Bodde, Derk. "The Attitude toward Science and Scientific Method in Ancient China." *T'ien Hsia Monthly* 2 (1936): 139–60.

———. "Evidence for 'Laws of Nature' in Chinese Thought." *Harvard Journal of Asiatic Studies* 20, no. 3/4 (1959): 709–27.

———. "Chinese 'Laws of Nature': A Reconsideration." *Harvard Journal of Asiatic Studies* 39, no. 1 (1979): 139–55.

———. *Chinese Thought, Society, and Science: The Intellectual and Social Background of Science and Technology in Pre-Modern China.* Honolulu: University of Hawaii Press, 1991.

Bol, Peter K. *"This Culture of Ours": Intellectual Transitions in T'ang and Sung China.* Stanford: Stanford University Press, 1992.

———. *Neo-Confucianism in History.* Harvard East Asian Monographs 307. Cambridge, MA: Harvard University Asia Center, 2008.

Boncompagni, Baldassare, ed. *Scritti di Leonardo Pisano matematico del secolo decimoterzo. I. Il Liber abbaci di Leonardo Pisano.* Roma: Tipografia delle Scienze Matematiche e Fisiche, 1857.

Bono, James J. *The Word of God and the Languages of Man: Interpreting Nature in Early Modern Science and Medicine.* Science and Literature. Madison: University of Wisconsin Press, 1995.

Boodberg, Peter A. "Philological Notes on Chapter One of the Lao Tzu." *Harvard Journal of Asiatic Studies* 20 (1957): 598–618.

Borrel, Jean (c. 1492–c. 1572?). *Ioan. Buteonis Logistica: quae & arithmetica vulgò dicitur, in libros quinque digesta quorum index summatim habetur in tergo eiusdem, ad locum Vitruuij corruptum restitutio, qui est de proportione lapidum mittendorum ad balistae foramen, libro decimo.* Lugduni: apud Gulielmum Rouillium . . . , 1559.

Bos, Henk J. M. *Redefining Geometrical Exactness: Descartes' Transformation of the Early Modern Concept of Construction.* Sources and Studies in the History of Mathematics and Physical Sciences. New York: Springer, 2001.

Bourdieu, Pierre. *Outline of a Theory of Practice.* Translated by Richard Nice. Cambridge Studies in Social Anthropology 16. Cambridge: Cambridge University Press, 1977. Originally published as *Esquisse d'une théorie de la pratique* (Genève: Droz, 1972).

———. "The Field of Cultural Production, or: The Economic World Reversed." *Poetics* 12 (1983): 311–56.

———. *In Other Words: Essays Towards a Reflexive Sociology.* Translated by Matthew Adamson. Stanford, CA: Stanford University Press, 1990. Originally published as *Choses dites* (Paris: Éditions de Minuit, 1987).

———. *The Logic of Practice.* Translated by Richard Nice. Stanford, CA: Stanford University Press, 1990. Originally published as *Le sens pratique* (Paris: Éditions de Minuit, 1980).

———. *Language and Symbolic Power.* Translated by Gino Raymond and Matthew Adamson. Cambridge, MA: Harvard University Press, 1991. Originally published as *Ce que parler veut dire: L'économie des échanges linguistiques* (Paris: Fayard, 1982).

Bourdieu, Pierre, and Jean-Claude Passeron. *Reproduction: In Education, Society and Culture.* Translated by Richard Nice. Sage Studies in Social and Educational Change 5. London: Sage Publications, 1977. Originally published as *La reproduction: Éléments pour une théorie du système d'enseignement* (Paris: Éditions de Minuit, 1970).

Brandom, Robert. *Making It Explicit: Reasoning, Representing, and Discursive Commitment.* Cambridge, MA: Harvard University Press, 1994.

———. *Between Saying and Doing: Towards an Analytic Pragmatism.* Oxford: Oxford University Press, 2010.

Bray, Francesca. *Technology and Gender: Fabrics of Power in Late Imperial China.* Berkeley: University of California Press, 1997.

Bray, Francesca, Vera Dorofeeva-Lichtmann, and Georges Métailié, eds. *Graphics and Text in the Production of Technical Knowledge in China: The Warp and the Weft.* Sinica Leidensia 79. Leiden: Brill, 2007.

Bray, Francesca, and Georges Métailié. "Who Was the Author of the *Nongzhen quanshu*?" In Jami et al., *Statecraft and Intellectual Renewal*, 322–59.

Bretelle-Establet, Florence, ed. *Looking at It from Asia: The Processes that Shaped the Sources of History of Science.* Boston Studies in the Philosophy of Science 265. New York: Springer, 2010.

Brockey, Liam Matthew. *Journey to the East: The Jesuit Mission to China, 1579–1724.* Cambridge, MA: Belknap Press of Harvard University Press, 2007.

Brokaw, Cynthia J. *The Ledgers of Merit and Demerit: Social Change and Moral Order in Late Imperial China.* Princeton, NJ: Princeton University Press, 1991.

Brokaw, Cynthia J., and Kai-wing Chow, eds. *Printing and Book Culture in Late Imperial China.* Studies on China 27. Berkeley: University of California Press, 2005.

Brook, Timothy. *Praying for Power: Buddhism and the Formation of Gentry Society in Late-Ming China.* Harvard-Yenching Institute Monograph Series 38. Cambridge, MA: Harvard University Press, 1994.

———. *The Confusions of Pleasure: Commerce and Culture in Ming China.* Berkeley: University of California Press, 1998.

———. "Xu Guangqi in His Context: The World of the Shanghai Gentry." In Jami et al., *Statecraft and Intellectual Renewal*, 72–98.

———. *The Chinese State in Ming Society.* Critical Asian Scholarship. London: RoutledgeCurzon, 2005.

Brooks, E. Bruce, and A. Taeko Brooks. *The Original Analects: Sayings of Confucius and His Successors; A New Translation and Commentary.* Translations from the Asian Classics. New York: Columbia University Press, 1998.

Brower, Reuben A., ed. *On Translation.* Harvard Studies in Comparative Literature 23. Reprint, New York: Oxford University Press, 1966. Cambridge, MA: Harvard University Press, 1959.

Buck, David D. "Forum on Universalism and Relativism in Asian Studies: Editor's Introduction." *Journal of Asian Studies* 50, no. 1 (February 1991): 29–34.

Bulliet, Richard W. *The Case for Islamo-Christian Civilization.* New York: Columbia University Press, 2004.

Burke, Peter, ed. *New Perspectives on Historical Writing.* 2nd ed. University Park: Pennsylvania State University Press, 2001.

Burnett, Charles, Jan P. Hogendijk, Kim Plofker, and Michio Yano, eds. *Studies in the History of the Exact Sciences in Honour of David Pingree.* Islamic Philosophy, Theology, and Science. Texts and Studies 54. Leiden: Brill, 2004.

Busard, Hubert L. L. *Latin Translation of the Arabic Version of Euclid's Elements Commonly Ascribed to Gerard of Cremona.* Leiden: E. J. Brill, 1984.

———. *The Mediaeval Latin Translation of Euclid's Elements: Made Directly from the Greek.* Boethius 15. Stuttgart: F. Steiner Verlag Wiesbaden, 1987.

Butler, Judith. *Excitable Speech: A Politics of the Performative*. New York: Routledge, 1997.

Butterfield, Herbert. *The Whig Interpretation of History*. London: G. Bell, 1931.

———. *The Origins of Modern Science, 1300–1800*. London: G. Bell, 1949.

Bynon, Theodora. *Historical Linguistics*. Cambridge Textbooks in Linguistics. Cambridge: Cambridge University Press, 1977.

Cajori, Florian. *A History of Mathematics*. 2nd ed. New York: Macmillan, 1919.

Calinescu, Matei. *Five Faces of Modernity: Modernism, Avant-Garde, Decadence, Kitsch, Postmodernism*. Durham, NC: Duke University Press, 1987.

Callon, Michel. "What Does It Mean to Say That Economics Is Performative?" In *Do Economists Make Markets? On the Performativity of Economics*, edited by Donald A. MacKenzie, Fabian Muniesa, and Lucia Siu. Princeton, NJ: Princeton University Press, 2007.

Canguilhem, Georges. *Ideology and Rationality in the History of the Life Sciences*. Cambridge, MA: MIT Press, 1988. Originally published as *Idéologie et rationalité dans l'histoire des sciences de la vie: Nouvelles études d'histoire et de philosophie des sciences* (Paris: J. Vrin, 1977).

Cañizares-Esguerra, Jorge. *Nature, Empire, and Nation: Explorations of the History of Science in the Iberian World*. Stanford, CA: Stanford University Press, 2006.

Carnap, Rudolf. *Foundations of Logic and Mathematics*. International Encyclopedia of Unified Science, 1, no. 3. Chicago: University of Chicago Press, 1939.

Cassirer, Ernst. *The Philosophy of Symbolic Forms*. New Haven: Yale University Press, 1996.

Certeau, Michel de. *The Practice of Everyday Life*. Translated by Steven Rendall. Berkeley: University of California Press, 1984.

———. *Heterologies: Discourse on the Other*. Translated by Brian Massumi. Theory and History of Literature 17. Minneapolis: University of Minnesota Press, 1986.

———. *Culture in the Plural*. Translated by Tom Conley. Minneapolis: University of Minnesota Press, 1997.

Chakrabarty, Dipesh. *Provincializing Europe: Postcolonial Thought and Historical Difference*. Princeton Studies in Culture/Power/History. Princeton, NJ: Princeton University Press, 2000.

Chan, Wing-tsit. "Neo-Confucianism and Chinese Scientific Thought." *Philosophy East and West* 6, no. 4 (1957): 309–32.

Chao, Yüan-ling. *Medicine and Society in Late Imperial China: A Study of Physicians in Suzhou, 1600–1850*. Asian Thought and Culture 61. New York: Peter Lang, 2009.

Chatterjee, Partha. *The Nation and Its Fragments: Colonial and Postcolonial Histories*. Princeton Studies in Culture/Power/History. Princeton, NJ: Princeton University Press, 1993.

Chemla, Karine. "Étude du livre Reflets des mesures du cercle sur la mer de Li Ye (1248)." Thèse de mathématiques, Université Paris XIII, 1982.

Chemla, Karine. "Similarities between Chinese and Arabic Mathematical Writings: (1) Root Extraction." *Arabic Sciences and Philosophy: A Historical Journal* 4 (1994): 207–66.

———. "What Is at Stake in Mathematical Proofs from Third-Century China?" *Science in Context* 10, no. 2 (1997): 227–51.

———. "Generality above Abstraction: The General Expressed in Terms of the Paradigmatic in Mathematics in Ancient China." *Science in Context* 16, no. 3 (2003): 413–58.

Chemla, Karine, and Guo Shuchun. *Les neuf chapitres: Le classique mathématique de la Chine ancienne et ses commentaires.* Paris: Dunod, 2004.

Chen, Cheng-Yih. *Early Chinese Work in Natural Science: A Re-Examination of the Physics of Motion, Acoustics, Astronomy and Scientific Thoughts.* Hong Kong: Hong Kong University Press, 1996.

Chen, Jiang-Ping Jeff. "The Evolution of Transformation Media in Spherical Trigonometry in 17th- and 18th-Century China, and Its Relation to 'Western Learning.'" *Historia Mathematica* 37 (2010): 62–109.

Ch'en, Kenneth. "Matteo Ricci's Contribution to, and Influence on, Geographical Knowledge in China." *Journal of the American Oriental Society* 59, no. 3 (1939): 325–59.

Chia, Lucille. *Printing for Profit: The Commercial Publishers of Jianyang, Fujian (11th–17th Centuries).* Harvard-Yenching Institute Monograph Series 56. Cambridge, MA: Harvard University Asia Center, 2002.

Ch'ien, Edward T. *Chiao Hung and the Restructuring of Neo-Confucianism in the Late Ming.* Neo-Confucian Studies. New York: Columbia University Press, 1986.

Cho, Gene J. *The Discovery of Musical Equal Temperament in China and Europe in the Sixteenth Century.* Studies in the History and Interpretation of Music 93. Lewiston, NY: E. Mellen Press, 2003.

Chomsky, Noam. *Knowledge of Language: Its Nature, Origin, and Use.* Convergence. New York: Praeger, 1986.

Chow, Kai-wing. *The Rise of Confucian Ritualism in Late Imperial China: Ethics, Classics, and Lineage Discourse.* Stanford, CA: Stanford University Press, 1994.

Chow, Rey. "Introduction: On Chineseness as a Theoretical Problem." *boundary 2* 25, no. 3 (1998): 1–24.

———. *The Protestant Ethic and the Spirit of Capitalism.* New York: Columbia University Press, 2002.

Chu, Ping-yi. "Trust, Instruments, and Cross-Cultural Scientific Exchanges: Chinese Debate over the Shape of the Earth, 1600–1800." *Science in Context* 12, no. 3 (1999): 385–411.

———. "Remembering Our Grand Tradition: The Historical Memory of the Scientific Exchanges between China and Europe, 1600–1800." *History of Science* 41 (2003): 193–215.

———. "Scientific Texts in Contest, 1600–1800." In Bretelle-Establet, *Looking at It from Asia,* 141–66.

Clagett, Marshall. *The Science of Mechanics in the Middle Ages.* Wisconsin University Publications in Medieval Science 4. Madison: University of Wisconsin Press, 1959.

Clavius, Christoph, SJ (1538–1612). *Christophori Clavii Bambergensis e Societate Iesv epitome arithmetica practicae.* Romae: Ex Typographia Dominici Basae, 1583.

———. *Euclidis Elementorum libri XV: Accessit XVI; De solidorum regularium cuiuslibet intra quodibet comparatione; Omnes perspicuis demonstrationibus accuratique Scholiis illustati; ac multarum rerum accessione locupletati.* 1591.

Clifford, James. *The Predicament of Culture: Twentieth-Century Ethnography, Literature, and Art.* Cambridge, MA: Harvard University Press, 1988.

Clifford, James, and George E. Marcus, eds. *Writing Culture: The Poetics and Politics of Ethnography: A School of American Research Advanced Seminar.* Berkeley: University of California Press, 1986.

Clunas, Craig. *Superfluous Things: Material Culture and Social Status in Early Modern China.* Cambridge: Polity Press, 1991.

———. "Text, Representation and Technique in Early Modern China." In *History of Science, History of Text,* edited by Karine Chemla, 107–21. Boston: Kluwer Academic, 2004.

Cohen, I. Bernard, ed. *Puritanism and the Rise of Modern Science: The Merton Thesis.* Edited, with an introduction by I. Bernard Cohen, with the assistance of K. E. Duffin and Stuart Strickland. New Brunswick, NJ: Rutgers University Press, 1990.

Cohen, Paul A. *Discovering History in China: American Historical Writing on the Recent Chinese Past.* Studies of the East Asian Institute. New York: Columbia University Press, 1984.

———. Review of *China and the Christian Impact: A Conflict of Cultures,* by Jacques Gernet. *Harvard Journal of Asiatic Studies* 47, no. 2 (1987): 674–83.

Cohen, Paul J. "The Independence of the Continuum Hypothesis." *Proceedings of the National Academy of Sciences of the United States of America* 50, no. 6 (1963): 1143–48.

———. "The Independence of the Continuum Hypothesis, II." *Proceedings of the National Academy of Sciences of the United States of America* 51, no. 6 (1964): 105–110.

———. *Set Theory and the Continuum Hypothesis.* New York: W. A. Benjamin, 1966.

Collins, H. M. *Changing Order: Replication and Induction in Scientific Practice.* Beverly Hills, CA: Sage Publications, 1985.

Corry, Leo. *Modern Algebra and the Rise of Mathematical Structures.* 2nd ed. Basel: Birkhäuser Verlag, 2004.

Creager, Angela N. H., Elizabeth Lunbeck, and M. Norton Wise, eds. *Science without Laws: Model Systems, Cases, Exemplary Narratives.* Science and Cultural Theory. Durham, NC: Duke University Press, 2007.

Criveller, Gianni. *Preaching Christ in Late Ming China: The Jesuits' Presentation of Christ from Matteo Ricci to Giulio Aleni.* Variétés Sinologiques 86. Taipei: Taipei Ricci Institute, 1997.

Crombie, Alistair C. *Robert Grosseteste and the Origins of Experimental Science, 1100–1700.* Oxford: Clarendon Press, 1958.

———. "The Significance of Medieval Discussions of Scientific Method for the Scientific Revolution." In *Critical Problems in the History of Science,* edited by Marshall Clagett, 79–101. Madison: University of Wisconsin Press, 1959.

———, ed. *Scientific Change: Historical Studies in the Intellectual, Social and Technical Conditions for Scientific Discovery and Technical Invention, from Antiquity to the Present.* London: Heinemann, 1963.

———. Review of *Rise of Early Modern Science,* by Toby E. Huff. *Journal of Asian Studies* 53, no. 4 (1994): 1213–14.

———. *Styles of Scientific Thinking in the European Tradition: The History of Argument and Explanation Especially in the Mathematical and Biomedical Sciences and Arts.* 3 vols. London: Duckworth, 1994.

———. *Science, Art and Nature in Medieval and Modern Thought.* London: Hambledon Press, 1996.

Cronin, Vincent. *The Wise Man from the West.* London: Rupert Hart-Davis, 1955.

Crossley, Pamela Kyle, Helen F. Siu, and Donald S. Sutton, eds. *Empire at the Margins: Culture, Ethnicity, and Frontier in Early Modern China.* Studies on China 28. Berkeley: University of California Press, 2006.

Cullen, Charles G. *Matrices and Linear Transformations.* 2nd ed. Dover Books on Advanced Mathematics. New York: Dover, 1990.

Cullen, Christopher. *Astronomy and Mathematics in Ancient China: The Zhou Bi Suan Jing.* Cambridge: Cambridge University Press, 1996.

Cuomo, Serafina. *Pappus of Alexandria and the Mathematics of Late Antiquity.* Cambridge: Cambridge University Press, 2000.

———. *Ancient Mathematics.* London: Routledge, 2001.

Dalen, Benno van. "Islamic and Chinese Astronomy under the Mongols: A Little-Known Case of Transmission." In Dold-Samplonius et al., *From China to Paris,* 327–56.

Dardess, John W. *Confucianism and Autocracy: Professional Elites in the Founding of the Ming Dynasty.* Berkeley: University of California Press, 1983.

———. *Ming China, 1368–1644: A Concise History of a Resilient Empire.* Critical Issues in History. World and International History. Lanham, MD: Rowman & Littlefield, 2011.

Daston, Lorraine, and Katharine Park. *Wonders and the Order of Nature, 1150–1750.* New York: Zone Books, 1998.

Dauben, Joseph W. *Abraham Robinson: The Creation of Nonstandard Analysis; A Personal and Mathematical Odyssey.* Princeton, NJ: Princeton University Press, 1995.

———. "Ancient Chinese Mathematics: The *Jiu Zhang Suan Shu* vs. Euclid's *Elements*; Aspects of Proof and the Linguistic Limits of Knowledge." *International Journal of Engineering Science* 36, no. 12 (1998): 1339–59.

———. "Chinese Mathematics." In Katz and Imhausen, *Mathematics of Egypt, Mesopotamia, China, India, and Islam,* 187–384.

———. "*Suan Shu Shu*: A Book on Numbers and Computations; English Translation with Commentary." *Archive for History of Exact Science* 62, no. 2 (2008): 91–178.

Dauben, Joseph W., and Christoph J. Scriba, eds. *Writing the History of Mathematics: Its Historical Development.* Science Networks Historical Studies 27. Boston: Birkhäuser, 2002.

Davidson, Donald. "On the Very Idea of a Conceptual Scheme." *Proceedings and Addresses of the American Philosophical Association* 47 (1974): 5–20.

———. *Inquiries into Truth and Interpretation.* New York: Oxford University Press, 1984.

———. *Essays on Actions and Events.* 2nd ed. Oxford: Clarendon Press, 2001.

Davis, Martin, ed. *The Undecidable: Basic Papers on Undecidable Propositions, Unsolvable Problems and Computable Functions.* Mineola, NY: Dover Publications, 2004.

de Bary, Wm. Theodore. *Neo-Confucian Orthodoxy and the Learning of the Mind-and-Heart.* New York: Columbia University Press, 1981.

———. *The Liberal Tradition in China.* New York: Columbia University Press, 1983.

———. *Learning for One's Self: Essays on the Individual in Neo-Confucian Thought.* New York: Columbia University Press, 1991.

de Bary, Wm. Theodore, Irene Bloom, Wing-tsit Chan, Joseph Adler, and Richard John Lufrano, eds. *Sources of Chinese Tradition.* 2nd ed. 2 vols. Introduction to Asian Civilization. New York: Columbia University Press, 1999–2000.

de Bary, Wm. Theodore, and the Conference on Ming Thought, eds. *Self and Society in Ming Thought.* New York: Columbia University Press, 1970.

de Bary, Wm. Theodore, and the Conference on Seventeenth-Century Chinese Thought, eds. *The Unfolding of Neo-Confucianism.* Studies in Oriental Culture 10. New York: Columbia University Press, 1975.

de Man, Paul. *Allegories of Reading: Figural Language in Rousseau, Nietzsche, Rilke, and Proust.* New Haven, CT: Yale University Press, 1979.

———. *Blindness and Insight: Essays in the Rhetoric of Contemporary Criticism.* Theory and History of Literature 7. Minneapolis: University of Minnesota Press, 1983.

De Weerdt, Hilde. *Competition over Content: Negotiating Standards for the Civil Service Examinations in Imperial China (1127–1279).* Harvard East Asian Monographs 289. Cambridge, MA: Harvard University Asia Center, 2007.

De Young, Gregg. "The Arabic Textual Traditions of Euclid's Elements." *Historia Mathematica* 11, no. 2 (1984): 147–60.

Deane, Thatcher E. "Instruments and Observation at the Imperial Astronomical Bureau during the Ming Dynasty." *Osiris,* 2nd ser., 9 (1994): 127–40.

Dear, Peter. *Discipline & Experience: The Mathematical Way in the Scientific Revolution.* Science and Its Conceptual Foundations. Chicago: University of Chicago Press, 1995.

Dear, Peter. *Revolutionizing the Sciences: European Knowledge and Its Ambitions,
1500–1700.* Princeton, NJ: Princeton University Press, 2001.

———. *The Intelligibility of Nature: How Science Makes Sense of the World.*
Chicago: University of Chicago Press, 2006.

Deleuze, Gilles, and Félix Guattari. *A Thousand Plateaus: Capitalism and
Schizophrenia.* Translated by Brian Massumi. Minneapolis: University of
Minnesota Press, 1987. Originally published as *Mille plateaux*, vol. 2 of *Cap-
italisme et schizophrénie* (Paris: Éditions de Minuit, 1980).

d'Elia, Pasquale M., SJ. *Fonti Ricciane: Documenti originali concernenti Matteo
Ricci e la storia delle prime relazioni tra l'Europa e la Cina (1579–1615).* 3 vols.
Roma: Libreria dello Stato, 1942–1949.

———. "Presentazione della prima traduzione Cinese di Euclide." *Monumenta
serica* 15, no. 1 (1956): 161–202.

———. *Galileo in China: Relations through the Roman College between Galileo
and the Jesuit Scientist-Missionaries (1610–1640).* Translated by Rufus Suter
and Matthew Sciascia. Cambridge, MA: Harvard University Press, 1960.

———. "Recent Discoveries and New Studies (1938–60) on the World Map in
Chinese of Father Matteo Ricci S.J." *Monumenta Serica* 20 (1961): 82–164.

Demiéville, Paul, and Martin Faigel. "The First Philosophic Contacts between
Europe and China." *Diogenes* 15 (June 1967): 75–103.

Derrida, Jacques. *Edmund Husserl's "Origin of Geometry": An Introduction.* Translated by John P. Leavey, Jr. Lincoln: University of Nebraska Press, 1978. Originally published as *Introduction à "L'origine de la géométrie" de Husserl* (Paris:
Presses Universitaires de France, 1962).

———. *Writing and Difference.* Translated by Alan Bass. Chicago: University of
Chicago, 1978. Originally published as *L'écriture et la différence* (Paris: Éditions du Seuil, 1967).

———. "The Supplement of Copula: Philosophy *Before* Linguistics." In Harari,
Textual Strategies.

———. *Dissemination.* Translated by Barbara Johnson. Chicago: University of
Chicago Press, 1981. Originally published as *La dissémination* (Paris: Éditions du Seuil, 1972).

———. *Margins of Philosophy.* Translated by Alan Bass. Chicago: University of
Chicago, 1982. Originally published as *Marges de la philosophie* (Paris: Éditions de Minuit, 1972).

Detlefsen, Michael. "On an Alleged Refutation of Hilbert's Program Using Gödel's
First Incompleteness Theorem." *Journal of Philosophical Logic* 19 (1990):
343–77.

———. "What Does Gödel's Second Theorem Say?" *Philosophia Mathematica* 9
(2001): 37–71.

Di Cosmo, Nicola, Allen J. Frank, and Peter B. Golden, eds. *The Cambridge History
of Inner Asia: The Chinggisid Age.* Cambridge: Cambridge University Press,
2009.

Dirlik, Arif. "The Postcolonial Aura: Third World Criticism in the Age of Global
Capitalism." *Critical Inquiry* 20, no. 2 (1994): 328–56.

———. "Chinese History and the Question of Orientalism." *History and Theory* (1996): 96–118.

———. "Is There History after Eurocentrism? Globalism, Postcolonialism, and the Disavowal of History." *Cultural Critique*, no. 42 (1999): 1–34.

Dold-Samplonius, Yvonne, Joseph W. Dauben, Menso Folkerts, and Benno van Dalen, eds. *From China to Paris: 2000 Years Transmission of Mathematical Ideas*. Boethius: Texte und Abhandlungen zur Geschichte der Mathematik und der Naturwissenschaften. Stuttgart: Franz Steiner Verlag Wiesbaden, 2002.

Dreyfus, Hubert L., and Paul Rabinow. *Michel Foucault: Beyond Structuralism and Hermeneutics*. 2nd ed. Chicago: University of Chicago Press, 1983.

Duara, Prasenjit. *Rescuing History from the Nation: Questioning Narratives of Modern China*. Chicago: University of Chicago Press, 1995.

Dubs, Homer. "The Failure of the Chinese to Produce Philosophical Systems." *T'oung Pao* 26 (1929): 96–109.

Dudink, Ad. "Opposition to Western Science and the Nanjing Persecution." In Jami et al., *Statecraft and Intellectual Renewal*, 191–224.

———. "The Image of Xu Guangqi as Author of Christian Texts." In Jami et al., *Statecraft and Intellectual Renewal*, 99–154.

———. "Xu Guangqi's Career: An Annotated Chronology." In Jami et al., *Statecraft and Intellectual Renewal*, 399–411.

———. "The Chinese Christian Texts in the Zikawei Collection in Shanghai: A Preliminary and Partial List." *Sino-Western Cultural Relations Journal* 33 (2011): 1–41.

Duhem, Pierre. *The Aim and Structure of Physical Theory*. Princeton, NJ: Princeton University Press, 1991. Originally published as *La théorie physique: Son objet, et sa structure* (Paris: Chevalier & Rivière, 1906).

Dunne, George H. *Generation of Giants: The Story of the Jesuits in China in the Last Decades of the Ming Dynasty*. Notre Dame, IN: University of Notre Dame Press, 1962.

Dupré, Sven, and Christoph Herbert Lüthy, eds. *Silent Messengers: The Circulation of Material Objects of Knowledge in the Early Modern Low Countries*. Low Countries Studies on the Circulation of Natural Knowledge 1. Berlin: LIT, 2011.

Durkheim, Émile, and Marcel Mauss. *Primitive Classification*. University of Chicago Press, 1963. Originally published as "De quelques formes primitives de classification, contribution à l'étude des représentations collectives," *Année Sociologique* 6 (1903): 1–72.

Ebrey, Patricia Buckley. *Confucianism and Family Rituals in Imperial China: A Social History of Writing about Rites*. Princeton, NJ: Princeton University Press, 1991.

Elison, George. *Deus Destroyed: The Image of Christianity in Early Modern Japan*. Harvard East Asian Monographs 141. Cambridge, MA: Council on East Asian Studies, Harvard University, 1988.

Elman, Benjamin A. *From Philosophy to Philology: Intellectual and Social Aspects of Change in Late Imperial China.* Harvard East Asian Monographs 110. Cambridge, MA: Harvard University Press, 1984.

———. "Imperial Politics and Confucian Societies in Late Imperial China: The Hanlin and Donglin Academies." *Modern China* 15, no. 4 (1989): 379–418.

———. *Classicism, Politics, and Kinship: The Ch'ang-Chou School of New Text Confucianism in Late Imperial China.* Berkeley: University of California Press, 1990.

———. *A Cultural History of Civil Examinations in Late Imperial China.* Berkeley: University of California Press, 2000.

———. "Naval Warfare and the Refraction of China's Self-Strengthening Reforms into Scientific and Technological Failure, 1865–1895." *Modern Asian Studies* 38, no. 2 (2004): 283–326.

———. *On Their Own Terms: Science in China, 1550–1900.* Cambridge, MA: Harvard University Press, 2005.

Elman, Benjamin A., and Alexander Woodside, eds. *Education and Society in Late Imperial China, 1600–1900.* Berkeley: University of California, 1993.

Elvin, Mark. *The Pattern of the Chinese Past: A Social and Economic Interpretation.* Stanford, CA: Stanford University Press, 1973.

———, ed. "Symposium: The Work of Joseph Needham." *Past and Present* 87, no. 1 (May 1980): 17–53.

Engelfriet, Peter M. "The Chinese Euclid and Its European Context." In *L'Europe en Chine: Interactions scientifiques, religieuses et culturelles aux XVIIe et XVIIIe siècles,* edited by Catherine Jami and Hubert Delahaye. Mémoires de L'Institut des hautes études chinoises 34. Collège de France, Institut des hautes études chinoises, 1993.

———. "Euclid in China: A Survey of the Historical Background of the First Chinese Translation of Euclid's Elements (Jihe Yuanben; Beijing, 1607), an Analysis of the Translation, and a Study of Its Influence up to 1723." Ph.D. dissertation, Leiden, 1996.

———. *Euclid in China: The Genesis of the First Chinese Translation of Euclid's Elements, Books I–VI (Jihe Yuanben, Beijing, 1607) and Its Reception up to 1723.* Sinica Leidensia 40. Boston: Brill, 1998.

Engelfriet, Peter, and Siu Man-Keung. "Xu Guangqi's Attempts to Integrate Western and Chinese Mathematics." In Jami et al., *Statecraft and Intellectual Renewal,* 279–310.

Eoyang, Eugene. "A Taste for Apricots: Approaches to Chinese Fiction." In *Chinese Narrative: Critical and Theoretical Essays,* edited by Andrew Plaks, 53–69. Princeton: Princeton University Press, 1977.

Evans, James. *The History and Practice of Ancient Astronomy.* New York: Oxford University Press, 1998.

Fabian, Johannes. *Time and the Other: How Anthropology Makes Its Object.* New York: Columbia University Press, 1983.

Fairbank, John King, and Merle Goldman. *China: A New History.* Enl. ed. Cambridge, MA: Belknap Press of the Harvard University Press, 1998.

Falkenhausen, Lothar von. *Suspended Music: Chime-Bells in the Culture of Bronze Age China.* Berkeley: University of California Press, 1993.

Fan, Fa-ti. *British Naturalists in Qing China: Science, Empire, and Cultural Encounter.* Cambridge, MA: Harvard University Press, 2004.

Farmer, Edward L. *Early Ming Government: The Evolution of Dual Capitals.* Harvard East Asian Monographs 66. Cambridge, MA: East Asian Research Center, Harvard University, 1976.

Fatoohi, L. J., and F. R. Stephenson. "Accuracy of Lunar Eclipse Observations Made by Jesuit Astronomers in China." *Journal for the History of Astronomy* 27, no. 1 (1996): 61–67.

Feingold, Mordechai. *Jesuit Science and the Republic of Letters.* Transformations. Cambridge, MA: MIT Press, 2003.

Ferguson, Niall. *Civilization: The West and the Rest.* 1st American ed. New York: Penguin Press, 2011.

Feyerabend, Paul. "Explanation, Reduction, and Empiricism." In *Scientific Explanation, Space, and Time,* edited by Herbert Feigl and Grover Maxwell, 28–97. Minnesota Center for Philosophy of Science. Minneapolis: University of Minnesota Press, 1962.

———. *Farewell to Reason.* London: Verso, 1987.

———. *Against Method.* London: Verso, 1988.

Findlen, Paula. *Possessing Nature: Museums, Collecting, and Scientific Culture in Early Modern Italy.* Studies on the History of Society and Culture 20. Berkeley: University of California Press, 1994.

Firth, J. R. *Papers in Linguistics: 1934–1951.* London: Oxford University Press, 1957.

———. "Linguistic Analysis and Translation." In *Selected Papers of J. R. Firth, 1952–59,* 75–83.

———. *Selected Papers of J. R. Firth, 1952–59.* Edited by F. R. Palmer. Indiana University Studies in the History and Theory of Linguistics. Bloomington: Indiana University Press, 1968.

Fischer, Michael M. J. *Anthropological Futures.* Experimental Futures: Technological Lives, Scientific Arts, Anthropological Voices. Durham, NC: Duke University Press, 2009.

Fleck, Ludwik. *Genesis and Development of a Scientific Fact.* Translated by Fred Bradley and Thaddeus J. Trenn. Chicago: University of Chicago Press, 1979. Originally published as *Entstehung und Entwicklung einer wissenschaftlichen Tatsache: Einführung in die Lehre vom Denkstil und Denkkollectiv* (Basel: Schwabe, 1935).

Folkerts, Menso. *Essays on Early Medieval Mathematics: The Latin Tradition.* Variorum Collected Studies Series. Brookfield, VT: Ashgate Variorum, 2003.

———. *The Development of Mathematics in Medieval Europe: The Arabs, Euclid, Regiomontanus.* Variorum Collected Studies Series. Aldershot, Hampshire: Ashgate Variorum, 2006.

Foucault, Michel. *The Order of Things: An Archaeology of the Human Sciences.* World of Man. New York: Pantheon Books, 1970. Originally published as *Les mots et les choses: Une archéologie des sciences humaines* (Paris: Gallimard, 1966).

Foucault, Michel. *The Archaeology of Knowledge & the Discourse on Language,* translated by A. M. Sheridan Smith. World of Man. New York: Pantheon Books, 1972. Originally published as *L'archeologie du savoir* (Paris: Gallimard, 1969); *L'ordre du discours* (Paris: Gallimard, 1971).

———. *The Birth of the Clinic: An Archaeology of Medical Perception.* Translated by A. M. Sheridan Smith. World of Man. New York: Vintage Books, 1975. Originally published as *Naissance de la clinique: Une archéologie du regard médical* (Paris: Presses universitaires de France, 1963).

———. *The Foucault Reader.* Edited by Paul Rainbow. Translated by Donald F. Bouchard and Sherry Simon. New York: Pantheon Books, 1984.

Fowler, D. H. *The Mathematics of Plato's Academy: A New Reconstruction.* 2nd ed. Oxford: Clarendon Press, 1999.

Frank, Andre Gunder. *Reorient: Global Economy in the Asian Age.* Berkeley: University of California Press, 1998.

Frege, Gottlob. *The Foundations of Arithmetic: A Logico-Mathematical Enquiry into the Concept of Number.* Translated by J. L. Austin. Oxford: Blackwell, 1950. Originally published as *Die Grundlagen der Arithmetik: eine logisch-mathematische Untersuchung über den Begriff der Zahl* (Breslau: Wilhelm Koebner, 1884).

———. "Begriffsschrift: A Formula Language Modeled upon that of Arithmetic, for Pure Thought." In Van Heijenoort, *From Frege to Gödel,* 5–82.

———. *Collected Papers on Mathematics, Logic, and Philosophy.* Edited by Brian McGuinness. Translated by Max Black. Oxford: B. Blackwell, 1984.

———. "On the Foundations of Geometry." In *Collected Papers on Mathematics, Logic, and Philosophy,* 293–340.

Friberg, Jöran. *A Remarkable Collection of Babylonian Mathematical Texts.* Sources and Studies in the History of Mathematics and the Physical Sciences. New York: Springer, 2007.

Fried, Michael N., and Sabetai Unguru. *Apollonius of Perga's Conica: Text, Context, Subtext.* Mnemosyne, Bibliotheca Classica Batava. Supplementum 222. Leiden: Brill, 2001.

Fu, Daiwie, ed. *The Challenging Relationship between Philosophy of Science and STS in East Asia.* A special issue of *East Asian Science, Technology and Society: An International Journal* 5, no. 1 (2011).

Fung, Yu-Lan. "Why China Has No Science—An Interpretation of the History and Consequences of Chinese Philosophy." *International Journal of Ethics* 32, no. 3 (April 1922): 237–63.

Funkenstein, Amos. *Theology and the Scientific Imagination from the Middle Ages to the Seventeenth Century.* Princeton, NJ: Princeton University Press, 1989.

Furth, Charlotte, Judith T. Zeitlin, and Ping-chen Hsiung, eds. *Thinking with Cases: Specialist Knowledge in Chinese Cultural History.* Honolulu: University of Hawai'i Press, 2007.

Gadamer, Hans-Georg. *Truth and Method*. 2nd ed. Translated by Joel Weinsheimer and Donald G. Marshall. New York: Crossroad, 1989. Originally published as *Wahrheit und Methode: Grundzüge einer philosophischen Hermeneutik* (Tübingen: Mohr, 1960).

Galison, Peter L. "Ten Problems in History and Philosophy of Science." *Isis* 99, no. 1 (2008): 111–24.

Galison, Peter L., and David J. Stump, eds. *The Disunity of Science: Boundaries, Contexts, and Power*. Writing Science. Stanford, CA: Stanford University Press, 1996.

Gardner, Daniel K. *Chu Hsi and the Ta-Hsueh: Neo-Confucian Reflection on the Confucian Canon*. Harvard East Asian Monographs 118. Cambridge, MA: Council on East Asian Studies, Harvard University, 1986.

Geertz, Clifford. *The Interpretation of Cultures: Selected Essays*. New York: Basic Books, 1973.

———. *Local Knowledge: Further Essays in Interpretive Anthropology*. New York: Basic Books, 1983.

———. "Anti Anti-Relativism." In Krausz, *Relativism*, 12–34.

———. "Local Knowledge and Its Limits." *Yale Journal of Criticism* 5, no. 2 (1991): 129–35.

Gellner, Ernest. *Spectacles & Predicaments: Essays in Social Theory*. Cambridge: Cambridge University Press, 1979.

———. "Relativism and Universals." In Hollis and Lukes, *Rationality and Relativism*, 181–200.

Gernet, Jacques. "Christian and Chinese Visions of the World in the Seventeenth Century." *Chinese Science* 4 (1980): 1–17.

———. *China and the Christian Impact: A Conflict of Cultures*. Translated by Janet Lloyd. New York: Cambridge University Press, 1985. Originally published as *Chine et christianisme: Action et réaction* (Paris: Gallimard, 1982).

———. "The Encounter between China and Europe." *Chinese Science* 11 (1993): 93–102.

———. *A History of Chinese Civilization*. 2nd ed. Translated by J. R. Foster and Charles Hartman. Cambridge: Cambridge University Press, 1996. Originally published as *Le monde chinois* (Paris: A. Colin, 1972).

———. "A Note on the Context of Xu Guangqi's Conversion." In Jami et al., *Statecraft and Intellectual Renewal*, 186–90.

Giles, Lionel. "Translations from the Chinese World Map of Father Ricci." *The Geographical Journal* 52, no. 6 (1918): 367–85.

Gillispie, Charles C. *The Edge of Objectivity: An Essay in the History of Scientific Ideas*. Princeton, NJ: Princeton University Press, 1960.

———, ed. *Dictionary of Scientific Biography*. New York: Scribner, 1970–1980.

Ginzburg, Carlo. *The Cheese and the Worms: The Cosmos of a Sixteenth-Century Miller*. Translated by John Tedeschi and Anne Tedeschi. Baltimore: Johns Hopkins University Press, 1980. Originally published as *Il formaggio e i vermi: Il cosmo di un mugnaio del '500* (Torino: G. Einaudi, 1976).

Ginzburg, Carlo. *Clues, Myths, and the Historical Method.* Translated by John Tedeschi and Anne C. Tedeschi. Baltimore: Johns Hopkins University Press, 1989. Originally published as *Miti, emblemi, spie: Morfologia e storia* (Turin: Giulio Einaudi, 1986).

———. "Clues: Roots of an Evidential Paradigm." Chap. 5 in *Clues, Myths, and the Historical Method*, 96–125.

———. "Checking the Evidence: The Judge and the Historian." *Critical Inquiry* 18, no. 1 (1991): 79–92.

———. "Microhistory: Two or Three Things that I Know about It." *Critical Inquiry* 20, no. 1 (1993): 10–35.

———. *Threads and Traces: True, False, Fictive.* Translated by Anne C. Tedeschi and John Tedeschi. Berkeley: University of California Press, 2012. Originally published as *Il filo e le tracce: Vero, falso, finto* (Bologna: Feltrinelli, 2006).

Ginzburg, Carlo, and Anna Davin. "Morelli, Freud, and Sherlock Holmes: Clues and Scientific Method." *History Workshop* 9 (Spring 1980): 5–36.

Gödel, Kurt. "On Formally Undecidable Propositions of *Principia Mathematica* and Related Systems." In Van Heijenoort, *From Frege to Gödel*, 596–616.

Golas, Peter J. "'Like Obtaining a Great Treasure': The Illustrations in Song Yingxing's the Exploitation of the Works of Nature." In Bray et al., *Graphics and Text in the Production of Technical Knowledge*, 569–614.

Goldberg, David Theo. *Racist Culture: Philosophy and the Politics of Meaning.* Oxford: Blackwell, 1993.

Golinski, Jan. *Making Natural Knowledge: Constructivism and the History of Science.* Chicago: University of Chicago Press, 2005.

Gombrich, E. H. "Eastern Inventions and Western Response." *Daedalus* 127, no. 1 (1998): 193–206.

Gooding, David, Trevor J. Pinch, and Simon Schaffer, eds. *The Uses of Experiment: Studies in the Natural Sciences.* Cambridge: Cambridge University Press, 1989.

Goodman, Howard L., and Anthony Grafton. "Ricci, the Chinese, and the Toolkits of Textualists." *Asia Major,* 3rd ser., 3, no. 2 (1990): 95–148.

Goodrich, L. Carrington, and Chaoying Fang, eds. *Dictionary of Ming Biography, 1368–1644.* 2 vols. New York: Columbia University Press, 1976.

Goody, Jack. *The Logic of Writing and the Organization of Society.* Studies in Literacy, Family, Culture, and the State. Cambridge: Cambridge University Press, 1986.

———. *The East in the West.* Cambridge: Cambridge University Press, 1996.

———. *The Theft of History.* Cambridge: Cambridge University Press, 2006.

———. *Renaissances: The One or the Many?* Cambridge: Cambridge University Press, 2010.

———. *The Eurasian Miracle.* Cambridge: Polity, 2010.

Grafton, Anthony. *Defenders of the Text: The Traditions of Scholarship in an Age of Science, 1450–1800.* Cambridge, MA: Harvard University Press, 1991.

———. *New Worlds, Ancient Texts: The Power of Tradition and the Shock of Discovery.* Cambridge, MA: Belknap Press of Harvard University Press, 1992.

Graham, A. C. "'Being' in Western Philosophy Compared with Shih/Fei Yu/Wu in Chinese Philosophy." *Asia Major* 7 (1959): 79–112.

———. "China, Europe and the Origin of Modern Science: Needham's *The Grand Titration*." In *Chinese Science: Explorations of an Ancient Tradition*, edited by Nathan Sivin and Shigeru Nakayama, 45–69. MIT East Asian Science Series. Cambridge, MA: MIT Press, 1973.

———. *Yin-Yang and the Nature of Correlative Thinking*. Occasional Paper and Monograph Series 6. Singapore: Institute of East Asian Philosophies, National University of Singapore, 1986.

———. *Disputers of the Tao: Philosophical Argument in Ancient China*. La Salle, IL: Open Court, 1989.

Granger, Gilles Gaston. *La théorie aristotélicienne de la science*. 22. Paris: Aubier Montaigne, 1976.

Grant, Edward. *The Foundations of Modern Science in the Middle Ages: Their Religious, Institutional, and Intellectual Contexts*. Cambridge: Cambridge University Press, 1996.

Grattan-Guinness, Ivor, ed. *Companion Encyclopedia of the History and Philosophy of the Mathematical Sciences*. Routledge Reference. New York: Routledge, 1994.

———. *The Search for Mathematical Roots, 1870–1940: Logics, Set Theories and the Foundations of Mathematics from Cantor through Russell to Gödel*. Princeton, NJ: Princeton University Press, 2000.

———. *Routes of Learning: Highways, Pathways, and Byways in the History of Mathematics*. Baltimore: Johns Hopkins University Press, 2009.

Gray, Jeremy J. *Ideas of Space: Euclidean, Non-Euclidean, and Relativistic*. 2nd ed. Oxford Science Publications. Oxford: Clarendon Press, 1989.

Gray, Jeremy J., and Karen Hunger Parshall, eds. *Episodes in the History of Modern Algebra (1800–1950)*. History of Mathematics 32. Providence, RI: American Mathematical Society, 2007.

Greenblatt, Stephen Jay. *Renaissance Self-Fashioning: From More to Shakespeare*. Chicago: University of Chicago Press, 1980.

———. *Shakespearean Negotiations: The Circulation of Social Energy in Renaissance England*. Berkeley: University of California Press, 1988.

Grice, Paul. *Studies in the Way of Words*. Cambridge, MA: Harvard University Press, 1989.

Günergun, Feza, and Dhruv Raina, eds. *Science between Europe and Asia: Historical Studies on the Transmission, Adoption and Adaptation of Knowledge*. Boston Studies in the Philosophy of Science 275. New York: Springer, 2011.

Guy, R. Kent. *The Emperor's Four Treasuries: Scholars and the State in the Late Ch'ien-Lung Era*. Cambridge: Council on East Asian Studies, Harvard University, 1987.

Hacking, Ian, ed. *Scientific Revolutions*. Oxford Readings in Philosophy. Oxford: Oxford University Press, 1981.

———. *Representing and Intervening: Introductory Topics in the Philosophy of Natural Science*. New York: Cambridge University Press, 1983.

Hacking, Ian. *Historical Ontology.* Cambridge, MA: Harvard University Press, 2002.

Hall, David L., and Roger T. Ames. "Chinese Philosophy." In *Routledge Encyclopedia of Philosophy.* London: Routledge, 1998.

Han Qi. "Astronomy, Chinese and Western: The Influence of Xu Guangqi's Views in the Early and Mid-Qing." In Jami et al., *Statecraft and Intellectual Renewal,* 360–79.

Hannah, John. "False Position in Leonardo of Pisa's Liber Abbaci." *Historia Mathematica* 34, no. 3 (2007): 306–32.

———. "Conventions for Recreational Problems in Fibonacci's Liber Abbaci." *Archive for History of Exact Sciences* 65, no. 2 (2010): 155–80.

Hansen, Chad. "Should the Ancient Masters Value Reason?" In *Chinese Texts and Philosophical Contexts: Essays Dedicated to Angus C. Graham,* edited by Henry Rosemont, Jr. La Salle, IL: Open Court, 1991.

Hanson, Marta E. *Speaking of Epidemics in Chinese Medicine: Disease and the Geographic Imagination in Late Imperial China.* Needham Research Institute Series. Milton Park: Routledge, 2011.

Harari, Josué V., ed. *Textual Strategies: Perspectives in Post-Structuralist Criticism.* Ithaca, NY: Cornell University Press, 1979.

Harding, Sandra G. *Is Science Multicultural? Postcolonialisms, Feminisms, and Epistemologies.* Race, Gender, and Science. Bloomington: Indiana University Press, 1998.

Harper, Donald John. *Early Chinese Medical Literature: The Mawangdui Medical Manuscripts.* Sir Henry Wellcome Asian Series. London: Kegan Paul International, 1998.

Harris, George L. "The Mission of Matteo Ricci, S. J.: A Case Study of an Effort at Guided Culture Change in China in the Sixteenth Century." *Monumenta Serica* 25, no. 1 (1966): 1–168.

Harris, Steven J. "Long-Distance Corporations, Big Sciences, and the Geography of Knowledge." *Configurations* 6, no. 2 (1998): 269–304.

Hart, Roger. "The Flight from Reason: Higher Superstition and the Refutation of Science Studies." In *Science Wars,* edited by Andrew Ross, 259–92. Durham, NC: Duke University Press, 1996.

———. "On the Problem of Chinese Science." In Biagioli, *Science Studies Reader,* 189–201.

———. "Translating the Untranslatable: From Copula to Incommensurable Worlds." In Liu, *Tokens of Exchange,* 45–73.

———. "Translating Worlds: Incommensurability and Problems of Existence in Seventeenth-Century China." *Positions: East Asia Cultures Critique* 7, no. 1 (1999): 95–128.

———. "Universals of Yesteryear: Hegel's Modernity in an Age of Globalization." In Hopkins, *Global History,* 66–97.

———. *The Chinese Roots of Linear Algebra.* Baltimore: Johns Hopkins University Press, 2011.

———. "Tracing Practices Purloined by the 'Three Pillars.'" *The Korean Journal for the History of Science* 34, no. 2 (2012): 287–358.

Hartwell, Robert M. "Historical Analogism, Public Policy, and Social Science in Eleventh- and Twelfth-Century China." *American Historical Review* 76, no. 3 (1971): 690–727.

Hashimoto, Keizo. *Hsu Kuang-Ch'i and Astronomical Reform: The Process of the Chinese Acceptance of Western Astronomy, 1629–1635.* Osaka: Kansai University Press, 1988.

Hashimoto, Keizo, Catherine Jami, and Lowell Skar, eds. *East Asian Science: Tradition and Beyond.* Osaka: Kansai University Press, 1995.

Hayot, Eric, Haun Saussy, and Steven G. Yao, eds. *Sinographies: Writing China.* Minneapolis: University of Minnesota Press, 2008.

Heath, Sir Thomas L., ed. *The Thirteen Books of Euclid's Elements.* 2nd ed. 3 vols. New York: Dover, 1956.

Heeffer, Albrecht. "Regiomontanus and Chinese Mathematics." *Philosophica* 82 (2008): 87–114.

Heidegger, Martin. *An Introduction to Metaphysics.* Translated by Ralph Manheim. New York: Anchor Books, 1961. Originally published as *Einführung in die Metaphysik* (Tübingen: M. Niemeyer, 1953).

Helmholtz, Hermann L. F. *On the Sensations of Tone as a Physiological Basis for the Theory of Music.* Cambridge: Cambridge University Press, 2009. Originally published as *Die Lehre von den Tonempfindungen als physiologische Grundlage für die Theorie der Musik* (Braunschweig: F. Vieweg, 1863).

Henderson, John B. "The Assimilation of the Exact Sciences into the Ch'ing Confucian Tradition." *Journal of Asian Affairs* 5, no. 1 (1980): 15–33.

Heng, Geraldine. *Empire of Magic: Medieval Romance and the Politics of Cultural Fantasy.* New York: Columbia University Press, 2003.

Henricks, Robert G., trans. *Lao-Tzu: Te-Tao Ching: A New Translation Based on the Recently Discovered Ma-Wang-Tui Texts.* New York: Ballantine Books, 1989.

Herrlich, Horst. *Axiom of Choice.* Lecture Notes in Mathematics 1876. Berlin: Springer, 2006.

Hilbert, David. *The Foundations of Geometry.* Translated by E. J. Townsend. La Salle, IL: Open Court, 1950. Reprint, Project Gutenberg eBook. Originally published as *Grundlagen der geometrie* (Leipzig: B. G. Teubner, 1899).

———. *David Hilbert's Lectures on the Foundations of Geometry, 1891–1902.* Edited by Michael Hallett and Ulrich Majer. Berlin: Springer, 2004.

———. *David Hilbert's Lectures on the Foundations of Mathematics and Physics, 1891–1933.* Edited by William Bragg Ewald, Michael Hallett, Ulrich Majer, and Wilfried Sieg. Berlin: Springer, 2004.

Hintikka, Jaakko, ed. *The Philosophy of Mathematics.* London: Oxford University Press, 1969.

Ho, Peng Yoke. *Chinese Mathematical Astrology: Reaching out to the Stars.* London: RoutledgeCurzon, 2003.

Hoe, Jock. *The Jade Mirror of the Four Unknowns by Zhu Shijie: An Early Fourteenth Century Mathematics Manual for Teaching the Derivation of Systems of*

Polynomial Equations in up to Four Unknowns; A Study. Christchurch, N.Z.: Mingming Bookroom, 2007.

Hoe, John [Jock]. *Les systèmes d'équations polynômes dans le Siyuan yujian (1303).* Mémoires de l'Institut des hautes études chinoises. Paris: Collège de France, l'Institut des hautes études chinoises, 1977.

Höfele, Andreas, and Werner von Koppenfels, eds. *Renaissance Go-Betweens: Cultural Exchange in Early Modern Europe.* Spectrum Literaturwissenschaft 2. Berlin: Walter de Gruyter, 2005.

Hoffman, Kenneth, and Ray Alden Kunze. *Linear Algebra.* 2nd ed. Englewood Cliffs, NJ: Prentice-Hall, 1971.

Hogendijk, Jan P., and A. I. Sabra. *The Enterprise of Science in Islam: New Perspectives.* Cambridge, MA: MIT Press, 2003.

Hollis, Martin, and Steven Lukes, eds. *Rationality and Relativism.* Cambridge, MA: MIT Press, 1982.

Hoon, Jun Yong. "Mathematics in Context: A Case in Early Nineteenth-Century Korea." *Science in Context* 19, no. 4 (2006): 475–512.

Hopkins, A. G., ed. *Globalization in World History.* New York: Norton, 2002.

———, ed. *Global History: Interactions between the Universal and the Local.* New York: Palgrave Macmillan, 2006.

Horiuchi, Annick. *Japanese Mathematics in the Edo Period (1600–1868): A Study of the Works of Seki Takakazu (?–1708) and Takebe Katahiro (1664–1739).* Translated by Silke Wimmer-Zagier. Science Networks Historical Studies 40. Basel: Birkhäuser, 2010. Originally published as *Les mathématiques japonaises à l'époque d'Edo (1600–1868): Une étude des travaux di Seki Takakazu (?–1708) et de Takebe Katahiro (1664–1739)* (Paris: J. Vrin, 1994).

Horn, Roger A., and Charles R. Johnson. *Matrix Analysis.* New York: Cambridge University Press, 1985.

———. *Topics in Matrix Analysis.* New York: Cambridge University Press, 1991.

Horng Wann-Sheng. "The Influence of Euclid's *Elements* on Xu Guangqi and His Successors." In Jami et al., *Statecraft and Intellectual Renewal*, 380–98.

Horton, Robin, and Ruth Finnegan, eds. *Modes of Thought: Essays on Thinking in Western and Non-Western Societies.* London: Faber, 1973.

Howland, Douglas. *Translating the West: Language and Political Reason in Nineteenth-Century Japan.* Honolulu: University of Hawai'i Press, 2002.

Hoyningen-Huene, Paul. *Reconstructing Scientific Revolutions: Thomas S. Kuhn's Philosophy of Science.* Chicago: University of Chicago Press, 1993.

Høyrup, Jens. *The Formation of a Myth: Greek Mathematics—Our Mathematics.* Filosofi og videnskabsteori på Roskilde Universitetscenter 3. Roskilde: Roskilde Universitetscenter, 1992.

———. "Leonardo Fibonacci and Abbaco Culture: A Proposal to Invert the Roles." *Revue d'histoire des mathèmatiques* 11, no. 1 (2005): 23–56.

———. *Jacopo Da Firenze's Tractatus Algorismi and Early Italian Abbacus Culture.* Science Networks. Historical Studies 34. Basel: Birkhäuser, 2007.

Hsia, Florence C. *Sojourners in a Strange Land: Jesuits and Their Scientific Missions in Late Imperial China.* Chicago: University of Chicago Press, 2009.

Hsia, R. Po-chia. *A Jesuit in the Forbidden City: Matteo Ricci 1552–1610*. New York: Oxford University Press, 2010.

Hsu, Elisabeth, ed. *Innovation in Chinese Medicine*. Needham Research Institute Studies 3. Cambridge: Cambridge University Press, 2001.

Hu Shih. "The Scientific Spirit and Method in Chinese Philosophy." *Philosophy East and West* 9, no. 1 and 2 (1959): 29–31.

———. "The Scientific Spirit and Method in Chinese Philosophy." In *The Chinese Mind: Essentials of Chinese Philosophy and Culture*, edited by Charles A. Moore, 104–31. Honolulu: University of Hawaii Press, 1967.

Hu, Minghui. "Provenance in Contest: Searching for the Origins of Jesuit Astronomy in Late Imperial China." *The International History Review* 24, no. 1 (March 2002): 1–36.

Hu, Mingjie. "Merging Chinese and Western Mathematics: The Introduction of Algebra and the Calculus in China, 1859–1903." PhD diss., Princeton University, 1998.

Huang Tsung-hsi. *The Records of Ming Scholars*. Edited by Julia Ching and Zhaoying Fang. Honolulu: University of Hawaii Press, 1987.

Huang Yi-Long. "Sun Yuanhua: A Christian Convert Who Put Xu Guangqi's Military Reform Policy into Practice." In Jami et al., *Statecraft and Intellectual Renewal*, 225–62.

Huang, Philip C. C. "Theory and the Study of Modern Chinese History: Four Traps and a Question." *Modern China* 24, no. 2 (April 1998): 183–208.

Huang, Ray. *1587, a Year of No Significance: The Ming Dynasty in Decline*. New Haven, CT: Yale University Press, 1981.

Huang, Xiang. "The Trading Zone Communication of Scientific Knowledge: An Examination of Jesuit Science in China (1582–1773)." *Science in Context* 18, no. 3 (2005): 393–427.

Hucker, Charles O. *The Ming Dynasty, Its Origins and Evolving Institutions*. Michigan Papers in Chinese Studies 34. Ann Arbor: Center for Chinese Studies, University of Michigan, 1978.

———. *A Dictionary of Official Titles in Imperial China*. Stanford, CA: Stanford University Press, 1985.

Huff, Toby E. *The Rise of Early Modern Science: Islam, China, and the West*. Cambridge: Cambridge University Press, 1993.

———. *Intellectual Curiosity and the Scientific Revolution: A Global Perspective*. Cambridge: Cambridge University Press, 2011.

Huntington, Samuel P. *The Clash of Civilizations and the Remaking of World Order*. New York: Simon & Schuster, 1996.

Husserl, Edmund. *The Crisis of European Sciences and Transcendental Phenomenology: An Introduction to Phenomenological Philosophy*. Translated by David Carr. Northwestern University Studies in Phenomenology & Existential Philosophy. Evanston, IL: Northwestern University Press, 1970. Originally published as *Die Krisis der europäischen Wissenschaften und die transzendentale Phänomenologie: eine Einleitung in die phänomenologische Philosophie* (Dordrecht: Nijhoff, 1954).

Iser, Wolfgang. *The Act of Reading: A Theory of Aesthetic Response*. Baltimore: Johns Hopkins University Press, 1978.

Isett, Christopher M. *State, Peasant, and Merchant in Qing Manchuria, 1644–1862*. Stanford, CA: Stanford University Press, 2007.

Jami, Catherine. "Western Influence and Chinese Tradition in an Eighteenth Century Chinese Mathematical Work." *Historia Mathematica* 15, no. 4 (1988): 311–31.

———. "Heavenly Learning, Statecraft, and Scholarship: The Jesuits and Their Mathematics in China." In Robson and Stedall, *Oxford Handbook of the History of Mathematics*.

Jami, Catherine, Gregory Blue, and Peter M. Engelfriet, eds. *Statecraft and Intellectual Renewal in Late Ming China: The Cross-Cultural Synthesis of Xu Guangqi (1562–1633)*. Sinica Leidensia 50. Leiden: Brill, 2001.

Jensen, Lionel M. *Manufacturing Confucianism: Chinese Traditions and Universal Civilization*. Durham, NC: Duke University Press, 1997.

Johnson, David G., Andrew J. Nathan, Evelyn Sakakida Rawski, and Judith A. Berling, eds. *Popular Culture in Late Imperial China*. Studies on China 4. Berkeley: University of California Press, 1985.

Jones, Matthew L. "Descartes's Geometry as Spiritual Exercise." *Critical Inquiry* 28, no. 1 (2001): 40–71.

———. *The Good Life in the Scientific Revolution: Descartes, Pascal, Leibniz, and the Cultivation of Virtue*. Chicago: University of Chicago Press, 2006.

Joseph, George Gheverghese. *The Crest of the Peacock: Non-European Roots of Mathematics*. 3rd ed. Princeton: Princeton University Press, 2011.

Katz, Victor J., and Annette Imhausen, eds. *The Mathematics of Egypt, Mesopotamia, China, India, and Islam: A Sourcebook*. Princeton, NJ: Princeton University Press, 2007.

Kelly, Edward Thomas. "The Anti-Christian Persecution of 1616–1617 in Nanking." PhD diss., Columbia University, 1971.

Kim, Sangkeun. *Strange Names of God: The Missionary Translation of the Divine Name and the Chinese Responses to Matteo Ricci's "Shangti" in Late Ming China, 1583–1644*. Studies in Biblical Literature. New York: P. Lang, 2004.

Kim, Yung Sik. *The Natural Philosophy of Chu Hsi (1130–1200)*. Memoirs of the American Philosophical Society 235. Philadelphia, PA: American Philosophical Society, 2000.

———. "The 'Why Not' Question of Chinese Science: The Scientific Revolution and Traditional Chinese Science." *East Asian Science, Technology, and Medicine* 22 (2004): 96–112.

———. "Confucian Scholars and Specialized Scientific and Technical Knowledge in Traditional China, 1000–1700: A Preliminary Overview." *East Asian Science, Technology and Society: An International Journal* 4, no. 2 (2010): 207–28.

Kim, Yung Sik, and Francesca Bray, eds. *Current Perspectives in the History of Science in East Asia*. Seoul, Korea: Seoul National University Press, 1999.

King, Gail. "The Family Letters of Xu Guangqi." *Ming Studies* 31 (1974): 1–41.

Kitagawa, Tomoko. "Samurai Culture and the Fashioning of Mathematics in Japan." Lecture presented at the Annual Meeting of the History of Science Society, Montréal, Quebec, November 4–7, 2010.

Kitcher, Philip. *The Nature of Mathematical Knowledge*. New York: Oxford University Press, 1983.

———. *The Advancement of Science: Science without Legend, Objectivity without Illusions*. New York: Oxford University Press, 1993.

Kleene, S. C. *Introduction to Metamathematics*. Princeton: Van Nostrand, 1952.

Kline, Morris. *Mathematics: The Loss of Certainty*. New York: Oxford University Press, 1980.

Knorr, Wilbur Richard. *The Evolution of the Euclidean Elements: A Study of the Theory of Incommensurable Magnitudes and Its Significance for Early Greek Geometry*. Synthese Historical Library. Boston: D. Reidel, 1975.

———. *Ancient Sources of the Medieval Tradition of Mechanics: Greek, Arabic, and Latin Studies of the Balance*. Monografia / Istituto e museo di storia della scienza 6. Firenze: Istituto e museo di storia della scienza, 1982.

———. *The Ancient Tradition of Geometric Problems*. Cambridge, MA: Birkhäuser Boston, 1985.

———. *Textual Studies in Ancient and Medieval Geometry*. Boston: Birkhäuser, 1989.

Knuth, Donald Ervin. *The Art of Computer Programming*. 2nd ed. Addison-Wesley Series in Computer Science and Information Processing. Reading, MA: Addison-Wesley, 1981.

Ko, Dorothy. *Teachers of the Inner Chambers: Women and Culture in Seventeenth-Century China*. Stanford, CA: Stanford University Press, 1994.

Koertge, Noretta, ed. *New Dictionary of Scientific Biography*. Detroit: Charles Scribner's Sons/Thomson Gale, 2008.

Koyré, Alexandre. *From the Closed World to the Infinite Universe*. Baltimore: Johns Hopkins University Press, 1957.

Krausz, Michael, ed. *Relativism: Interpretation and Confrontation*. Notre Dame, IN: University of Notre Dame Press, 1989.

Kripke, Saul A. *Naming and Necessity*. Cambridge, MA: Harvard University Press, 1980.

———. *Wittgenstein on Rules and Private Language: An Elementary Exposition*. Cambridge, MA: Harvard University Press, 1982.

Kuhn, Thomas S. *The Copernican Revolution: Planetary Astronomy in the Development of Western Thought*. Cambridge, MA: Harvard University Press, 1957.

———. *The Structure of Scientific Revolutions*. 2nd ed. International Encyclopedia of Unified Science. Chicago: University of Chicago Press, 1970.

———. "Second Thoughts on Paradigms." In *The Structure of Scientific Theories*, edited by Frederick Suppe, 459–82. Urbana: University of Illinois Press, 1974.

———. *The Road since* Structure*: Philosophical Essays, 1970–1993, with an Autobiographical Interview*. Edited by James Conant and John Haugeland. Chicago: University of Chicago Press, 2000.

Kuriyama, Shigehisa. *The Expressiveness of the Body and the Divergence of Greek and Chinese Medicine*. New York: Zone Books, 1999.

Kuttner, Fritz A. "Prince Chu Tsai-Yü's Life and Work: A Re-Evaluation of His Contribution to Equal Temperament Theory." *Ethnomusicology* 19, no. 2 (1975): 163–206.

Kyburg, Henry E., Jr. *Science & Reason*. New York: Oxford University Press, 1990.

Lacan, Jacques. *Écrits: A Selection*. Translated by Alan Sheridan. New York: Norton, 1977. Selected translation from *Écrits: Le champ freudien* (Paris: Éditions du Seuil: 1966).

Lach, Donald F., and Edwin J. Van Kley. *Asia in the Making of Europe*, Vol. 3, *A Century of Advance*, Book 4, *East Asia*. Chicago: University of Chicago Press, 1993.

Lackner, Michael, Iwo Amelung, and Joachim Kurtz, eds. *New Terms for New Ideas: Western Knowledge and Lexical Change in Late Imperial China*. Sinica Leidensia 52. Leiden: Brill, 2001.

Lakatos, Imre. *Proofs and Refutations: The Logic of Mathematical Discovery*. Edited by John Worrall and Elie Zahar. Cambridge: Cambridge University Press, 1976.

Lam, Joseph S. C. "Ritual and Musical Politics in the Court of Ming Shizong." In *Harmony and Counterpoint: Ritual Music in Chinese Context*, edited by Bell Yung, Evelyn S. Rawski, and Rubie S. Watson, 35–53. Stanford, CA: Stanford University Press, 1996.

———. *State Sacrifices and Music in Ming China: Orthodoxy, Creativity, and Expressiveness*. SUNY Series in Chinese Local Studies. Albany, NY: State University of New York Press, 1998.

Lam, Lay Yong. *A Critical Study of the Yang Hui Suan Fa: A Thirteenth-Century Chinese Mathematical Treatise*. Singapore: Singapore University Press, 1977.

Lam, Lay-Yong, and Tian-Se Ang. "Li Ye and His 'Yi Gu Yan Duan' (Old Mathematics in Expanded Sections)." *Archive for History of Exact Sciences* 29, no. 3 (1984): 237–66.

———. *Fleeting Footsteps: Tracing the Conception of Arithmetic and Algebra in Ancient China*. Rev. ed. River Edge, NJ: World Scientific, 2004.

Lancashire, Douglas, and Peter Kuo-chen Hu, SJ. "Introduction." In Ricci, *True Meaning of the Lord of Heaven*.

Latour, Bruno. *Science in Action: How to Follow Scientists and Engineers through Society*. Cambridge, MA: Harvard University Press, 1987.

———. *We Have Never Been Modern*. Translated by Catherine Porter. Cambridge, MA: Harvard University Press, 1993. Originally published as *Nous n'avons jamais été modernes: Essais d'anthropologie symétrique* (Paris: La Découverte, 1991).

———. *Reassembling the Social: An Introduction to Actor-Network-Theory*. Clarendon Lectures in Management Studies. Oxford: Oxford University Press, 2005.

Latour, Bruno, and Steve Woolgar. *Laboratory Life: The Social Construction of Scientific Facts.* Sage Library of Social Research 80. Beverly Hills, CA: Sage Publications, 1979.

Lattis, James M. *Between Copernicus and Galileo: Christoph Clavius and the Collapse of Ptolemaic Cosmology.* Chicago: University of Chicago Press, 1994.

Lau, D. C. "Meng-Tzu." In Loewe, *Early Chinese Texts.*

Le Roy Ladurie, Emmanuel. *Montaillou: The Promised Land of Error.* New York: G. Braziller, 1978. Originally published as *Montaillou, village occitan de 1294 à 1324* (Paris: Gallimard, 1975).

Lenoir, Timothy, ed. *Inscribing Science: Scientific Texts and the Materiality of Communication.* Stanford, CA: Stanford University Press, 1998.

Leung, Angela Ki Che. *Leprosy in China: A History.* Studies of the Weatherhead East Asian Institute, Columbia University. New York: Columbia University Press, 2009.

Levenson, Joseph R. "'History' and 'Value': Tensions of Intellectual Choice in Modern China." In *Studies in Chinese Thought,* edited by Arthur F. Wright, 146–94. Comparative Studies of Cultures and Civilizations. Chicago: University of Chicago Press, 1953.

———. *The Problem of Intellectual Continuity.* Vol. 1 of *Confucian China and Its Modern Fate.* Berkeley: University of California Press, 1958.

Lewis, Albert C. "The Divine Truth of Mathematics and the Origins of Linear Algebra." *Theology and Science* 9, no. 1 (2011): 109–20.

Li Yan and Du Shiran. *Chinese Mathematics: A Concise History.* Translated by John N. Crossley and Anthony W.-C. Lun. Oxford: Clarendon Press, 1987.

Liang Qichao. *Intellectual Trends in the Ch'ing Period.* Translated by Immanuel C. Y. Hsü. Harvard East Asian Studies 2. Cambridge, MA: Harvard University Press, 1959.

Libbrecht, Ulrich. *Chinese Mathematics in the Thirteenth Century: The Shu-Shu Chiu-Chang of Ch'in Chiu-Shao.* MIT East Asian Series 1. Cambridge, MA: MIT Press, 1973.

Liew, Foon Ming. "Debates on the Birth of Capitalism in China during the Past Three Decades." *Ming Studies* 26 (1988): 61–75.

Lim, Jongtae. "The Introduction of Western Science and the Rationalization of Traditional Astrology: Reevaluating Yi Ik's 'On Field-Allocation.'" *Seoul Journal of Korean Studies* 17 (2004): 45–65.

———. "Locating a Center on the Surface of a Globe: Negotiating China's Position on the Spherical Earth in Seventeenth and Eighteenth-Century China and Korea." *Historia Scientiarum* 17, no. 3 (2008): 175–88.

Lindberg, David C., and Ronald L. Numbers, eds. *God and Nature: Historical Essays on the Encounter between Christianity and Science.* Berkeley: University of California Press, 1986.

———, eds. *When Science & Christianity Meet.* Chicago: University of Chicago Press, 2003.

Lindberg, David C., and Robert S. Westman, eds. *Reappraisals of the Scientific Revolution.* Cambridge: Cambridge University Press, 1990.

Lippiello, Tiziana, and Roman Malek, eds. *"Scholar from the West": Giulio Aleni S. J. (1582–1649) and the Dialogue between Christianity and China*. Monumenta Serica Monographs Series 42. Sankt Augustin: Monumenta Serica Institute, 1997.

Liu Dun. "400 Years of the History of Mathematics in China: An Introduction to the Major Historians of Mathematics since 1592." *Historia Scientiarum* 4, no. 2 (1994): 103–11.

———. "A Homecoming Stranger: Transmission of the Method of Double-False-Position and the Story of Hiero's Crown." In Dold-Samplonius et al., *From China to Paris*, 157–66.

Liu Dun and Joseph W. Dauben. "China." In Dauben and Scriba, *Writing the History of Mathematics*, 297–306.

Liu, Lydia H. *Translingual Practice: Literature, National Culture, and Translated Modernity—China, 1900–1937*. Stanford, CA: Stanford University Press, 1995.

———. "The Question of Meaning-Value in the Political Economy of the Sign." In Liu, *Tokens of Exchange*, 13–41.

———, ed. *Tokens of Exchange: The Problem of Translation in Global Circulations*. Post-Contemporary Interventions. Durham, NC: Duke University Press, 1999.

———. *The Clash of Empires: The Invention of China in Modern World Making*. Cambridge, MA: Harvard University Press, 2004.

Lloyd, G. E. R. *Adversaries and Authorities: Investigations into Ancient Greek and Chinese Science*. Ideas in Context. Cambridge: Cambridge University Press, 1996.

———. *The Ambitions of Curiosity: Understanding the World in Ancient Greece and China*. Cambridge: Cambridge University Press, 2002.

———. *Ancient Worlds, Modern Reflections: Philosophical Perspectives on Greek and Chinese Science and Culture*. New York: Oxford University Press, 2004.

———. *Disciplines in the Making: Cross-Cultural Perspectives on Elites, Learning, and Innovation*. New York: Oxford University Press, 2009.

Lloyd, G. E. R., and Nathan Sivin. *The Way and the Word: Science and Medicine in Early China and Greece*. New Haven: Yale University Press, 2002.

Loewe, Michael, ed. *Early Chinese Texts: A Bibliographical Guide*. Early China Special Monograph Series 2. Berkeley: Society for the Study of Early China, Institute of East Asian Studies, University of California, Berkeley, 1993.

Long, Pamela O. *Openness, Secrecy, Authorship: Technical Arts and the Culture of Knowledge from Antiquity to the Renaissance*. Baltimore: Johns Hopkins University Press, 2001.

Lü, Lingfeng. "Eclipses and the Victory of European Astronomy in China." *East Asian Science, Technology, and Medicine* 27 (2007): 127–45.

Lukes, Steven. "Some Problems about Rationality." In *Rationality*, edited by Bryan R. Wilson, 194–213. Evanston: Harper & Row, 1970.

Lyons, John. *Semantics*. 2 vols. Cambridge: Cambridge University Press, 1977.

Major, John S., et al., trans. *The Huainanzi: A Guide to the Theory and Practice of Government in Early Han China.* Translations from the Asian Classics. New York: Columbia University Press, 2010.

Mancosu, Paolo. "On the Status of Proofs by Contradiction in the Seventeenth Century." *Synthese* 88, no. 1 (1991): 15–41.

———. "Aristotelian Logic and Euclidean Mathematics: Seventeenth-Century Developments of the 'Quaestio de Certitudine Mathematicarum.'" *Studies in History and Philosophy of Science* 23, no. 2 (1992): 241–65.

———. *Philosophy of Mathematics and Mathematical Practice in the Seventeenth Century.* New York: Oxford University Press, 1996.

———, ed. *From Brouwer to Hilbert: The Debate on the Foundations of Mathematics in the 1920s.* New York: Oxford University Press, 1998.

———, ed. *The Philosophy of Mathematical Practice.* Oxford: Oxford University Press, 2008.

Mancosu, Paolo, Klaus Frovin Jørgensen, and Stig Andur Pedersen, eds. *Visualization, Explanation and Reasoning Styles in Mathematics.* Synthese Library 327. Dordrecht: Springer, 2005.

Marcus, George E., and Michael M. J. Fischer. *Anthropology as Cultural Critique: An Experimental Moment in the Human Sciences.* 2nd ed. Chicago: University of Chicago Press, 1999.

Martinez, Alberto A. *Science Secrets: The Truth about Darwin's Finches, Einstein's Wife, and Other Myths.* Pittsburgh, PA: University of Pittsburgh Press, 2011.

Martzloff, Jean-Claude. "La compréhension chinoise des méthodes démonstratives euclidiennes au cours du XVIIe siècle et au début du XVIIIe." In *Actes du IIe colloque international de sinologie: Les rapports entre la Chine et l'Europe au temps des lumières,* 125–43. Paris: Les Belles Letters, 1980.

———. "La géométrie euclidienne selon Mei Wending." *Historia Scientiarum* 21 (1981): 27–42.

———. "Matteo Ricci's Mathematical Works and Their Influence." In *International Symposium on Chinese-Western Cultural Interchange in Commemoration of the 400th Anniversary of the Arrival of Matteo Ricci, S. J. In China. Taibei, Sept. 11–16, 1983,* 889–95. Taibei: Furen daxue chubanshe, 1983.

———. "Eléments de réflexion sur les réactions chinoises à la géométrie euclidienne à la fin du XVIIe siècle: Le *Jihe lunyue* de Du Zhigeng vu principalement à partir de la préface de l'auteur et de deux notices bibliographiques rédigées par des lettres illustres." *Historia Mathematica* 20, no. 2 (1993): 160–79.

———. "Space and Time in Chinese Texts of Astronomy and Mathematical Astronomy in the Seventeenth and Eighteenth Centuries." *Chinese Science* 11 (1993): 66–92.

———. "Chinese Mathematics." In Grattan-Guinness, *Companion Encyclopedia,* 93–103.

———. *A History of Chinese Mathematics.* New York: Springer, 2006. Corrected second printing of the first (1997) English edition. Originally published as *Histoire des mathématiques chinoises* (Paris: Masson, 1988).

Martzloff, Jean-Claude. *Le calendrier chinois: Structure et calculs (104 av. J.-C.-1644); Indétermination céleste et réforme permanente; La construction chinoise officielle du temps quotidien discret à partir d'un temps mathématique caché, linéaire et continu.* Sciences, techniques et civilizations du Moyen Âge à l'aube des lumières 11. Paris: Honoré Champion, 2009.

———. "Review of *The Chinese Roots of Linear Algebra*, by Roger Hart." *Zentralblatt MATH* (2011).

McKirahan, Richard D. *Principles and Proofs: Aristotle's Theory of Demonstrative Science.* Princeton, NJ: Princeton University Press, 1992.

McNeill, William H. "World History and the Rise and Fall of the West." *Journal of World History* 9, no. 2 (1998): 215–36.

Menegon, Eugenio. *Ancestors, Virgins, and Friars: Christianity as a Local Religion in Late Imperial China.* Harvard-Yenching Institute Monograph Series 69. Cambridge, MA: Harvard University Press, 2009.

Meng, Yue. "Hybrid Science versus Modernity: The Practice of the Jiangnan Arsenal, 1867–1904." *East Asian Science, Technology and Medicine* 16 (1999): 13–52.

Merleau-Ponty, Maurice. "Phenomenology and the Sciences of Man." In *The Primacy of Perception*, edited by James Edie, translated by John Wild, 43–95. Evanston: Northwestern University Press, 1964. Originally published as *Les sciences de l'homme et la phénoménologie* (Paris: Centre de Documentation Universitaire, 1958).

———. *Signs.* Translated by Richard C. McCleary. Northwestern University Studies in Phenomenology and Existential Philosophy. Evanston, IL: Northwestern University Press, 1964.

Meskill, John Thomas. *Academies in Ming China: A Historical Essay.* Monographs of the Association for Asian Studies 39. Tucson: University of Arizona Press, 1982.

Mikami, Yoshio. *The Development of Mathematics in China and Japan.* 2nd ed. Reprint, New York: Chelsea, 1974.

Montelle, Clemency. *Chasing Shadows: Mathematics, Astronomy, and the Early History of Eclipse Reckoning.* Johns Hopkins Studies in the History of Mathematics. Baltimore: Johns Hopkins University Press, 2011.

Moore, Gregory H. *Zermelo's Axiom of Choice: Its Origins, Development, and Influence.* New York: Springer-Verlag, 1982.

Morrow, Glenn R., trans. *A Commentary on the First Book of Euclid's Elements.* Princeton, NJ: Princeton University Press, 1970.

Mostern, Ruth. *"Dividing the Realm in Order to Govern": The Spatial Organization of the Song State (960–1276 CE).* Harvard-Yenching Institute Monograph Series 73. Cambridge: Harvard University Asia Center, 2011.

Mote, Frederick W., and Denis Twitchett, eds. *The Ming Dynasty, 1368–1644.* 2 vols. The Cambridge History of China. New York: Cambridge University Press, 1988, 1998.

Mueller, Ian. *Philosophy of Mathematics and Deductive Structure in Euclid's Elements.* Cambridge, MA: MIT Press, 1981.

Muir, Edward, and Guido Ruggiero, eds. *Microhistory and the Lost Peoples of Europe*. Translated by Eren Branch. Baltimore: Johns Hopkins University Press, 1991.

Muir, Thomas. *The Theory of Determinants in the Historical Order of Development*. 4 vols. London: Macmillan, 1906–1923. Reprint, 4 vols. bound as 2, New York: Dover Publications, 1960.

Mullaney, Thomas S. *Coming to Terms with the Nation: Ethnic Classification in Modern China*. Asia: Local Studies/Global Themes 18. Berkeley: University of California Press, 2011.

Mungello, David E. *Leibniz and Confucianism: The Search for Accord*. Honolulu: University Press of Hawaii, 1977.

———. *Curious Land: Jesuit Accommodation and the Origins of Sinology*. Studia Leibnitiana Supplementa 25. Stuttgart: Steiner, 1985.

———. *The Great Encounter of China and the West, 1500–1800*. 4th ed. Critical Issues in World and International History. Lanham, MD: Rowman & Littlefield, 2013. First ed. published in 1999.

Murata, Tamotsu. "Wallis' 'Arithmetica Infinitorium' and Takebe's 'Tetsujutsu Sankei': What Underlies Their Similarities and Dissimilarities?" *Historia Scientiarum: International Journal of the History of Science Society of Japan* 19 (1980): 77–100.

Murdoch, John E. "The Medieval Euclid: Salient Aspects of the Translations of the Elements by Adelard of Bath and Campanus of Novara." *Revue de synthèse* 89 (1968): 67–94.

———. "Euclid: Transmission of the Elements." In Gillispie, *Dictionary of Scientific Biography*, 4:437–59.

———. *Antiquity and the Middle Ages*. New York: Scribner, 1984.

Nader, Laura, ed. *Naked Science: Anthropological Inquiry into Boundaries, Power, and Knowledge*. New York: Routledge, 1996.

Nagel, Ernest, and James Roy Newman. *Gödel's Proof*. Rev. ed. Edited by Douglas R. Hofstadter. New York: New York University Press, 2001.

Nakayama, Shigeru. *A History of Japanese Astronomy: Chinese Background and Western Impact*. Harvard-Yenching Institute Monograph Series 18. Cambridge, MA: Harvard University Press, 1969.

———. "Periodization of the East Asian History of Science." *Revue de Synthese* 4, nos. 3–4 (1988): 375–79.

Nappi, Carla. *The Monkey and the Inkpot: Natural History and Its Transformations in Early Modern China*. Cambridge, MA: Harvard University Press, 2009.

Naquin, Susan. *Peking: Temples and City Life, 1400–1900*. Philip E. Lilienthal Book. University of California Press, 2000.

Narasimha, Roddam. "The Indian Half of Needham's Question: Some Thoughts on Axioms, Models, Algorithms, and Computational Positivism." *Interdisciplinary Science Reviews* 28, no. 1 (2003): 1–13.

Needham, Joseph. *Human Law and the Laws of Nature in China and the West*. London: Oxford University Press, 1951.

Needham, Joseph, ed. *Science and Civilisation in China.* 7 vols. Published in 27 physical vols. Cambridge: Cambridge University Press, 1954–2008.

———. *History of Scientific Thought.* Vol. 2 of *Science and Civilisation in China.* Cambridge: Cambridge University Press, 1956.

———. *Chinese Astronomy and the Jesuit Mission: An Encounter of Cultures.* China Society Occasional Papers 10. London: China Society, 1958.

———. *Mathematics and the Sciences of the Heavens and the Earth.* Vol. 3 of *Science and Civilisation in China.* Cambridge: Cambridge University Press, 1959.

———. "Poverties and Triumphs of the Chinese Scientific Tradition." In *Grand Titration,* 14–54.

———. *The Grand Titration: Science and Society in East and West.* London: George Allen & Unwin, 1969.

———. *Science in Traditional China: A Comparative Perspective.* Cambridge, MA: Harvard University Press, 1981.

Netz, Reviel. *The Shaping of Deduction in Greek Mathematics: A Study in Cognitive History.* Ideas in Context 51. Cambridge: Cambridge University Press, 1999.

———. *The Transformation of Mathematics in the Early Mediterranean World: From Problems to Equations.* Cambridge: Cambridge University Press, 2004.

———. *Ludic Proof: Greek Mathematics and the Alexandrian Aesthetic.* Cambridge: Cambridge University Press, 2009.

Neugebauer, Otto. *A History of Ancient Mathematical Astronomy.* 3 vols. Studies in the History of Mathematics and Physical Sciences 1. New York: Springer-Verlag, 1975.

Niranjana, Tejaswini. *Siting Translation: History, Post-Structuralism, and the Colonial Context.* Berkeley: University of California Press, 1992.

Northrop, Filmer S. C. "The Complementary Emphases of Eastern Intuitive and Western Scientific Philosophy." In *Philosophy—East and West,* edited by Charles A. Moore, 168–234. Princeton, NJ: Princeton University Press, 1946.

———. *The Meeting of East and West: An Inquiry Concerning World Understanding.* New York: Macmillan, 1946.

O'Brien, Patrick K. "The Needham Question Updated: A Historiographical Survey and Elaboration." *History of Technology* 29 (2009): 7–28.

Oh, Young Sook. "Ways of Calculation in Late Choson: A Case Study of Choe Sok-Chong's 崔錫鼎 Kusuryak 九數略." Presented at the First Templeton Conference on Science and Religion in East Asia, "Science and Christianity in the Encounter of Confucian East Asia with West: 1600–1800," December 15–17, 2011, Seoul National University.

O'Malley, John W. *The Jesuits: Cultures, Sciences, and the Arts, 1540–1773.* Toronto: University of Toronto Press, 1999.

Peirce, Charles S. *Philosophical Writings of Peirce.* New York: Dover Publications, 1955. Republication of *The Philosophy of Peirce: Selected Writings* (New York: Harcourt, Brace, 1940).

Perdue, Peter C. *Exhausting the Earth: State and Peasant in Hunan, 1500–1850.* Harvard East Asian Monographs 130. Cambridge, MA: Council on East Asian Studies, Harvard University, 1987.

———. *China Marches West: The Qing Conquest of Central Eurasia.* Cambridge, MA: Belknap Press of Harvard University Press, 2005.

Peterson, Willard J. *Bitter Gourd: Fang I-Chih and the Impetus for Intellectual Change.* New Haven, CT: Yale University Press, 1979.

———. "Calendar Reform Prior to the Arrival of Missionaries at the Ming Court." *Ming Studies* 21 (1986): 45–61.

———. "Why Did They Become Christians? Yang T'ing-Yun, Li Chih-Tsao, and Hsu Kuang-Ch'i." In Ronan and Oh, *East Meets West,* 129–52.

———. "Confucian Learning in Late Ming Thought." In Mote and Twitchett, *Ming Dynasty, Part 2,* 708–88.

Pickering, Andrew. *The Mangle of Practice: Time, Agency, and Science.* Chicago: University of Chicago Press, 1995.

Pingree, David. "Astronomy and Astrology in India and Iran." *Isis* 54, no. 2 (June 1963): 229–46.

———. "Hellenophilia Versus the History of Science." *Isis* 83, no. 4 (1992): 554–63.

Pingree, David, Gherardo Gnoli, and Antonio Panaino. *Kayd: Studies in History of Mathematics, Astronomy and Astrology in Memory of David Pingree.* Serie orientale Roma 102. Roma: Istituto Italiano per l'Africa e l'Oriente, 2009.

Plaks, Andrew. "Pa-Ku Wen 八股文." In *Indiana Companion to Traditional Chinese Literature,* edited by William Nienhauser, 641–43. Bloomington: Indiana University Press, 1986.

———. "The Prose of Our Time." In *The Power of Culture: Studies in Chinese Cultural History,* edited by Willard J. Peterson, Andrew H. Plaks, and Yu Yingshi, 206–17. Hong Kong: Chinese University Press, 1994.

Plofker, Kim. *Mathematics in India.* Princeton, NJ: Princeton University Press, 2009.

Pomeranz, Kenneth. *The Great Divergence: China, Europe, and the Making of the Modern World Economy.* Princeton, NJ: Princeton University Press, 2000.

Prakash, Gyan. *Another Reason: Science and the Imagination of Modern India.* Princeton, NJ: Princeton University Press, 1999.

Pratt, Keith. "Art in the Service of Absolutism: Music at the Courts of Louis XIV and the Kangxi Emperor." *Seventeenth Century* 7, no. 1 (1992): 83–110.

Price, Derek J. de Solla. *Science since Babylon.* New Haven, CT: Yale University Press, 1961.

Putnam, Hilary. "Mathematics without Foundations." In Benacerraf and Putnam, *Philosophy of Mathematics,* 295–314.

Qian, Wen-yuan. *The Great Inertia: Scientific Stagnation in Traditional China.* Dover, NH: Croom Helm, 1985.

Qu Wanli. *A Catalogue of the Chinese Rare Books in the Gest Collection of the Princeton University Library.* Taibei: Yiwen yinshuguan, 1974.

Quine, Willard Van Orman. "Meaning and Translation." In Brower, *On Translation,* 148–72.

Quine, Willard Van Orman. "Two Dogmas of Empiricism." In *From a Logical Point of View: 9 Logico-Philosophical Essays*, 2nd ed. Cambridge, MA: Harvard University Press, 1961. Originally published in *The Philosophical Review* 60 (1951): 20-43.

———. "On the Very Idea of a Third Dogma." In *Theories and Things*. Cambridge, MA: Harvard University Press, 1981.

———. "Truth by Convention." In Benacerraf and Putnam, *Philosophy of Mathematics*, 329–54.

———. "2 Dogmas in Retrospect." *Canadian Journal of Philosophy* 21, no. 3 (1991): 265–74.

Rafael, Vicente L. *Contracting Colonialism: Translation and Christian Conversion in Tagalog Society under Early Spanish Rule*. Ithaca, NY: Cornell University Press, 1988.

———. *The Promise of the Foreign: Nationalism and the Technics of Translation in the Spanish Philippines*. Durham, NC: Duke University Press, 2005.

Ragep, F. J., Sally P. Ragep, and Steven John Livesey, eds. *Tradition, Transmission, Transformation: Proceedings of Two Conferences on Pre-Modern Science Held at the University of Oklahoma*. Collection de travaux de l'Académie internationale d'histoire des sciences 37. Leiden: Brill, 1996.

Raj, Kapil. *Relocating Modern Science: Circulation and the Construction of Knowledge in South Asia and Europe, 1650–1900*. Houndmills, Basingstoke, Hampshire: Palgrave Macmillan, 2007.

Renn, Jürgen, ed. *Galileo in Context*. Cambridge: Cambridge University Press, 2001.

Rheinberger, Hans-Jörg. *On Historicizing Epistemology: An Essay*. Translated by David Fernbach. Cultural Memory in the Present. Stanford, CA: Stanford University Press, 2010. Originally published as *Historische Epistemologie zur Einführung* (Junius Verlag, 2007).

Ricci, Matteo, SJ. *China in the Sixteenth Century: The Journals of Matthew Ricci, 1583–1610*. Translated by Louis J. Gallagher, SJ. New York: Random House, 1953. Translation of Trigault's 1615 Latin version of the Ricci commentaries, titled *De Christiana expeditione apud Sinas suscepta ab Societate Jesu*.

———. *The True Meaning of the Lord of Heaven (T'ien-Chu Shih-I)*. Translated by Douglas Lancashire and Peter Kuo-chen Hu, SJ. Series 1: Jesuit Primary Sources in English Translations 6. Edited by Edward J. Malatesta. St. Louis, MO: Institute of Jesuit Sources, 1985.

———. *On Friendship: One Hundred Maxims for a Chinese Prince*. Translated by Timothy James Billings. New York: Columbia University Press, 2009.

Ricci, Matteo, SJ, and Pietro Tacchi Venturi. *Opere storiche*. Macerata: Premiato stab. tip. F. Giorgetti, 1911–13.

Ricoeur, Paul. *Interpretation Theory: Discourse and the Surplus of Meaning*. Fort Worth: Texas Christian University Press, 1976.

———. *Time and Narrative*. Translated by Kathleen Blamey and David Pellauer. Chicago: University of Chicago Press, 1984–1988.

———. *Memory, History, Forgetting*. Chicago: University of Chicago Press, 2004.

Robinson, Kenneth. *A Critical Study of Chu Tsai-Yü's Contribution to the Theory of Equal Temperament in Chinese Music*. Sinologica Coloniensia 9. With additional notes by Erich F. W. Altwein and a preface by Joseph Needham. Wiesbaden: Steiner, 1980.

Robson, Eleanor. *Mathematics in Ancient Iraq: A Social History*. Princeton, NJ: Princeton University Press, 2008.

Robson, Eleanor, and Jacqueline A. Stedall, eds. *The Oxford Handbook of the History of Mathematics*. Oxford: Oxford University Press, 2009.

Rochberg, Francesca. *The Heavenly Writing: Divination, Horoscopy, and Astronomy in Mesopotamian Culture*. Cambridge: Cambridge University Press, 2004.

Rogaski, Ruth. *Hygienic Modernity: Meanings of Health and Disease in Treaty-Port China*. Berkeley: University of California Press, 2004.

Rommevaux, Sabine, Ahmed Djebbar, and Bernard Vitrac. "Remarques sur l'histoire du texte des Éléments d'Euclide." *Archive for History of Exact Sciences* 55, no. 3 (2001): 221–95.

Ronan, Charles E., and Bonnie B. C. Oh, eds. *East Meets West: The Jesuits in China, 1582–1773*. Chicago: Loyola University Press, 1988.

Rorty, Richard. "Is Science a Natural Kind?" In *Objectivity, Relativism, and Truth: Philosophical Papers*, 24–30. Cambridge: Cambridge University Press, 1991.

Rosenthal, Jean-Laurent, and R. Bin Wong. *Before and Beyond Divergence: The Politics of Economic Change in China and Europe*. Cambridge, MA: Harvard University Press, 2011.

Rotman, Brian. *Ad Infinitum: The Ghost in Turing's Machine; Taking God out of Mathematics and Putting the Body Back In; An Essay in Corporeal Semiotics*. Stanford, CA: Stanford University Press, 1993.

———. *Mathematics as Sign: Writing, Imagining, Counting*. Stanford, CA: Stanford University Press, 2000.

Rowbotham, Arnold H. *Missionary and Mandarin: The Jesuits at the Court of China*. Berkeley: University of California Press, 1942.

Rusk, Bruce. "Not Written in Stone: Ming Readers of the *Great Learning* and the Impact of Forgery." *Harvard Journal of Asiatic Studies* 66, no. 1 (2006): 189–231.

Russell, Bertrand. *The Problem of China*. London: Allen & Unwin, 1922.

———. "Chinese and Western Civilization Contrasted." In *The Basic Writings of Bertrand Russell, 1903–1959*, edited by Robert E. Egner and Lester E. Denonn, 195–209. New York: Simon & Schuster, 1961.

———. *An Essay on the Foundations of Geometry*. Cambridge: Cambridge University Press, 2012. First published in 1897.

Sachsenmaier, Dominic. *Global Perspectives on Global History: Theories and Approaches in a Connected World*. Cambridge: Cambridge University Press, 2011.

Said, Edward W. *Orientalism*. New York: Pantheon Books, 1978.

———. "The Text, the World, the Critic." In Harari, *Textual Strategies*.

———. "Orientalism Reconsidered." *Race & Class* 27, no. 2 (1985): 1–15.

Said, Edward W. *Culture and Imperialism*. New York: Knopf, 1993.

Saliba, George. *A History of Arabic Astronomy: Planetary Theories during the Golden Age of Islam*. New York University Studies in Near Eastern Civilization 19. New York: New York University Press, 1994.

———. *Islamic Science and the Making of the European Renaissance*. Transformations. Cambridge, MA: MIT Press, 2007.

Saraiva, Luís, ed. *History of Mathematical Sciences: Portugal and East Asia II; Scientific Practices and the Portuguese Expansion in Asia (1498–1759)*. Hackensack, NJ: World Scientific, 2004.

Saraiva, Luís, and Catherine Jami, eds. *History of Mathematical Sciences: Portugal and East Asia III; The Jesuits, the Padroado and East Asian Science (1552–1773)*. Hackensack, NJ: World Scientific, 2008.

Saussure, Ferdinand de. *Course in General Linguistics*. Translated by Roy Harris. London: Duckworth, 1983. Originally published as *Cours de linguistique générale* (Paris: Payot, 1916).

———. *Writings in General Linguistics*. Oxford: Oxford University Press, 2006. Originally published as *Écrits de linguistique générale* (Paris: Gallimard, 2002).

Saussy, Haun. *The Problem of a Chinese Aesthetic*. Meridian: Crossing Aesthetics. Stanford: Stanford University Press, 1993.

———. "Always Multiple Translation, or, How the Chinese Language Lost Its Grammar." In Liu, *Tokens of Exchange*, 107–23.

———. *Great Walls of Discourse and Other Adventures in Cultural China*. Harvard East Asian Monographs 212. Cambridge, MA: Harvard University Asia Center, 2001.

Schäfer, Dagmar. *The Crafting of the 10,000 Things: Knowledge and Technology in Seventeenth-Century China*. Chicago: University of Chicago Press, 2011.

Schaffer, Simon. "Glass Works: Newton's Prisms and the Uses of Experiment." In Gooding et al., *Uses of Experiment*, 67–104.

———. "Self Evidence." *Critical Inquiry* 18, no. 2 (1992): 327–62.

Schaffer, Simon, Lissa Roberts, Kapil Raj, and James Delbourgo, eds. *The Brokered World: Go-Betweens and Global Intelligence, 1770–1820*. Uppsala Studies in History of Science 35. Sagamore Beach, MA: Science History Publications, 2009.

Schemmel, Matthias. *The English Galileo: Thomas Harriot's Work on Motion as an Example of Preclassical Mechanics*. Boston Studies in the Philosophy of Science 268. New York: Springer, 2008.

Schleicher, August. *Die Sprachen Europas in systematischer Uebersicht*. Bonn: H.B. König, 1850.

Schwartz, Benjamin I. *The World of Thought in Ancient China*. Cambridge, MA: Belknap Press of Harvard University Press, 1985.

Searle, John R. *Speech Acts: An Essay in the Philosophy of Language*. Cambridge: Cambridge University Press, 1969.

———. "How Performatives Work." *Linguistics and Philosophy* 12, no. 5 (1989): 535–58.

Seidenberg, Abraham. "The Ritual Origin of Geometry." *Archive for History of Exact Sciences* 1 (1960): 488–527.

———. "Did Euclid's Elements, Book I, Develop Geometry Axiomatically?" *Archive for History of Exact Sciences* 14 (1974): 263–95.

———. "The Origin of Mathematics." *Archive for History of Exact Sciences* 18, no. 4 (1978): 301–42.

Selin, Helaine. *Mathematics across Cultures: The History of Non-Western Mathematics.* Science Across Cultures 2. Dordrecht: Kluwer Academic, 2000.

Serre, Denis. *Matrices: Theory and Applications.* Graduate Texts in Mathematics 216. New York: Springer, 2002.

Serres, Michel. *Hermes: Literature, Science, Philosophy.* Edited by Josué V. Harari and David F. Bell. Baltimore: Johns Hopkins University Press, 1983.

Sesiano, Jacques. "The Appearance of Negative Solutions in Mediaeval Mathematics." *Archive for History of Exact Sciences* 32, no. 2 (1985): 105–50.

Sha, Xin Wei. "Whitehead's Poetical Mathematics." *Configurations* 13, no. 1 (2007): 77–94.

Shapin, Steven. *The Scientific Revolution.* Chicago: University of Chicago Press, 1996.

———. *Never Pure: Historical Studies of Science as If It Was Produced by People with Bodies, Situated in Time, Space, Culture, and Society, and Struggling for Credibility and Authority.* Baltimore: Johns Hopkins University Press, 2010.

Shapin, Steven, and Simon Schaffer. *Leviathan and the Air-Pump: Hobbes, Boyle, and the Experimental Life.* Princeton, NJ: Princeton University Press, 1985.

Sheldon, Wilmon Henry. "Main Contrasts between Eastern and Western Philosophy." In *Essays in East-West Philosophy: An Attempt at World Philosophical Synthesis,* edited by Charles A. Moore, 288–97. Honolulu: University of Hawaii Press, 1951.

Shen Kangshen, Anthony W.-C. Lun, and John N. Crossley. *The Nine Chapters on the Mathematical Art: Companion and Commentary.* New York: Oxford University Press, 1999.

Shi, Yunli. "Eclipse Observations Made by Jesuit Astronomers in China: A Reconsideration." *Journal for the History of Astronomy* 31, no. 2 (2000): 135–47.

———. "A Note on the Islamic Influence on the Astronomical Instrumentation of the Choson Dynasty." *Historia Scientiarum* 13 (2003): 33–41.

———. "The Korean Adaptation of the Chinese-Islamic Astronomical Tables." *Archive for History of Exact Sciences* 57 (2003): 25–60.

Shin, Minchol. "Jiang Yong's Study on the Gougu 句股 Method and His Correlative Cosmology." Presented at the First Templeton Conference on Science and Religion in East Asia, "Science and Christianity in the Encounter of Confucian East Asia with West: 1600–1800," December 15–17, 2011, Seoul National University.

Sigler, Laurence E. *Fibonacci's Liber Abaci: A Translation into Modern English of Leonardo Pisano's Book of Calculation.* Sources and Studies in the History of Mathematics and Physical Sciences. New York: Springer, 2002.

Sivasundaram, Sujit. "Sciences and the Global: On Methods, Questions and Theory." *Isis* 101, no. 1 (March 2010): 146–58.

Sivin, Nathan. "On China's Opposition to Western Science during the Late Ming and Early Ch'ing." *Isis* 56 (1965): 201–5.

———. *Chinese Alchemy: Preliminary Studies.* Harvard Monographs in the History of Science 1. Cambridge, MA: Harvard University Press, 1968.

———. "Shen Kua." In Gillispie, *Dictionary of Scientific Biography*, s.v.

———. "Wang Hsi-Shan." In Gillispie, *Dictionary of Scientific Biography*, s.v.

———. "Copernicus in China." In *Colloquia Copernica II: Études sur l'audience de la théorie héliocentrique*, 63–122. Vol. II. Warsaw, 1973.

———. "Why the Scientific Revolution Did Not Take Place in China—or Didn't It?" *Chinese Science* 5 (1982): 45–66.

———. "Max Weber, Joseph Needham, Benjamin Nelson: The Question of Chinese Science." In *Civilizations East and West: A Memorial Volume for Benjamin Nelson*, edited by Eugene Victor Walter, 37–49. Atlantic Highlands, NJ: Humanities Press, 1985.

———. "Science and Medicine in Chinese History." In *Heritage of China: Contemporary Perspectives on Chinese Civilization*, edited by Paul S. Ropp, 164–96. Berkeley: University of California Press, 1990.

———. *Science in Ancient China: Researches and Reflections.* Variorum Collected Studies Series. Brookfield, VT: Ashgate, 1995.

———. "Selected, Annotated Bibliography of the History of Chinese Science: Sources in Western Languages." In *Science in Ancient China*.

Sivin, Nathan, Kiyoshi Yabuuchi, and Shigeru Nakayama. *Granting the Seasons: The Chinese Astronomical Reform of 1280, with a Study of Its Many Dimensions and a Translation of Its Records; Shou Shih Li Cong Kao.* Sources and Studies in the History of Mathematics and Physical Sciences. New York: Springer, 2009.

Skinner, Quentin. "Meaning and Understanding in the History of Ideas." *History and Theory* 8 (1969): 3–53.

———. "'Social Meaning' and the Explanation of Social Action." In *Philosophy, Politics and Society*, edited by Peter Laslett, W. G. Runciman, and Quentin Skinner, 136–57. Series 4. Oxford: Basil Blackwell, 1972.

———. "Some Problems in the Analysis of Political Thought and Action." *Political Theory* 2, no. 3 (1974): 277–303.

Slaughter, M. M. *Universal Languages and Scientific Taxonomy in the Seventeenth Century.* Cambridge: Cambridge University Press, 1982.

Smith, Joanna Handlin. *The Art of Doing Good: Charity in Late Ming China.* Berkeley: University of California Press, 2009.

Smith, Pamela H. *The Body of the Artisan: Art and Experience in the Scientific Revolution.* Chicago: University of Chicago Press, 2004.

Smith, Pamela H., and Paula Findlen, eds. *Merchants & Marvels: Commerce, Science and Art in Early Modern Europe.* New York, NY: Routledge, 2002.

Smith, Pamela H., and Benjamin Schmidt, eds. *Making Knowledge in Early Modern Europe: Practices, Objects, and Texts, 1400–1800.* Chicago: University of Chicago Press, 2007.

Smith, Paul J., and Richard von Glahn, eds. *The Song-Yuan-Ming Transition in Chinese History.* Harvard East Asian Monographs 221. Cambridge, MA: Harvard University Asia Center, 2003.

Smith, Richard J. *Chinese Maps: Images of "All Under Heaven."* New York: Oxford University Press, 1996.

Song Yingxing, E-tu Zen Sun, and Shiou-chuan Sun. *Chinese Technology in the Seventeenth Century: T'ien-Kung K'ai-Wu.* Mineola, NY: Dover Publication, 1997. Originally published: University Park: Pennsylvania State University, 1966.

Spence, Jonathan D. *To Change China: Western Advisers in China, 1620–1960.* Boston: Little, Brown, 1969.

———. *The Memory Palace of Matteo Ricci.* New York: Viking Penguin, 1984.

Spence, Jonathan D., and John E. Wills, Jr., eds. *From Ming to Ch'ing: Conquest, Region, and Continuity in Seventeenth-Century China.* New Haven, CT: Yale University Press, 1979.

Spiesser, Maryvonne. "Problèmes linéaires dans le Compendy de la praticque des nombres de Barthélemy de Romans et Mathieu Préhoude (1471): Une approche nouvelle basée sur des sources proches du Liber abbaci de Léonard de Pise." *Historia Mathematica* 27, no. 4 (2000): 362–83.

Spivak, Gayatri Chakravorty. *A Critique of Postcolonial Reason: Toward a History of the Vanishing Present.* Cambridge, MA: Harvard University Press, 1999.

Standaert, Nicolas. *Yang Tingyun, Confucian and Christian in Late Ming China: His Life and Thought.* Sinica Leidensia 19. Leiden: E. J. Brill, 1988.

———, ed. *Handbook of Christianity in China.* 2 vols. Handbook of Oriental Studies 15. Leiden: Brill, 2001.

———. "Xu Guangqi's Conversion as a Multifaceted Process." In Jami et al., *Statecraft and Intellectual Renewal,* 170–85.

———. "Erik Zürcher's Study of Christianity in Seventeenth-Century China: An Intellectual Portrait." *China Review International* 15, no. 4 (2008): 476–502.

———. *The Interweaving of Rituals: Funerals in the Cultural Exchange between China and Europe.* Seattle: University of Washington Press, 2008.

Standaert, Nicolas, and Adrianus Dudink, eds. *Forgive Us Our Sins: Confession in Late Ming and Early Qing China.* Monumenta Serica Monograph Series 55. Sankt Augustin: Institut Monumenta Serica, 2006.

Steele, John M. "Predictions of Eclipse Times Recorded in Chinese History." *Journal for the History of Astronomy* 29 (1998): 275–85.

———. *Observations and Predictions of Eclipse Times by Early Astronomers.* Archimedes 4. Dordrecht: Kluwer Academic Publishers, 2000.

Steele, John M., and F. R. Stephenson. "Astronomical Evidence for the Accuracy of Clocks in Pre-Jesuit China." *Journal for the History of Astronomy* (1998): 35–48.

Stephenson, F. R., and L. J. Fatoohi. "Accuracy of Solar Eclipse Observations Made by Jesuit Astronomers in China." *Journal for the History of Astronomy* 26 (1995): 227–36.

Stewart, G. W. *Introduction to Matrix Computations.* New York: Academic Press, 1973.

———. *Matrix Algorithms.* Philadelphia: Society for Industrial and Applied Mathematics, 1998.

Stone, Richard. "Scientists Fete China's Supreme Polymath." *Science* 318 (November 2, 2007): 733.

Strang, Gilbert. *Linear Algebra and Its Applications.* 4th ed. Belmont, CA: Thomson, Brooks/Cole, 2006.

Sun Xiaochun. "On the Star Catalogue and Atlas of Chongzhen Lishu." In Jami et al., *Statecraft and Intellectual Renewal,* 311–21.

Swerdlow, Noel M., ed. *Ancient Astronomy and Celestial Divination.* Dibner Institute Studies in the History of Science and Technology. Cambridge, MA: MIT Press, 1999.

Swerdlow, Noel M., and Otto Neugebauer. *Mathematical Astronomy in Copernicus's* De Revolutionibus. 2 vols. Studies in the History of Mathematics and Physical Sciences 10. New York: Springer, 1984.

Swetz, Frank. *The Sea Island Mathematical Manual: Surveying and Mathematics in Ancient China.* University Park: Pennsylvania State University Press, 1992.

———. *Legacy of the Luoshu: The 4,000 Year Search for the Meaning of the Magic Square of Order Three.* Chicago: Open Court, 2002.

———. *Mathematical Expeditions: Exploring Word Problems across the Ages.* Baltimore: Johns Hopkins University Press, 2012.

Swetz, Frank J., and Ang Tian Se. "A Brief Chronological and Bibliographic Guide to the History of Chinese Mathematics." *Historia Mathematica* 11 (1984): 39–56.

Takeuti, Gaisi, and Wilson M. Zaring. *Introduction to Axiomatic Set Theory.* 2nd ed. Graduate Texts in Mathematics 1. New York: Springer-Verlag, 1982.

Tambiah, Stanley Jeyaraja. *Magic, Science, Religion, and the Scope of Rationality.* Cambridge: Cambridge University Press, 1990.

Tarski, Alfred. "What is Elementary Geometry?" In Hintikka, *Philosophy of Mathematics,* 164–75.

Teich, Mikulas, and Robert Young, eds. *Changing Perspectives in the History of Science: Essays in Honour of Joseph Needham.* Boston: D. Reidel, 1973.

Teiser, Stephen F. *The Scripture on the Ten Kings and the Making of Purgatory in Medieval Chinese Buddhism.* Kuroda Institute Studies in East Asian Buddhism 9. Honolulu: University of Hawaii Press, 1994.

Toomer, G. J. *Ptolemy's Almagest.* New York: Springer-Verlag, 1984.

———. *Eastern Wisedome and Learning: The Study of Arabic in Seventeenth-Century England.* Oxford: Clarendon Press, 1996.

Tropfke, Johannes, Kurt Vogel, Karin Reich, and Helmuth Gericke. *Geschichte der Elementarmathematik, Band 1, Arithmetik und Algebra.* 4th ed. 3 vols. New York: Walter de Gruyter, 1980.

Tseng, Lillian Lan-ying. *Picturing Heaven in Early China*. Cambridge, MA: Harvard University Asia Center for the Harvard-Yenching Institute, 2011.

Tsu, Jing. *Failure, Nationalism, and Literature: The Making of Modern Chinese Identity, 1895–1937*. Stanford, CA: Stanford University Press, 2005.

Tymoczko, Thomas, ed. *New Directions in the Philosophy of Mathematics: An Anthology*. Boston: Birkhäuser, 1986.

Unguru, Sabetai. "On the Need to Rewrite the History of Greek Mathematics." *Archive for History of Exact Sciences* 15, no. 1 (1975): 67–114.

Van Brummelen, Glen. *The Mathematics of the Heavens and the Earth: The Early History of Trigonometry*. Princeton, NJ: Princeton University Press, 2009.

Van Heijenoort, Jean, ed. *From Frege to Gödel: A Source Book in Mathematical Logic, 1879–1931*. Source Books in the History of the Sciences. Cambridge, MA: Harvard University Press, 1967.

Van Helden, Albert. *Measuring the Universe: Cosmic Dimensions from Aristarchus to Halley*. Chicago: University of Chicago Press, 1985.

Vogel, Hans Urlich, and Gunter Dux, eds. *Concepts of Nature: A Chinese-European Cross-Cultural Perspective*. Conceptual History and Chinese Linguistics 1. With an overview and introduction by Mark Elvin. Leiden: Brill, 2010.

Volkov, Alexei. "Mathematics and Mathematics Education in Traditional Vietnam." In Robson and Stedall, *Oxford Handbook of the History of Mathematics*, 153–76.

Voloshinov, V. N. *Marxism and the Philosophy of Language*. Translated by Ladislav Matejka and I. R. Titunik. Studies in Language 1. Cambridge, MA: Harvard University Press, 1973.

von Glahn, Richard. *The Sinister Way: The Divine and the Demonic in Chinese Religious Culture*. Berkeley: University of California Press, 2004.

von Staden, Heinrich. *Herophilus: The Art of Medicine in Early Alexandria: Edition, Translation, and Essays*. Cambridge: Cambridge University Press, 1989.

———. "Liminal Perils: Early Roman Receptions of Greek Medicine." In Ragep, Ragep, and Livesey, *Tradition, Transmission, Transformation*.

Wakeman, Frederic E., Jr. *The Great Enterprise: The Manchu Reconstruction of Imperial Order in Seventeenth-Century China*. 2 vols. Berkeley: University of California Press, 1985.

Waltner, Ann Beth. *Getting an Heir: Adoption and the Construction of Kinship in Late Imperial China*. Honolulu: University of Hawaii Press, 1990.

Wang Yangming. *Instructions for Practical Living, and Other Neo-Confucian Writing*. Translated by Wing-tsit Chan. Records of Civilization: Sources and Studies 68. New York: Columbia University Press, 1963.

Wang, Aihe. *Cosmology and Political Culture in Early China*. Cambridge: Cambridge University Press, 2000.

Wang, Dewei, and Wei Shang, eds. *Dynastic Crisis and Cultural Innovation: From the Late Ming to the Late Qing and Beyond*. Harvard East Asian Monographs 249. Cambridge, MA: Harvard University Asia Center, 2005.

Wang, Xiaochao. *Christianity and Imperial Culture: Chinese Christian Apologetics in the Seventeenth Century and Their Latin Patristic Equivalent.* Studies in Christian Mission 20. Leiden: Brill, 1998.

Wardy, Robert. *Aristotle in China: Language, Categories, and Translation.* Needham Research Institute Studies 2. Cambridge: Cambridge University Press, 2000.

Watson, Burton, trans. *Chuang Tzu: Basic Writings.* Translations from the Asian Classics. New York: Columbia University Press, 1964.

Weber, Max. *The Religion of China: Confucianism and Taoism.* Translated by Hans H. Gerth. Glencoe, IL: Free Press, 1951. Originally published as "Konfuzianismus und Taoismus," in *Gesammelte Aufsätze zur Religionssoziologie*, vol. 1 (Tübingen: Mohr, 1922).

Westman, Robert S. "Proof, Poetics, and Patronage: Copernicus's Preface to De Revolutionibus." In Lindberg and Westman, *Reappraisals of the Scientific Revolution*, 167–206.

White, Hayden V. *Metahistory: The Historical Imagination in Nineteenth-Century Europe.* Baltimore: Johns Hopkins University Press, 1973.

——. *The Content of the Form: Narrative Discourse and Historical Representation.* Baltimore: Johns Hopkins University Press, 1990.

Whitehead, Alfred North, and Bertrand Russell. *Principia Mathematica.* 3 vols. Cambridge: Cambridge University Press, 1910–1913.

Whitney, William Dwight. *Language and the Study of Language: Twelve Lectures on the Principles of Linguistic Science.* London: N. Trübner, 1867.

Whorf, Benjamin Lee. *Language, Thought, and Reality: Selected Writings.* Technology Press Books in the Social Sciences. Cambridge, MA: MIT Press, 1959.

——. "The Punctual and Segmentative Aspects of Verbs in Hopi." In *Language, Thought, and Reality*, 51–56.

Wigner, Eugene P. "The Unreasonable Effectiveness of Mathematics in the Natural Sciences." In *Symmetries and Reflections: Scientific Essays of Eugene P. Wigner*, edited by W. J. Moore and M. Scriven, 222–37. Bloomington: Indiana University Press, 1967. First published in *Communications in Pure and Applied Mathematics* 13, no. 1 (February 1960).

Wills, John E. *China and Maritime Europe, 1500–1800: Trade, Settlement, Diplomacy, and Missions.* Cambridge: Cambridge University Press, 2011.

Wilson, Thomas A., ed. *On Sacred Grounds: Culture, Society, Politics, and the Formation of the Cult of Confucius.* Harvard East Asian Monographs 217. Cambridge, MA: Harvard University Asia Center, 2002.

Wise, M. Norton, ed. *Growing Explanations.* Durham, NC: Duke University Press, 2004.

Witek, John W., SJ, ed. *Ferdinand Verbiest (1623–1688): Jesuit Missionary, Scientist, Engineer and Diplomat.* Monumenta Serica Monograph Series 30. Nettetal: Steyler Verlag, 1994.

Wittfogel, Karl August. *Oriental Despotism: A Comparative Study of Total Power.* New Haven, CT: Yale University Press, 1957.

Wittgenstein, Ludwig. *Philosophical Investigations.* Translated by G. E. M. Anscombe. Oxford: Blackwell, 1958. First published posthumously in 1953 as *Philosophische Untersuchungen.*

———. *On Certainty.* Edited by G. E. M. Anscombe and G. H. von Wright. Translated by Denis Paul and G. E. M. Anscombe. New York: Harper & Row, 1969. Translation of *Über Gewißheit.*

———. *Remarks on the Foundations of Mathematics.* Translated by G. E. M. Anscombe. Cambridge, MA: MIT Press, 1978. Originally published in Oxford, 1956, in English translation with German text.

———. *Remarks on Frazer's "Golden Bough."* Translated by A. C. Miles and Rush Rhees. Atlantic Heights, NJ: Humanities Press, 1979. Translation of *Bemerkungen über Frazers Golden bough,* written in about 1931.

———. *Tractatus Logico-Philosophicus.* Translated by David Pears and Brian McGuinness. London: Routledge, 2001.

Wolpert, Lewis. *The Unnatural Nature of Science.* Cambridge, MA: Harvard University Press, 1994.

Wong, David B. "Three Kinds of Incommensurability." In Krausz, *Relativism,* 140–58.

Wong, George C. "China's Opposition to Western Science during the Late Ming and Early Ch'ing." *Isis* 54 (1963): 29–49.

Wong, R. Bin. *China Transformed: Historical Change and the Limits of European Experience.* Ithaca, NY: Cornell University Press, 1997.

Wright, Arthur F. Review of *Science and Civilisation in China,* Vol. 2, *History of Scientific Thought,* by Joseph Needham. *American Historical Review* 62, no. 4 (1957): 918–20.

Wright, David. *Translating Science: The Transmission of Western Chemistry into Late Imperial China, 1840–1900.* Sinica Leidensia 48. Leiden: Brill, 2000.

Wu, Kuang-ming. "Counterfactuals, Universals, and Chinese Thinking," review of *The Linguistic Shaping of Thought: A Study in the Impact of Language on Thinking in China and the West* by Alfred Bloom. *Philosophy East and West* 37, no. 1 (1987): 84–94.

Xu, Yibao. "The First Chinese Translation of the Last Nine Books of Euclid's *Elements* and Its Source." *Historia Mathematica* 32, no. 1 (2005): 4–32.

Yabuuti, Kiyosi. *Une histoire des mathématiques chinoises.* Translated by Kaoru Baba and Catherine Jami. Regards sur la science. Paris: Belin, 2000.

Young, John D. *East-West Synthesis: Matteo Ricci and Confucianism.* Hong Kong: Centre of Asian Studies, University of Hong Kong, 1980.

———. *Confucianism and Christianity: The First Encounter.* Hong Kong: Hong Kong University Press, 1983.

Yü, Chün-fang. *The Renewal of Buddhism in China: Chu-Hung and the Late Ming Synthesis.* IASWR Series. New York: Columbia University Press, 1981.

Yu, Pauline, Peter Bol, Stephen Owen, and Willard Peterson, eds. *Ways with Words: Writing about Reading Texts from Early China.* Studies on China 24. Berkeley: University of California Press, 2000.

Zanasi, Margherita. *Saving the Nation: Economic Modernity in Republican China.* Chicago: University of Chicago Press, 2006.

Zhang, Baichun, and Jürgen Renn, eds. *Transformation and Transmission: Chinese Mechanical Knowledge and the Jesuit Intervention.* Preprint (Max-Planck-Institut für Wissenschaftsgeschichte) 313. Berlin: Max-Planck-Institut für Wissenschaftsgeschichte, 2006.

Zhang, Fuzhen. *Matrix Theory: Basic Results and Techniques.* Universitext. New York: Springer, 1999.

Zhang, Longxi. *Mighty Opposites: From Dichotomies to Differences in the Comparative Study of China.* Stanford, CA: Stanford University Press, 1998.

Zhang, Qiong. "About God, Demons, and Miracles: The Jesuit Discourse on the Supernatural in Late Ming China." *Early Science and Medicine* 4, no. 1 (February 1999): 1–36.

———. "Demystifying Qi: The Politics of Cultural Translation and Interpretation in the Early Jesuit Mission to China." In Liu, *Tokens of Exchange,* 74–106.

———. "Matteo Ricci's World Maps in Late Ming Discourse of Exotica." *Horizons: Seoul Journal of Humanities* 1, no. 2 (December 2010): 215–50.

Zilsel, Edgar. *The Social Origins of Modern Science.* Edited by Diederick Raven, Wolfgang Krohn, and Robert S. Cohen. Boston Studies in the Philosophy of Science 200. Boston: Kluwer Academic Publishers, 2000. Republication of essays first published in the early 1940s.

Zürcher, Erik. "The Jesuit Mission in Fujian in Late Ming Times: Levels of Response." In *Development and Decline of Fukien Province in the 17th and 18th Centuries,* edited by E. B. Vermeer, 417–57. Sinica Leidensia. Leiden: Brill, 1990.

———. "Xu Guangqi and Buddhism." In Jami et al., *Statecraft and Intellectual Renewal,* 155–69.

———, trans. *Kouduo Richao: Li Jiubiao's Diary of Oral Admonitions: A Late Ming Christian Journal.* Monumenta Serica Monograph Series, 56/1–2. Sankt Augustin: Institut Monumenta Serica, 2007.

———. *The Buddhist Conquest of China: The Spread and Adaptation of Buddhism in Early Medieval China.* 3rd ed. Sinica Leidensia 11. Leiden: Brill, 2007. First edition published in 1959.

Index

abacus, 82–83, 105, 110, 124–25, 128, 130, 135
"Advent and Triumph of the Watermill" (Bloch), 15
Against Method (Feyerabend), 61n46
algebra: Celestial Origin (*tian yuan*), 25, 32, 78, 83, 85–90, 96–102, 105–6, 128–29; Four Origin, 78n4, 85–87; linear, 19–20, 110, 132–34, 137–38, 148–49, 158, 180, 261–62
Almagest (Ptolemy), 209
Ames, Roger, 111n88
Analects (*Lunyu*; Confucius), 24, 114, 187nn55–56, 224–25, 228–29, 233n129
Ancestors, Virgins, and Friars (Menegon), 23
Anderson, Benedict, 6
anti-Christian persecution, 245, 249–50
Archaeology of Knowledge and the Discourse on Language (Foucault), 68n73
Archimedes, 47n62
Aristotle, 18n22, 20, 38, 45, 46n55, 57–58
astronomy, 30, 41–42, 45–46, 83, 117
Averroës, 45
Axiom of Choice, 208nn43–44
Axiom of Foundation, 208n42
axiomatics, 37–38, 206–9

back substitution, 148–58, 159, 167
Bacon, Francis, 37
Bai Shangshu, 90n56, 101n75
Banach-Tarski Paradox, 208n44
"'Being' in Western Philosophy" (Graham), 56
Benveniste, Emile, 54–59, 61, 63, 66
Biagioli, Mario, 14, 21, 27, 68
bilingualism, 16, 68n75
Bishop, John, 35–36
Bloch, Marc, 15

Bloom, Alfred, 49, 57
Blue, Gregory, 22
Bol, Peter K., 79n5
Boodberg, Peter, 56–57, 58n32
Borrel, Jean, 192
Bourdieu, Pierre, 13
Boyle, Robert, 27
Brahe, Tycho, 46n55
Brockey, Liam, 22
Brooks, A. Taeko, 24
Brooks, E. Bruce, 24
Buddhism, 79, 225–26, 257–58; and Jesuits, 71–72, 75, 231, 258–61; in Xu's writings, 220, 223, 228, 237–40, 243n187, 246–49

calculus, 209
calendrics, 117, 119–20, 245, 257
Cambridge History of China, 22
Cao Daxian, 214
capitalism, 33, 35, 39n25, 80
Cassirer, Ernst, 59
Catholicism, 20, 22–23, 196, 198, 237–40. *See also* Jesuit missionaries
Ce yuan hai jing. See *Sea Mirror of Circle Measurement*
Celestial Origin. *See under* algebra
Cheng Dawei, 82n24, 84n30; *Comprehensive Source of Mathematical Methods*, 78, 83n27, 102–10, 130, 181–82, 186, 190
Chi'en, Edward, 212
China: and first encounter with the West, 1, 52, 257–58; science in, 17n18, 31, 37, 39, 198; thought in, 11, 13, 21, 56–57; and the West, 2–4, 11, 16n16, 28–29, 31, 36, 51–57, 66–68, 74–75, 261–62. *See also particular dynasties*

367